Lecture Notes in Computer Sc

T0230299

Commenced Publication in 1973
Founding and Former Series Editors:
Gerhard Goos, Juris Hartmanis, and Jan van Leeu

James F. Peters   Andrzej Skowron (Eds.)

Transactions on Rough Sets IV

 Springer

Volume Editors

James F. Peters
University of Manitoba
Department of Electrical and Computer Engineering
15 Gillson Street, ENGR 504, Winnipeg, MB R3T 5V6, Canada
E-mail: jfpeters@ee.umanitoba.ca

Andrzej Skowron
Warsaw University
Institute of Mathematics
Banacha 2, 02-097 Warsaw, Poland
E-mail: skowron@mimuw.edu.pl

Library of Congress Control Number: 2005935117

CR Subject Classification (1998): F.4.1, F.1, I.2, H.2.8, I.5.1, I.4

ISSN   0302-9743
ISBN-10   3-540-29830-4 Springer Berlin Heidelberg New York
ISBN-13   978-3-540-29830-4 Springer Berlin Heidelberg New York

Springer is a part of Springer Science+Business Media

springeronline.com

© Springer-Verlag Berlin Heidelberg 2005
Printed in Germany

Typesetting: Camera-ready by author, data conversion by Scientific Publishing Services, Chennai, India
Printed on acid-free paper   SPIN: 11574798   06/3142   5 4 3 2 1 0

Preface

Volume IV of the Transactions on Rough Sets (TRS) introduces a number of new advances in the theory and application of rough sets. Rough sets and approximation spaces were introduced more than 30 years ago by Zdzisław Pawlak. These advances have profound implications in a number of research areas such as the foundations of rough sets, approximate reasoning, artificial intelligence, bioinformatics, computational intelligence, cognitive science, intelligent systems, data mining, machine intelligence, and security. In addition, it is evident from the papers included in this volume that the foundations and applications of rough sets is a very active research area worldwide. A total of 16 researchers from 7 countries are represented in this volume, namely, Canada, India, Norway, Sweden, Poland, Russia and the United States of America. Evidence of the vigor, breadth and depth of research in the theory and applications of rough sets can be found in the 10 articles in this volume.

Prof. Pawlak has contributed a treatise on the philosophical underpinnings of rough sets. In this treatise, observations are made about the Cantor notion of a set, antinomies arising from Cantor sets, the problem of vagueness (especially, *vague (imprecise)* concepts), fuzzy sets, rough sets, fuzzy vs. rough sets as well as logic and rough sets. Among the many vistas and research directions suggested by Prof. Pawlak, one of the most fruitful concerns the model for a rough membership function, which was incarnated in many different forms since its introduction by Pawlak and Skowron in 1994. Recall, here, that Prof. Pawlak introduced approximation spaces in the context of rough sets during the early 1980s. Later, the model for rough membership provided a basis for a model for rough inclusion in generalized approximation spaces introduced by Skowron and Stepaniuk during the early 1990s.

In addition, this volume includes seven papers that explore the theory of rough sets, and two papers that present new applications of rough sets. New developments in rough set theory are represented by papers that investigate a framework for reasoning with rough sets utilizing extended logic programs (Aida Vitória), optimization of decision trees (Igor V. Chikalov, Mikhail Ju. Moshkov, and Maria S. Zelentsova), fuzzy set and rough set approaches to dealing with missing data (Dan Li, Jitender Deogun, William Spaulding, and Bill Shuart), generalization of the indiscernibility relation as an aid to dealing with incompletely specified decision tables (Jerzy W. Grzymała-Busse), deterministic and non-deterministic decision tree complexity in the context of both finite and infinite information systems (Mikhail Ju. Moshkov), analogy-based reasoning in classifier construction (Arkadiusz Wojna), and incremental learning and evaluation of structures of rough decision tables (Wojciech Ziarko). In addition, two papers in this volume introduce new applications of rough sets, namely, supervised learning in the gene ontology (Herman Midelfart) and the design of an intrusion detection system (Sanjay Rawat, V.P. Gulati, and Arun K. Pujari).

This issue of the TRS was made possible thanks to the laudable efforts of a great many generous persons and organizations. We express our thanks to the many anonymous reviewers for their heroic efforts in providing detailed reviews of the articles in this issue of the TRS. The editors and authors of this volume also extend an expression of gratitude to Alfred Hofmann, Ursula Barth, Christine Günther and the LNCS staff at Springer for their support in making this volume of the TRS possible. In addition, the editors of this volume extend their thanks to Marcin Szczuka for his consummate skill and care in the compilation of this volume. The editors of this volume have been supported by the Ministry for Scientific Research and Information Technology of the Republic of Poland, Research Grant No. 3T11C00226, and the Natural Sciences and Engineering Research Council of Canada (NSERC) Research Grant 185986.

July 2005   James F. Peters
                                                             Andrzej Skowron

LNCS Transactions on Rough Sets

This journal subline has as its principal aim the fostering of professional exchanges between scientists and practitioners who are interested in the foundations and applications of rough sets. Topics include foundations and applications of rough sets as well as foundations and applications of hybrid methods combining rough sets with other approaches important for the development of intelligent systems.

The journal includes high-quality research articles accepted for publication on the basis of thorough peer reviews. Dissertations and monographs up to 250 pages that include new research results can also be considered as regular papers. Extended and revised versions of selected papers from conferences can also be included in regular or special issues of the journal.

Table of Contents

Regular Papers

Dissertations and Monographs

A Treatise on Rough Sets

Zdzisław Pawlak

[1] Institute for Theoretical and Applied Informatics,
Polish Academy of Sciences, Bałtycka 5, 44-100 Gliwice, Poland
[2] Warsaw School of Information Technology, Newelska 6, 01-447 Warsaw, Poland
zpw@ii.pw.edu.pl

The central problem of our age is how to act decisively in the absence of certainty.
Bertrand Russell (1950). An Inquiry into Meaning and Truth.
George Allen and Unwin, London;
W. W. Norton, New York

Abstract. This article presents some general remarks on rough sets and their place in general picture of research on vagueness and uncertainty - concepts of utmost interest, for many years, for philosophers, mathematicians, logicians and recently also for computer scientists and engineers particularly those working in such areas as AI, computational intelligence, intelligent systems, cognitive science, data mining and machine learning. Thus this article is intended to present some philosophical observations rather than to consider technical details or applications of rough set theory. Therefore we also refrain from presentation of many interesting applications and some generalizations of the theory.

Keywords: Sets, fuzzy sets, rough sets, antinomies, vagueness.

1 Introduction

In this article we are going to give some general remarks on rough sets and their place in general picture of research on vagueness and uncertainty - concepts of utmost interest, for many years, for philosophers, mathematicians, logicians and recently also for computer scientists and engineers particularly those working in such areas as AI, computational intelligence, intelligent systems, cognitive science, data mining and machine learning. Thus this article is intended to present some philosophical observations rather than to consider technical details or applications of rough set theory. Therefore we also refrain from presentation of many interesting applications and some generalizations of the theory.

We start our consideration in Section 2 with general comments on classical notion of a set, formulated by Georg Cantor [8] over one hundred years ago. Next, we discuss briefly in Section 3 a source of basic discomfort of classical set theory, namely the antinomies, which shocked the foundation of mathematics

J.F. Peters and A. Skowron (Eds.): Transactions on Rough Sets IV, LNCS 3700, pp. 1–17, 2005.

and the ways out of this embarrassment. Further, the notion of vagueness and its role in mathematics, as formulated by Gottlob Frege [12] are briefly discussed in Section 4. The basic notions concerning fuzzy sets and rough sets are presented in Sections 5 and 6, respectively. The contrast between fuzzy membership [51] and rough membership [31] is briefly considered in Section 7. Than we discuss the notions of fuzzy set [51] and rough set [28,30] as certain formalizations of vagueness. A brief comparison of both notions close this section.

We conclude our deliberation in Section 8 with brief discussion of deductive, inductive and common sense reasoning and the role of rough sets has played in these kinds of inference.

2   Sets

The notion of a set is the basic one of mathematics. All mathematical structures refer to it.

The definition of this notion and the creation of set theory are due to German mathematician Georg Cantor (1845-1918), who laid the foundations of contemporary set theory about 100 years ago. The original, intuitive definition of the Cantor's notion of the set [8] is given below:

"Unter einer "Manningfaltigkeit" oder "Menge" verstehe ich nämlich allgemein jedes Viele, welches sich als Eines denken lässt, d.h. jeden Inbegriff bestimmter Elemente, welcher durch ein Gesetz zu einem Ganzen verbunden werden kann."

Thus according to Cantor a set is a collection of any objects, which according to some law can be considered as a whole. As one can see the notion is very intuitive and simple.

All mathematical objects, e.g., relations, functions, numbers, etc. are some kind of sets. In fact set theory is needed in mathematics to provide rigor.

The notion of a set is not only fundamental for the whole mathematics but it also plays an important role in natural language. We often speak about sets (collections) of various objects of interest such as, collection of books, paintings and people.

The intuitive meaning of a set according to some dictionaries is the following:

"A number of things of the same kind that belong or are used together."

Webster's Dictionary

"Number of things of the same kind, that belong together because they are similar or complementary to each other."

The Oxford English Dictionary

Thus a set is a collection of things which are somehow related to each other but the nature of this relationship is not specified in these definitions.

In fact, these definitions are due to the original definition given by Cantor.

3   Antinomies

Well! I'have seen often a cat without a grin, thought Alice; but a grin without a cat!
Lewis Carroll (1994). Alice's Adventures in Wonderland.
Penguin Books, London

In 1903 the renown English philosopher Bertrand Russell (1872-1970) observed [37] that the intuitive notion of a set given by Cantor leads to logical *antinomies* (contradictions), i.e., set theory was contradictory (there also exist other kinds of antinomies - we refrain from considering them here). A logical antinomy, for the sake of simplicity called antinomy in the remaining part of this paper, arises when after carrying on a correct logical reasoning we come to a contradiction, i.e., to the propositions A and *non-A*, which is not allowed in logic.

As an example let us discuss briefly the so-called Russell's antinomy. Consider the set X containing all the sets Y, which are not the elements of themselves. If we assume that X is its own element then X, by definition, cannot be its own element; while if we assume that X is not its own element then, according to the definition of the set X, it must be its own element. Thus while applying each assumption we obtain contradiction.

The above antinomy is often illustrated with the example of a barber, who got the instruction, that he could only shave all the men who did not shave themselves. Then a question arises if he may shave himself or not. If we assume that the barber shaves himself then, according to the instruction, he may not shave himself. But when we assume that he does not shave himself then, according to the instruction, he should shave himself. Thus we have run across an antinomy.

Another well known antinomy, called the power-set antinomy, goes as follows: consider (infinite) set X of all sets. Thus X is the greatest set. Let Y denote the set of all subsets of X. Obviously Y is greater then X, because the cardinality of the family of all subsets of a given set is always greater the cardinality of the set of all its elements. For example, if $X = \{1, 2, 3\}$ then $Y = \{\emptyset, \{1\}, \{2\}, \{3\}, \{1, 2\}, \{1, 3\}, \{2, 3\}, \{1, 2, 3\}\}$, where $\emptyset$ denotes the empty set. Hence, X is not the greatest set as assumed and we have arrived at contradiction.

Antinomies show that a set cannot be a collection of arbitrary elements, as was stipulated by Cantor.

One could think that antinomies are ingenuous logical play, but it is not so. They question the essence of logical reasoning. That is why there have been attempts to "repair" Cantor's theory for over 100 years or to substitute another set theory for it but the results have not been good so far. Is then all mathematics based on doubtful foundations?

As a remedy for this defect several improvements of set theory have been proposed. For example,

- Axiomatic set theory (Zermello and Fraenkel, 1904);
- Theory of types (Whitehead and Russell, 1910);
- Theory of classes (v. Neumann, 1920).

All of these improvements consist in restrictions put on objects which can form a set. Such restrictions are expressed by properly chosen axioms, which say how a set can be built. They are called, in contrast to Cantors' intuitive set theory, axiomatic set theories.

Instead of improvements of Cantors' set theory by its axiomatization, some mathematicians proposed escape from classical set theory by creating a completely new idea of a set, which would free the theory from antinomies. Some of them are listed below.

- Mereology (Leśniewski, 1915, [19]);
- Alternative set theory (Vopenka, 1970, [49]);
- "Penumbral" set theory (Apostoli and Kanada, 1999, [1]).

No doubt the most interesting proposal was given by Polish logician Stanisław Leśniewski, who proposed instead of membership relation between elements and sets, employed in classical set theory, the relation of "being a part". In his set theory, called *mereology*, this relation is a fundamental one [19].

None of the three mentioned above "new" set theories were accepted by mathematicians. However, Leniewski's mereology attracted some attention of philosophers and recently also researchers in computer science (see, e.g., [9,33,43]).

The problem of finding an alternative to classical set theory has failed to be solved until now.

Basic concept of mathematics, the set, leads to antinomies, i.e., it is contradictory. How is it is then possible that mathematics is so successful and can be applied almost everywhere – that bridges are not collapsing, air-planes are not falling down and man has landed on the moon?

The deficiency of sets, mentioned above, has rather philosophical than practical meaning, since sets used practically in mathematics are free from the above discussed faults. Antinomies are associated with very "artificial" sets constructed in logic but not found in sets used in "everyday" mathematics. That is why we can use mathematics safely.

4   Vagueness

> *Besides known and unknown what else is three?*
> *Harold Pinter (1965). The Homecoming.*
> *Methuen, London*

Another issue discussed in connection with the notion of a set is vagueness. Mathematics requires that all mathematical notions (including set) must be exact, otherwise precise reasoning would be impossible. However, philosophers [17,18,36,38] and recently computer scientists [21,23,24,41] as well as other researchers have become interested in *vague* (imprecise) concepts.

In classical set theory a set is uniquely determined by its elements. In other words, this means that every element must be uniquely classified as belonging to the set or not. That is to say the notion of a set is a *crisp* (precise) one. For example, the set of odd numbers is crisp because every number is either odd or even.

In contrast to odd numbers, the notion of a beautiful painting is vague, because we are unable to classify uniquely all paintings into two classes: beautiful and not beautiful. Some paintings cannot be decided whether they are beautiful or not and thus they remain in the doubtful area. Thus *beauty* is not a precise but a vague concept.

Almost all concepts we are using in natural language are vague. Therefore common sense reasoning based on natural language must be based on vague concepts and not on classical logic. This is why vagueness is important for philosophers and recently also for computer scientists. Interesting discussion of this issue can be found in [36].

The idea of vagueness can be traced back to ancient Greek philosophers Eubulides (ca. 400BC) who first formulated so called sorites (Bald Man or Heap) paradox. The paradox goes as follows: suppose a man has 100 000 hair on his head. Removing on hair from his head surely cannot make him bald. Repeating this step we arrive at the conclusion the a man without any hair is not bald. Similar reasoning can be applied to a hip stones.

Vagueness is usually associated with the boundary region approach (i.e., existence of objects which cannot be uniquely classified relative to a set or its complement) which was first formulated in 1893 by the father of modern logic, German logician, Gottlob Frege (1848-1925). He wrote:

"Der Begrieff muss scharf begrenzt sein. Einem unscharf begrenzten Begriff würde ein Bezirk ensprechen, der nicht überall ein scharfe Grentzlinie hätte, sondern stellenweise gantz verschwimmend in die Umgebung übergine" [12].

Thus according to Frege:

"The concept must have a sharp boundary. To the concept without a sharp boundary there would correspond an area that had not a sharp boundary-line all around."

It means mathematics must use crisp, not vague concepts, otherwise it would be impossible to reason precisely. Summing up, vagueness is

- not allowed in mathematics;
- interesting for philosophy;
- necessary for computer science.

5   Fuzzy Sets

There is nothing new under the sun.
Ecclesiates 1:9

At the same time, independently of mathematicians' and philosophers' investigations, engineers became interested in the notion of a set. It turned out that many practical problems could not be formulated and solved by means of classical Cantor's notion of a set.

In 1965 Lotfi Zadeh, Professor of University of Berkely, proposed a different notion of a set, in which elements can belong to a set to some extent and not definitively, as it is in case of the classical set theory. This proposal turned out applicable in many domains and initiated extensive research in fuzzy set theory, what became the name of Zadeh's theory [43].

In his approach an element can belong to a set to a degree k $(0 \leq k \leq 1)$, in contrast to classical set theory, where an element must definitely belong or not to a set. For example, in classical set theory one can say that someone is definitely ill or healthy, whereas in the fuzzy set theory language we can say that someone is ill (or healthy) at the 60 percent level (i.e., in degree 0.6).

Let us observe that the definition of fuzzy set involves more advanced mathematical concepts – real numbers and functions – whereas in classical set theory the notion of a set is used as a fundamental notion of whole mathematics and is used to derive any other mathematical concepts, e.g., numbers and functions. Consequently fuzzy set theory cannot replace classical set theory, because, in fact, the theory is needed to define fuzzy sets.

Fuzzy membership function has the following properties:

a) $\mu_{U-X}(x) = 1 - \mu_X(x)$ for any $x \in U$;
b) $\mu_{X \cup Y}(x) = max(\mu_X(x), \mu_Y(x))$ for any $x \in U$;
c) $\mu_{X \cap Y}(x) = min(\mu_X(x), \mu_Y(x))$ for any $x \in U$.

That means that the membership of an element to the union and intersection of sets is uniquely determined by its membership to constituent sets. This is a very nice property and allows very simple operations on fuzzy sets, which is a very important feature both theoretically and practically.

Several generalizations of this basic approach to concept approximation are presented in the literature (see, e.g., [14,42,44,45,50]).

Let us stress once more that classical set is a primitive notion and is defined intuitively or axiomatically. Fuzzy sets are defined by employing the fuzzy membership function, which involves advanced mathematical structures, numbers and functions. Thus it cannot play an analogous role in mathematics similar to that played by the classical concept of a set, which is used to define numbers and functions.

Fuzzy set theory can be perceived as new model of vagueness. The theory and its applications developed very extensively over the past four decades and attracted attention of engineers, logicians, mathematicians and philosophers worldwide.

6   Rough Sets

Data! data! data!
Sir Artur Conan Doyle (1994). The Adventures of Sherlock Holmes.
Penguin Books, London

Rough set theory, proposed by the author in 1982 [28,30], is still another approach to vagueness.

Rough set theory expresses vagueness not by means of membership, but by employing a boundary region of a set. If the boundary region of a set is empty it means that the set is crisp, otherwise the set is rough (inexact). A nonempty boundary region of a set means that our knowledge about the set is not sufficient to define the set precisely.

In a manner similar to fuzzy set theory, rough set theory it is not an alternative to classical set theory but it is embedded in it. Rough set theory can be viewed as a specific implementation of Frege's idea of vagueness, i.e., imprecision in this approach is expressed by a boundary region of a set, and not by a partial membership, as in fuzzy set theory.

The rough set concept can be defined quite generally by means of topological operations, *interior* and *closure*, called *approximations*. At the onset of an introduction to rough sets, it was observed that the key to the presented approach is provided by the exact mathematical formulation of the concept of approximative (rough) equality of sets in a given approximation space [28]. In [30], an approximation space is represented by the pair (U, R), where U is a universe of objects, and $R \subseteq U \times U$ is an indiscernibility relation defined by an attribute set. The relation R is an equivalence relation. Let $[x]_R$ denote an equivalence class of an element $x \in U$ under the indiscernibility relation R, where $[x]_R = \{y \in U \mid xRy\}$.

In this context, R-approximations of any set $X \subseteq U$ are based on the exact (crisp) containment of sets. Then set approximations are defined as follows:

- $x \in U$ belongs with certainty to the R-lower approximation of $X \subseteq U$, if $[x]_R \subseteq X$.
- $x \in U$ belongs with certainty to the complement set of $X \subseteq U$, if $[x]_R \subseteq U - X$.
- $x \in U$ belongs with certainty to the R-boundary region of $X \subseteq U$, if $[x]_R \cap X \neq \oslash$ and $[x]_R \cap (U - X) \neq \oslash$.

Generalized approximation spaces were introduced in [42]. A *generalized approximation space* is a system $GAS = (U, I, \nu)$ where

- U is a non-empty set of objects, and $\mathcal{P}(U)$ is the powerset of U;
- $I : U \to \mathcal{P}(U)$ is an uncertainty function such that $x \in I(x)$ for any $x \in U$;
- $\nu : \mathcal{P}(U) \times \mathcal{P}(U) \to [0, 1]$ denotes rough inclusion

The uncertainty function I defines for every object x a set of similarly defined objects. In effect, I defines a neighborhood of every sample element x belonging to the universe U (see, e.g., [32]). The rough inclusion function ν computes the degree of overlap between two subsets of U. Let $\mathcal{P}(U)$ denote the powerset of U. In general, rough inclusion $\nu : \mathcal{P}(U) \times \mathcal{P}(U) \to [0, 1]$ can be defined in terms of the relationship between two sets, e.g., by

$$\nu(X, Y) = \begin{cases} \frac{|X \cap Y|}{|Y|}, & \text{if } Y \neq \emptyset \\ 1 \quad , & \text{otherwise} \end{cases}$$

for any $X, Y \subseteq U$.

From practical point of view it is better to define basic concepts of this theory in terms of data. Therefore we will start our considerations from a data set called an *information system*. An information system is a data table containing rows labeled by objects of interest, columns labeled by attributes and entries of the table are attribute values. For example, a data table can describe a set of patients in a hospital. The patients can be characterized by some attributes, like *age, sex, blood pressure, body temperature*, etc. With every attribute a set of its values is associated, e.g., values of the attribute age can be *young, middle*, and *old*. Attribute values can be also numerical. In data analysis the basic problem we are interested in is to find patterns in data, i.e., to find relationship between some set of attributes, e.g., we might be interested whether *blood pressure* depends on *age and sex*.

Let us describe this problem more precisely. Suppose we are given a finite, non-empty set of objects U called the *universe* and a set of *attributes* A, describing objects of the universe in terms of *attribute values*. Let X be a subset of U and B a subset of A. We want to characterize the set X in terms of attributes B. To this end we will need the basic concepts of rough set theory given below.

- The *lower approximation* of a set X with respect to B is the set of all objects, which can be for *certain* classified as X using B (are *certainly* X in view of B).
- The *upper approximation* of a set X with respect to B is the set of all objects which can be *possibly* classified as X using B (are *possibly* X in view of B).
- The *boundary region* of a set X with respect to B is the set of all objects, which can be classified neither as X nor as not-X using B.

Now we are ready to give the definition of rough sets.

- Set X is *crisp* (exact with respect to B), if the boundary region of X is empty.
- Set X is *rough* (inexact with respect to B), if the boundary region of X is nonempty.

Thus a set is *rough* (imprecise) if it has nonempty boundary region; otherwise the set is crisp (precise). This is exactly the idea of vagueness proposed by Frege.

Let us observe that the definition of rough sets refers to data (knowledge), and is *subjective*, in contrast to the definition of classical sets, which is in some sense an *objective* one.

The approximations and the boundary region can be defined more precisely. To this end we need some additional notation.

Every subset of attributes B determines an equivalence relation on U. This relation will be referred to as an *indiscernibility relation*. The equivalence class determined by an element x and the set of attributes B will be denoted $B(x)$. The indiscernibility relation in a certain sense describes our lack of knowledge about the universe. Equivalence classes of the indiscernibility relation, called *granules* generated by the set of attributes B, represent an elementary portion of knowledge we are able to perceive in terms of available data. Thus in view of the data we are unable, in general, to observe individual objects but we are forced to reason only about the accessible granules of knowledge (see, e.g., [27,30,35].

Formal definitions of approximations and the boundary region are as follows:

– *B-lower approximation* of X

$$B_*(X) = \bigcup_{x \in U} \{B(x) : B(x) \subseteq X\};$$

– *B-upper approximation* of X

$$B^*(X) = \bigcup_{x \in U} \{B(x) : B(x) \cap X \neq \emptyset\};$$

– *B-boundary region* of X

$$BN_B(X) = B^*(X) - B_*(X).$$

As we can see from the definition approximations are expressed in terms of granules of knowledge. The lower approximation of a set is union of all granules determined by the set of attributes B which are entirely included in the set; the upper approximation is union of all granules which have non-empty intersection with the set; the boundary region of set is the difference between the upper and the lower approximation.

Thus the definition of rough set also requires advanced mathematical concepts (relations) and consequently, similarly as fuzzy set, cannot replace classical concept of a set.

Several generalizations of the above approach have been proposed in the literature (see, e.g., [14,27,42,44,45,50]). In particular, in some of these approaches the set inclusion to a degree is used instead of the exact inclusion. It is worthwhile to mention that the set inclusion to a degree has been considered by Lukasiewicz [20] in studies on assigning fractional truth values to logical formulas.

Different aspects of vagueness in the roughs set framework are discussed, e.g., in [21,24,36,41].

Our knowledge about the approximated concepts is often partial and uncertain [15]. For example, the concept approximation should be constructed from examples and counter examples of objects for the concepts [16]. Hence, the concept approximations constructed from a given sample of objects is extended, using inductive reasoning, on unseen so far objects. The rough set approach for dealing with concept approximation under such partial knowledge is presented, e.g., in [44]. Moreover, the concept approximations should be constructed under dynamically changing environments [41]. This leads to a more complex situation when the boundary regions are not crisp sets what is consistent with the postulate of the higher order vagueness, considered by philosophers (see, e.g., [17]). It is worthwhile to mention that is has been also developed a rough set approach to approximation of compound concepts that we are unable to approximate using the traditional methods [7,47]. The approach is based on hierarchical learning and ontology approximation [27,22,43,5] (see Section 8). Approximation of concepts in distributed environments is discussed in [40]. A survey of algorithmic methods for concept approximation based on rough sets and boolean reasoning in presented in [39].

7   Fuzzy Versus Rough

In order to compare both concepts, fuzzy and rough sets we let us observe that rough sets can be also defined employing, instead of approximation, *rough membership function* [31]

$$\mu_X^B : U \to <0,1>,$$

where

$$\mu_X(x) = \frac{card(B(x) \cap X)}{card(X)},$$

and $card(X)$ denotes the cardinality of X.

The rough membership function expresses a conditional probability that x belongs to X given B and can be interpreted as a degree that x belongs to X in view of knowledge about x expressed by B. This means that the definition reflects a subjective knowledge about elements of the universe, in contrast to classical definition of a set.

It can be shown that the rough membership function has the following properties [31]:

1) $\mu_X^B(x) = 1$ *iff* $x \in B_*(X)$;
2) $\mu_X^B(x) = 0$ *iff* $x \in U - B^*(X)$;
3) $0 < \mu_X^B(x) < 1$ *iff* $x \in BN_B(X)$;
4) $\mu_{U-X}^B(x) = 1 - \mu_X^B(x)$ *for any* $x \in U$;
5) $\mu_{X \cup Y}^B(x) \geq max(\mu_X^B(x),\ \mu_Y^B(x))$ *for any* $x \in U$;
6) $\mu_{X \cap Y}^B(x) \leq min(\mu_X^B(x),\ \mu_Y^B(x))$ *for any* $x \in U$.

From the properties it follows that the rough membership differs essentially from the fuzzy membership, for properties 5) and 6) show that the membership for union and intersection of sets, in general, cannot be computed – as in the case of fuzzy sets – from their constituents membership. Thus formally the rough membership is more general from fuzzy membership. Moreover, the rough membership function depends on an available knowledge (represented by attributes from B). Besides, the rough membership function, in contrast to fuzzy membership function, has a probabilistic flavor.

Let us also mention that rough set theory, in contrast to fuzzy set theory, clearly distinguishes two very important concepts, vagueness and uncertainty, very often confused in the AI literature. Vagueness is the property of sets and can be described by approximations, whereas uncertainty is the property of elements of a set and can be expressed by the rough membership function.

Both fuzzy and rough set theory represent two different approaches to vagueness. Fuzzy set theory addresses *gradualness* of knowledge, expressed by the fuzzy membership – whereas rough set theory addresses *granularity* of knowledge, expressed by the indiscernibility relation. A nice illustration of this difference has been given by Dider Dubois and Henri Prade [11] in the following example. In image processing fuzzy set theory refers to gradualness of gray level, whereas rough set theory is about the size of pixels.

Consequently, both theories are not competing but are rather complementary. In particular, the rough set approach provides tools for approximate construction of fuzzy membership functions. The rough-fuzzy hybridization approach proved to be successful in many applications (see, e.g., [25,26]).

Interesting discussion of fuzzy and rough set theory in the approach to vagueness can be found in [36].

Finally, let us observe that fuzzy set and rough set theory are not a remedy for classical set theory difficulties.

8   Logic and Rough Sets

> *Reality, or the world we all know, is only a description.*
> *Carlos Castaneda (1972). Journey to Ixtlan: The lesson of Don Juan.*
> *Simon & Schuster, New York*

The father of contemporary logic is a German mathematician Gottlob Frege (1848-1925). He thought that mathematics should not be based on the notion of set but on the notions of logic. He created the first axiomatized logical system but it was not understood by the logicians of those days.

In the thirties of the previous century a rapid development of logic took place, to which Polish logicians contributed to a large extent, in particular Alfred Tarski (1901-1983).

Development of computers and their applications stimulated logical research and widened their scope.

When we speak about logic we generally mean *deductive logic*. It gives us tools designed for deriving true propositions from other true propositions. Deductive reasoning always leads to true conclusions. The theory of deduction has well established generally accepted theoretical foundations. Deductive reasoning is the main tool used in mathematical reasoning and found no application beyond it.

Rough set theory has contributed to some extent to various kind of deductive reasoning. Particularly, rough set methodology contributed essentially to modal logics, many valued logic, intuitionistic logic and others (see, e.g., [3]). A summary of this research can be found in [33] and interested reader is advised to consult this volume.

In natural sciences (e.g., in physics) *inductive reasoning* is of primary importance. The characteristic feature of such reasoning is that it does not begin from axioms (expressing general knowledge about the reality) like in deductive logic, but some partial knowledge (examples) about the universe of interest are the starting point of this type of reasoning, which are generalized next and they constitute the knowledge about wider reality than the initial one. In contrast to deductive reasoning, inductive reasoning does not lead to true conclusions but only to probable (possible) ones. Also in contrast to the logic of deduction, the logic of induction does not have uniform, generally accepted, theoretical foundations as yet, although many important and interesting results have been obtained, e.g., concerning statistical and computational learning and others.

Verification of validity of hypotheses in the logic of induction is based on experiment rather than the formal reasoning of the logic of deduction. Physics is the best illustration of this fact.

The research on inductive logic have a few centuries', long history and outstanding English philosopher John Stuart Mill (1806-1873) is considered its father.

The creation of computers and their innovative applications essentially contributed to the rapid growth of interest in inductive reasoning. This domain develops very dynamically thanks to computer science. Machine learning, knowledge discovery, reasoning from data, expert systems and others are examples of new directions in inductive reasoning. It seems that rough set theory is very well suited as a theoretical basis for inductive reasoning. Basic concepts of this theory fit very well to represent and analyze knowledge acquired from examples, which can be next used as starting point for generalization. Besides, in fact rough set theory has been successfully applied in many domains to find patterns in data (data mining) and acquire knowledge from examples (learning from examples). Thus, rough set theory seems to be another candidate as a mathematical foundation of inductive reasoning [5,22,44].

The most interesting from computer science point of view is *common sense* reasoning. We use this kind of reasoning in our everyday life, and examples of such kind of reasoning we face in news papers, radio TV etc., in political, economic etc., debates and discussions.

The starting point to such reasoning is the knowledge possessed by the specific group of people (*common knowledge*) concerning some subject and intuitive methods of deriving conclusions from it. We do not have here possibilities of resolving the dispute by means of methods given by deductive logic (reasoning) or by inductive logic (experiment). So the best known methods of solving the dilemma is voting, negotiations or even war. See e.g., Gulliver's Travels [46], where the hatred between Tramecksan (High-Heels) and Slamecksan (Low-Heels) or disputes between Big-Endians and Small-Endians could not be resolved without a war.

These methods do not reveal the truth or falsity of the thesis under consideration at all. Of course, such methods are not acceptable in mathematics or physics. Nobody is going to solve by voting, negotiations or declare a war – the truth of Fermat's theorem or Newton's laws.

Reasoning of this kind is the least studied from the theoretical point of view and its structure is not sufficiently understood, in spite of many interesting theoretical research in this domain [13]. The meaning of common sense reasoning, considering its scope and significance for some domains, is fundamental and rough set theory can also play an important role in it but more fundamental research must be done to this end [43].

In particular, the rough truth introduced in [29] and studied, e.g., in [2,4] seems to be important for investigating commonsense reasoning in the rough set framework.

Let us consider a simple example. In the considered decision table we assume $U = Birds$ is a set of birds that are described by some condition attributes from a set A. The decision attribute is a binary attribute $Flies$ with possible values yes if the given bird flies and no, otherwise. Then, we define the set of abnormal birds by $Ab_A(Birds) = A_*(\{x \in Birds : Flies(x) = no\})$. Hence, we have, $Ab_A(Birds) = Birds - A^*(\{x \in Birds : Flies(x) = yes\})$ and $Birds - Ab_A(Birds) = A^*(\{x \in Birds : Flies(x) = yes\})$. It means that for normal birds it is consistent, with knowledge represented by A, to assume that they can fly, i.e., it is possible that they can fly. One can optimize $Ab_A(Birds)$ using A to obtain minimal boundary region in the approximation of $\{x \in Birds : Flies(x) = no\}$.

It is worthwhile to mention that in [10] has been presented an approach combining the rough sets with nonmonotonic reasoning. There are distinguished some basic concepts that can be approximated on the basis of sensor measurements and more complex concepts that are approximated using so called transducers defined by first order theories constructed overs approximated concepts. Another approach to commonsense reasoning has been developed in a number of papers (see, e.g., [35,43,22,27,5]). The approach is based on an ontological framework for approximation. In this approach approximations are constructed for concepts and dependencies between the concepts represented in a given ontology expressed, e.g., in natural language. Still another approach combining rough sets with logic programming is discussed in [48].

To recapitulate, the characteristics of the three above mentioned kinds of reasoning are given below:

1. deductive:
    - reasoning method: axioms and rules of inference;
    - applications: mathematics;
    - theoretical foundations: complete theory;
    - conclusions: true conclusions from true premises;
    - hypotheses verification: formal proof.
2. inductive:
    - reasoning method: generalization from examples;
    - applications: natural sciences (physics);
    - theoretical foundation: lack of generally accepted theory;
    - conclusions: not true but probable (possible);
    - hypotheses verification - experiment.
3. common sense:
    - reasoning method based on common sense knowledge with intuitive rules of inference expressed in natural language;
    - applications: every day life, humanities;

- theoretical foundation: lack of generally accepted theory;
- conclusions obtained by mixture of deductive and inductive reasoning based on concepts expressed in natural language, e.g., with application of different inductive strategies for conflict resolution (such as voting, negotiations, cooperation, war) based on human behavioral patterns;
- hypotheses verification - human behavior.

9   Conclusions

Basic concept of mathematics, the set, leads to antinomies, i.e., it is contradictory.

The deficiency of sets, has philosophical rather than practical meaning, since sets used in mathematics are free from the above discussed faults. Antinomies are associated with very "artificial" sets constructed in logic but not found in sets used in mathematics. That is why we can use mathematics safely.

Fuzzy set and rough set theory are two different approaches to vagueness and are not remedy for classical set theory difficulties.

Fuzzy set theory addresses gradualness of knowledge, expressed by the fuzzy membership - whereas rough set theory addresses granularity of knowledge, expressed by the indiscernibility relation.

From practical point of view both theories are not competing but are rather complementary.

Summing up:

- The notion of classical set is fundamental for whole mathematics and is necessary to provide rigor in mathematics.
- Non-classical sets (fuzzy and rough) cannot replace classical sets - for their definitions need classical set theory (i.e., more advanced mathematical concepts, real numbers, functions and relations).
- The classical sets, lead to antinomies.
- The deficiency of classical sets has rather philosophical than practical meaning, since sets used in everyday mathematics are free from antinomies.
- Non-classical sets (fuzzy and rough) are not remedy for classical set theory difficulties but are two different approaches to vagueness.

Summary

In this paper a brief discussion on the rough set concept and its place in various ideas of sets is presented. The article is not intended to serve as an introduction to rough set theory but is rather meant to give some philosophical background underlining the theory.

Acknowledgments

Thanks are due to Professor Andrzej Skowron and Professor Jim Peters for critical remarks.

References

1. P. Apostoli, A. Kanda (1999). *Parts of the Continuum: Towards a Modern Ontology of Sciences*. Technical Reports in Philosophical Logic **96**(1), **97**(1), Revised March, 1999, The University of Toronto, Department of Philosophy.
2. M. Banerjee, M.K. Chakraborty (1994). Rough Consequence and Rough Algebra. In: W.P. Ziarko (ed.), *Rough Sets, Fuzzy Sets and Knowledge Discovery, Proc. Int. Workshop on Rough Sets and Knowledge Discovery (RSKD'93)*, Workshops in Computing, Springer-Verlag & British Computer Society, London, Berlin, 196–207.
3. M. Banerjee, M.K. Chakraborty (2004). Algebras from Rough Sets. In: [27], 157–188.
4. M. Banerjee (2004). Rough Truth, Consequence, Consistency and Belief Revision. In: S. Tsumoto et. al. (eds.), *Proc. 4th Int. Conf. On Rough Sets and Current Trends in Computing (RSCTC2004)*, Uppsala, Sweden, June 2004, LNAI **3066**, Springer-Verlag, Heidelberg, 95–102.
5. J.G. Bazan, J.F. Peters, A. Skowron (2005). Behavioral Pattern Identification Through Rough Set Modelling. *Proceedings of RSFDGrC 2005*, Regina, Canada, September 1–3, 2005, LNCS, Springer, Heidelberg (to appear).
6. M.K. Chakraborty, M. Banerjee (1993). Rough Consequence. *Bull. Polish Acad. Sc. (Math.)*, **41**(4): 299–304.
7. L. Breiman (2001). Statistical Modeling: The Two Cultures. *Statistical Science* **16**(3): 199–231.
8. G. Cantor (1883). *Grundlagen einer allgemeinen Mannigfaltigkeitslehre*. Leipzig, 1883.
9. R. Casati, A. Varzi (1999). *Parts and Places. The Structures of Spatial Representation, Cambridge, MA, and London: MIT Press, Bradford Books*.
10. P. Doherty, W. Łukaszewicz, A. Skowron, A. Szałas (2005). *Knowledge Engineering: A Rough Set Approach*. Springer, Heidelberg (to appear).
11. D. Dubois, H. Prade (1991). Foreword. In: Zdzisław Pawlak, *Rough Sets, Theoretical Aspects of Reasoning about Data*, Kluwer Academic Publishers, Dordrecht.
12. G. Frege (1893). *Grundlagen der Arithmetik* **2**, Verlag von Herman Pohle, Jena.
13. D.M. Gabbay, C.J. Hogger, J.A. Robinson (eds.) (1994). *Handbook of Logic in Aretificial Intelligence and Logic Programming: Volume 3: Nonmonotonic Reasoning and Uncertain Reasoning*, Calderon Press, Oxford.
14. S. Greco, B. Matarazzo, R. Słowiński, (2001). Rough Set Theory for Multicriteria Decision Analysis. *European Journal of Operational Research* **129**(1): 1–47.
15. J.W. Grzymała-Busse (1990). *Managing Uncertainty in Expert Systems*. Kluwer Academic Publishers, Norwell, MA.
16. J. Friedman, T. Hastie, R. Tibshirani (2001). *The Elements of Statistical Learning: Data Mining, Inference, and Prediction*. Springer-Verlag, Heidelberg.
17. R. Keefe (2000). *Theories of Vagueness*. Cambridge Studies in Philosophy, Cambridge, UK.
18. R. Keefe, P. Smith (eds.) (1997). *Vagueness: A Reader*. MIT Press, Massachusetts, MA.
19. S. Leśniewski (1929). Grungzüge eines neuen Systems der Grundlagen der Mathematik. *Fundamenta Matemaicae* **14**: 1–81.
20. J. Łukasiewicz (1913). Die Logischen grundlagen der Wahrscheinlichkeitsrechnung. Kraków, 1913, In: L. Borkowski (ed.), *Jan Łukasiewicz - Selected Works*, North Holland Publishing Company, Amsterdam, London, Polish Scientific Publishers, Warsaw, 1970.

21. S. Marcus (1998). The Paradox of the Heap of Grains, in Respect to Roughness, Fuzziness and Negligibility. In L.Polkowski, A. Skowron (eds.) *First International Conference on Rough Sets and Current Trends in Computing* (RSCTC'98), Warsaw, Poland, June 22-26, 1998, LNAI **1424**, Springer-Verlag, Heidelberg, 19–23.

22. S. Hoa Nguyen, J. Bazan, A. Skowron, and H. Son Nguyen (2004). Layered Learning for Concept Synthesis. *Transactions on Rough Sets I: Journal Subline*, LNCS **3100**, Springer, Heidelberg, 187–208.

23. E. Orłowska (1984). Semantics of Vague Conepts. In: G. Dorn, P. Weingartner (eds.), *Foundation of Logic and Linguistics*, Plenum Press, New York, 465–482.

24. E. Orłowska (1987). *Reasoning about Vague Concepts. Bull. Polish Acad. Sci., Math.* **35**, 643–652.

25. S.K. Pal, A. Skowron (1999). *Rough Fuzzy Hybridization: A New Trend in Decision-Making.* Springer-Verlag, Singapore.

26. S.K. Pal, P. Mitra (2004). *Pattern Recognition Algorithms for Data Mining.* CRC Press, Boca Raton, Florida.

27. S.K. Pal, L. Polkowski, A. Skowron (Eds.) (2004). *Rough-Neural Computing: Techniques for Computing with Words.* Springer, Heidelberg, 2004.

28. Z. Pawlak (1982). Rough Sets. *Int. J. of Information and Computer Sciences* **11**(5), 341–356.

29. Z. Pawlak (1987). Rough Logic. *Bull. Polish. Acad. Sci., Tech.* **35**(5-6), 253–258.

30. Z. Pawlak (1991). *Rough Sets: Theoretical Aspects of Reasoning about Data.* System Theory, Knowledge Engineering and Problem Solving **9**, Kluwer Academic Publishers, Dordrecht.

31. Z. Pawlak, A. Skowron (1994). Rough Membership Functions. In: R. R. Yager, M. Fedrizzi, J. Kacprzyk (eds.), *Advances in the Dempster-Schafer Theory of Evidence*, John Wiley and Sons, New York, 251–271.

32. J.F. Peters, A, Skowron, P. Synak, S. Ramanna (2003). Rough Sets and Information Granulation. In: Bilgic, T., Baets, D., Kaynak, O. (eds.), *Tenth International Fuzzy Systems Association World Congress IFSA*, Istanbul, Turkey, June 30-July 2, 2003, LNAI **2715**, Springer-Verlag, Heidelberg, 370–377.

33. L. Polkowski (2002). *Rough Sets: Mathematical Foundations.* Physica-Verlag, Heidelberg.

34. L. Polkowski, A. Skowron (1996). Rough Mereology: A New Paradigm for Approximate Reasoning. *International Journal of Approximate Reasoning* **15**, 333–365.

35. L. Polkowski, A. Skowron (2001). Rough Mereological Calculi of Granules: A Rough Set Approach to Computation. *Computational Intelligence* **17**, 472–492.

36. S. Read (1995). *Thinking about Logic. An Introduction to the Philosophy of Logic.* Oxford University Press, Oxford, New York.

37. B. Russell (1903). *The Principles of Mathematics.* London, George Allen & Unwin Ltd., 1st Ed. 1903 (2nd Edition in 1937).

38. B. Russell (1940). *An Inquiry into Meaning and Truth.* George Allen and Unwin, London; W.W. Norton, New York.

39. A. Skowron (2000). Rough Sets in KDD. (plenary lecture) In: Z. Shi, B. Faltings, and M. Musen (eds.), *16-th World Computer Congress (IFIP'2000): Proceedings of Conference on Intelligent Information Processing (IIP'2000)*, Publishing House of Electronic Industry, Beijing, 1–17.

40. A. Skowron (2004). Approximate Reasoning in Distributed Environments. In: Zhong, N., Liu, J. (Eds.), *Intelligent Technologies for Information Analysis*, Springer, Heidelberg, 433-474.

41. A. Skowron (2005). Rough Sets and Vague Concepts. *Fundamenta Informaticae* **64**(1-4): 417–431.

42. A. Skowron, J. Stepaniuk (1996). Tolerance Approximation Spaces. *Fundamenta Informaticae* **27**(2-3): 245–253.
43. A. Skowron, J. Peters (2003). Rough Sets: Trends and Challenges (plenary talk). In: G. Wang, Q. Liu, Y.Y. Yao, A. Skowron (eds.), *Proceedings of the Ninth International Conference on Rough Sets, Fuzzy Sets, Data Mining and Granular Computing (RSFDGrC'2003)*, Chongqing, China, May 26-29, 2003, LNAI **2639**, 25–34.
44. A. Skowron, R. Swiniarski, P. Synak (2005). Approximation Spaces and Information Granulation. *Transactions on Rough Sets III: Journal Subline*, LNCS **3400**, Springer, Heidelberg 2005, 175–189.
45. R. Słowiński, D. Vanderpooten (1997). Similarity Relation as a Basis for Rough Approximations. In: P. Wang (ed.), *Advances in Machine Intelligence and Soft Computing* **4**, Duke University Press, 17-33.
46. J. Swift (1726). *Gulliver's Travels into Several Remote Nations of the World.* London, M, DCC, XXVI.
47. V. Vapnik (1998). *Statistical Learning Theory.* John Wiley & Sons, New York, NY.
48. A. Vitória (2005). A Framework for Reasoning with Rough Sets. Licentiate Thesis, Linköping University 2004. *Transactions on Rough Sets IV: Journal Subline*, LNCS, Springer, Heidelberg (to appear).
49. P. Vopenka (1979). *Mathematics in the Alternative Set Theory.* Teubner, Leipzig.
50. W. Ziarko (1993). Variable Precision Rough Set Model. *Journal of Computer and System Sciences* **46**, 39–59.
51. L.A. Zadeh (1965). Fuzzy Sets. *Information and Control* **8**, 338–353.

On Optimization of Decision Trees

Igor V. Chikalov[1], Mikhail Ju. Moshkov[2], and Maria S. Zelentsova[3]

[1] Intel Labs, Russian Research Center,
30, Turgenev Str., Nizhny Novgorod 603950, Russia
igor.chikalov@intel.com
[2] Institute of Computer Science, University of Silesia,
39, Będzińska St., Sosnowiec, 41-200, Poland
moshkov@us.edu.pl
[3] Faculty of Computing Mathematics and Cybernetics of Nizhny,
Novgorod State University, 23, Gagarina Av., Nizhny Novgorod 603950, Russia
zelentsova_mary@mail.ru

Abstract. In the paper algorithms are considered which allow to consecutively optimize decision trees for decision tables with many-valued decisions relatively different complexity measures such as number of nodes, weighted depth, average weighted depth, etc. For decision tables over an arbitrary infinite restricted information system [5] these algorithms have (at least for the three mentioned measures) polynomial time complexity depending on the length of table description. For decision tables over one of such information systems experimental results of decision tree optimization are described.

Keywords: Decision trees, complexity measures, optimization.

1 Introduction

Decision trees are widely used in different applications as algorithms for task solving and as a way of knowledge representation. Problems of decision tree optimization are very complicated.

We study decision tables with many-valued decisions. A finite nonempty set of decisions is attached to each row of such table. For a given row it is required to find a decision from the set attached to the row. To this end we can choose an arbitrary attribute (column) and ask for the value at intersection of this column and the considered row. As algorithms for this problem solving we use decision trees, and as complexity measures for decision trees we consider number of nodes, weighted depth, average weighted depth, etc.

Many problems studied in rough set theory [7, 8] can be reduced to decision tables with many-valued decisions. Let us consider a finite decision system. Such system is specified by a number of conditional attributes that divide the universe into domains on which these attributes have fixed values. Our aim is to find the value of a decision attribute using only values of conditional attributes. Consider an arbitrary domain. If the decision attribute is constant on this domain then we can find the exact value of the decision attribute on objects from the domain

J.F. Peters and A. Skowron (Eds.): Transactions on Rough Sets IV, LNCS 3700, pp. 18–36, 2005.
© Springer-Verlag Berlin Heidelberg 2005

using only values of conditional attributes. Otherwise, it is impossible to find exact value of the decision attribute for all objects from the domain.

In order to minimize the number of mispredictions the most frequent decision should be reported as answer that is a value of the decision attribute assigned to maximal number of objects in the domain. Note that multiple choice of most frequent decision is possible in some cases. The considered problem of the search of a most frequent decision can be reduced to a decision table with many-valued decisions. In this table rows correspond to domains and each row is labeled by the set of most frequent decisions for the corresponding domain.

In the paper an algorithm is considered which for a given decision table with many-valued decisions constructs a graph describing the set of all reduced decision trees. Also an algorithm of decision tree optimization relative to one of complexity measures is described. The possibilities are discussed for consecutive optimization of decision trees concerning various measures. Some results of the paper generalize the results of [1, 2, 4–6] on the case of decision tables with many-valued decisions.

Note that the considered algorithms have polynomial time complexity depending on the length of table description for decision tables over so-called infinite restricted information systems [5]. In the paper we consider one of such information systems in detail. For considered example an effective implementation of the tree optimization algorithm is described. Potential of sequential optimization, temporal and spatial complexity of the problem are illustrated by results of computational experiments.

2 Reduced Decision Trees

Consider a *decision table* T depicted in Fig. 1.

Here $f_1, \ldots, f_n$ are names of columns (attributes); $w_1, \ldots, w_n$ are natural numbers (weights of columns) each of which could be interpreted as time of computation of the corresponding attribute value; $D_1, \ldots, D_m$ are sets of decisions corresponding to rows each of which is a nonempty finite set of nonnegative integers; $\pi_1, \ldots, \pi_m$ are natural numbers which are interpreted as "probabilities" of rows; δ_{ij} are numbers from the set $E_k = \{0, 1, \ldots, k-1\}$, $k \geq 2$, which are interpreted as values of attributes (we assume that the rows $(\delta_{11}, \ldots, \delta_{1n}), \ldots, (\delta_{m1}, \ldots, \delta_{mn})$ are pairwise different).

Denote $C(T) = \bigcap_{i=1}^{m} D_i$. Elements from the set $C(T)$ will be called *common decisions* of the table T.

w_1 ... w_n	
f_1 ... f_n	
δ_{11} ... δ_{1n}	D_1 π_1
...	... ...
δ_{m1} ... δ_{mn}	D_m π_m

Fig. 1

Let $f_{i_1}, \ldots, f_{i_t} \in \{f_1, \ldots, f_n\}$ and $a_1, \ldots, a_t \in E_k$. Denote by $T(f_{i_1}, a_1) \ldots$ (f_{i_t}, a_t) the sub-table of the table T which consists of all rows of T that on the intersection with columns $f_{i_1}, \ldots, f_{i_t}$ have numbers $a_1, \ldots, a_t$ respectively. Such nonempty tables (including the table T) will be called *separable sub-tables* of the table T. Denote by $S(T)$ the set of separable sub-tables of the table T.

We will fix some natural form of *decision table description* such that the numbers w_i, π_i, δ_{ij} and numbers from sets D_i will be given by binary representation. We will say later about the *length* of decision table description. We will assume that for any table T and for any separable sub-table Θ of the table T the length of the sub-table Θ description is at most the length of the table T description.

One can associate with the table T a game of two players: the first player chooses a row of the table, and the second one must find a number (decision) from the set of decisions corresponding to this row. For this purpose the second player can choose a column (an attribute) and ask for the value at intersection of the chosen row and this column.

One can interpret decision trees, which we will consider, as strategies of the second player. During the work of a decision tree the chosen row will be localized in lesser and lesser separable sub-tables. The process finishes when the obtained separable sub-table will have a common decision.

More formally, a *decision tree* for the table T is a finite directed tree with the root in which each terminal node is labeled by a decision (a number from the set $D_1 \cup \ldots \cup D_m$), each nonterminal node is labeled by an attribute from the set $\{f_1, \ldots, f_n\}$, and for each nonterminal node edges issuing from this node are labeled by some pairwise different numbers from E_k. Let v be an arbitrary node of the considered decision tree. Let us define a sub-table $T(v)$ of the table T. If v is the root then $T(v) = T$. Let v be not the root, and in the path from the root to v nodes be labeled by attributes $f_{i_1}, \ldots, f_{i_t}$ and edges be labeled by numbers $a_1, \ldots, a_t$ respectively. Then $T(v) = T(f_{i_1}, a_1) \ldots (f_{i_t}, a_t)$. It is required that for each row r_j of the table T there exists a terminal node v of the tree such that r_j belongs to the sub-table $T(v)$, and v is labeled by a decision from the set D_j attached to the row r_j.

Denote by $E(T)$ the set of attributes (columns of the table T) having at least two distinct values in the rows of T. For $f_i \in E(T)$ let $E(T, f_i)$ be the set of numbers occurred in the column f_i.

Among decision trees for the table T we select *reduced* decision trees. Suppose that during the work of a decision tree Γ for the table T we came to a node v of the tree and localized the considered row in the separable sub-table $\Theta = T(v)$. Let the table Θ have a common decision. Then the considered node of the decision tree Γ is terminal and is labeled by a common decision from the set $C(\Theta)$. Let the table Θ have no common decisions. Then the node v is labeled by an attribute $f_i \in E(\Theta)$. If $E(\Theta, f_i) = \{a_1, \ldots, a_t\}$ then t edges issue from the node v, and these edges are labeled by numbers $a_1, \ldots, a_t$ respectively.

3 Representation of the Set of Reduced Decision Trees

Consider an algorithm $\mathcal{A}$ that constructs the graph $\Delta(T)$ representing in some sense the set of all reduced decision trees for the table T. Nodes of this graph are some separable sub-tables of the table T. During each step we treat exactly one node and mark this node by the symbol *. We begin with the graph, which consists of one node T, and will finish when all nodes of the graph are treated.

Let the algorithm have performed p steps. The step $(p+1)$ will be performed as follows: if in the considered graph all nodes are treated then the work of the algorithm is finished, and the considered graph is $\Delta(T)$. Let the graph have nodes which are not treated. Choose a node that is not treated. Let this node be a table Θ. If Θ has common decisions then we mark the considered node by the set $C(\Theta)$ of common decisions of Θ, mark it by the symbol * and pass to the step $(p+2)$. Let Θ have no common decisions. For each $f_i \in E(\Theta)$ we draw from the node Θ a bundle of edges. Let $E(\Theta, f_i) = \{a_1, \ldots, a_t\}$. Then we draw t edges from Θ, and mark these edges by pairs $(f_i, a_1), \ldots, (f_i, a_t)$ respectively. These edges enter nodes $\Theta(f_i, a_1), \ldots, \Theta(f_i, a_t)$. If some of these nodes are not in the graph then we add these nodes to the graph. We mark the node Θ by the symbol * and pass to the step $(p+2)$.

It is not difficult to prove the following statement.

Proposition 1. *For any decision table T the algorithm $\mathcal{A}$ constructs the graph $\Delta(T)$ and performs at most $|S(T)|+1$ steps. The time of the algorithm $\mathcal{A}$ work is bounded from above by a polynomial on the number $|S(T)|$ of separable sub-tables of the table T and on the length of the table T description.*

Now for each node of the graph $\Delta(T)$ we describe the set of decision trees corresponding to it. It is clear that $\Delta(T)$ is a directed acyclic graph. A node of such graph will be called *terminal* if there are no edges which issue from this node. We will "move" from terminal nodes, which are labeled by sets of numbers (sets of common decisions), to the node T. Let Θ be a node which is labeled by the set $C(\Theta)$ of common decisions. Then the set of trivial decision trees depicted in Fig. 2, $d \in C(\Theta)$, corresponds to the considered node. Let

Fig. 2

Θ be a node (table) which have no common decisions. Let $f_i \in E(\Theta)$ and $E(\Theta, f_i) = \{a_1, \ldots, a_t\}$. Let $\Gamma_1, \ldots, \Gamma_t$ be decision trees from sets corresponding to the nodes $\Theta(f_i, a_1), \ldots, \Theta(f_i, a_t)$. Then the decision tree depicted in Fig. 3 belongs to the set of decision trees which corresponds to the node Θ. All such decision trees belong to the considered set. This set does not contain any other decision trees. We denote by $D(\Theta)$ the set of decision trees corresponding to the node Θ.

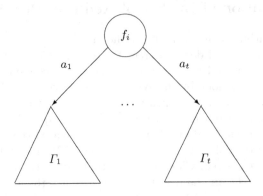

Fig. 3

The following proposition shows that the graph $\Delta(T)$ represents all reduced decision trees for the table T.

Proposition 2. *Let T be a decision table and Θ be a node in the graph $\Delta(T)$. Then the set $D(\Theta)$ coincides with the set of all reduced decision trees for the table Θ.*

Proof. Prove the proposition by induction on nodes in the graph $\Delta(T)$.

For each terminal node Θ only trivial reduced decision trees exist depicted in Fig. 2, where $d \in C(\Theta)$. The set $D(\Theta)$ contains all these trees and does not contain any other trees.

Let Θ be a nonterminal node and the statement of proposition hold for all its descendants. Consider an arbitrary decision tree $\Gamma \in D(\Theta)$. Obviously, Γ contains more than one node. Let the root of Γ be labeled by the attribute f_i and the edges issuing from root be labeled by the numbers $a_1, \ldots, a_t$. For $j = 1, \ldots, t$ we denote by Γ_j the decision tree (a sub-tree of the tree Γ) connected to the root of Γ with the edge labeled by the number a_j. From definition of the set $D(\Theta)$ it follows that f_i is contained in the set $E(\Theta)$, $E(\Theta, f_i) = \{a_1, \ldots, a_t\}$ and for $j = 1, \ldots, t$ the decision tree Γ_j belongs to the set $D(\Theta(f_i, a_j))$. According to the inductive hypothesis, the tree Γ_j is a reduced decision tree for the table $\Theta(f_i, a_j)$. Using these facts one can show that the tree Γ is a reduced decision tree for the table Θ.

Now consider an arbitrary reduced decision tree Γ for the table Θ. Using definition of reduced decision tree one can show that the root of Γ is labeled by an attribute f_i from the set $E(\Theta)$, $|E(\Theta, f_i)|$ edges issue from the root and these edges are labeled by numbers from the set $E(\Theta, f_i)$, and the sub-trees whose roots are nodes, which these edges enter, are reduced decision trees for corresponding descendants of the node Θ. Then, according to definition of the set $D(\Theta)$ and to inductive hypothesis, the tree Γ belongs to the set $D(\Theta)$. $\qquad\square$

4   Proper Subgraphs of the Graph $\Delta(T)$

Let us introduce the notion of a proper subgraph of the graph $\Delta(T)$. For each node of the graph $\Delta(T)$, which is not terminal, we can remove any but not all bundles that issue from the node. Denote the obtained subgraph by G. Such subgraphs will be called *proper subgraphs* of the graph $\Delta(T)$. It is clear that the set of nodes of G coincides with the set of nodes of $\Delta(T)$. It is clear also that all terminal nodes of G are terminal nodes of the graph $\Delta(T)$.

Now for each node of the graph G we describe the set of decision trees corresponding to it. We will "move" from terminal nodes, which are labeled by sets of numbers (sets of common decisions), to the node T. Let Θ be a node which is labeled by the set $C(\Theta)$ of common decisions. Then the set of trivial decision trees depicted in Fig. 2, $d \in C(\Theta)$, corresponds to the considered node. Let Θ be a node (table) which have no common decisions. Let $f_i \in E(\Theta)$, $E(\Theta, f_i) = \{a_1, \ldots, a_t\}$ and there exists a bundle of edges leaving Θ that are labeled by pairs $(f_i, a_1), \ldots, (f_i, a_t)$ respectively. Let $\Gamma_1, \ldots, \Gamma_t$ be decision trees from sets corresponding to the nodes $\Theta(f_i, a_1), \ldots, \Theta(f_i, a_t)$. Then the decision tree depicted in Fig. 3 belongs to the set of decision trees which corresponds to the node Θ. All such decision trees belong to the considered set. This set does not contain any other decision trees. We denote by $D_G(\Theta)$ the set of decision trees corresponding to the node Θ of the graph G.

5   Complexity Measures

We will consider complexity measures which are given in the following way: values of a considered complexity measure ψ, which are nonnegative integers, are defined by induction on pairs (T, Γ), where T is a decision table and Γ is a decision tree for T. Let Γ be a decision tree represented in Fig. 2. Then $\psi(T, \Gamma) = \psi^0$ where ψ^0 is a nonnegative integer. Let Γ be a decision tree depicted in Fig. 3. Then

$$\psi(T, \Gamma) = F(\pi(T), w_i, \psi(T(f_i, a_1), \Gamma_1), \ldots, \psi(T(f_i, a_t), \Gamma_t)) \ .$$

Here $\pi(T)$ is the sum of "probabilities" attached to rows of the table T, w_i is the weight of the column f_i and $F(\pi, w, \psi_1, \psi_2, \ldots)$ is a computable operator which transforms the considered tuple of nonnegative integers into a nonnegative integer. Note that the number of variables $\psi_1, \psi_2, \ldots$ is not bounded from above. So the complexity measure ψ is defined by the pair (ψ^0, F).

The considered complexity measure will be called *monotone* if for any natural i, t, $1 \leq i \leq t - 1$, and any nonnegative integers $a, b, c_1, \ldots, c_t, d_1, \ldots, d_t$ the inequality $F(a, b, c_1, \ldots, c_t) \geq \max\{c_1, \ldots, c_t\}$ holds, the equality $F(a, b, c_1, \ldots, c_i, c_{i+1}, \ldots, c_t) = F(a, b, c_1, \ldots, c_{i+1}, c_i, \ldots, c_t)$ holds, the inequality $F(a, b, c_1, \ldots, c_{t-1}) \leq F(a, b, c_1, \ldots, c_t)$ holds if $t \geq 2$, and from inequalities $c_1 \leq d_1, \ldots, c_t \leq d_t$ the inequality $F(a, b, c_1, \ldots, c_t) \leq F(a, b, d_1, \ldots, d_t)$ follows.

The considered complexity measure will be called *strongly monotone* if it is monotone and for any natural t and any nonnegative integers $a, b, c_1, \ldots, c_t, d_1, \ldots, d_t$

from inequalities $a > 0, b > 0, c_1 \leq d_1, \ldots, c_t \leq d_t$ and inequality $c_i < d_i$, which is true for some $i \in \{1, \ldots, t\}$, the inequality $F(a, b, c_1, \ldots, c_t) < F(a, b, d_1, \ldots, d_t)$ follows.

The following proposition, which is simple corollary of results from [6], describes some properties of linear combination of monotone complexity measures.

Proposition 3. *Let $\varphi_1, \ldots, \varphi_n$ be monotone complexity measures defined by pairs $(\varphi_1^0, F_1), \ldots, (\varphi_n^0, F_n)$, and $\alpha_1, \ldots, \alpha_n$ be natural numbers. Consider complexity measure ψ defined by a pair (ψ^0, F), where $\psi^0 = \sum_{i=1}^n \alpha_i \varphi_i^0$ and $F = \sum_{i=1}^n \alpha_i F_i$. Then the following statements hold:*

a) ψ is a monotone complexity measure;

b) if there exists a number $j \in \{1, \ldots, n\}$ such that φ_j is a strongly monotone complexity measure then ψ is a strongly monotone complexity measure.

Now we take a closer view of some complexity measures.

Number of nodes: $\psi(T, \Gamma)$ is the number of nodes in decision tree Γ. For this complexity measure $\psi^0 = 1$ and $F(\pi, w, \psi_1, \ldots, \psi_t) = 1 + \sum_{i=1}^t \psi_i$. This measure is strongly monotone.

Weighted depth: we attach a weight to each path from the root to a terminal node of tree, which is equal to the sum of weights of attributes attached to nodes of the path. Then $\psi(T, \Gamma)$ is the maximal weight of a path from the root to a terminal node of Γ. For this complexity measure $\psi^0 = 0$ and $F(\pi, w, \psi_1, \ldots, \psi_t) = w + \max\{\psi_1, \ldots, \psi_t\}$. This measure is monotone.

Average weighted depth: for an arbitrary row $\bar{\delta}$ of the table T we denote by $\pi(\bar{\delta})$ its "probability" and by $w(\bar{\delta})$ we denote the weight of the path from the root to a terminal node of Γ which accepts $\bar{\delta}$ (it means that $\bar{\delta}$ belongs to the sub-table $T(v)$ where v is the terminal node of the considered path). Then $\psi(T, \Gamma) = \sum_{\bar{\delta}} w(\bar{\delta})\pi(\bar{\delta})$ where we take the sum over all rows $\bar{\delta}$ of the table T. For this complexity measure $\psi^0 = 0$ and $F(\pi, w, \psi_1, \ldots, \psi_t) = w\pi + \sum_{i=1}^t \psi_i$. This measure is strongly monotone.

The following proposition shows that for any monotone complexity measure among reduced decision trees at least one is optimal in sense of this measure.

Proposition 4. *Let T be a decision table and ψ be a monotone complexity measure. Then there exists a reduced decision tree for T that is optimal relative to the complexity measure ψ.*

Proof. Let Γ be an optimal decision tree for T relative to the complexity measure ψ. We will modify the tree Γ in order to obtain a reduced decision tree. The algorithm sequentially considers nonterminal nodes of the tree Γ. Let u be the current node and f_i be an attribute assigned to the node u. The algorithm tries to apply the following rules to this node.

1. Let $C(T(u)) \neq \emptyset$. Then remove all the descendants of u and mark u by a number $r \in C(T(u))$ instead of the attribute f_i.

2. Let the set $E(T(u), f_i)$ contain only one number a. Denote by Γ_a the decision tree whose root the edge, issuing from u and labeled by a, enters. Then replace the sub-tree whose root is u with Γ_a.

3. For each edge issuing from u and labeled by a number that does not contain in $E(T(u), f_i)$ remove this edge and all descendant nodes and edges.

Since each node is considered at most once, the work of the algorithm is finished after a finite number of steps. Denote the resulted decision tree by $\hat{\Gamma}$. One can show that $\hat{\Gamma}$ is a reduced decision tree for T. Obviously, the applied transformation does not increase the complexity and, thus, $\hat{\Gamma}$ is also optimal.   □

We will say that $\psi = (\psi^0, F)$ is a *strongly polynomial* complexity measure if it satisfies the following conditions:

a) there exists a polynomial such that for any decision table T and for any reduced decision tree Γ for T the length of binary representation of the number $\psi(T, \Gamma)$ is bounded from above by the value of this polynomial on the length of the table T description;

b) the operator F has polynomial time complexity.

Proposition 5. *Number of nodes, weighted depth and average weighted depth are strongly polynomial complexity measures.*

Proof. Let $\psi_1 = (\psi_1^0, F_1)$ be number of nodes, $\psi_2 = (\psi_2^0, F_2)$ be weighted depth and $\psi_3 = (\psi_3^0, F_3)$ be average weighted depth. Consider the decision table T depicted in Fig. 1. Let Γ be a reduced decision tree for T. It is clear that at least two edges issue from each nonterminal node of Γ. Using this fact it is not difficult to prove that the number of nonterminal nodes in Γ is at most the number of terminal nodes. Taking into account that Γ is a reduced decision tree one can show that the number of terminal nodes in Γ is at most m where m is the number of rows in T. Therefore $\psi_1(T, \Gamma) \leq 2m$. Since Γ is a reduced decision tree, nodes in any path from the root of Γ to a terminal node are labeled by pairwise different attributes. Therefore the weight of each such path is at most $\sum_{i=1}^{n} w_i$. From this fact it follows that $\psi_2(T, \Gamma) \leq \sum_{i=1}^{n} w_i$ and $\psi_3(T, \Gamma) \leq (\sum_{i=1}^{n} w_i) \times (\sum_{i=1}^{m} \pi_i)$. Using the obtained inequalities one can show that for each $i \in \{1, 2, 3\}$ there exists a polynomial such that for any decision table T and for any reduced decision tree Γ for T the length of binary representation of the number $\psi_i(T, \Gamma)$ is bounded from above by the value of this polynomial on the length of the table T description. It is clear that the operators F_1, F_2 and F_3 have polynomial time complexities. Therefore ψ_1, ψ_2 and ψ_3 are strongly polynomial complexity measures.   □

6   Procedure of Optimization

Let G be a proper subgraph of the graph $\Delta(T)$, and ψ be a complexity measure defined by the pair (ψ^0, F). Describe an algorithm $\mathcal{B}(\psi)$ which transforms the graph G into a proper subgraph G_ψ of G.

We begin from terminal nodes and move to the node T. We attach a number to each node, and possibly remove some bundles of edges, which start in the considered node. During each step of the algorithm we treat exactly one node of the graph G. We attach the number ψ^0 to each terminal node. Consider a node Θ, which is not terminal, and a bundle of edges, which starts in this node. Let edges be labeled by pairs $(f_i, a_1), \ldots, (f_i, a_t)$, and edges enter to nodes $\Theta(f_i, a_1), \ldots, \Theta(f_i, a_t)$, to which numbers $\psi_1, \ldots, \psi_t$ are attached already. Then we attach to the considered bundle the number $F(\pi(\Theta), w_i, \psi_1, \ldots, \psi_t)$.

Among numbers attached to bundles starting in Θ we choose the minimal number p and attach it to the node Θ. We remove all bundles starting in Θ to which numbers are attached that are greater than p. When all nodes will be treated we obtain a graph. Denote this graph by G_ψ.

It is clear that G_ψ is a proper subgraph of the graph $\Delta(T)$. As it was done previously, for any node Θ of G_ψ we denote by $D_{G_\psi}(\Theta)$ the set of decision trees corresponding to Θ in the graph G_ψ.

Proposition 6. *Let $\psi = (\psi^0, F)$ be a strongly polynomial complexity measure, T be a decision table, and G be a proper subgraph of the graph $\Delta(T)$. Then the algorithm $\mathcal{B}(\psi)$ constructs the proper subgraph G_ψ of the graph G and performs at most $|S(T)|$ steps. The time of the algorithm $\mathcal{B}(\psi)$ work is bounded from above by a polynomial on $|S(T)|$ and the length of the table T description.*

Proof. It is clear that the number of the algorithm $\mathcal{B}(\psi)$ steps coincides with the number of nodes in the graph G, and the number of nodes in the graph G is at most $|S(T)|$. Therefore the algorithm $\mathcal{B}(\psi)$ performs at most $|S(T)|$ steps.

Taking into account that ψ is a strongly polynomial complexity measure we conclude that there exists a polynomial Q such that for any decision table T' and for any reduced decision tree Γ for T' the length of binary representation of the number $\psi(T', \Gamma)$ is bounded from above by the value of Q on the length of the table T' description. It is not difficult to prove by induction on nodes of G that for each node Θ of the graph G the number, attached to Θ during the work of the algorithm $\mathcal{B}(\psi)$, is the complexity (relative to the complexity measure ψ) of a decision tree Γ from $D_G(\Theta)$. It is clear that $D_G(\Theta) \subseteq D(\Theta)$. Using Proposition 2 we conclude that Γ is a reduced decision tree for the table Θ. Therefore for each node Θ of the graph G the length of binary representation of the number, attached to Θ during the work of the algorithm $\mathcal{B}(\psi)$, is at most the value of the polynomial Q on the length of the table T description. Using this fact and taking into account that the operator F has polynomial time complexity one can show that the time of the algorithm $\mathcal{B}(\psi)$ work is bounded from above by a polynomial on $|S(T)|$ and the length of the table T description. $\square$

7   Results for Monotone Complexity Measures

Let T be a decision table and ψ be a monotone complexity measure. Let G be a proper subgraph of $\Delta(T)$ and Θ be an arbitrary node in G. We will denote by $D_{\psi,G}(\Theta)$ the subset of $D_G(\Theta)$ containing all decision trees having minimal complexity relative to ψ, i.e. $D_{\psi,G}(\Theta) = \{\hat{\Gamma} : \hat{\Gamma} \in D_G(\Theta), \psi(\Theta, \hat{\Gamma}) = \min\{\psi(\Theta, \Gamma) : \Gamma \in D_G(\Theta)\}\}$.

The following theorem shows that optimized graph G_ψ describes only optimal decision trees in sense of chosen complexity measure.

Theorem 1. *Let T be a decision table and ψ be a monotone complexity measure defined by the pair (ψ^0, F). Let G be a proper subgraph of $\Delta(T)$ and Θ be an arbitrary node in the graph G. Then $D_{G_\psi}(\Theta) \subseteq D_{\psi,G}(\Theta)$.*

Preface proof of the theorem by the following lemma.

Lemma 1. *Let T be a decision table and ψ be a monotone complexity measure defined by the pair (ψ^0, F). Let G be a proper subgraph of $\Delta(T)$, Θ be an arbitrary node in the graph G and p be a number assigned to the node Θ by the algorithm $\mathcal{B}(\psi)$. Then for each decision tree Γ from the set $D_{G_\psi}(\Theta)$ the equality $\psi(\Theta, \Gamma) = p$ holds.*

Proof. Prove the considered statement by induction on nodes in the graph G. For each terminal node Θ only trivial reduced decision trees exist depicted in Fig. 2, and the statement of lemma obviously holds for Θ. Let now Θ be a nonterminal node and the statement of lemma hold for all descendants of Θ. Consider an arbitrary decision tree $\Gamma \in D_{G_\psi}(\Theta)$. Let the root of Γ be labeled by the attribute f_i and the edges issuing from the root be labeled by the numbers $a_1, \ldots, a_t$. For $j = 1, \ldots, t$ denote by Γ_j the decision tree (sub-tree of Γ) connected to the root with the edge labeled by the number a_j. Let for $j = 1, \ldots, t$ the node $\Theta(f_i, a_j)$ be labeled by the number p_j. According to the inductive hypothesis, the equality $\psi(\Theta(f_i, a_j), \Gamma_j) = p_j$ holds for $j = 1, \ldots, t$. According to the algorithm $\mathcal{B}(\psi)$ description, $p = F(\pi(\Theta), w_i, p_1, \ldots, p_t)$. From the definition of the complexity measure ψ it follows that $\psi(\Theta, \Gamma) = F(\pi(\Theta), w_i, \psi(\Theta(f_i, a_1), \Gamma_1), \ldots, \psi(\Theta(f_i, a_t), \Gamma_t))$. Using the three last equalities we obtain $\psi(\Theta, \Gamma) = p$. Since Γ is an arbitrary tree from $D_{G_\psi}(\Theta)$, all the trees in $D_{G_\psi}(\Theta)$ have the same complexity p. $\square$

Proof (of Theorem 1). The statement of the theorem will be proved by induction on nodes of the graph G. Let Θ be a terminal node. Then the set $D_{G_\psi}(\Theta)$ contains only trees depicted in Fig. 2, $d \in C(\Theta)$, and these trees, obviously, belong to $D_{\psi,G}(\Theta)$. So the statement of the theorem holds for the node Θ.

Let now Θ be a nonterminal node in G and the statement of the theorem hold for any descendant of Θ in the graph G. Let the number p be assigned to the node Θ by the algorithm $\mathcal{B}(\psi)$. Lemma 1 implies that all decision trees in $D_{G_\psi}(\Theta)$ have the same complexity p. Consider an arbitrary decision tree Γ from the set $D_{\psi,G}(\Theta)$. From the definition of the set $D_{\psi,G}(\Theta)$ it follows that $\psi(\Gamma, \Theta) \leq p$.

To prove the statement of the theorem we need to show that $\psi(\Gamma, \Theta) = p$. Let the root of Γ be labeled by the attribute f_i. Since Γ is a reduced decision tree, f_i is contained in the set $E(\Theta)$. Let $E(\Theta, f_i) = \{a_1, \ldots, a_t\}$. Then t edges issue from the root and these edges are labeled by numbers $a_1, \ldots, a_t$. For $j = 1, \ldots, t$ denote by Γ_j the sub-tree of Γ that is connected to the root with the edge labeled by a_j. It is clear that Γ_j is contained in the set $D_G(\Theta(f_i, a_j))$. Let p_j be the number assigned to the node $\Theta(f_i, a_j)$ during the process of the algorithm $\mathcal{B}(\psi)$ work. Taking into account that the statement of the theorem holds for the node $\Theta(f_i, a_j)$ and using Lemma 1 we obtain $\psi(\Gamma_j, \Theta(f_i, a_j)) \geq p_j$. Let w_i be the weight of the attribute f_i. From the description of the algorithm $\mathcal{B}(\psi)$

it follows that $F(\pi(\Theta), w_i, p_1, \ldots, p_t) \geq p$. Since ψ is a monotone complexity measure, we have $\psi(\Gamma, \Theta) = F(\pi(\Theta), w_i, \psi(\Gamma_1, \Theta(f_i, a_1)), \ldots, \psi(\Gamma_t, \Theta(f_i, a_t))) \geq F(\pi(\Theta), w_i, p_1, \ldots, p_t)$. The equality $\psi(\Gamma, \Theta) = p$ follows from the two last inequalities and the inequality $\psi(\Gamma, \Theta) \leq p$. $\qquad\square$

8   Results for Strongly Monotone Complexity Measures

Theorem 2. *Let T be a decision table and ψ be a strongly monotone complexity measure. Let G be a proper subgraph of $\Delta(T)$ and Θ be an arbitrary node in the graph G. Then the set $D_{G_\psi}(\Theta)$ coincides with the set $D_{\psi,G}(\Theta)$.*

Proof. Since ψ is strongly monotone, ψ is monotone. Using Theorem 1 obtain $D_{G_\psi}(\Theta) \subseteq D_{\psi,G}(\Theta)$. Let us prove that for an arbitrary tree $\Gamma \in D_{\psi,G}(\Theta)$ the tree Γ belongs to the set $D_{G_\psi}(\Theta)$. The induction on the nodes of G will be used. If Θ is a terminal node then, as it is not difficult to show, $D_{G_\psi}(\Theta) = D_{\psi,G}(\Theta) = D_G(\Theta)$. Therefore the statement of the theorem holds for the node Θ. Let Θ be a nonterminal node and the statement of theorem hold for all descendants of Θ. Let the root of the tree Γ be labeled by the attribute f_i. Since Γ is a reduced decision tree, f_i is contained in the set $E(\Theta)$. Let $E(\Theta, f_i) = \{a_1, \ldots, a_t\}$. Then t edges issuing from the root are labeled by the numbers $a_1, \ldots, a_t$. For $j = 1, \ldots, t$ denote by Γ_j the sub-tree that is connected to the root with the edge labeled by a_j. Since ψ is a strongly monotone complexity measure and $\Gamma \in D_{\psi,G}(\Theta)$, the tree Γ_j belongs to the set $D_{\psi,G}(\Theta(f_i, a_j))$. Since the statement of the theorem holds for the node $\Theta(f_i, a_j)$, the tree Γ_j belongs to the set $D_{G_\psi}(\Theta(f_i, a_j))$ for $j = 1, \ldots, t$. Consider the bundle of edges in the graph $\Delta(T)$ that leave the node Θ and are labeled by the pairs $(f_i, a_1), \ldots, (f_i, a_t)$. Since $\Gamma \in D_{\psi,G}(\Theta)$, these edges were not removed by the algorithm $\mathcal{B}(\psi)$. Then, according to the definition of the set $D_{G_\psi}(\Theta)$, the tree Γ belongs to this set. $\qquad\square$

9   Possibilities of Consecutive Optimization

Let the graph $\Delta(T)$ be constructed for a decision table T by the algorithm $\mathcal{A}$. Let ψ_1 and ψ_2 be strongly monotone complexity measures. Apply the algorithm $\mathcal{B}(\psi_1)$ to the graph $\Delta(T)$. As a result we obtain the proper subgraph $(\Delta(T))_{\psi_1}$ of the graph $\Delta(T)$. Denote this subgraph by G_1. According to Proposition 2 and Theorem 2, the set of decision trees corresponding to the node T of this graph coincides with the set of all reduced decision trees for the table T, which have minimal complexity relative to ψ_1. Denote this set by D_1. Using Proposition 4 we conclude that decision trees from D_1 are optimal relative to ψ_1 not only among reduced decision trees but also among all decision trees for the table T.

Apply the algorithm $\mathcal{B}(\psi_2)$ to the graph G_1. As a result we obtain the proper subgraph $(G_1)_{\psi_2}$ of the graph $\Delta(T)$. Denote this subgraph by G_2. The set of decision trees corresponding to the node T of this graph coincides with the set of all decision trees from D_1, which have minimal complexity relative to ψ_2. It is possible to continue this process of consecutive optimization concerning various criteria.

If ψ_2 is a monotone complexity measure then according to Theorem 1 the set of decision trees, corresponding to the node T of the graph G_2, is a subset of the set of all decision trees from D_1, which have minimal complexity relative to ψ_2.

10   Decision Tables over Information Systems

Let A be a nonempty set, F be a nonempty set of functions from A to $E_k = \{0, \ldots, k-1\}$, and $f \not\equiv \text{const}$ for any $f \in F$. Functions from F will be called *attributes*, and the pair $U = (A, F)$ will be called a *k-valued information system*. Let $f_1, \ldots, f_m \in F$. We will say that the table T depicted in Fig. 1 is a *decision table over the information system U* if the set $\{(\delta_{11}, \ldots, \delta_{1n}), \ldots, (\delta_{m1}, \ldots, \delta_{mn})\}$ of the table T rows coincides with the set of tuples $(\delta_1, \ldots, \delta_n) \in E_k^n$ such that the system of equations

$$\{f_1(x) = \delta_1, \ldots, f_n(x) = \delta_n\} \tag{1}$$

is compatible (has a solution) on the set A. The number of attributes (columns) n in the table T will be called the *dimension* of the table T and will be denoted by $\dim T$. Denote by $T(U)$ the set of decision tables over U. Consider the function

$$C_U(n) = \max\{|S(T)| : T \in T(U), \dim T \leq n\}$$

which characterizes the maximal number of separable sub-tables depending on the number of columns in decision tables over U.

Let ψ be a strongly polynomial complexity measure. Using Propositions 1 and 6 we conclude that for tables over U time complexity of the algorithms $\mathcal{A}$ and $\mathcal{B}(\psi)$ is bounded from above by some polynomials on the length of table description if the function $C_U(n)$ is bounded from above by a polynomial on n. Now we consider the criterion of the function $C_U(n)$ to be bounded from above by a polynomial.

A system of equations of the kind (1) will be called a *system of equations over U*. Two systems of equations are called *equivalent* if they have the same set of solutions on A. A compatible system of equations will be called *uncancellable* if each its proper subsystem is not equivalent to the system. Let r be a natural number. An information system U will be called *r-restricted* if each uncancellable system of equations over U consists of at most r equations. An information system U will be called *restricted* if it is r-restricted for some natural r. The following proposition was proved in [5].

Proposition 7. *Let $U = (A, F)$ be a k-valued information system. Then the following statements hold:*

a) if U is r-restricted information system then $C_U(n) \leq (nk)^r + 1$ for any natural n;

b) if U is not a restricted information system then $C_U(n) \geq 2^n$ for any natural n.

The following example was considered in [5].

Example 1. Denote by P the set of all points in a plane. Consider an arbitrary straight line l, which divides the plane into positive and negative open half-planes and the line l. Assign a function $f : P \to E_2$ to the straight line l. The function f takes the value 1 if a point is situated in the positive half-plane, and f takes the value 0 if a point is situated in the negative half-plane or in the line l. Denote by F an infinite set of functions, which correspond to some straight lines in the plane. Consider two cases.

1) Functions from the set F correspond to t infinite classes of parallel straight lines. One can show that the information system U is $2t$-restricted.

2) Functions from the set F correspond to all straight lines on the plane. One can show that the information system U is not restricted.

It is possible to consider not only decision tables over an information system U but also sub-tables of such tables. A table T' will be called a *sub-table* of a decision table T if T' can be obtained from T by removal of some rows. It is clear that T is a sub-table of T. Denote by $\mathcal{T}^*(U)$ the set of sub-tables of decision tables over U. Consider the function

$$C_U^*(n) = \max\{|S(T)| : T \in \mathcal{T}^*(U), \dim T \le n\} .$$

One can show that $C_U^*(n) = C_U(n)$ for any natural n. Using this equality and Propositions 1, 6 and 7 it is not difficult to prove the following statement.

Proposition 8. *Let U be a restricted information system and ψ be a strongly polynomial complexity measure. Then for sub-tables of decision tables over U time complexity of the algorithms $\mathcal{A}$ and $\mathcal{B}(\psi)$ is bounded from above by some polynomials on the length of sub-table description.*

In the next sections we will consider sub-tables of decision tables over 4-restricted 2-valued information system $U = (P, F)$, where P is the set of points in a plane, and functions from F correspond to all straight lines in the plane each of which is parallel to a coordinate axis.

11   Problem of Classification of Points in a Plane

In this section we consider the following problem. Let A be a finite set of black and white points in the plane. For a given point from A one should recognize the color of this point using values of attributes each of which corresponds to a straight line that is parallel to a coordinate axis. The problem could be represented in form of a decision table that is in fact a sub-table of some decision table over information system $U = (P, F)$ described at the end of the previous section. Since U is restricted, Proposition 8 implies that there is a polynomial algorithm for consecutive optimization of decision trees for the considered problem relatively various complexity measures. In this section we consider a software implementation of such algorithm.

Similar problem is search for a set of lines of minimal cardinality that divides plane into regions such that there is no region containing both black and white points. The latter problem is proven to be NP-hard in [3].

Let us describe a decision table representing the problem of color recognition for points from A. Denote XOY coordinate system in the plane and n the number of points in A. Assume for simplicity that all the points are pairwise different on each coordinate. Project all points to both coordinate axes and denote by $x_1, \ldots, x_n$ and $y_1, \ldots, y_n$ coordinates of points in the ascending order. For $i = 1, \ldots, n - 1$ denote by l_i^x the vertical line passing by the point $((x_i + x_{i+1})/2, 0)$ and denote by l_i^y the horizontal line passing by the point $(0, (y_i + y_{i+1})/2)$. Denote by T the decision table, in which columns correspond to the lines $l_1^x, \ldots, l_{n-1}^x, l_1^y, \ldots, l_{n-1}^y$, all probabilities and attribute weights are equal to 1 and each decision set D_i assigned to a row is equal to $\{0\}$ if the color of the corresponding point is white, and $\{1\}$ otherwise.

The above-mentioned general algorithm of decision tree optimization is applicable to T. Now we describe an effective implementation of the algorithm that uses specific properties of the table T.

First the procedure of graph traversal will be described. Then two algorithms will be considered that optimize a subgraph of $\Delta(T)$ and build a single decision tree described by a subgraph of $\Delta(T)$ respectively.

Let us precede algorithm description by some auxiliary notions. Choose numbers x_0, x_{n+1}, y_0, y_{n+1} such that $x_0 < x_1$, $x_{n+1} > x_n$, $y_0 < y_1$, $y_{n+1} > y_n$. Denote by l_0^x, l_n^x the vertical lines passing by the points $(x_0, 0)$, $(x_{n+1}, 0)$ and l_0^y, l_n^y the horizontal lines passing by the points $(0, y_0)$, $(0, y_{n+1})$. One can see that for any separable sub-table τ of T its set of rows corresponds to all points from A in the rectangle bounded by lines $l_t^x, l_u^x, l_v^y, l_w^y$ for some numbers $0 \leq t < u \leq n$, $0 \leq v < w \leq n$. Then each sub-table could be identified by an unique quartet of numbers. One can see that for sub-table τ identified by a quartet $\langle t, u, v, w \rangle$ the set $E(\tau)$ contains attributes $l_{t+1}^x, l_{t+2}^x, \ldots, l_{u-1}^x, l_{v+1}^y, l_{v+2}^y, \ldots, l_{w-1}^y$ and does not contain other attributes that allows to quickly restore set $E(\tau)$ by the sub-table identifier.

Describe data structure used for storing of a graph $\Delta(T)$ and its sub-graphs. Balance between time and space complexity forces us to avoid keeping edges of $\Delta(T)$. Instead for each node the edges are rebuilt upon request by a simple procedure. Thus $\Delta(T)$ is stored in an associative list C where key is a quartet of numbers identifying a sub-table. After q consecutive steps of optimization the data associated to each record of C is a vector of q values. For $i = 1, \ldots, q$ denote by ψ_i complexity measure used at the i-th step of optimization and G_i the resulted sub-graph of $\Delta(T)$. Let τ be a separable sub-table of T and $\langle z_1, \ldots, z_q \rangle$ the corresponding associated data vector. Then for $i = 1 \ldots, q$ the value z_i is the minimal complexity (relatively to ψ_i) of a decision tree for τ described by graph G_i. Let f_l be an attribute from $E(\tau)$ and $\langle z_1^0, \ldots, z_q^0 \rangle$, $\langle z_1^1, \ldots, z_q^1 \rangle$ associated data vectors for $\tau(f_l, 0)$, $\tau(f_l, 1)$ respectively (if $\tau(f_l, \delta)$ is terminal then assume that $\langle z_1^\delta, \ldots, z_q^\delta \rangle$ is $\langle \psi_1^0, \ldots, \psi_q^0 \rangle$). To check whether the split of τ by f_l remain in the graph G_q one should check the equalities $z_i = F_i(\pi(\tau), w_l, z_i^0, z_i^1)$ for $i = 1, ..., q$. Violation at least one equality means that the split was removed by optimization procedure.

Further a graph traversal procedure is described. The procedure is used by several algorithms such as consecutive optimization of $\Delta(T)$ and counting the number of trees described by a subgraph of $\Delta(T)$. Using this procedure each algorithm can be described as specific *Callback()* procedure performed at examining of a split.

11.1   Procedure of Graph Traversal

The graph traversal procedure extends the associated data vector for all records in C by a flag indicating that the sub-table was already processed. The procedure uses a stack O for storing identifiers of "open" sub-tables (for which processing is started but not finished yet). Each record of O also contains a spilt counter that iterates among splits from the set $E(\tau)$.

Step 0. Push the identifier $\langle 0, n, 0, n \rangle$ to O. Clear "processed" flag in all records of C. Proceed to the step 1.

Step i.
If O is empty then the algorithm finishes its work. Otherwise let t be an identifier at the top of O, τ sub-table having identifier t and l the split counter for t. Let $E(\tau) = \{f_1, \ldots, f_m\}$.
If $l > m$ then mark the record $\langle t, l \rangle$ in C as "processed", pop the record from O and proceed to the next step. Otherwise check whether the split f_l was removed at previous optimization step.
If it is true increase l and proceed to the next step. Otherwise call *Callback($\langle t, l \rangle$)*. If for $\delta = 0$ or $\delta = 1$ the table $\tau(f_l, \delta)$ is nonterminal and corresponding node is not marked in C as "processed" (add new record to C if there is no record with such identifier) then push identifier of $\tau(f_l, \delta)$ to O. Increase l if no records were pushed to O. Proceed to the next step.

11.2   Optimization of Sub-graph $\Delta(T)$

Let q optimization steps have been already done and ψ_{q+1} be the complexity measure for the current step. Extend associated data vector for all records in C by z_{q+1} value and set this value to be infinite. The *Callback* procedure for a sub-table τ and a split f_l updates the value z_{q+1} if both sub-tables $\tau(f_l, 0)$ and $\tau(f_l, 1)$ are either terminal or corresponding records are marked as "processed". For $\delta = 0, 1$ let z_{q+1}^δ be the last element of associated data vector for $\tau(f_l, \delta)$ or ψ_{q+1}^0 if the sub-table is terminal. Then z_{q+1} is assigned the value $\min(z_{q+1}, \psi_{q+1}(\pi(\tau), w_l, z_{q+1}^0, z_{q+1}^1))$.

11.3   Building of a Single Decision Tree Described by a Sub-graph

We assume that the optimization procedure was applied to $\Delta(T)$ at least once. Denote by G the decision tree being constructed. Denote by Y the number of nodes added to the tree. Associate with each record in O the number of the last node added to the decision tree. At start of algorithm add a single node to G and associate its number to the record on the top of O. Describe a single step of the algorithm.

If C is empty then work of algorithm is finished and G is the resulted tree. Else let t be the record at the top of O, τ the corresponding sub-table and m the associated node in G. First the algorithm finds a split that was not removed at optimization using the procedure described above. Once split is found the algorithm assigns the corresponding attribute f_l to the node m, adds two nodes $(Y + 1)$ and $(Y + 2)$ to the tree and connects them to m with edges labeled with 0 and 1 respectively. The record t is popped from O. Then for $\delta = 0, 1$ the algorithm checks whether the table $\tau(f_l, \delta)$ is terminal. If it is true the common decision is assigned to the node $(Y + 1 + \delta)$. Otherwise the corresponding record is pushed to C. The algorithm proceed to the next step.

12   Experimental Results

In this section experimental results are presented that characterize complexity of the stated optimization problem and computational effectiveness of the algorithm. The algorithm described above was implemented as a part of a research software system that allows for consecutive optimization of decision trees against multiple complexity measures. The system counts number of nodes in the graph $\Delta(T)$ and total number of reduced decision trees for the problem as well as minimal value of complexity measure and number of optimal decision trees at each step of optimization. The system is capable of processing problems of size up to 200 points at a desktop PC.

In experiments we considered a class of problems where both coordinates and color of points were randomly chosen from an uniform distribution. Decision trees were optimized by three complexity measures that are depth, average depth and number of nodes. We studied dependence on the number of points for the following parameters:

- the number of nodes in the graph $\Delta(T)$;
- the number of reduced decision trees;
- the number of optimal decision trees for given complexity measure;
- the minimal value of complexity measure;
- execution time (measured at a desktop PC with Pentium 4 1.4GHz CPU and 512MB of RAM).

Table 1 shows results of computational experiments. One can see that the number of optimal decision trees for a monotone complexity measure (depth) is larger than for strongly monotone ones (average depth and number of nodes). Also it should be noted the big number of optimal decision trees that enables rich possibilities for a consecutive optimization.

Figure 4 shows dependence of execution time of the algorithm on the number of points and its polynomial approximation.

Table 1. Dependency of several problem characteristics on number of points (mean values of 10 experiments are shown). The columns contain (left to right) number of points in the problem, number of nodes in the graph $\Delta(T)$, number of reduced decision trees described by $\Delta(T)$, minimal value and number of optimal reduced decision trees for each of three above-mentioned complexity measures, execution time.

# of pts.	# of nodes	# of trees	# of nodes trees	value	Depth trees	value	Av. Depth trees	value	Time, sec.
10	$1.5 \cdot 10^2$	$2.5 \cdot 10^7$	$1.0 \cdot 10^2$	7.6	$1.6 \cdot 10^3$	2.5	8.8	2.1	0.12
20	$1.7 \cdot 10^3$	$1.0 \cdot 10^{19}$	$9.3 \cdot 10^3$	13.6	$5.9 \cdot 10^8$	3.3	$2.4 \cdot 10^2$	2.8	0.22
30	$7.3 \cdot 10^3$	$9.9 \cdot 10^{31}$	$2.4 \cdot 10^6$	19.6	$2.7 \cdot 10^9$	4.0	$1.9 \cdot 10^3$	3.3	0.71
40	$2.3 \cdot 10^4$	$4.7 \cdot 10^{45}$	$1.1 \cdot 10^9$	24.7	$3.5 \cdot 10^{21}$	4.1	$1.0 \cdot 10^5$	3.7	3.0
50	$5.1 \cdot 10^4$	$1.0 \cdot 10^{60}$	$6.5 \cdot 10^{10}$	30.4	$3.3 \cdot 10^{25}$	4.8	$1.0 \cdot 10^7$	3.9	9.3
60	$1.0 \cdot 10^5$	$7.7 \cdot 10^{74}$	$2.7 \cdot 10^{14}$	36.3	$9.8 \cdot 10^{26}$	5.0	$7.6 \cdot 10^7$	4.2	24.7
70	$1.9 \cdot 10^5$	$4.2 \cdot 10^{90}$	$6.2 \cdot 10^{16}$	40.8	$7.9 \cdot 10^{26}$	5.0	$4.0 \cdot 10^9$	4.3	56.6
80	$3.2 \cdot 10^5$	$6.6 \cdot 10^{105}$	$7.8 \cdot 10^{17}$	46.1	$2.0 \cdot 10^{26}$	5.0	$3.5 \cdot 10^{10}$	4.5	124.8
90	$5.1 \cdot 10^5$	$3.7 \cdot 10^{122}$	$2.2 \cdot 10^{22}$	51.7	$9.6 \cdot 10^{58}$	5.3	$3.4 \cdot 10^{12}$	4.7	242.0
100	$7.7 \cdot 10^5$	$3.5 \cdot 10^{139}$	$4.2 \cdot 10^{22}$	57.0	$1.9 \cdot 10^{62}$	5.7	$2.0 \cdot 10^{16}$	4.7	454.0

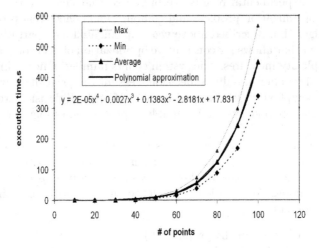

Fig. 4. Dependence of execution time on the number of points (maximal, minimal and average values over 10 experiments are shown)

When sequential optimization is performed one can be interested in finding a decision tree that is optimal according to all considered complexity measures. However it is often impossible. For example consider a multicriteria optimization of decision trees by depth and average depth. Draw each decision tree described by $\Delta(T)$ as a point in a plane whose coordinates are values of corresponding complexity measures. Consider south-west frontier of the convex hull for the resulted set of points, and select points that belongs to this line. The selected points correspond to decision trees that have undominated characteristics. Sequential optimization is capable of finding the leftmost and the rightmost points

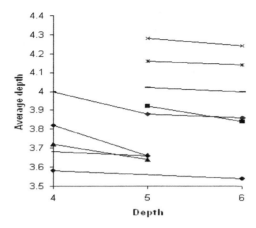

Fig. 5. Relation between depth and average depth of decision tree resulted in consecutive optimization (pairs of points showing different optimization sequence for the same problem are connected by line)

depending on order of applying the complexity measures. Figure 5 shows results of sequential optimization for several problems, each of which contains 50 points. One can see that even for small dimensionality of a problem there are decision trees with undominated characteristics. Thus the order in which optimization measures are applied impacts the result.

13   Conclusion

We have described algorithms which allow to optimize decision trees consecutively concerning such complexity measures as number of nodes, weighted depth, average weighted depth. These algorithms may be useful for a detailed investigation of the decision tree set for relatively small decision tables. We have considered examples of such investigations for decision tables connected with the problem of recognition of point color in the plane.

References

1. Chikalov, I.V.: Algorithm for constructing of decision trees with minimal number of nodes. Proceedings of the Second International Conference on Rough Sets and Current Trends in Computing. Banff, Canada (2000) 107–111
2. Chikalov, I.V.: Algorithm for constructing of decision trees with minimal average depth. Proceedings of the Eighth International Conference on Information Processing and Management of Uncertainty in Knowledge-based Systems, Vol. 1. Madrid, Spain (2000) 376–379
3. Chlebus, B.S., Nguyen, S.H.: On finding optimal discretization for two attributes. Proceedings of the First International Conference on Rough Sets and Current Trends in Computing. Warsaw, Poland. Lecture Notes in Artificial Intelligence **1424**, Springer-Verlag (1998) 537–544

4. Moshkov, M.Ju., Chikalov, I.V.: On effective algorithms for construction of decision trees. Proceedings of the Twelfth International Conference Problems of Theoretical Cybernetics, Part 2. Nizhny Novgorod, Russia (1999) 165 (in Russian)
5. Moshkov, M.Ju., Chikalov, I.V.: On algorithm for constructing of decision trees with minimal depth. Fundamenta Informaticae **41**(3) (2000) 295–299
6. Moshkov, M.Ju., Chikalov, I.V.: Consecutive optimization of decision trees concerning various complexity measures. Fundamenta Informaticae **61**(2) (2004) 87–96
7. Pawlak, Z.: Rough Sets – Theoretical Aspects of Reasoning about Data. Kluwer Academic Publishers, Dordrecht, Boston, London, 1991
8. Skowron, A., Rauszer, C.: The discernibility matrices and functions in information systems. Intelligent Decision Support. Handbook of Applications and Advances of the Rough Set Theory. Edited by R. Slowinski. Kluwer Academic Publishers, Dordrecht, Boston, London (1992) 331–362

Dealing with Missing Data: Algorithms Based on Fuzzy Set and Rough Set Theories*

Dan Li[1], Jitender Deogun[1], William Spaulding[2], and Bill Shuart[2]

[1] Department of Computer Science & Engineering,
University of Nebraska-Lincoln, Lincoln NE 68588-0115
[2] Department of Psychology, University of Nebraska-Lincoln,
Lincoln NE 68588-0308

Abstract. Missing data, commonly encountered in many fields of study, introduce inaccuracy in the analysis and evaluation. Previous methods used for handling missing data (e.g., deleting cases with incomplete information, or substituting the missing values with estimated mean scores), though simple to implement, are problematic because these methods may result in biased data models. Fortunately, recent advances in theoretical and computational statistics have led to more flexible techniques to deal with the missing data problem. In this paper, we present missing data imputation methods based on clustering, one of the most popular techniques in Knowledge Discovery in Databases (KDD). We combine clustering with soft computing, which tends to be more tolerant of imprecision and uncertainty, and apply fuzzy and rough clustering algorithms to deal with incomplete data. The experiments show that a hybridization of fuzzy set and rough set theories in missing data imputation algorithms leads to the best performance among our four algorithms, i.e., crisp K-means, fuzzy K-means, rough K-means, and rough-fuzzy K-means imputation algorithms.

Keywords: Missing data, imputation, K-means clustering, fuzzy sets, rough sets, rough-fuzzy hybridization.

1 Introduction

The problem of missing (or incomplete) data is relatively common in many fields, and it may have different causes such as equipment malfunction, unavailability of equipment, refusal of respondents to answer certain questions, etc. The overall result is that the observed data cannot be analyzed because of the incompleteness of the data. The most appropriate method for handling missing or incomplete data depends upon how data items became missing. There are three types of missing data mechanisms [1]. First, data are *missing completely at random* (MCAR). This means the probability that an observation x_i is missing is

---

* This work was supported, in part, by a grant from NSF (EIA-0091530), a cooperative agreement with USADA FCIC/RMA (2IE08310228), and an NSF EPSCOR Grant (EPS-0091900).

J.F. Peters and A. Skowron (Eds.): Transactions on Rough Sets IV, LNCS 3700, pp. 37–57, 2005.

unrelated to the value of x_i or to the value of any other variables. In this case, the pattern of data missingness is unpredictable. Second, data are *missing at random* (MAR), and the missingness does not depend on the value of x_i after controlling for another variable. In other words, records with incomplete data differ from records with complete data, but the pattern of data missingness is traceable or predictable from other variables in the database rather than being due to the specific variable on which the data are missing. For example, people who are at higher anxiety level might be less likely to report their income, thus the level of anxiety will be related to the reported income. However, if within anxious patients the probability of reported income was unrelated to income level, then the data would be considered MAR. If data are MCAR or MAR, we say that the missingness is *ignorable*. The third type of missing data is that the missingness is *non-ignorable*. This means the pattern of data missingness is non-random and it cannot be predicted from other variables in the database.

This paper deals with missing data in two applications. First, we are developing a Geospatial Decision Support System (GDSS), with an initial focus on drought risk management [2]. Data are collected at automated weather stations and some data items are missing because of malfunction or unavailability of equipment. This type of missing data is unintended and uncontrolled by the researchers, and data are missing completely at random. Interpolation methods can be employed to handle incomplete datasets [3, 4, 5]. Most conventional interpolation methods such as Kriging, are more suitable for handling data when the data are distributed with relatively high density and the regional conditions are almost homogeneous. However environmental databases must cover a variety of areas with different natural/socio economic conditions, and the distribution of point-based data available for interpolation tends to be much biased. For these reasons, many conventional interpolation methods cannot be effectively utilized to solve missing data problem.

The second application focuses on a psychotherapy study to understand information processing, judgment and decision making in psychiatric diagnosis and rehabilitation. Here, data are missing at random because respondents refuse to answer certain questions or because the answers were not recorded. This paper focuses on situations in which some information is missing from an individual case rather than the total lack of response to a diagnosis.

A number of researchers over the last several decades have investigated techniques for dealing with missing data [6, 7, 8, 9, 1, 10, 11, 12, 13]. Methods for handling missing data can be divided into three categories. The first is *ignoring and discarding data*. *Listwise deletion* and *pairwise deletion* are two widely used methods in this category [7]. The second group includes the methods based on *parameter estimation*, which uses variants of the *Expectation-Maximization* algorithm to estimate parameters in the presence of missing data [6]. The third category is *imputation*, which denotes the process of filling in the missing values in a dataset by some plausible values based on information available in the dataset [10].

Among imputation methods, there are many approaches varying from simple methods such as mean imputation, to some more robust and intricate methods based on the analysis of the relationships among attributes. For example, in hot deck imputation, the missing data are replaced by other cases with the same (or similar) characteristics. These common characteristics are derived from auxiliary variables, e.g., age, gender, race, or education degree, whose values are available from the cases to be imputed. Generally, there are two steps in hot deck imputation [14]. First, data are partitioned into several clusters based on certain similarity metric, and each instance with missing data is associated with one of the clusters. Second, by calculating the mean of the attribute within a cluster, the complete cases in the cluster are used to fill in the missing values.

One of the most well known clustering algorithms is the K-means method [15], which takes the number of desirable clusters, K, as an input parameter, and outputs a partition consisting of K clusters on a set of objects. Conventional clustering algorithms are normally crisp. However, in reality, an object sometimes could be assigned to more than one cluster. Therefore, a fuzzy membership function can be applied to the K-means clustering, which models the degree of an object belonging to a cluster. Additionally, the theory of rough set has emerged as a major method for managing uncertainty in many domains, and has proved to be a useful tool in a variety of KDD processes. The theories of fuzzy set and rough set present the basic idea of soft computing. The soft computing paradigm is to exploit the tolerance for imprecision, uncertainty and partial truth to achieve tractability, robustness and low solution cost [16]. The use of soft computing techniques in missing data imputation presents the major difference of our approach from that presented in [14].

In our earlier paper [17], we developed two missing data imputation algorithms. The first algorithm was based on original crisp K-means clustering, and the second algorithm integrated the concept of fuzzy logic into K-means clustering. This paper is an extended version of that paper and presents two more missing data imputation algorithms based on rough set theory. The details of these four algorithms are introduced in Section 2. Experiments and analysis are presented in Section 3. Finally, concluding remarks and directions for future research are presented in Section 4.

2   Missing Data Imputation Algorithms

In this section, we review previous research related to missing data imputation based on K-means clustering. Particularly, missimg data imputation approaches based on *Fuzzy K-means Clustering* as well as *Rough K-means Clustering* are important in the context of our research.

2.1   Missing Data Imputation with K-Means Clustering

A fundamental problem in missing data imputation is to fill in missing information about an object based on the knowledge of other information about the object [18]. As one of the most popular techniques in data mining, the clustering method facilitates the process of solving this problem. Given a set of objects, the overall objective of clustering is to divide the dataset into groups based on similarity of objects and to minimize the intra-cluster dissimilarity. In K-means clustering, the intra-cluster dissimilarity is measured by the summation of distances between the objects and the centroid of the cluster they are assigned to. A cluster centroid represents the mean value of the objects in a cluster. A number of different distance functions, e.g., Euclidean distance, Cosine-based distance, can be used.

Given a set of N objects $X = \{x_1, x_2, ..., x_N\}$ where each object has S attributes, we use x_{ij} ($1 \leq i \leq N$ and $1 \leq j \leq S$) to denote the value of attribute j in object x_i. Object x_i is called a *complete* object, if $\{x_{ij} \neq \phi \mid \forall\ 1 \leq j \leq S\}$, and an *incomplete* object, if $\{x_{ij} = \phi \mid \exists\ 1 \leq j \leq S\}$, and we say object x_i has a missing value on attribute j. For any incomplete object x_i, we use $R = \{j \mid x_{ij} \neq \phi, 1 \leq j \leq S\}$ to denote the set of attributes whose values are available, and these attributes are called *reference* attributes. Our objective is to obtain the values of non-reference attributes for the incomplete objects. By K-means clustering method, we divide dataset X into K clusters, and each cluster is represented by the centroid of the set of objects in the cluster. Let $V = \{v_1, v_2, ..., v_K\}$ be the set of K clusters, where v_k ($1 \leq k \leq K$) represents the centroid of cluster k. Note that v_k is also a vector in an S-dimensional space. We use $d(v_k, x_i)$ to denote the distance between centroid v_k and object x_i.

Algorithm. *K-means-imputation(X, K, ε)*
1) Initialization -- randomly select K complete objects from X as centroids;
2) Assign each object (complete or incomplete) in X to the closest cluster centroid;
3) Recompute the centroid of each cluster;
4) Repeat steps 2 & 3, until $\sum_{k=1}^{K}\sum_{i=1}^{N} d(v_k, x_i) < \varepsilon$;
5) For each incomplete object, apply *nearest neighbor* algorithm to fill in all the non-reference attributes.

Fig. 1. K-means Clustering for Missing Data Imputation

Figure 1 shows the algorithm for missing data imputation with K-means clustering method. The algorithm can be divided into three processes. First (Step 1), randomly select K complete data objects as K centroids. Rather than random selection, an alternative is to choose the first centroid as the object that is most central to the dataset, and then pick other $(k-1)$ centroids one by one in such a way that each one is most dissimilar to all the objects that have already been selected. This makes the initial K centroids evenly distributed. Second (Steps 2 to 4), iteratively modify the partition to reduce the sum of the distances for each object from the centroid of the cluster to which the object belongs. The process terminates once the summation of distances is less than a user-specified threshold ε. The last process (Step 5) is to fill in all the non-reference attributes for each incomplete object based on the cluster information. Data objects that belong to the same cluster are taken as nearest neighbors of each other. The missing data are replace by Inverse Distance Weighted (IDW) approach based on the available data values from nearest neighbors.

Generalized L_P norm distance [19] is selected to measure the distance between a centroid and a data object in the cluster,

$$d(v_k, x_i) = \left(\sum_{j=1}^{S} |x_{i,j} - v_{k,j}|^p \right)^{1/p}.$$ (1)

The Euclidean distance is the L_2 distance and the Manhattan distance is the L_1 distance. Another distance metric is the Cosine-based distance which is calculated from Cosine Similarity,

$$d(v_k, x_i) = e^{-Sim(v_k, x_i)}, \quad \text{where:} \quad Sim(v_k, x_i) = \frac{\sum_{j=1}^{S} x_{i,j} * v_{k,j}}{\sqrt{\sum_{j=1}^{S} x_{i,j}^2 \sum_{j=1}^{S} v_{k,j}^2}}.$$ (2)

The distance functions are normalized for two reasons. First, the distances can be calculated only from the values of reference attributes, but for incomplete objects, the number of reference attributes is different. Second, each attribute (either numerical or categorical) has a different domain and the distance functions do not make sense without normalization. Because the domain of each attribute is already known in our application domains, we employ the min-max method to normalize the input data sets.

2.2   Missing Data Imputation with Fuzzy K-Means Clustering

Now, the original K-means clustering method is extended to a fuzzy version to impute missing data. The reason for applying the fuzzy approach is that fuzzy clustering provides a better tool when the clusters are not well-separated, as is sometimes the case in missing data imputation. Moreover, the original K-means clustering may be trapped in local minimum if the initial points are not selected properly. However, continuous membership values in fuzzy clustering make the resulting algorithms less susceptible to get stuck in local minimum [20].

In fuzzy clustering, each data object x_i has a membership function which describes the degree that this data object belongs to certain cluster v_k. The membership function is:

$$U(v_k, x_i) = \frac{d(v_k, x_i)^{-2/(m-1)}}{\sum_{j=1}^{K} d(v_j, x_i)^{-2/(m-1)}},$$ (3)

where $m > 1$ is the fuzzifier and $\sum_{j=1}^{K} U(v_j, x_i) = 1$ for any data object x_i $(1 \leq i \leq N)$ [21]. Now, the cluster centroids cannot be calculated simply by the mean values. Instead, the calculation a each cluster centroid needs to consider the membership degree of each data object. The formula for cluster centroid computation is:

$$v_k = \frac{\sum_{i=1}^{N} U(v_k, x_i) * x_i}{\sum_{i=1}^{N} U(v_k, x_i)}. \tag{4}$$

Because there are unavailable data in incomplete objects, the fuzzy K-means approach uses only reference attributes to compute the cluster centroids.

Figure 2 shows the algorithm for missing data imputation with fuzzy K-means clustering method. This algorithm has three processes which are the same as *K-means-imputation*. In the initialization process (Steps 1 & 2), the algorithm picks K centroids which are evenly distributed to avoid local minimum situation. The second process (Steps 3 to 5), iteratively updated membership functions and centroids until the overall distance meets the user-specified distance threshold ε. In this process, a data object cannot be assigned to a concrete cluster represented by a cluster centroid (as did in the basic K-mean clustering algorithm), because each data object belongs to all K clusters with different membership degrees. Finally (Step 6), the algorithm replaces non-reference attributes for each incomplete object.

Algorithm. *Fuzzy-K-means-imputation(X, K, ε)*
1) Compute the most centered complete object and select it as the first centroid, i.e.,
$$v_1 = \min_{1 \leq i \leq N} \sum_{j=1}^{N} d(x_i, x_j);$$
2) Select other $(K - 1)$ complete objects as centroids such that each one is most dissimilar to all the centroids that have already been selected, i.e.,
$$\text{for } (2 \leq i \leq K) \quad \{v_i = \max_{1 \leq j \leq N, x_j \notin V} (\min_{1 \leq k \leq K, v_k \in V} d(x_j, v_k))\};$$
3) Compute the membership function $U(v_k, x_i)$ using Equation (3) for each $1 \leq k \leq K$, and $1 \leq i \leq N$;
4) Recompute centroid v_k using Equation (4);
5) Repeat steps 3 & 4, until $\sum_{k=1}^{K} \sum_{i=1}^{N} U(v_k, x_i) d(v_k, x_i) < \varepsilon$;
6) Fill in all the non-reference attributes for each incomplete data object.

Fig. 2. Fuzzy K-means Clustering for Missing Data Imputation

Algorithm *K-means-imputation* fills in missing data by a nearest neighbor algorithm which takes the data points belonging to the same cluster as nearest neighbors. However, in *Fuzzy-K-means-imputation*, the nearest neighbors are not available, because clusters are not well-separated with regard to the fuzzy concept. *Fuzzy-K-means-imputation* replaces non-reference attributes for each incomplete data object x_i based on the information about membership degrees and the values of cluster centroids,

$$x_{i,j} = \sum_{k=1}^{K} U(x_i, v_k) * v_{k,j}, \text{ for any non-refence attribute } j \notin R. \tag{5}$$

2.3 Missing Data Imputation with Rough K-Means Clustering

This section presents a missing data imputation algorithm based on rough set theory. Theories of rough set and fuzzy set are distinct generalizations of set theory [22, 23, 16]. A fuzzy set allows a membership value between 0 and 1 which describes the degree that an object belongs to a set. Based on rough set theory, a pair of upper and lower bound approximations are used to describe a reference set. Given an arbitrary set X, the lower bound $\underline{A}(X)$ is the union of all elementary sets, which are subsets of X. The upper bound $\overline{A}(X)$ is the union of all elementary sets that have a non-empty intersection with X [22]. In other words, elements in the lower bound of X definitely belong to X, while elements in the upper bound of X may or may not belong to X.

Algorithm. `Rough-Assignment`(x_i, K, θ)
1) Find the cluster centroid v_k to which the data object x_i has the minimum distance, i.e. $v_k = \min d(v_{k'}, x_i)$ for all $1 \leq k' \leq K$;
2) x_i is assigned to the lower and upper bounds of cluster v_k, i.e. $x_i \in \underline{A}(v_k)$ and $x_i \in \overline{A}(v_k)$, if $d(v_{k'}, x_i) - d(v_k, x_i) > \theta$ for all $1 \leq k' \leq K$, and $k' \neq k$; otherwise
3) x_i is assigned to the upper bounds of clusters v_k and $v_{k'}$, i.e. $x_i \in \overline{A}(v_k)$ and $x_i \in \overline{A}(v_{k'})$, if $d(v_{k'}, x_i) - d(v_k, x_i) \leq \theta$ for any $1 \leq k' \leq K$, and $k' \neq k$.

Fig. 3. Data Object Assignment in Rough K-means algorithm

In the original crisp K-means clustering algorithm, data objects are grouped into the same cluster if they are close to each other and each data object belongs to only one cluster. In the rough K-means algorithm, each cluster is represented by two sets which include all the data objects that approximate its lower bound and upper bound, respectively. Different from crisp K-means method, in rough K-means, a data object may exist in the upper bound of one or more clusters. One of the most important issues in the rough K-means clustering is how to assign each data object into the lower or upper bound of one or more clusters.

In the crisp K-means algorithm, data objects are assigned to different clusters simply based on the distances between data objects and cluster centroids. The rough K-means clustering still uses distance metrics defined earlier to determine cluster membership, but the process is more complicated because each cluster is represented by both the lower and upper bound approximations. This process is shown in Figure 3. A new parameter, θ, is introduced which is used to control the similarity among the data objects belonging to a common upper bound of a cluster. Algorithm *Rough-Assignment* shows that in rough K-means clustering, each data object can only belong to the lower bound of one cluster, but it may exist in the upper bound of one or more clusters.

Another important modification in rough K-means clustering is the computation of cluster centroids. Each cluster is represented by two sets, the lower bound approximation and the upper bound approximation. Both sets are used to re-compute the value of a cluster centroid [24]:

$$
v_k = \begin{cases} \dfrac{\sum\limits_{x_i \in \underline{A}(v_k)} x_i}{|\underline{A}(v_k)|} \times W_{lower} + \dfrac{\sum\limits_{x_i \in (\overline{A}(v_k) - \underline{A}(v_k))} x_i}{|\overline{A}(v_k) - \underline{A}(v_k)|} \times W_{upper}, \\[2em] \qquad\qquad\qquad\qquad\qquad\qquad\quad \text{if } |\overline{A}(v_k)| \neq |\underline{A}(v_k)|, \qquad (6) \\[2em] \dfrac{\sum\limits_{x_i \in \underline{A}(v_k)} x_i}{|\underline{A}(v_k)|}, \qquad\qquad\qquad\qquad\qquad\qquad \text{otherwise.} \end{cases}
$$

In this equation, there are two more parameters, W_{lower} and W_{upper}, which are used to control the relative importance of lower and upper bound approximations. For the purpose of normalization, the equation does not use the weight function in the second case. This is different from the equation given in [24]. Generally, $W_{lower} + W_{upper} = 1$ and $W_{lower} \geq W_{upper}$, based on the definitions of lower and upper bounds in rough set theory. If a cluster includes an incomplete data object, only the reference attributes of the data object are used for centroid computation.

Overall, the major difference between rough K-means and crisp K-means imputation methods lies in the second process. For rough imputation algorithm, each data object is assigned to the lower or upper bound of one or more clusters based on *Rough-Assignment* process and re-computed the centroid for each cluster based on Equation 6.

The imputation methods applied to crisp K-means and fuzzy K-means clusterings cannot be applied to rough K-means clustering. In crisp K-means clustering, a data object only belongs to one cluster, and in fuzzy K-means clustering, a data object belongs to all K clusters with different membership degrees. However, in rough K-means clustering, an incomplete data object either exists in the lower bound of one cluster (also in the upper bound of this cluster) or exists in the upper bounds of two or more clusters. Equation 7 shows how we deal with these two different situations:

$$
x_i = \begin{cases}
\dfrac{\displaystyle\sum_{x_j \in \underline{A}(v_k)} x_j}{|x_j|} \times W_{lower} + \dfrac{\displaystyle\sum_{x_j \in (\overline{A}(v_k) - \underline{A}(v_k))} x_j}{|x_j|} \times W_{upper}, \\[2ex]
\quad \text{if } x_i \in \underline{A}(v_k) \text{ for any } 1 \le k \le K, \text{ and } x_j \text{ is a complete object,} \quad (7) \\[2ex]
\dfrac{\displaystyle\sum_{x_j \in \overline{A}(v_k)} x_j}{|x_j|}, \text{ if } x_i \notin \underline{A}(v_{k'}) \text{ for all } 1 \le k' \le K.
\end{cases}
$$

For fuzzy K-means imputation, the computation of a non-reference attribute is based on the values of cluster centroids and the information about membership degrees. This is feasible because each cluster includes all data objects, and the cluster centroids, in turn, are calculated based on all data points. For rough K-means imputation, to make the algorithm more accurate, the value of an incomplete data object is computed based on the values of data objects (rather than cluster centroids) that are in the same cluster as the imputed data object. Moreover, two weight parameters, W_{lower} and W_{upper}, are used if the imputed data object exists in both the lower and upper bounds of a cluster.

2.4 Missing Data Imputation with Rough-Fuzzy K-Means Clustering

There are ongoing efforts to integrate fuzzy logic with rough set theory for dealing with uncertainty arising from inexact or incomplete information [25, 26, 20]. In this section, we present a rough-fuzzy hybridization method to capture the intrinsic uncertainty involved in cluster analysis. In this hybridization, fuzzy sets help handle ambiguity in input data, while rough sets represent each cluster with lower and upper approximations. In rough K-means clustering, a data object either exists in the lower bound of one cluster or exists in the upper bounds of two or more clusters. To deal with the uncertainty involved in lower and upper bound approximations,the rough K-means clustering assigns a data object to the lower bounds of two or more clusters. At the same time, each data object belongs to the upper bounds of all clusters with different membership degrees. This drives the main idea of the rough-fuzzy K-means clustering algorithm.

Figure 4 shows the algorithm for data object assignment in rough-fuzzy cluster-ing. From the description of the algorithm, each data object may be assigned to the lower bound of one or more clusters depending on the value of distance, and each object is assigned to the upper bound of every cluster. Therefore, each data object x_i has two membership functions which describe the degrees that this data object belongs to the lower and upper bounds of certain cluster v_k. The membership functions are defined in Equations 8 and 9.

Here, $\underline{U}(v_k, x_i)$ denotes the membership degree that data object x_i belongs to the lower bound of cluster v_k ($\underline{U}(v_k, x_i) = 0$, if $x_i \notin \underline{A}(v_k)$), and $\overline{U}(v_k, x_i)$ denotes the membership degree that data object x_i belongs to the upper bound of cluster v_k. $\sum_{k=1}^{K} \underline{U}(v_k, x_i) = \sum_{k=1}^{K} \overline{U}(v_k, x_i) = 1$ for any data object x_i ($1 \le i \le N$).

Algorithm. `Rough-Fuzzy-Assignment`(x_i, K, θ)
1) Find the cluster centroid v_k to which the data object x_i has the minimum distance, i.e. $v_k = \min d(v_{k'}, x_i)$ for all $1 \le k' \le K$;
2) Assign x_i to the lower bound of cluster v_k, i.e. $x_i \in \underline{A}(v_k)$;
3) Assign x_i to the lower bounds of clusters $v_{k'}$, i.e. $x_i \in \underline{A}(v_{k'})$, if there exists $1 \le k' \le K$, and $k' \ne k$ such that $d(v_{k'}, x_i) - d(v_k, x_i) \le \theta$;
4) Assign x_i to the upper bound of each cluster, i.e. $x_i \in \overline{A}(v_k)$, for all $1 \le k \le K$.

Fig. 4. Data Object Assignment in Rough-Fuzzy Algorithm

$$\underline{U}(v_k, x_i) = \frac{d(v_k, x_i)^{-2/(m-1)}}{\displaystyle\sum_{x_i \in \underline{A}(v_j)} d(v_j, x_i)^{-2/(m-1)}}. \tag{8}$$

$$\overline{U}(v_k, x_i) = \frac{d(v_k, x_i)^{-2/(m-1)}}{\displaystyle\sum_{j=1}^{K} d(v_j, x_i)^{-2/(m-1)}}. \tag{9}$$

To accommodate the properties of fuzzy and rough sets, we combine Equations 4 and 6 into a new formula to calculate cluster centroids in rough-fuzzy clustering algorithm, as shown in Equations 10.

$$v_k = \begin{cases} \dfrac{\displaystyle\sum_{x_i \in \underline{A}(v_k)} \underline{U}(v_k, x_i) * x_i}{\displaystyle\sum_{x_i \in \underline{A}(v_k)} \underline{U}(v_k, x_i)} \times W_{lower} + \dfrac{\displaystyle\sum_{x_i \notin \underline{A}(v_k)} \overline{U}(v_k, x_i) * x_i}{\displaystyle\sum_{x_i \notin \underline{A}(v_k)} \overline{U}(v_k, x_i)} \times W_{upper}, \\ \qquad\qquad\qquad\qquad\qquad\qquad\qquad\qquad \text{if } |\overline{A}(v_k)| \ne |\underline{A}(v_k)|; \\[2em] \dfrac{\displaystyle\sum_{i=1}^{N} \overline{U}(v_k, x_i) * x_i}{\displaystyle\sum_{i=1}^{N} \overline{U}(v_k, x_i)}, \qquad\qquad\qquad\qquad\qquad\quad \text{otherwise.} \end{cases} \tag{10}$$

The computation of a non-reference attribute for an incomplete data object is based on two parts considering both lower and upper approximations of a cluster, as shown in Equation 11. Because $\sum_{k=1}^{K} \underline{U}(v_k, x_i) = \sum_{k=1}^{K} \overline{U}(v_k, x_i) = 1$ and $W_{lower} + W_{upper} = 1$, this computation formula is well normalized.

$$x_i = \sum_{k=1,\ x_i \in \underline{A}(v_k)}^{K} \frac{\sum_{j=1,\ x_j \in \underline{A}(v_k)}^{N} \underline{U}(v_k, x_i) * x_j}{|x_j|} \times W_{lower} +$$

$$\sum_{k=1}^{K} \frac{\sum_{j=1,\ x_j \notin \underline{A}(v_k)}^{N} \overline{U}(v_k, x_i) * x_j}{|x_j|} \times W_{upper}.$$

(11)

The rough-fuzzy K-means imputation algorithm is shown in Figure 5.

Algorithm. *Rough-Fuzzy-K-means-imputation(X, K, ε)*
1) Select K initial data objects as cluster centroids;
2) Assign each data object x_i in X to the appropriate lower and upper bounds with Algorithm *Rough-Fuzzy-Assignment*;
3) Compute the membership functions $\underline{U}(v_k, x_i)$ and $\overline{U}(v_k, x_i)$ using Equation (8) and (9) for each $1 \le k \le K$;
4) Recompute cluster centroid v_k using Equation (10);
5) Repeat steps 2, 3 & 4, until distance threshold $ε$ is satisfied;
6) Fill in all the non-reference attributes using Equation (11) for each incomplete data object.

Fig. 5. Rough-Fuzzy Clustering for Missing Data Imputation

3 Experiments and Analysis

Two types of experiments are designed. First, the algorithms are evaluated based on complete datasets which are subsets of real-life databases without incomplete data objects. The overall objective of the experiments is to find the best value for each of the parameters (e.g,. missing percentage, the fuzzifier value, and the number of clusters, etc.). Second, the algorithms are evaluated based on real-life datasets with missing values. The best parameter values discovered earlier are used in this process. There are two types of real-life datasets. One is weather databases for drought risk management. Weather data are collected at auto-mated weather stations in Nebraska. These weather stations serve as long-term reference sites to search for key patterns among climatic events. The other type of data is the Integrated Psychological Therapy (IPT) outcome databases for psychotherapy study. A common property in these two types of datasets is that missing data are present either due to the malfunction (or unavailability) of equipment or caused by the refusal of respondents. The experimental results shown in this section are based on the monthly weather data in Clay Center, NE, from 1950-1999. The dataset includes ten fields. Because each data attribute has different domain, to test the algorithms meaningfully, the dataset is first normalized so that all the data values are between 0 and 100. The experiments are based on the following input parameters: distance metric = Manhattan distance; the number of cluster $K = 7$; the fuzzifier $m = 1.2$; the percentage of missing data = 5%; the distance threshold $\theta = 1$; $W_{lower} = 0.9$; and $W_{upper} = 0.1$.

We evaluate the quality of algorithms based on cross validation resampling method. Each algorithm is tested ten times and each time a sample is randomly divided into two subsets, test set and training set. The test results are validated by comparing across sub-samples. The Root Mean Squared Error (RMSE) is selected to compare the prediction value with the actual value of a test instance. RMSE error analysis metric is defined as follows:

$$RMSE = \sqrt{\frac{\sum_{i=1}^{n} |F_i - f_i|^2}{n}},$$

where n is the total number of test points, F_i are the estimated data values, and f_i are the actual data values. Note that the RMSE is much biased because it exaggerates the prediction error of test cases in which the prediction error is larger than others. However, from another point of view, if the RMSE number is significantly greater than zero, it means that there are test cases in which the prediction value is significantly greater or less than the actual value. Therefore, sensitivity of RMSE number is useful in highlighting test cases in which prediction value is significantly lower or higher.

3.1   Experiments on Complete Datasets

The performance of the four K-means imputation algorithms is evaluated and analyzed from two aspects. First, the experiments show the influence of the missing percentage. Second, the experiments test various input parameters (i.e., distance metrics, the value of fuzzifier m, and cluster number K, etc.), and conclude with the best values. The evaluation of these two aspects is based on complete datasets, which are subsets of real-life datasets without incomplete data objects.

3.1.1   Percentage of Missing Data

Figure 6 summarizes the results for varying percentages of missing values in the test cases. Besides the four K-means imputation algorithms, the experiments also test a widely used missing data imputation algorithm, mean substitution. There are four observations from Figure 6:

1. As the percentage of missing values increases, the overall error also increases considering all of these five algorithms. This is reasonable because we lose more useful information when the amount of missing data increases.

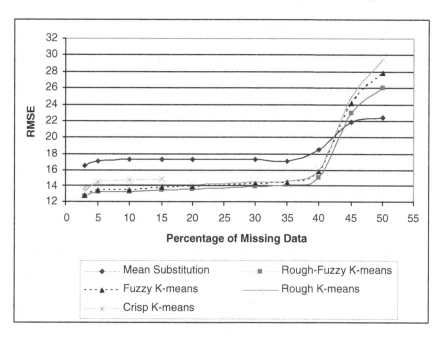

Fig. 6. RMSE for Varying Percentages of Missing Values

2. When the missing percentage is less than or equal to 40%, rough-fuzzy K-means algorithm provides the best results, while the performance of mean substitution imputation algorithm is the worst.

3. When the missing percentage is greater than 15%, the curve for the crisp K-means algorithm terminates, as shown in Figure 6. This occurs because for any incomplete data object, when filling in the values for its non-reference attributes, the algorithm needs to have the values on these attributes from other data objects which are within the same cluster as this incomplete object. However, it is possible that all the data objects within the same cluster have a common non-reference attribute. In this case, the nearest neighbor algorithm used for K-means imputation will not work. This will not happen in fuzzy or rough algorithms. In the fuzzy imputation algorithm, the final imputation process is based on the centroid information and the membership degrees. These two kinds of information are always available for computation. In the rough imputation algorithm, an incomplete data object may belong to two or more clusters, and the information on lower and upper bounds for a given cluster makes the computation flexible and feasible to deal with uncertainty.

4. There is a sharp increase in the value of RMSE when the missing percentage is greater than 40% considering the four K-means imputation methods. The mean substitution approach outperforms the four imputation algorithms when the missing percentage is greater than 45%. This indicates that the four K-means algorithms cannot properly discover the similarity among data objects when there are too many missing values.

3.1.2 Effects of Input Parameters

Distance Metrics. The experiments are designed to evaluate the four missing data imputation algorithms by testing on different input parameters. First, the experiments test three distance metrics, Euclidean distance, Manhattan distance, and Cosine-based distance, as shown in Equations 1 and 2. Figure 7 presents the influence of these metrics. The performance of the four imputation algorithms is shown in four different groups. Considering all of these four algorithms, Manhattan distance provides the best performance while the Cosine-based distance metric is the worst.

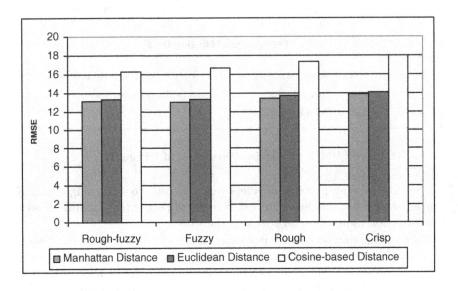

Fig. 7. RMSE for Varying Distance Metrics

Values of Fuzzifier in Fuzzy Algorithms. The experiments test the effect of the value of fuzzifier in the fuzzy and the rough-fuzzy K-means imputation methods. Because fuzzifier is a parameter only in the fuzzy imputation algorithms, as shown in Figure 8, the RMSE in the crisp K-means and the rough K-means clustering methods does not change much as the value of m changes. However, for the fuzzy algorithms, the change in performance is obvious, and the best value of m is 1.2 for both the fuzzy and the rough-fuzzy algorithms. When the value of fuzzifier goes to 1.5, the crisp K-means algorithm outperforms the fuzzy K-means and the rough-fuzzy K-means methods. This indicates that selecting a proper parameter value is important for system performance. Moreover, the experimental results are consistent with the recommendation in [21], which suggested a value between 1 and 1.5 for m.

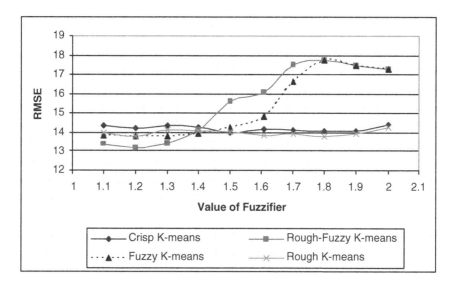

Fig. 8. RMSE for Varying the Value of Fuzzifier

Number of Clusters. Now, the experiments test the influence of the number of clusters, K. The value of K is varied from 4 to 11. Figure 9 shows the performance of the algorithms when there are 6000 data items in the test dataset. From the figure, the best value of K is 7 for all four algorithms. It is worth mentioning that for $K = 4$, the crisp K-means algorithm is the best one among all four algorithms. This is because the smaller number of clusters have fewer centroids. This, in turn, limits the possible variance in the imputed data values for the other three K-means imputation algorithms. On the other hand, when the number of clusters is small, the number of data objects in each cluster increases. This provides more information for the basic K-means algorithm when nearest neighbor algorithm is applied to estimate missing values.

Weights of Lower and Upper Bounds in Rough Algorithms. The rough K-means and rough-fuzzy K-means imputation algorithms introduce two weight parameters, W_{lower} and W_{upper}. These two parameters correspond to the relative importance of lower and upper bounds in rough set theory. Figure 10 presents the performance of rough K-means and rough-fuzzy K-means algorithms as we change the value of W_{upper}. (Because $W_{lower} + W_{upper} = 1$, Figure 10 does not show the value of W_{lower} in Figure 10.) As the value of W_{upper} increases, the RMSE of these two algorithms also increases. This is reasonable because the elements in the lower bound of a cluster definitely belong to the cluster, while the elements in the upper bound of a cluster may or may not belong to the cluster. The weight function has stronger influence on the rough K-means imputation algorithm than on the rough-fuzzy K-means algorithm, because fuzzy sets help handle ambiguity in cluster information.

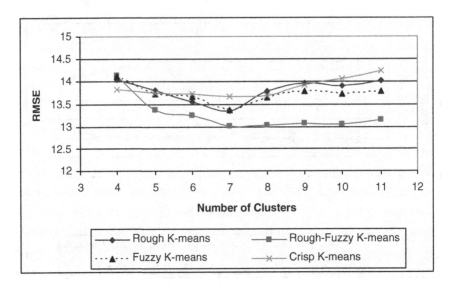

Fig. 9. RMSE for Varying the Number of Clusters

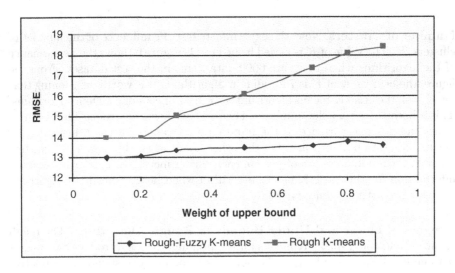

Fig. 10. RMSE for Varying the Weight of Lower and Upper Bounds

Distance Threshold in Rough Algorithms. The rough K-means and rough-fuzzy K-means algorithms use distance threshold to control the similarity between data objects that belong to the same upper (for rough imputation algorithms) or lower (for rough-fuzzy imputation algorithm) bound of a cluster. The experiments test the effect of distance threshold by setting the weight of the lower bound to two different values (0.8 and 0.5 respectively). As can be seen in Figure 11, the RMSE increases as the value of distance threshold increases.

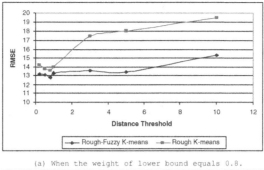

(a) When the weight of lower bound equals 0.8.

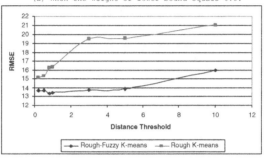

(b) When the weight of lower bound equals 0.5.

Fig. 11. RMSE for Varying the Value of Distance Threshold

This occurs because the greater distance threshold results in less similarity between data objects in a given cluster. The experiments present the best performance when the distance threshold equals 0.8. When the threshold is less than
0.8, the performance of the two algorithms slightly decreases because the smaller
threshold reduces the number of data objects in the upper or lower bound of a
cluster. This, in turn, compromises the possible benefit we should gain based on
rough set theory.

3.2   Experiments on Real-Life Datasets with Missing Values

The previous experiments are based on test datasets, which are subsets of real-
life datasets without incomplete data objects. However, in reality, the datasets
include incomplete data objects, and we do not have actual data values on the
non-reference attributes for an incomplete data object. Therefore, the algorithms
cannot be evaluated simply based on the root mean square error. To solve this
problem, the experiments are designed in this way: 1) Initially fill the missing
data with one of the algorithms and get a complete dataset. 2) From this new
dataset, randomly remove a certain percentage data which have actual data
values in original dataset. 3) Different imputation algorithms are applied to
estimate these missing data values. 4) The RMSE is computed for each algorithm
based on estimated data values and actual data values.

Table 1. Experiments on Actual Datasets with 3% Missing

Fill actual data with	Initially fill missing data with				
	Mean Sub	Crisp K-m	Rough K-m	Fuzzy K-m	R-F K-m
Mean Substitution	16.22	13.72	13.48	13.45	13.14
Crisp K-means	16.43	13.35	12.92	12.81	12.57
Rough K-means	16.61	13.27	12.51	12.26	12.13
Fuzzy K-means	16.70	13.39	12.40	12.16	12.07
R-F K-means	16.65	13.35	12.29	12.14	12.04

Table 2. Experiments on Actual Datasets with 10% Missing

Fill actual data with	Initially fill missing data with				
	Mean Sub	Crisp K-m	Rough K-m	Fuzzy K-m	R-F K-m
Mean Substitution	17.19	14.43	13.57	13.53	13.31
Crisp K-means	17.24	14.15	13.42	13.21	13.03
Rough K-means	17.34	14.11	13.23	13.03	12.87
Fuzzy K-means	17.29	14.21	13.24	12.95	12.86
R-F K-means	17.30	14.32	13.27	12.94	12.85

In addition to the four K-means imputation algorithms, the mean substitution method is also implemented. Tables 1 and 2 compare the five algorithms when the percentage of missing data is 3% and 10% respectively. We make three observations from these two tables:

1. The experimental results are mainly determined by the algorithm which we initially select to estimate the actual missing data. This explains why the root mean square errors in each column in Table 1 and 2 have similar values. Based on this observation, once we fix the algorithm which is initially used, no matter which algorithm we later choose to estimate missing data, there is no much difference among the five algorithms.

2. The four K-means imputation algorithms provide better results than the widely used mean substitution imputation approach and the three algorithms based on soft computing (i.e. rough, fuzzy and rough-fuzzy K-means imputation methods) are better than crisp K-means imputation algorithm. Among the three algorithms, the rough-fuzzy algorithm is the best. From the experiments, comparing the rough-fuzzy algorithm with the mean substitution algorithm, the percentage of improvement is between 18% and 27%, and the improvement is between 5% and 12% when we compare the rough-fuzzy imputation algorithm with the crisp K-means algorithm. This shows that the hybridization of rough set theory and fuzzy set theory takes advantages of both theories and improves the performance comparing with simple fuzzy or rough algorithm.

3. The performance of all of these algorithms decreases as the percentage of missing data increases comparing Table 1 with Table 2. This is consistent with previous experimental results.

4   Conclusion

Analysis and estimation of incomplete data is an increasingly important issue in many fields of study. This paper investigates missing data imputation techniques with the aim of constructing robust algorithms. Traditional clustering algorithms, e.g., K-means clustering, which are normally crisp, have been widely used in hot deck imputation. However, the "crispness" property makes the algorithms less practical, because an object could be assigned to more than one cluster. Integrating fuzzy logic into K-mean clustering helps solve the "crispness" because the fuzzy membership function models the membership degree of an object in a cluster. Rough set theory has emerged as a major method for managing uncertainty in many domains and has proved to be a useful tool in KDD. Based on fuzzy set theory and rough set theory, this paper presents three imputation algorithms, fuzzy K-means, rough K-means, and rough-fuzzy K-means. The experimental results demonstrate the strength of these methods compared with crisp K-means imputation approach. We evaluate the performance of the algorithms based on the RMSE. The experiments discover that the crisp K-means algorithm outperforms the mean substitution method, which is a simple and common approach for missing data imputation. The experiments also show that the overall performance of the rough-fuzzy K-means method is the best one among the four K-means imputation methods. For these experiments, we test the performance of the algorithms based on various input parameters and find the best value for each parameter. The experiments are based on the datasets where data are numerical. For future work, a more sophisticated algorithm needs to be developed to deal with categorical data.

References

1. Little, R.J., Rubin, D.B.: Statistical Analysis with Missing Data. Wiley, New York (1987)
2. Harms, S., Li, D., Deogun, J.S., Tadesse, T.: Efficient rule discovery in a geo-spatial desicion support system. In: Proceedings of the Second National Conference on Digital Government. (2002) 235–241
3. Li, D., Deogun, J.: Spatio-temporal association mining for un-sampled sites. In: Proceedings of the 14th International Symposium on Methodologies for Intelligent Systems (ISMIS'03), Maebashi City, Japan (2003) 478–485
4. Li, D., Deogun, J., Harms, S.: Interpolation techniques for geo-spatial association rule mining. In: Proceedings of the 9th International Conference on Rough Sets, Fuzzy Sets, Data Mining and Granular Computing, Chongqing, China (2003) 573–580
5. Li, D., Deogun, J.S.: Interpolation models for spatio-temporal association mining. Fundamenta Informaticae **59** (2004) 153–172

6. Dempster, A.P., Laird, N.M., Rubin, D.B.: Maximum likelihood from incomplete data via the EM algorithm. Journal of the Royal Statistical Society Series B **39** (1977) 1–38

7. Gary, K., Honaker, J., Joseph, A., Scheve, K.: Listwise deletion is evil: What to do about missing data in political science (2000) http://GKing.Harvard.edu.

8. Grzymala-Busse, J.W.: Rough set strategies to data with missing attribute values. In: Proceedings of the Workshop on Foundations and New Directions in Data Mining, the third IEEE International Conference on Data Mining, Melbourne, FL (2003) 56–63

9. Grzymala-Busse, J.W.: Data with missing attribute values: Generalization of indiscernibility relation and rule induction. Transactions on Rough Sets **1** (2004) 78–95

10. Myrtveit, I., Stensrud, E., Olsson, U.H.: Analyzing data sets with missing data: an empirical evaluation of imputation methods and likelihood-based methods. IEEE Transactions on Software Engineering **27** (2001) 999–1013

11. Roth, P.: Missing data: A conceptual review for applied psychologists. Personnel Psychology **47** (1994) 537–560

12. Schafer, J.L.: Analysis of Incomplete Multivariate Data. Chapman & Hall/CRC (1997)

13. Weiss, S.M., Indurkhya, N.: Decision-rule solutions for data mining with missing values. In: IBERAMIA-SBIA. (2000) 1–10

14. Fujikawa, Y., Ho, T.: Cluster-based algorithms for dealing with missing values. In: Proceedings of Advances in Knowledge Discovery and Data Mining, 6th Pacific-Asia Conference, (PAKDD). (2002) 535–548

15. Hartigan, J., Wong, M.: Algorithm AS136: A k-means clustering algorithm. Applied Statistics **28** (1979) 100–108

16. Zadeh, L.: Fuzzy sets. Information and Control **8** (1965) 338–353

17. Li, D., Deogun, J.S., Spaulding, W., Shuart, B.: Towards missing data imputation: A study of fuzzy k-means clustering method. In: S. Tsumoto, R. Slowinski, J. Komorowski, J.W. Grzymala-Busse (Eds.), Proceedings of the 4th International Conference on Rough Sets and Current Trends in Computing.Lecture Notes in Artificial Intelligence 3066, Uppsala, Sweden, Springer-Verlag Berlin (2004) 573–579

18. Yager, R.R.: Using fuzzy methods to model nearest neighbor rules. IEEE Transactions on Systems, Man and Cybernetics, Part B **32** (2002) 512–525

19. Akleman, E., Chen, J.: Generalized distance functions. In: Proceedings of the '99 International Conference on Shape Modeling. (1999) 72–79

20. Joshi, A., Krishnapuram, R.: Robust fuzzy clustering methods to support web mining. In: Proc. Workshop in Data Mining and knowledge Discovery, SIGMOD. (1998) 15-1 – 15-8

21. Krishnapuram, R., Joshi, A., Nasraoui, O., Yi, L.: Low-complexity fuzzy relational clustering algorithms for web mining. IEEE Transactions on Fuzzy Systems **9** (2001) 595–607

22. Pawlak, Z.: Rough sets. International Journal of Computer and Information Sciences **11** (1982) 341–356

23. Peters, J., Borkowski, M.: K-means indiscernibility over pixels. In: S. Tsumoto, R. Slowinski, J. Komorowski, J.W. Grzymala-Busse (Eds.), Proceedings of the 4th International Conference on Rough Sets and Current Trends in Computing, Lecture Notes in Artificial Intelligence 3066, Uppsala, Sweden, Springer-Verlag Berlin (2004) 580–585

24. Lingras, P., Yan, R., West, C.: Comparison of conventional and rough k-means clustering. In: Proc. of the 9th Intl Conf. on Rough Sets, Fuzzy Sets, Data Mining, and Granular Computing, Chongqing, China (2003) 130–137
25. Asharaf, S., Murty, M.N.: An adaptive rough fuzzy single pass algorithm for clustering large data sets. Pattern Recognition **36** (2003) 3015–3018
26. Banerjee, M., Mitra, S., , Pal, S.K.: Rough fuzzy mlp: Knowledge encoding and classification. IEEE Trans. Neural Networks **9** (1998) 1203–1216

Characteristic Relations for Incomplete Data: A Generalization of the Indiscernibility Relation

Jerzy W. Grzymala-Busse

Department of Electrical Engineering and Computer Science,
University of Kansas, Lawrence, KS 66045, USA
Institute of Computer Science,
Polish Academy of Sciences, 01-237 Warsaw, Poland
Jerzy@ku.edu
http://lightning.eecs.ku.edu/index.html

Abstract. This paper shows that attribute-value pair blocks, used for many years in rule induction, may be used as well for computing indiscernibility relations for completely specified decision tables. Much more importantly, for incompletely specified decision tables, i.e., for data with missing attribute values, the same idea of attribute-value pair blocks is a convenient tool to compute characteristic sets, a generalization of equivalence classes of the indiscernibility relation, and also characteristic relations, a generalization of the indiscernibility relation. For incompletely specified decision tables there are three different ways lower and upper approximations may be defined: singleton, subset and concept. Finally, it is shown that, for a given incomplete data set, the set of all characteristic relations for the set of all congruent decision tables is a lattice.

1 Introduction

An idea of an attribute-value pair block, used for many years in rule induction algorithms such as LEM2 [4], may be applied not only for computing indiscernibility relations for completely specified decision tables but also for computing characteristic relations for incompletely specified decision tables. A characteristic relation is a generalization of the indiscernibility relation.

Using attribute-value pair blocks for completely specified decision tables, equivalence classes of the indiscernibility relation are computed first, then the indiscernibility relation is defined from such equivalence classes. Similarly, for incompletely specified decision tables, attribute-value pair blocks, defined in a slightly modified way, are used to compute characteristic sets, then characteristic relations are computed from these sets.

Decision tables are incomplete mainly for two reasons. First, an attribute value is lost, i. e., it was recorded but currently is unavailable. Second, the original value was irrelevant and as such not recorded and the case was classified on the basis of remaining attribute values. Such missing attribute values will be called "do not care" conditions.

J.F. Peters and A. Skowron (Eds.): Transactions on Rough Sets IV, LNCS 3700, pp. 58–68, 2005.
© Springer-Verlag Berlin Heidelberg 2005

Initially, decision tables with all missing attribute values that are lost were studied, within rough set theory, in [8], where two algorithms for rule induction from such data were presented. This approach was studied later, see, e.g., [15] and [16] where the indiscernibility relation was generalized to describe such incompletely specified decision tables.

The first attempt to study "do not care" conditions using rough set theory was presented in [3], where a method for rule induction was introduced in which missing attribute values were replaced by all values from the domain of the attribute. "Do not care" conditions were also studied later, see, e.g., [9] and [10], where the indiscernibility relation was again generalized, this time to describe incomplete decision tables with "do not care" conditions.

In this paper we will assume that the same incomplete decision table may have missing attribute values of both types—lost attribute values and "do not care" conditions.

For a given completely specified decision table and concept, the lower and upper approximations of the concept are unique, though they may be defined in a few different ways [11] and [12]. For an incomplete decision table, lower and upper approximations of the concept may be defined in a few different ways, but—in general—the approximations of different types differ. In this paper we will discuss three different lower and upper approximations, called singleton, subset, and concept approximations [5]. Singleton lower and upper approximations were studied in [9], [10], [15] and [16]. As it was observed in [4], concept lower and upper approximations should be used for data mining. Note that similar three definitions of lower and upper approximations, though not for incomplete decision tables, were studied in [2], [13], [17], [18] and [19].

The last topic of the paper is studying the class of congruent incomplete decision tables, i.e., tables with the same set of all cases, the same attribute set, the same decision, and the same corresponding specified attribute values. Two congruent decision tables may differ only by missing attribute values (some of them are lost attribute values the others are "do not care" conditions). A new idea of a signature, a vector of all missing attribute values, is introduced. There is a one-to-one correspondence between signatures and congruent decision tables. The paper includes also the Homomorphism Theorem showing that the defined operation on characteristic relations is again a characteristic relation for some congruent decision table. For a given incomplete decision table, the set of all characteristic relations for the set of all congruent decision tables is a lattice.

A preliminary version of this paper was presented at the Fourth International Conference on Rough Sets and Current Trends in Computing, Uppsala, Sweden, June 15, 2004 [6].

2 Blocks of Attribute-Value Pairs, Characteristic Sets, and Characteristic Relations

An example of a decision table is presented in Table 1. Rows of the decision table represent *cases*, while columns represent *variables*. The set of all cases is denoted by U. In Table 1, $U = \{1, 2, ..., 7\}$. Independent variables are called *attributes* and a dependent variable is called a *decision* and is denoted by d. The set of all attributes will be denoted by A. In Table 1, $A = \{Age, Hypertension, Complications\}$. Any decision table defines a function ρ that maps the direct product of U and A into the set of all values. For example, in Table 1, $\rho(1, Age) = 20..29$. Function ρ describing Table 1 is completely specified (total). A decision table with completely specified function ρ will be called *completely specified*, or, simpler, *complete*.

Table 1. A complete decision table

Case	Age	Hypertension	Complications	Delivery
		Attributes		**Decision**
1	20..29	no	none	fullterm
2	20..29	yes	obesity	preterm
3	20..29	yes	none	preterm
4	20..29	no	none	fullterm
5	30..39	yes	none	fullterm
6	30..39	yes	alcoholism	preterm
7	40..50	no	none	fullterm

Rough set theory, see, e.g., [11] and [12], is based on the idea of an indiscernibility relation, defined for complete decision tables. Let B be a nonempty subset of the set A of all attributes. The indiscernibility relation $IND(B)$ is a relation on U defined for $x, y \in U$ as follows

$$(x, y) \in IND(B) \quad if \ and \ only \ if \ \rho(x, a) = \rho(y, a) \ for \ all \ a \in B.$$

The indiscernibility relation $IND(B)$ is an equivalence relation. Equivalence classes of $IND(B)$ are called *elementary sets* of B and are denoted by $[x]_B$. For example, for Table 1, elementary sets of $IND(A)$ are $\{1, 4\}$, $\{2\}$, $\{3\}$, $\{5\}$, $\{6\}$, $\{7\}$. The indiscernibility relation $IND(B)$ may be computed using the idea of blocks of attribute-value pairs. Let a be an attribute, i.e., $a \in A$ and let v be a value of a for some case. For complete decision tables if $t = (a, v)$ is an attribute-value pair then a *block* of t, denoted $[t]$, is a set of all cases from U that for attribute a have value v. For Table 1,

[(Age, 20..29)] = {1, 2, 3, 4},
[(Age, 30..39)] = {5, 6},
[(Age, 40..50)] = {7},
[(Hypertension, no)] = {2, 3, 5, 6},
[(Hypertension, yes)] = {1, 4, 7},
[(Complications, none)] = {1, 3, 4, 5, 7},
[(Complications, obesity)] = {2}, and
[(Complications, alcoholism)] = {6}.

The indiscernibility relation $IND(B)$ is known when all elementary sets of $IND(B)$ are known. Such elementary sets of B are intersections of the corresponding attribute-value pairs, i.e., for any case $x \in U$,

$$[x]_B = \cap\{[(a,v)]|a \in B, \rho(x,a) = v\}.$$

We will illustrate the idea how to compute elementary sets of B for Table 1 where $B = A$:

$[1]_A = [4]_A = [(Age, 20..29)] \cap [(Hypertension, no)] \cap [(Complications, none)] = \{1, 4\}$,

$[2]_A = [(Age, 20..29)] \cap [(Hypertension, yes)] \cap [(Complications, obesity)] = \{2\}$,

$[3]_A = [(Age, 20..29)] \cap [(Hypertension, yes)] \cap [(Complications, none)] = \{3\}$,

$[5]_A = [(Age, 30..39)] \cap [(Hypertension, yes)] \cap [(Complications, none)] = \{5\}$,

$[6]_A = [(Age, 30..39)] \cap [(Hypertension, yes)] \cap [(Complications, alcohol)] = \{6\}$, and

$[7]_A = [(Age, 40..50)] \cap [(Hypertension, no)] \cap [(Complications, none)] = \{7\}$.

A decision table with an incompletely specified (partial) function ρ will be called *incompletely specified*, or *incomplete*. For the rest of the paper we will assume that all decision values are specified, i.e., they are not missing. Also, we will assume that all missing attribute values are denoted either by "?" or by "*", lost values will be denoted by "?", "do not care" conditions will be denoted by "*". Additionally, we will assume that for each case at least one attribute value is specified. Incomplete decision tables are described by characteristic relations instead of indiscernibility relations. Also, elementary sets are replaced by characteristic sets. An example of an incomplete table is presented in Table 2.

Table 2. An incomplete decision table

| Case | | Attributes | | Decision |
	Age	Hypertension	Complications	Delivery
1	?	*	none	fullterm
2	20..29	yes	obesity	preterm
3	20..29	yes	none	preterm
4	20..29	no	none	fullterm
5	30..39	yes	?	fullterm
6	*	yes	alcoholism	preterm
7	40..50	no	?	fullterm

For incomplete decision tables the definition of a block of an attribute-value pair must be modified. If for an attribute a there exists a case x such that

$\rho(x, a) = ?$, i.e., the corresponding value is lost, then the case x should not be included in any block $[(a, v)]$ for all values v of attribute a. If for an attribute a there exists a case x such that the corresponding value is a "do not care" condition, i.e., $\rho(x, a) = *$, then the corresponding case x should be included in all blocks $[(a, v)]$ for every possible value v of attribute a. This modification of the definition of the block of attribute-value pair is consistent with the interpretation of missing attribute values, lost and "do not care" condition. Thus, for Table 2

$[(\text{Age}, 20..29)] = \{2, 3, 4, 6\}$,
$[(\text{Age}, 30..39)] = \{5, 6\}$,
$[(\text{Age}, 40..50)] = \{6, 7\}$,
$[(\text{Hypertension}, \text{no})] = \{1, 4, 7\}$,
$[(\text{Hypertension}, \text{yes})] = \{1, 2, 3, 5, 6\}$,
$[(\text{Complications}, \text{none})] = \{1, 3, 4\}$,
$[(\text{Complications}, \text{obesity})] = \{2\}$,
$[(\text{Complications}, \text{alcoholism})] = \{6\}$.

We define a *characteristic set* $KB(x)$ as the intersection of blocks of attribute-value pairs (a, v) for all attributes a from B for which $\rho(x, a)$ is specified and $\rho(x, a) = v$. For Table 2 and $B = A$,

$K_A(1) = \{1, 3, 4\}$,
$K_A(2) = \{2, 3, 4, 6\} \cap \{1, 2, 3, 5, 6\} \cap \{2\} = \{2\}$,
$K_A(3) = \{2, 3, 4, 6\} \cap \{1, 2, 3, 5, 6\} \cap \{1, 3, 4\} = \{3\}$,
$K_A(4) = \{2, 3, 4, 6\} \cap \{1, 4, 7\} \cap \{1, 3, 4\} = \{4\}$,
$K_A(5) = \{5, 6\} \cap \{1, 2, 3, 5, 6\} = \{5, 6\}$,
$K_A(6) = \{1, 2, 3, 5, 6\} \cap \{6\} = \{6\}$, and
$K_A(7) = \{6, 7\} \cap \{1, 4, 7\} = \{7\}$.

The characteristic set $KB(x)$ may be interpreted as the smallest set of cases that are indistinguishable from x using all attributes from B and using a given interpretation of missing attribute values. Thus, $K_A(x)$ is the set of all cases that cannot be distinguished from x using all attributes. The *characteristic relation* $R(B)$ is a relation on U defined for $x, y \in U$ as follows:

$$(x, y) \in R(B) \ if \ and \ only \ if \ y \in K_B(x).$$

We say that $R(B)$ is *implied* by its characteristic sets $K_B(x)$, $x \in U$. The characteristic relation $R(B)$ is reflexive but—in general—does not need to be symmetric or transitive. Also, the characteristic relation $R(B)$ is known if we know characteristic sets $K(x)$ for all $x \in U$. In our example, $R(A) = \{(1, 1),$ $(1, 3)$, $(1, 4)$, $(2, 2)$, $(3, 3)$, $(4, 4)$, $(5, 5)$, $(5, 6)$, $(6, 6)$, $(7, 7)\}$. The most convenient way to define the characteristic relation is through the characteristic sets. Nevertheless, the characteristic relation $R(B)$ may be defined independently of characteristic sets in the following way:

$$(x, y) \in R(B) \ if \ and \ only \ if \ \rho(x, a) = \rho(y, a) \ or \ \rho(x, a) = * \ or \rho(y, a) = * \ for$$
$$all \ a \in B \ such \ that \ \rho(x, a) \neq ?.$$

3   Lower and Upper Approximations

For completely specified decision tables lower and upper approximations are defined on the basis of the indiscernibility relation. Any finite union of elementary sets, associated with B, will be called a *B-definable set*. Let X be any subset of the set U of all cases. The set X is called a *concept* and is usually defined as the set of all cases defined by a specific value of the decision. In general, X is not a B-definable set. However, set X may be approximated by two B-definable sets, the first one is called a *B-lower approximation* of X, denoted by $\underline{B}X$ and defined as follows

$$\{x \in U | [x]_B \subseteq X\}.$$

The second set is called a B-upper approximation of X, denoted by $\overline{B}X$ and defined as follows

$$\{x \in U | [x]_B \cap X \neq \emptyset.$$

The above shown way of computing lower and upper approximations, by constructing these approximations from singletons x, will be called the *first method*. The B-lower approximation of X is the greatest B-definable set, contained in X. The B-upper approximation of X is the smallest B-definable set containing X.

As it was observed in [12], for complete decision tables we may use a *second method* to define the B-lower approximation of X, by the following formula

$$\cup\{[x]_B | x \in U, [x]_B \subseteq X\},$$

and the B-upper approximation of x may de defined, using the second method, by

$$\cup\{[x]_B | x \in U, [x]_B \cap X \neq \emptyset).$$

For incompletely specified decision tables lower and upper approximations may be defined in a few different ways. First, the definition of definability should be modified. Any finite union of characteristic sets of B is called a *B-definable set*. In this paper we suggest three different definitions of lower and upper approximations. Again, let X be a concept, let B be a subset of the set A of all attributes, and let $R(B)$ be the characteristic relation of the incomplete decision table with characteristic sets $K(x)$, where $x \in U$. Our first definition uses a similar idea as in the previous articles on incompletely specified decision tables [9], [10], [14], [15] and [16], i.e., lower and upper approximations are sets of singletons from the universe U satisfying some properties. Thus, lower and upper approximations are defined by analogy with the above first method, by constructing both sets from singletons. We will call these approximations *singleton*. A singleton B-lower approximation of X is defined as follows:

$$\underline{B}X = \{x \in U | K_B(x) \subseteq X\}.$$

A singleton B-upper approximation of X is

$$\overline{B}X = \{x \in U | K_B(x) \cap X \neq \emptyset\}.$$

In our example of the decision table presented in Table 2 let us say that $B = A$. Then the singleton A-lower and A-upper approximations of the two concepts: $\{1, 4, 5, 7\}$ and $\{2, 3, 6\}$ are:

$$\underline{A}\{1, 4, 5, 7\} = \{4, 7\},$$

$$\underline{A}\{2, 3, 6\} = \{2, 3, 6\},$$

$$\overline{A}\{1, 4, 5, 7\} = \{1, 4, 5, 7\},$$

$$\overline{A}\{2, 3, 6\} = \{1, 2, 3, 5, 6\}.$$

Note that the set $\overline{A}\{1, 4, 5, 7\} = \{1, 4, 5, 7\}$ is not A-definable (this set cannot be presented as a union of intersections of attribute-value pair blocks). The problem is caused by case 5. This case appears twice in the list of all blocks of attribute-value pairs, namely, in [(Age, 30..39)] and [(Hypertension, yes)]. However, both of these blocks contain also case 6. Hence any intersection of blocks of attribute value pairs, containing case 5, must also contain case 6. Thus, using intersection and union of blocks of attribute-value pairs we may construct the set $\{1, 4, 5, 6, 7\}$ but not the set $\{1, 4, 5, 7\}$. Therefore, singleton approximations are, in general, not A-definable, and, as such, are not useful for rule induction.

The second method of defining lower and upper approximations for complete decision tables uses another idea: lower and upper approximations are unions of elementary sets, subsets of U. Therefore we may define lower and upper approximations for incomplete decision tables by analogy with the second method, using characteristic sets instead of elementary sets. There are two ways to do this. Using the first way, a *subset* B-lower approximation of X is defined as follows:

$$\underline{B}X = \cup\{K_B(x) | x \in U, K_B(x) \subseteq X\}.$$

A *subset* B-upper approximation of X is

$$\overline{B}X = \cup\{K_B(x) | x \in U, K_B(x) \cap X \neq \emptyset\}.$$

Since any characteristic relation $R(B)$ is reflexive, for any concept X, singleton B-lower and B-upper approximations of X are subsets of the subset B-lower and B-upper approximations of X, respectively. For the same decision table, presented in Table 2, the subset A-lower and A-upper approximations are

$$\underline{A}\{1, 4, 5, 7\} = \{4, 7\},$$

$$\underline{A}\{2, 3, 6\} = \{2, 3, 6\},$$

$$\overline{A}\{1, 4, 5, 7\} = \{1, 3, 4, 5, 6, 7\},$$

$$\overline{A}\{2, 3, 6\} = \{1, 2, 3, 4, 5, 6\}.$$

The second possibility is to modify the subset definition of lower and upper approximation by replacing the universe U from the subset definition by a concept X. A *concept* B-lower approximation of the concept X is defined as follows:

$$\underline{B}X = \cup\{K_B(x)|x \in X, K_B(x) \subseteq X\}.$$

Obviously, the subset B-lower approximation of X is the same set as the concept B-lower approximation of X. A concept B-upper approximation of the concept X is defined as follows:

$$\overline{B}X = \cup\{K_B(x)|x \in X, KB(x) \cap X \neq \emptyset\} = \cup\{K_B(x)|x \in X\}.$$

The concept B-upper approximation of X is a subset of the subset B-upper approximation of X. Thus, concept upper approximations are more useful for rule induction than subset upper approximations. For the decision presented in Table 2, the concept A-lower and A-upper approximations are

$$\underline{A}\{1,4,5,7\} = \{4,7\},$$

$$\underline{A}\{2,3,6\} = \{2,3,6\},$$

$$\overline{A}\{1,4,5,7\} = \{1,3,4,5,6,7\},$$

$$\overline{A}\{2,3,6\} = \{2,3,6\}.$$

Note that for complete decision tables, all three definitions of lower approximations, singleton, subset and concept, coalesce to the same definition. Also, for complete decision tables, all three definitions of upper approximations coalesce to the same definition. This is not true for incomplete decision tables, as our example shows.

4   Congruent Decision Tables

In this section, for simplicity, all characteristic relations will be defined for the entire set A of attributes instead of its subset B. In addition, and the characteristic relation will be denoted by R instead of $R(A)$. Finally, in characteristic sets $K_A(x)$, the subscript A will be omitted.

Two decision tables with the same set U of all cases, the same attribute set A, the same decision d, and the same specified attribute values will be called *congruent*. Thus, two congruent decision tables may differ only by missing attribute values * and ?. Obviously, there is 2^n congruent decision tables, where n is the total number of all missing attribute values in a decision table.

To every incomplete decision table we will assign a *signature* of missing attribute values, a vector $(p_1, p_2, ..., p_n)$, where p_i is equal to either ? or *, the value taken from the incomplete decision table; $i = 1, 2, ..., n$, by scanning the decision table, row after row, starting from the top row, from left to right. Thus every consecutive missing attribute value should be placed as a component of the signature, where p_1 is the first missing attribute value, identified during scanning, and p_n is the last one. For Table 2, the signature is $(?, *, ?, *, ?)$. In the set of all congruent decision tables, a signature uniquely identifies the table and vice versa. On the other hand, congruent decision tables with different signatures may have the same characteristic relations. For example, tables congruent with Table 2, with signatures $(?, *, *, *, *)$ and $(*, ?, *, *, *)$, have the same characteristic relations. Two congruent decision tables that have the same characteristic relations will be called *indistinguishable*.

Let D_1 and D_2 be two congruent decision tables, let R_1 and R_2 be their characteristic relations, and let $K_1(x)$ and $K_2(x)$ be their characteristic sets for some $x \in U$, respectively. We say that $R_1 \leq R_2$ if and only if $K_1(x) \subseteq K_2(x)$ for all $x \in U$. For two congruent decision tables D_1 and D_2 we define a characteristic relation $R = R_1 \cdot R_2$ as implied by characteristic sets $K_1(x) \cap K_2(x)$. For two signatures p and q, $p \cdot q$ is defined as a signature r with $r_i(x) = *$ if and only if $p_i(x) = *$ and $q_i(x) = *$, otherwise $r_i(x) = ?$, $i = 1, 2, ..., n$.

Let $A = \{a_1, a_2, ..., a_k\}$. Additionally, let us define, for $x \in U$ and $a \in A$, the set $[(a, \rho(x, a))]^+$ in the following way: $[(a, \rho(x, a))]^+ = [(a, \rho(x, a))]$ if $\rho(x, a) \neq *$ and $\rho(x, a) \neq ?$ and $[(a, \rho(x, a))]^+ = U$ otherwise.

Lemma. For $x \in U$, the characteristic set $K(x) = \cap_{i=1}^{k} [(a_i, \rho(x, a_i))]^+$.

Proof. In the definition of $K(x)$, if $\rho(x, a) = *$ or $\rho(x, a) = ?$, the corresponding block $[(a, \rho(x, a))]$ is ignored. Additionally, by our assumption, for every $x \in U$ there exists an attribute $a \in A$ such that $\rho(x, a) \neq *$ and $\rho(x, a) \neq ?$.

Let D be an incomplete decision table and let p be the signature of D. Let ψ be a function that maps a signature p into a characteristic relation R of D.

Homomorphism Theorem. Let p and q be two signatures of congruent decision tables. Then $\psi(p \cdot q) = \psi(p) \cdot \psi(q)$, i.e., ψ is a homomorphism.

Proof. Let D_1, D_2 be two congruent decision tables with functions ρ_1 and ρ_2, signatures p and q, and characteristic relations R_1, R_2, respectively, where $\psi(p) = R_1$ and $\psi(q) = R_2$. Let D be a congruent decision table with function ρ and signature $p \cdot q$ and let $\psi(p \cdot q) = R$. Due to Lemma, for every $x \in U$

$$K_1(x) \cdot K_2(x) = (\cap_{i=1}^{k}[(a_i, \rho_1(x, a_i))]^+) \cap (\cap_{i=1}^{k}[(a_i, \rho_2(x, a_i))]^+) = \cap_{i=1}^{k}[(a_i, \rho_1(x, a_i))]^+ \cap [(a_i, \rho_2(x, a_i))]^+$$

If $\rho_j(x, a_i) \neq *$ and $\rho_j(x, a_i) \neq ?$ then $[(a_i, \rho_j(x, a_i))]$ contains $y \in U$ if and only if $\rho_j(y, a_i) = *$, $j = 1, 2$. Moreover, $[(a_i, \rho(x, a_i))]^+$ contains y if and only if $\rho_1(y, a_i) = *$ and $\rho_2(y, a_i) = *$. Thus, $K_1(x) \cdot K_2(x) = \cap_{i=1}^{k}[(a_i, \rho_1(x, a_i))]^+ = K(x)$.

Thus, $\psi(p) \cdot \psi(q)$ is the characteristic relation of a congruent decision table with the signature $p \cdot q$. For the set L of all characteristic relations for the set of all congruent decision tables, the operation $\cdot$ on relations is idempotent, commutative, and associative, therefore, L is a semilattice [1], p. 9. Moreover, L has a universal upper bound $\psi(*, *, ..., *)$ and its length is finite, so L is a lattice, see [1], p. 23. The second lattice operation, resembling addition, is defined directly from the diagram of a semilattice.

Let us define subset E of the set of all congruent decision tables as the set of tables with exactly one missing attribute value "?" and all remaining attribute values equal to "*". Let G be the set of all characteristic relations associated with the set E. The lattice L can be generated by G, i.e., every element of L can be expressed as $\psi(*, *, ...*)$ or as a product of some elements from G.

5   Conclusions

An attribute-value pair block is a very useful tool not only for dealing with completely specified decision tables but, much more importantly, also for incompletely specified decision tables. For completely specified decision tables attribute-value pair blocks provide for easy computation of equivalence classes of the indiscernibility relation. Similarly, for incompletely specified decision tables, attribute-value pair blocks make possible, by equally simple computations, determining characteristic sets and then, if necessary, characteristic relations.

For a given concept of the incompletely specified decision table, lower and upper approximations can be easily computed from characteristic sets—knowledge of characteristic relations is not required. Note that for incomplete decision tables there are three different approximations possible: singleton, subset and concept. The concept approximations are the best fit for the intuitive expectations for lower and upper approximations. Our last observation is that for a given incomplete decision table, the set of all characteristic relations for the set of all congruent decision tables is a lattice.

References

1. Birkhoff, G.: Lattice Theory. American Mathematical Society, Providence, RI (1940).
2. Greco, S., Matarazzo, S.B. and Slowinski, R.: Dealing with missing data in rough set analysis of multi-attribute and multi-criteria decision problems. In Decision Making: Recent developments and Worldwide Applications, ed. by S. H. Zanakis, G. Doukidis and Z. Zopounidis, Kluwer Academic Publishers, Dordrecht, Boston, London (2000) 295–316.
3. Grzymala-Busse, J.W.: On the unknown attribute values in learning from examples. Proc. of the ISMIS-91, 6th International Symposium on Methodologies for Intelligent Systems, Charlotte, North Carolina, October 16–19, 1991. Lecture Notes in Artificial Intelligence, vol. 542, Springer-Verlag, Berlin, Heidelberg, New York (1991) 368–377.
4. Grzymala-Busse, J.W.: LERS—A system for learning from examples based on rough sets. In Intelligent Decision Support. Handbook of Applications and Advances of the Rough Sets Theory, ed. by R. Slowinski, Kluwer Academic Publishers, Dordrecht, Boston, London (1992) 3–18.
5. Grzymala-Busse, J.W.: Rough set strategies to data with missing attribute values. Workshop Notes, Foundations and New Directions of Data Mining, the 3-rd International Conference on Data Mining, Melbourne, FL, USA, November 19–22, 2003, 56–63.

6. Grzymala-Busse, J.W.: Characteristic relations for incomplete data: A generalization of the indiscernibility relation. Proceedings of the RSCTC'2004, the Fourth International Conference on Rough Sets and Current Trends in Computing, Uppsala, Sweden, June 1–5, 2004. Lecture Notes in Artificial Intelligence 3066, Springer-Verlag 2004, 244–253.

7. Grzymala-Busse, J.W. and Hu, M.: A comparison of several approaches to missing attribute values in data mining. Proceedings of the Second International Conference on Rough Sets and Current Trends in Computing RSCTC'2000, Banff, Canada, October 16–19, 2000, 340–347.

8. Grzymala-Busse, J.W. and Wang, A.Y.: Modified algorithms LEM1 and LEM2 for rule induction from data with missing attribute values. Proc. of the Fifth International Workshop on Rough Sets and Soft Computing (RSSC'97) at the Third Joint Conference on Information Sciences (JCIS'97), Research Triangle Park, NC, March 2–5, 1997, 69–72.

9. Kryszkiewicz, M.: Rough set approach to incomplete information systems. Proceedings of the Second Annual Joint Conference on Information Sciences, Wrightsville Beach, NC, September 28–October 1, 1995, 194–197.

10. Kryszkiewicz, M.: Rules in incomplete information systems. *Information Sciences* **113** (1999) 271–292.

11. Pawlak, Z.: Rough Sets. *International Journal of Computer and Information Sciences* **11** (1982) 341–356.

12. Pawlak, Z.: Rough Sets. Theoretical Aspects of Reasoning about Data. Kluwer Academic Publishers, Dordrecht, Boston, London (1991).

13. Slowinski, R. and Vanderpooten, D.: A generalized definition of rough approximations based on similarity. *IEEE Transactions on Knowledge and Data Engineering* **12** (2000) 331–336.

14. Stefanowski, J.: Algorithms of Decision Rule Induction in Data Mining. Poznan University of Technology Press, Poznan, Poland (2001).

15. Stefanowski, J. and Tsoukias, A.: On the extension of rough sets under incomplete information. Proceedings of the 7th International Workshop on New Directions in Rough Sets, Data Mining, and Granular-Soft Computing, RSFDGrC'1999, Ube, Yamaguchi, Japan, November 8–10, 1999, 73–81.

16. Stefanowski, J. and Tsoukias, A.: Incomplete information tables and rough classification. *Computational Intelligence* **17** (2001) 545–566.

17. Yao, Y.Y.: Two views of the theory of rough sets in finite universes. *International J. of Approximate Reasoning* **15** (1996) 291–317.

18. Yao, Y.Y.: Relational interpretations of neighborhood operators and rough set approximation operators. *Information Sciences* **111** (1998) 239–259.

19. Yao, Y.Y.: On the generalizing rough set theory. Proc. of the 9th Int. Conference on Rough Sets, Fuzzy Sets, Data Mining and Granular Computing (RSFDGrC'2003), Chongqing, China, October 19–22, 2003, 44–51.

Supervised Learning in the Gene Ontology
Part I: A Rough Set Framework

Herman Midelfart

Department of Biology, Norwegian University of Science and Technology,
N-7491 TRONDHEIM, Norway
herman@bio.ntnu.no

Abstract. Prediction of gene function introduces a new learning problem where the decision classes associated with the objects (i.e., genes) are organized in a directed acyclic graph (DAG). Rough set theory, on the other hand, assumes that the classes are unrelated cannot handle this problem properly. To this end, we introduce a new rough set framework. The traditional decision system is extended into DAG decision system which can represent the DAG. From this system we develop several new operators, which can determine the known and the potential objects of a class and show how these sets can be combined with the usual rough set approximations. The properties of these operators are also investigated.

1   Introduction

A supervised learning algorithm is given a training set with a set of objects with known classes. From the training set, it produces a classifier that predicts the class of an object from some observable attributes of the object. When a classifier has been found it can later applied to classify objects with unknown class. An important assumption behind such an algorithm is that the classes are unordered and discrete. In many real life situations, however, this assumption does not hold. One particular example is prediction of gene function annotations.

Molecular biologists have sequenced the genome and identified the genes in many organisms, but they have only fragmented knowledge of the function of these genes, and this knowledge is scattered in the biological literature and various biological databases. Consequently, there is a need for databases that describe the function of the genes.

There are currently several efforts that develop such databases. One important effort has been driven by the Gene Ontology (GO) Consortium [2]. This consortium provides ontologies[1] that define structured vocabularies for describing the function of gene products and catalogs with annotations[2] for several

---

[1] The ontologies describe three different aspects of the function of a gene product: *Biological process*, which refers to the objective to which a gene or a gene product contributes, *molecular function*, which refers to the biochemical activity, i.e., what the gene product does on the biochemical level, and *cellular component*, which refers to the subcellular location where a gene product is active.

[2] An annotation associates a gene with a category in the ontology.

J.F. Peters and A. Skowron (Eds.): Transactions on Rough Sets IV, LNCS 3700, pp. 69–97, 2005.
© Springer-Verlag Berlin Heidelberg 2005

model organisms [3, 4]. However, the catalogs are developed manually, and the genome of an organism consists of thousands of genes. Developing annotations is therefore a formidable and time-consuming task. Several studies have been conducted where annotations have been predicted automatically with supervised learning methods. One example is Brown et al. [1], which used support vector machines on microarray data to predict 5 classes from the MIPS ontology [5]. A more recent attempt is Hvidsten et al. [9], which attempted to predict 16 classes from the process ontology of the GO consortium used Rough Sets [16].

The ontologies of the GO form directed acyclic graphs (DAGs) where the classes are related (The MIPS ontology is tree). In particular, a class subsumes its children classes so that a gene annotated to class c also belongs to the parents of c[3]. Therefore, a standard supervised learning may not be applied directly to this task. This problem was avoided in the mentioned studies by a selecting few unrelated classes manually. However, this is impractical when we want to develop a full classifier and will typically result in a sub-optimal solution.

In a previous paper [15] we therefore developed an algorithm for supervised learning in the Gene Ontology. However, this algorithm was not developed formally and was not very robust. In this paper we will present a formal rough set framework for modeling decision classes in a DAG. A part of this framework was described in [14], but without most of the details. Here we present the full details of the framework, which has been revised to large extend. We will introduce a DAG-decision system, which is an extension of the traditional decision system. This system captures the structure of the DAG by the means of a partial order. Using this partial order, we will develop several boundary sets, which allow us to determine whether an object belongs to a class or not, and show how to combine the standard rough set operators. Properties of these boundary sets and approximations will be investigated. We will also consider a special type of DAGs, which we call well-defined. These allow us to simplify the definition of the boundary sets and hence ease the implementation of these sets. In [13] we will show how the framework may be used in a supervised learning algorithm and introduce an new bottom-up algorithm. This algorithm will be evaluated on both experimental and artificial data sets.

2   What Makes Learning in a DAG Difficult?

It may not be apparent at first sight that learning in a DAG is a special problem that cannot be handled by an ordinary learning algorithm. In this section, we will discuss the issues that are introduced by the DAG and motivate the solutions that will be presented.

2.1   The Classes Are Related

The most obvious problem is the structure between the classes. An ordinary rule learning algorithm assumes that the classes are unrelated and will try to

---

[3] A full discussion of the semantics can be found in [12–ch. 6].

discern between related classes if it is applied on the DAG. This may result in very specific rules. Specific rules are, however, a problem since they cover few objects and are more likely to be based on artifacts (in the training data) that do not actually contribute to discerning the objects.

For example, an algorithm, which creates possible rules [7, 8], may create the rules shown in Figure 2 if it is trained on the decision system in Figure 1. These rules cover few objects, and it is possible to find a smaller set of more general rules if the structure of the ontology is considered. According to the ontology in Figure 1, intracellular protein traffic is a subclass of cell growth & maintenance so that the rules need not discriminate between them. The following rule would thus be created.

$$\langle \texttt{15m-30m, down} \rangle \wedge \langle \texttt{30m-1H, down} \rangle \rightarrow \langle \texttt{Process, cell growth} \rangle$$

This rule covers objects o_3, o_7, o_8, and o_9 and is more general than the corresponding rules found in Figure 2. So it is clearly an improvement on these rules. Still, objects o_3 and o_9 are labeled with subclasses of cell growth & maintenance. The predictions made by this rule will be less detailed than the original decision classes of objects o_3 and o_9. So some information is lost.

However, we may find more general rules for the subclasses as well. Objects o_8 and o_9 have almost the same information vector, and it is likely that object o_8 belongs to intracellular protein traffic. If we do not discern between related classes, we may learn a rule such as:

$$\langle \texttt{30m-1H, down} \rangle \wedge \langle \texttt{1H-2H, down} \rangle \rightarrow \langle \texttt{Process, intracellular p. traffic} \rangle$$

This rule covers more objects than the corresponding rules in Figure 2. The rules in the figure have the same number of descriptors as this rule. However, they try to discern between the objects by means of attribute OH-15m for which o_8 and o_9 have a different value. This difference is most likely an artifact due to noise such that the rules in Figure 2 may have a lower prediction power than the new rule. Furthermore, the rule gives a more detailed prediction for object o_8.

If the DAG structure is considered we may also resolve inconsistencies. Object o_7 is annotated with a superclass of the class of object o_3, and both objects are members of the same elementary set. Thus, we may assume that object o_7 belongs to the same class as object o_3 and remove the fifth rule.

Hence, an ordinary learner will try to find rules that discern between related classes. In order to do so it may choose noisy attributes for the descriptors in the rules and neglect other more relevant attributes. It will consequently be very sensitive to noise and may have a low predictive power. If we do not discriminate between related classes, on the other hand, we may obtain more general and more accurate rules.

2.2 The Detail Level of the Annotations Varies

Another problem that arises in the ontology is the detail level of the annotations. The existing biological knowledge about the functions of genes has a strongly varying detail level. Some annotations refer to leaf classes, but many annotations concern non-leaf classes as illustrated in Figure 1.

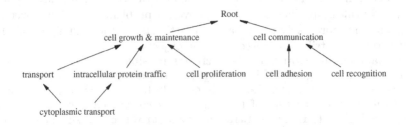

Object	0H-15m	15m-30m	30m-1H	1H-2H	2H-4H	Process Name
o_1	up	up	up	const	const	cell proliferation
o_2	up	up	up	down	down	cell proliferation
o_3	down	down	down	up	up	cell proliferation
o_4	up	up	up	down	down	cell adhesion
o_5	down	up	up	down	up	cell adhesion
o_6	const	const	const	down	down	cell adhesion
o_7	down	down	down	up	up	cell growth & maintenance
o_8	up	down	down	down	down	cell growth & maintenance
o_9	const	down	down	down	down	intracellular protein traffic
o_{10}	down	up	up	up	up	transport
o_{11}	down	down	up	up	up	cytoplasmic transport

Fig. 1. *A DAG-decision system* $\mathcal{A} = \langle U, A, d, \succcurlyeq \rangle$. The DAG is a small part of the process ontology (rev. 1.221 - 05-Feb-2001) from the Gene Ontology Consortium.

$$\langle \text{OH--15m, u} \rangle \wedge \langle \text{15m--30m, u} \rangle \qquad \rightarrow \langle \text{Process, cell prolif.} \rangle \qquad (o_1, o_2, o_4)$$
$$\langle \text{OH--15m, d} \rangle \wedge \langle \text{15m--30m, d} \rangle \qquad \rightarrow \langle \text{Process, cell prolif.} \rangle \qquad (o_3, o_7)$$
$$\langle \text{30m--1H, u} \rangle \wedge \langle \text{1H--2H, d} \rangle \qquad \rightarrow \langle \text{Process, cell adhesion} \rangle \qquad (o_4, o_5)$$
$$\langle \text{OH--15m, c} \rangle \wedge \langle \text{15m--30m, c} \rangle \qquad \rightarrow \langle \text{Process, cell adhesion} \rangle \qquad (o_6)$$
$$\langle \text{OH--15m, d} \rangle \wedge \langle \text{15m--30m, d} \rangle \qquad \rightarrow \langle \text{Process, cell growth} \rangle \qquad (o_3, o_7)$$
$$\langle \text{OH--15m, u} \rangle \wedge \langle \text{15m--30m, d} \rangle \qquad \rightarrow \langle \text{Process, cell growth} \rangle \qquad (o_8)$$
$$\langle \text{OH--15m, c} \rangle \wedge \langle \text{15m--30m, d} \rangle) \qquad \rightarrow \langle \text{Process, intra. p. traffic} \rangle \qquad (o_9)$$
$$\langle \text{15m--30m, u} \rangle \wedge \langle \text{30m--1H, u} \rangle \wedge \langle \text{1H--2H, u} \rangle \rightarrow \langle \text{Process, transport} \rangle \qquad (o_{10})$$
$$\langle \text{15m--30m, d} \rangle \wedge \langle \text{30m--1H, u} \rangle \wedge \langle \text{1H--2H, u} \rangle \rightarrow \langle \text{Process, cytop. transport} \rangle \qquad (o_{11})$$

Fig. 2. *Possible rules when the class relationships are ignored.* The values *up*, *down*, and *const* are represented as *u*, *d*, and *c*, respectively. The objects covered by each rules is displayed in parentheses.

The variation in the detail level means that we cannot avoid the structure of the ontology such that an ordinary learning algorithm can be applied directly. For example, if all of the annotations referred to leaf classes, we could reduce the problem by only using the leaf classes. However, since an annotation may refer to a non-leaf class, we would lose a lot of annotations if we tried. Moving all non-leaf annotations to the leaf classes is not an alternative, either. Most of the annotations in the leaf classes would belong to non-leaf classes, and these would be shared by many leaf classes. Discerning between these classes would thus be very hard.

Alternatively, we could create a "cut" through the ontology by selecting the most general classes, which contain genes (i.e., cell growth & maintenance and cell adhesion in Figure 1) and move the annotations from the subclasses to these classes. However, the details of the moved annotations would be lost.

Hence, the ontology cannot be reduced to a flat set of unrelated classes without losing whole annotations or annotation details. So the structure of the ontology must be considered during learning and prediction.

2.3 There Are Few Objects per Class

A rule learning algorithm needs a minimum number of objects in order to learn rules that accurately predict a class. The number of objects in an ordinary training set is therefore much larger than the number of classes. However, the number of genes in a microarray study may be of same magnitude as (or even smaller than) the number of classes in the process ontology. Hence, the number of objects that are available for each class will be very low.

One extreme example is the yeast genome that has about $6,000$ genes. Revision 2.577 of the process ontology, on the other hand, contains about $4,400$ different classes. There will consequently be only 1.34 objects/class.

The number of objects/class may be a little higher in practice. Some of classes in the ontology may not be relevant for a particular organism, and these may be removed. Moreover, a gene may be annotated to more than one class so that several classes may share the same objects. However, there may still be a lot of classes that need to be discerned, and the annotations may be spread throughout the whole ontology so that each class may have very few objects. Finding genuine distinguishing properties in the information vectors may be very hard even if we do *not* discriminate between the related classes.

Thus, it may be necessary to increase the number of objects in each class. This may be achieved by moving the objects from some related classes to a joint class. Rules for the joint class can then be learned from the total set of objects.

The annotations may be moved either upwards or downwards. Moving objects upwards is the simplest procedure and is always correct. An object x annotated with a class c belongs also to any of the superclasses of c. So, if a superclass of c is predicted for x, a correct prediction has been made even though the prediction is less detailed.

Moving objects downwards is more difficult. Details are gained in this case, but a decision must be made with regard to the subclass to which an annotation should be moved. The only available knowledge for making such a decision is the information vectors. However, objects having a similar information vectors are likely to belong to same class. It seems reasonable to move generally annotated objects to subclasses where the objects have similar information vectors. One way to do this would be to apply a similarity measure and compute the similarity of the information vectors. An object could then be moved to the subclass with the most similar objects according to this measure.

We would like to stress that we are not suggesting that we move all objects to the most general or the most specific classes. This idea was dismissed in Section 2.2, and we are not reconsidering it. We are only proposing that the objects are moved to some of the related classes so that the precision and the detail level of the predictions are optimized.

2.4   There Is a Trade-Off Between Detail Level and Precision

Moving annotations downwards is obviously preferable since more detailed predictions are obtained. However, it may not be possible to get satisfactory precision by moving genes downwards, and better results may be obtained by moving the objects upwards. There is, in fact, a trade-off between the detail level and the precision of the predictions:

- The number of classes that need to be separated increases as one moves downwards in the ontology. The number of classes immediately below the root is quite small, and it is therefore quite easy to predict these classes. However, the number of leaf classes is large (The process ontology, rev. 2.577 has 2, 725 leaf classes), and it may be very difficult to separate all of them. Thus, the difficulty of the learning problem increases with the detail level of the predictions.
- The number of objects is independent of the number of classes and will remain the same even if objects are moved. This means that the number of objects per class will decrease as we move downwards. So, not only must a larger number of classes be separated, but each class will have fewer objects that set it apart from the rest.
- The available information for discerning between the objects, i.e., the information vectors, remains the same as well. Hence, the ability to discern between the objects does not change, while more classes must be separated when objects are moved downwards. In particular, the information vectors may describe only properties of general classes, and it may not be possible to discern between detailed classes without additional data. The information vectors may consequently be insufficient for discerning between the more detailed classes.

Thus, it may not be possible to have both details and precision. There is a trade-off where the precision of the predictions increases as the detail level is reduced, and vice versa. In this trade-off, details should be sacrificed for precision since a classifier giving detailed, but inaccurate predictions is quite useless. However, no more details should be given away than what is absolutely necessary. Otherwise, we could just use the solution from Section 2.2 where only the most general classes were selected.

3   A *More General* Order on the DAG

In order to solve the problems that we just have discussed we need a framework that allows us to represent the DAG structure. We begin the development of this framework by defining an ordering relation on the DAG and show that this relation corresponds to a partial order. In the following sections we will introduce two concepts — the neighborhood and the complement of a class — before we introduce the DAG decision system in Section 6.

Definition 1 (More general). *Let $G = \langle V_d, E \rangle$ be a DAG where V_d is a set of decision classes and $E \subseteq V_d \times V_d$ is a set of edges in the graph. The relation $c \succcurlyeq e$ denotes that c is* more general *than e $(c, e \in V_d)$ and is defined on G such that*

1. *$c \succcurlyeq c$ for all $c \in V_d$*
2. *$c \succcurlyeq e$ for all $\langle e, c \rangle \in E$*
3. *for all $c, f \in V_d$, $c \succcurlyeq f$, if there is an $e \in V_d$ such that $c \succcurlyeq e$ and $e \succcurlyeq f$.*

Definition 2 (Less general). *$c \preccurlyeq e$ denotes that c is* less general *than e and holds iff $e \succcurlyeq c$*

$\succcurlyeq$ (and $\preccurlyeq$) is a partial order as stated by the following proposition.

Proposition 1. *$\succcurlyeq$ is a partial order on V_d.*

Proof. $\succcurlyeq$ is reflexive and transitive by case 1 and 3, respectively, and it remains to show that $\succcurlyeq$ is anti-symmetric. Assume that $\succcurlyeq$ is not anti-symmetric. Then, there are some classes c and e in V_d such that $c \succcurlyeq e$ and $e \succcurlyeq c$, but $c \neq e$. $c \succcurlyeq e$, together with case 2 and 3, implies that there is a path $\langle e, v_1, \ldots, v_k, c \rangle$ from e to c in G. Similarly, there must be a path $\langle c, w_1, \ldots, w_l, e \rangle$ from c to e (since $e \succcurlyeq c$). Then, there is a cycle $\langle c, w_1, \ldots, w_k, e, v_1, \ldots, v_l, c \rangle$ in the graph, and this results in a contradiction. Hence, $\succcurlyeq$ is anti-symmetric if G is acyclic. This means that $\succcurlyeq$ must be a partial order.

The following relations can be derived from $\succcurlyeq$.

Definition 3 (Strictly more general). *Given a partial order $\succcurlyeq$ on a set of classes V_d, we say that c is* strictly more general *than e $(c, e \in V_d)$, denoted as $c \succ e$, iff $c \succcurlyeq e$ and $c \neq e$.*

Two classes are related if one is more general than the other.

Definition 4 (Related classes). *Let $\succcurlyeq$ be a partial order on a set of classes V_d. The classes $c, e \in V_d$ are* related, *denoted as $c \approx e$, iff $c \succcurlyeq e$ or $e \succcurlyeq c$. If $c \not\approx e$, we say that c and e are* unrelated.

Notice that $\approx$ is reflexive and symmetric, but not transitive. In the following, we will also these definitions.

Definition 5 (Classes).

- *A class c is a* superclass *of e if $c \succ e$. A class c is an* immediate *superclass of e if it is a superclass of e and there is no $f \in V_d$ such that $c \succ f \succ e$.*

- *A class c is a* subclass *of e if e ≻ c. A class c is an* immediate *subclass of e if it is a subclass of e and there is no f ∈ V_d such that e ≻ f ≻ c.*
- *A* leaf *class c has no subclasses. Hence, for all e ∈ V_d, c ⊁ e.*
- *A non-leaf class c has at least one subclass. Hence, there is some e ∈ V_d such that c ≻ e.*
- *A* root *class c has no superclasses. Hence, for all e ∈ V_d, e ⊁ c holds for c.*

4   The Neighborhood of a Class

The neighborhood of a class c consists of the classes that are related to c. The known and the potential sets that we will develop in Section 7 are defined through the classes in their neighborhood. We will therefore define special sets that denote theses classes. Sometimes we need to consider only a part of the neighborhood such as the classes above or below a class. Hence, we introduce three different sets: The above set, the below set, and the related set.

Definition 6. *Let $C \subseteq V_d$ be a set of classes. Then the classes above, below or related to C are defined as follows:*

- ***Above set:*** $[C]^{\succeq} = \{e \in V_d \mid for\ some\ c \in C, e \succcurlyeq c\}$
- ***Below set:*** $[C]^{\preceq} = \{e \in V_d \mid for\ some\ c \in C, c \succcurlyeq e\}$
- ***Related set:*** $[C]^{\approx} = \{e \in V_d \mid for\ some\ c \in C, c \approx e\}$

The sets are defined for set of classes rather than a single class, but may obviously be used for single classes as well. For example, the set of classes above class d is $[\{d\}]^{\succeq} = \{e \in V_d \mid e \succcurlyeq d\}$. Sometimes we will abuse this notation and write $[c]^{\succeq}$ instead of $[\{c\}]^{\succeq}$ when our intention is the later. However, we follow the convention that sets of classes are denoted with capital letters, and single classes are denoted with non-capital letters. Hence, it will always be clear whether we refer to neighborhood of a class or a set of classes.

The above, below, and related sets have several useful properties that we will apply in our proofs.

Lemma 1

a) $C \cap [D]^{\succeq} = \emptyset \iff [C]^{\preceq} \cap D = \emptyset$
b) $C \cap [D]^{\preceq} = \emptyset \iff [C]^{\succeq} \cap D = \emptyset$
c) $C \cap [D]^{\approx} = \emptyset \iff [C]^{\approx} \cap D = \emptyset$

Proof.

a) $C \cap [D]^{\succeq} = \emptyset \iff$ not (for some $e \in C, e \in [D]^{\succeq}$)
$\iff$ not (for some $e \in C$, and for some $f \in D, e \succcurlyeq f$)
$\iff$ not (for some $f \in D$, and for some $e \in C, e \succcurlyeq f$)
$\iff$ not (for some $f \in D, f \in [C]^{\preceq}$)
$\iff [C]^{\preceq} \cap D = \emptyset$

b) Follows from directly from the first case and the commutativity of $\cup$ (i.e., $[A]^{\succcurlyeq} \cap B = \emptyset \Leftrightarrow A \cap [B]^{\succcurlyeq} = \emptyset$ is equivalent to $B \cap [A]^{\succcurlyeq} = \emptyset \Leftrightarrow [B]^{\succcurlyeq} \cap A = \emptyset$).

c) Proven in the same manner as the first case.

Lemma 2

a) $[C \cup D]^{\succcurlyeq} = [C]^{\succcurlyeq} \cup [D]^{\succcurlyeq}$
b) $[C \cup D]^{\preccurlyeq} = [C]^{\preccurlyeq} \cup [D]^{\preccurlyeq}$
c) $[C \cup D]^{\approx} = [C]^{\approx} \cup [D]^{\approx}$

Proof. Follows directly from the definitions.

Lemma 3. *If $C \subseteq D$,*

a) $[C]^{\succcurlyeq} \subseteq [D]^{\succcurlyeq}$
b) $[C]^{\preccurlyeq} \subseteq [D]^{\preccurlyeq}$
c) $[C]^{\approx} \subseteq [D]^{\approx}$

Proof. The first case is proven as follows: Since $C \subseteq D$, we have that $D = C \cup (D - C)$. Hence, $[D]^{\succcurlyeq} = [C \cup (D - C)]^{\succcurlyeq} = [C]^{\succcurlyeq} \cup [D - C]^{\succcurlyeq} \supseteq [C]^{\succcurlyeq}$ by Lemma 2a. The last two cases can be proven in the same manner.

The DAG induces a subset inclusion order on the above and below sets. If e is above f, every class above e is also above f. Moreover, every class below f is also below e. This property is captured by the following lemma.

Lemma 4

a) $e \succcurlyeq f$ iff $[e]^{\succcurlyeq} \subseteq [f]^{\succcurlyeq}$
b) $e \preccurlyeq f$ iff $[e]^{\preccurlyeq} \subseteq [f]^{\preccurlyeq}$
c) $A \cap [e]^{\preccurlyeq} \neq \emptyset$ iff $[e]^{\succcurlyeq} \subseteq [A]^{\succcurlyeq}$
d) $A \cap [e]^{\succcurlyeq} \neq \emptyset$ iff $[e]^{\preccurlyeq} \subseteq [A]^{\preccurlyeq}$

Proof.

a) If $e \succcurlyeq f$, then for any $g \in [e]^{\succcurlyeq}$ we have that $g \succcurlyeq e \succcurlyeq f$. Consequently, $g \in [e]^{\succcurlyeq}$ implies $g \in [f]^{\succcurlyeq}$ so that $[e]^{\succcurlyeq} \subseteq [f]^{\succcurlyeq}$. If $[e]^{\succcurlyeq} \subseteq [f]^{\succcurlyeq}$ then $e \in [f]^{\succcurlyeq}$ since $e \in [e]^{\succcurlyeq}$. Hence, $e \succcurlyeq f$.
b) This is proven in the same manner as case a.
c) If $A \cap [e]^{\preccurlyeq} \neq \emptyset$, there is an $f \in A$ such that $e \succcurlyeq f$. Case a implies that $[e]^{\succcurlyeq} \subseteq [f]^{\succcurlyeq}$, and Lemma 3a implies that $[f]^{\succcurlyeq} \subseteq [A]^{\succcurlyeq}$ (since $\{f\} \subseteq A$). Hence, $[e]^{\succcurlyeq} \subseteq [f]^{\succcurlyeq} \subseteq [A]^{\succcurlyeq}$.
   If $[e]^{\succcurlyeq} \subseteq [A]^{\succcurlyeq}$, e must be in $[A]^{\succcurlyeq}$ (since $e \in [e]^{\succcurlyeq}$). This means that $[A]^{\succcurlyeq} \cap \{e\} \neq \emptyset$. Then $A \cap [e]^{\preccurlyeq} \neq \emptyset$ follows by Lemma 1a.
d) This is proven in the same manner as case c.

5   The Complement of a Class

The complement of a class c is the set of classes that should be separated from
c by a learning algorithm. In an ordinary (flat) decision system, we want to
separate each class from the other classes in V_d. A learning algorithm will thus
try to discern the objects of c (i.e., the objects in X_c) from the objects of $V_d - \{c\}$
(i.e., the objects in $\bigcup_{e \in V_d - \{c\}} X_e = U - X_c$). Hence, the complement of c may
be defined as $V_d - \{c\}$ in this case.

This definition is not appropriate for a DAG-decision system. The DAG defines relationships between the classes, and these relationships create implicit
annotations. For example, an object does not only belong to the class(es) to
which it is annotated, it also belongs to the superclasses of the class(es). Thus,
we define the complement of a class to be the set of classes that are unrelated
to the class.

Definition 7 (Complement). *For a set of classes $C \subseteq V_d$, the complement,
denoted by $\sim C$, is the set of classes that are unrelated to any class in C:*

$$\sim C = \{e \in V_d \mid \text{for all } c \in C, \ e \not\approx c\}$$

Note that we define the complement on a set of classes just as we did with the
above, below, and related sets since the complement of a single class is a special
case of the complement of a set of classes (i.e, $\sim \{c\} = \{e \in V_d \mid e \not\approx c\}$). Just as
before we may abuse the notation slightly and write $\sim c$ instead of $\sim \{c\}$ when
we consider single classes.

The complement $\sim C$ has several interesting properties. In particular, it is
equal to the intersection of the single class complements.

Lemma 5. $\sim C = \left(\bigcap_{c \in C} \sim c \right)$ *if we assume that* $\left(\bigcap_{c \in C} \sim c \right) = V_d$ *when* $C = \emptyset$.

Proof. $\sim C = \{e \in V_d \mid \text{for all } c \in C, \ e \not\approx c\} = \{e \in V_d \mid \text{for all } c \in C, \ e \in \sim c\}$
$= \left(\bigcap_{c \in C} \sim c \right)$

Moreover, it has the following properties on unions and intersections of classes.

Lemma 6. $\sim(C \cup D) = \sim C \cap \sim D$

Proof. Follows directly from Lemma 5.

Lemma 7. $\sim(C \cap D) \supseteq \sim C \cup \sim D$

Proof. For the set C, we have $C = (C \cap D) \cup (C - D)$. Then it follows from
Lemma 6 that $\sim C = \sim(C \cap D) \cap \sim(C - D)$. This means that $\sim C \subseteq \sim(C \cap D)$
since $A \cap B \subseteq A$ holds for any sets A and B. A similar argument holds for D.
Hence, $\sim D \subseteq \sim(C \cap D)$. Then $\sim C \cup \sim D \subseteq \sim(C \cap D)$ holds since for any sets
A, A', and B, we have that $A \subseteq B$ and $A' \subseteq B$ implies $A \cup A' \subseteq B$.

The complement is also closely connected to the related set that was defined in
Section 4.

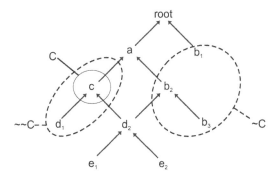

Fig. 3. *The double complement problem:* $C \neq \sim\sim C$

Lemma 8. $\sim C = V_d - [C]^{\approx}$

Proof. $\sim C = \{e \in V_d \mid$ for all $c \in C, e \not\approx c\} = V_d - \{e \in V_d \mid$ for some $c \in C, c \approx e\} = V_d - [C]^{\approx}$

The standard complement for sets obeys the law of double negation so that $C = V_d - (V_d - C)$. Unfortunately, it is not true for the complement of a class such that $\sim\sim C = V_d - [\sim C]^{\approx} = V_d - [V_d - [C]^{\approx}]^{\approx} \neq C$. For example, let $C = \{c\}$ in Figure 3. Then $\sim C = \{b_1, b_2, b_3\}$ and $\sim\sim C = \{c, d_1\}$. C is nonetheless a subset of $\sim\sim C$.

Lemma 9. $C \subseteq \sim\sim C$

Proof. $[C]^{\approx} \cap \sim C = \emptyset \xleftarrow{\text{Lem 2c}} C \cap [\sim C]^{\approx} = \emptyset \xleftarrow{\text{Lem 8}} C \cap (V_d - \sim\sim C) = \emptyset$ $\xleftarrow{\text{Lem 23}} C \subseteq \sim\sim C$

Even though $\sim\sim C \subseteq C$ does not hold, we may prove that $\sim\sim\sim C \subseteq \sim C$.

Lemma 10. $\sim C = \sim\sim\sim C$

Proof. Lemma 9 implies that $\sim C \subseteq \sim\sim\sim C$. Hence, we will only prove that $\sim\sim\sim C \subseteq \sim C$. From Lemma 9, we have that $C \subseteq \sim\sim C$, and $C \subseteq \sim\sim C \xrightarrow{\text{Lem 3 c}}$ $[C]^{\approx} \subseteq [\sim\sim C]^{\approx} \xrightarrow{\text{Lem 22}} V_d - [\sim\sim C]^{\approx} \subseteq V_d - [C]^{\approx} \xleftarrow{\text{Lem 8}} \sim\sim\sim C \subseteq \sim C$.

We may also prove a weak subset relation between $\sim\sim C$ and C.

Lemma 11. $[e]^{\approx} \cap \sim\sim C \neq \emptyset$ *iff* $[e]^{\approx} \cap C \neq \emptyset$

Proof. Lemmas 8 and 10 implies that

$$[\sim\sim C]^{\approx} = V_d - (V_d - [\sim\sim C]^{\approx}) = V_d - \sim\sim\sim C = V_d - \sim C = [C]^{\approx}$$

By this property and Lemma 1c, it follows that $[e]^{\approx} \cap \sim\sim C \neq \emptyset \iff \{e\} \cap [\sim\sim C]^{\approx} \neq \emptyset \iff \{e\} \cap [C]^{\approx} \neq \emptyset \iff [e]^{\approx} \cap C \neq \emptyset$.

In Lemma 24 (in the appendix) states that if $A \cap B \neq \emptyset$ and we replace A (or B) with a superset A' (or B'), the intersection $A' \cap B$ ($A \cap B'$) will still be empty. This allows us to prove the following properties on C and $\sim\sim C$.

Lemma 12

a) $[e]^{\preccurlyeq} \cap C \neq \emptyset$ implies $[e]^{\approx} \cap C \neq \emptyset$
b) $[e]^{\succcurlyeq} \cap C \neq \emptyset$ implies $[e]^{\approx} \cap C \neq \emptyset$
c) $[e]^{\preccurlyeq} \cap C \neq \emptyset$ implies $[e]^{\preccurlyeq} \cap \sim\sim C \neq \emptyset$
d) $[e]^{\succcurlyeq} \cap C \neq \emptyset$ implies $[e]^{\succcurlyeq} \cap \sim\sim C \neq \emptyset$

Proof. These properties follow from Lemma 24 since $[e]^{\preccurlyeq} \subseteq [e]^{\approx}$, $[e]^{\succcurlyeq} \subseteq [e]^{\approx}$, and $C \subseteq \sim\sim C$ (Lemma 9).

The above set $[e]^{\succcurlyeq}$ and the below set $[e]^{\preccurlyeq}$ of a class e are related through the complement so that if some class in $\sim\sim C$ is in the above set (the below set) then no class in $\sim C$ is in the below set (the above set), and vice versa. Note this result is due to a special property of the complement and does not hold for C.

Lemma 13

a) $[e]^{\preccurlyeq} \cap \sim\sim C \neq \emptyset$ implies $[e]^{\succcurlyeq} \cap \sim C = \emptyset$
b) $[e]^{\succcurlyeq} \cap \sim\sim C \neq \emptyset$ implies $[e]^{\preccurlyeq} \cap \sim C = \emptyset$
c) $[e]^{\succcurlyeq} \cap \sim C = \emptyset$ implies $[e]^{\approx} \cap \sim\sim C \neq \emptyset$
d) $[e]^{\preccurlyeq} \cap \sim C = \emptyset$ implies $[e]^{\approx} \cap \sim\sim C \neq \emptyset$
e) $[e]^{\approx} \cap \sim C = \emptyset$ implies $[e]^{\preccurlyeq} \cap \sim\sim C \neq \emptyset$
f) $[e]^{\approx} \cap \sim C = \emptyset$ implies $[e]^{\succcurlyeq} \cap \sim\sim C \neq \emptyset$

Proof.

a) $[e]^{\preccurlyeq} \cap \sim\sim C \neq \emptyset \xleftarrow{\text{Lem 4c}} [e]^{\succcurlyeq} \subseteq [\sim\sim C]^{\succcurlyeq} \xrightarrow{[D]^{\succcurlyeq} \subseteq [D]^{\approx}} [e]^{\succcurlyeq} \subseteq [\sim\sim C]^{\approx}$
$\xleftarrow{\text{Lem 23}} [e]^{\succcurlyeq} \cap (V_d - [\sim\sim C]^{\approx}) = \emptyset \xleftarrow{\text{Lem 8}} [e]^{\succcurlyeq} \cap \sim\sim\sim C = \emptyset \xleftarrow{\text{Lem 10}}$
$[e]^{\succcurlyeq} \cap \sim C = \emptyset$
b) Proven in the same manner as a.
c) $[e]^{\succcurlyeq} \cap \sim C = \emptyset \xleftarrow{\text{Lem 8 \& 23}} [e]^{\succcurlyeq} \subseteq [C]^{\approx} \xrightarrow{e \in [e]^{\succcurlyeq}} e \in [C]^{\approx} \Longleftrightarrow$
$\{e\} \cap [C]^{\approx} \neq \emptyset \xleftarrow{\text{Lem 1c}} [e]^{\approx} \cap C \neq \emptyset \xrightarrow{\text{Lem 11}} [e]^{\approx} \cap \sim\sim C \neq \emptyset$
d) Proven in the same manner as c.
e) $[e]^{\approx} \cap \sim C = \emptyset \xrightarrow{\text{Lem 1c}} \{e\} \cap [\sim C]^{\approx} = \emptyset \xleftarrow{\text{Lem 8 \& 23}} \{e\} \subseteq \sim\sim C \Longleftrightarrow$
$\{e\} \cap \sim\sim C \neq \emptyset \xrightarrow{\text{Lem 24b}} [e]^{\preccurlyeq} \cap \sim\sim C \neq \emptyset$ (since $\{e\} \subseteq [e]^{\preccurlyeq}$)
f) Proven in the same manner as e.

Theorem 1. $[e]^{\preccurlyeq} \cap \sim\sim C \neq \emptyset$ iff $[e]^{\succcurlyeq} \cap \sim C = \emptyset$

Proof. Lemma 13a states that $[e]^{\preccurlyeq} \cap \sim\sim C \neq \emptyset$ implies $[e]^{\succcurlyeq} \cap \sim C = \emptyset$. So, we need only to prove that $[e]^{\succcurlyeq} \cap \sim C = \emptyset$ implies $[e]^{\preccurlyeq} \cap \sim\sim C \neq \emptyset$.

$[e]^{\succcurlyeq} \cap \sim C = \emptyset \xrightarrow{\text{Lem 13c}} [e]^{\approx} \cap \sim\sim C \neq \emptyset \xleftarrow{\text{By def}} ([e]^{\preccurlyeq} \cup [e]^{\succcurlyeq}) \cap \sim\sim C \neq \emptyset$
$\xleftarrow{\text{Lem 25}} [e]^{\preccurlyeq} \cap \sim\sim C \neq \emptyset$ or $[e]^{\succcurlyeq} \cap \sim\sim C \neq \emptyset$

The first part of this disjunction is our goal. So, it remains only to prove that $[e]^{\preceq} \cap \sim\sim C \neq \emptyset$ also follows from $[e]^{\succeq} \cap \sim C = \emptyset$ when $[e]^{\succeq} \cap \sim\sim C \neq \emptyset$. Lemma 13b states that $[e]^{\preceq} \cap \sim C = \emptyset$ follows from $[e]^{\succeq} \cap \sim\sim C \neq \emptyset$. Hence, both $[e]^{\preceq} \cap \sim C = \emptyset$ and $[e]^{\succeq} \cap \sim C = \emptyset$ are true.

$$[e]^{\preceq} \cap \sim C = \emptyset \text{ and } [e]^{\succeq} \cap \sim C = \emptyset \xleftarrow{\text{Lem 25}} ([e]^{\preceq} \cup [e]^{\succeq}) \cap \sim C = \emptyset$$

$$\xleftarrow{\text{By def}} [e]^{\approx} \cap \sim C = \emptyset \xrightarrow{\text{Lem 13a}} [e]^{\preceq} \cap \sim\sim C \neq \emptyset$$

Theorem 2. $[e]^{\succeq} \cap \sim\sim C \neq \emptyset$ iff $[e]^{\preceq} \cap \sim C = \emptyset$

Proof. The theorem is proven in the same manner as Theorem 1.

6   DAG-Decision Systems

A DAG-decision system is an extension of a decision system and is defined on the partial ordered that was introduced in Section 3.

Definition 8 (DAG-decision system). *Let $\mathcal{A} = \langle U, A, d, \succeq \rangle$ denote a DAG-decision system where*

- *U is a non-empty finite set of (observable) objects, called the universe.*
- *A is a set of conditional attributes describing the objects. Each attribute $a \in A$ is a function $a : U \to V_a$ where V_a is a set of values that an object may have for a.*
- *d is the decision attribute, which is not in A. It is a function $d : U \to V_d$ where V_d is a set of decision classes.*
- *$\succeq$ is a partial order on the classes in V_d where $p \succeq r$ denotes that p is more general than r $(p, r \in V_d)$.*

This definition allows us to formalize the additional information provided by the ontology. However, gene function prediction introduces also another problem: A gene may be annotated with more than one class. Thus, we need to represent and predict multiple classes for each gene.

One way to achieve this would be redefined the decision attribute so that it maps an object to set of classes rather than a single class. Another would be to consider the annotations (i.e., tuples of a gene and a class) as the objects. The last approach is more convenient for a number of reasons. First, the standard rough set approximations may be applied directly without any modification since each object has only one decision class in this case. Second, we may use the standard rough set approximations to determine, which genes belong to a set D of classes. This set of genes cannot be defined exactly, since the genes have more than one class. However, it can be described by a lower and an upper approximation where the lower contains the genes only labeled with classes from D, while the upper also contains the genes labeled with classes from both D and the complement of D.

Thus, we consider an object to be a tuple $\langle x, d \rangle$ where x is a gene in G and d is a decision class from V_d so that $U \subseteq G \times V_d$. The decision attribute is defined as $d(\langle x, d' \rangle) = d'$. We define an indiscernibility relation $IND(G)$ as follows.

$$IND(G) = \{\langle x, y \rangle \in U \times U \mid x = \langle z, d_x \rangle \text{ and } y = \langle z, d_y \rangle\}$$

This partitions the universe into a collection of elementary sets that correspond to the genes in G. The generalized decision defined on $IND(G)$ is $\partial_G(x) = \{d'(y) \mid y \in [x]_G\}$ and represents the classes which are associated with gene $[x]_G$. Moreover, $IND(G)$ creates a finer partition than $IND(A)$ (the partition created by the attributes) so that we may apply $\overline{A}$ and $\underline{A}$ on top of $\overline{G}$ and $\underline{G}$. For example, we have that $\overline{A}X = \overline{A}\,\overline{G}X$ and $\underline{A}X = \underline{A}\,\underline{G}X$. More details on this issue may be found in [12].

7   Ambiguities in the DAG

In set theory, the membership of all objects in the universe is assumed to be known. An object y is either a member of a set X or it is not. This is also the starting point in rough set theory, but here an elementary set may belong to both X and the complement $U - X$. In this case, we say that the membership is inconsistent. Hence, rough set theory distinguishes between three different ways that an element can be related to a set: *in*, *not in*, and *inconsistent*.

We may also recognize a fourth membership category; it may be *unknown* whether an object belongs to a set or not. This situation arises in the DAG as shown in Figure 4 for class c. The objects of the superclasses a and *root* may belong to c or some of the classes that are unrelated to c, i.e., b_1, b_2, and b_3. However, we do not know which class. The membership of these objects is in other words unknown with respect to c.

This means that a framework for a DAG-decision system must be able to represent objects whose memberships are unknown. Since rough set theory does not have such facilities, we must extend it. We do this by employing the same strategy as rough set theory and define a *lower* and an *upper* boundary for

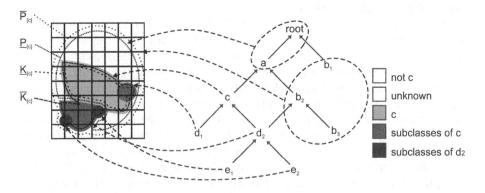

Fig. 4. Discernibility of the class c

each class. The lower boundary K_c, which is also called the *known* set, basically consists of the objects that belong to the class c or to the subclasses of c. Note that the objects of the subclasses are included in this set since they are also members of c. The upper boundary P_c, which is also called the *potential* set, is a superset of K_c. It contains the objects of the superclasses of c in addition to the objects of K_c. Using these two sets, we can represent the unknown objects with regard to c as the difference $P_c - K_c$.

Formally, we could try to define the lower and the upper boundaries as follows:

$$K_c = \{x \in U \mid c \succcurlyeq d(x)\} \qquad (= \{x \in U \mid [d(x)]^{\succcurlyeq} \cap \{c\} \neq \emptyset\})$$
$$P_c = \{x \in U \mid c \approx d(x)\} \qquad (= \{x \in U \mid [d(x)]^{\approx} \cap \{c\} \neq \emptyset\})$$

For a set of classes $C \subseteq V_d$, these boundaries could also be defined as the union of the corresponding boundaries of each class, i.e.,

$$K_C = \bigcup_{c \in C} K_c = \{x \in U \mid [d(x)]^{\succcurlyeq} \cap \{c\} \neq \emptyset, \text{ for some } c \in C\} \text{ and}$$
$$P_C = \bigcup_{c \in C} P_c = \{x \in U \mid [d(x)]^{\approx} \cap \{c\} \neq \emptyset, \text{ for some } c \in C\}$$

Unfortunately, these definitions are complicated for several reasons.

One problem is that these boundaries are not complementary. We would like K_C to be complementary to $P_{\sim C}$ (i.e., $K_C = U - P_{\sim C}$) and P_C to be complementary to $K_{\sim C}$ (i.e., $P_C = U - K_{\sim C}$). Unfortunately, $K_C = U - P_{\sim C}$ does not hold.

Example 1. Assume that $C = \{d_1, d_2\}$ in Figure 4. Then K_C consists of the objects in d_1, d_2, e_1, and e_2. The complement $\sim C$ is, in this case, equal to $\{b_1, b_3\}$, and $P_{\sim C}$ consists of the objects in *root*, a, b_1, b_2, and b_3. However, the objects in c do not occur in either set.

The reason that the objects of the class c do not occur in K_C and $P_{\sim C}$ in the last example is that c has no other immediate subclasses than d_1 and d_2. All of its subclasses are related to C so that no class in the complement $\sim C$ is related to c. Its objects are therefore not included in $P_{\sim C}$. Moreover, $c \notin C$ so that its objects are not in K_C, either.

Notice that this issue is not specific to a DAG. It occurs in a tree as well. The problem also arises when a class has only a single child.

Example 2. Assume that the edge from d_2 to c has been removed in Figure 4. In this case, d_1 is a lonely child of c. If $C = \{d_1\}$, then $\sim C = \{b_1, b_2, b_3, d_2, e_1, e_2\}$ and the objects of c are in neither K_C nor $P_{\sim C}$.

We may correct this problem by changing the definition of either K_C or $P_{\sim C}$. However, it would be counter-intuitive to define $P_{\sim C}$ so that it included the objects of c in Example 1. Class c is not a superclass of any class in $\sim C$, and its objects are not potential members of these classes. On the other hand, if we assume that the objects of c must belong to either d_1 or d_2, we *know* that the objects of c belong to C since C includes both classes. Hence, K_C should include objects from a class *if none of its subclasses are in* $\sim C$. More formally, object x should be in K_C if x satisfies

$$[d(x)]^{\preccurlyeq} \cap \sim C = \emptyset \qquad \text{(Cond. 1)}$$

Note that this condition subsumes the condition in K_C (as shown by the following lemma). Thus, the objects in K_C will also satisfy Cond. 1, and we may replace the condition in the definition of K_C with Cond. 1.

Lemma 14. $[e]^{\succcurlyeq} \cap \{c\} \neq \emptyset$, *for some* $c \in C$ *implies* $[e]^{\preccurlyeq} \cap \sim C = \emptyset$

Proof. $[e]^{\succcurlyeq} \cap \{c\} \neq \emptyset$, for some $c \in C$ $\xleftrightarrow{\text{Lem 25}}$ $[e]^{\succcurlyeq} \cap C \neq \emptyset$ $\xrightarrow{\text{Lem 12d}}$ $[e]^{\succcurlyeq} \cap \sim\sim C \neq \emptyset$ $\xleftrightarrow{\text{Th 2}}$ $[e]^{\preccurlyeq} \cap \sim C = \emptyset$

Remark 1. Observe that when we assume that the objects of c belong to $K_{\{d_1,d_2\}}$, we make an assumption similar to the Closed World Assumption in Logic Programming. We have that $d_1 \to c$ and $d_2 \to c$ since the objects of d_1 and d_2 also belong to c. This means that $d_1 \vee d_2 \to c$. If the objects of c belong to $K_{\{d_1,d_2\}}$, then $d_1 \vee d_2 \leftarrow c$. So, c is equivalent to $d_1 \vee d_2$ in this case.

There is another problem with the boundary sets. A class may have several immediate superclasses in a DAG. This means that some objects may be included in the (lower and upper) boundaries of both a set C and the complement $\sim C$.

Example 3. Consider Figure 4 again. If $C = \{c\}$, we have $\sim C = \{b_1, b_2, b_3\}$. The classes d_2, e_1, and e_2 are subclasses of c and b_2. Hence, they are below a class in C and a class in $\sim C$. Their objects will consequently be in both K_C and $K_{\sim C}$ (and P_C and $P_{\sim C}$).

The DAG introduces in this way a new type of inconsistency, which we will call DAG-inconsistency. Thus, we have to define upper and lower approximations of K_C and P_C. The upper approximations contain all objects from the subclasses of the classes in C, just as before. The lower approximations, on the other hand, do not contain the objects from subclasses, which introduce this kind of inconsistency.

A DAG-inconsistency occurs when a subclass of a class in C also has another superclass in $\sim C$. Hence, the objects of a class should only be included in the lower approximations if *there are no superclasses in* $\sim C$. Formally, object x should be included in the lower approximations if

$$[d(x)]^{\succcurlyeq} \cap \sim C = \emptyset \qquad \text{(Cond. 2)}$$

Note that it is necessary to define this condition for a set of classes since a DAG-inconsistency that occurs for a single class may disappear if a set is considered.

Example 4. Let $C' = \{c, b_2\}$ (in Figure 4), which is a superset of C in Example 3. In this case, $\sim C' = \{b_1\}$, and the objects of d_2, e_1, and e_2 belong only to $K_{C'}$ and $P_{C'}$ and not to $K_{\sim C'}$ and $P_{\sim C'}$. So the DAG-inconsistency that occurred for c in Example 3 has vanished.

We may define the following upper and lower approximations of the boundary sets:

- $\underline{K}_C = \{x \in U \mid \text{Cond. 1 and Cond. 2}\}$
- $\overline{K}_C = \{x \in U \mid \text{Cond. 1}\}$
- $\underline{P}_C = \{x \in U \mid [d(x)]^{\approx} \cap C \neq \emptyset \text{ and Cond. 2}\}$
- $\overline{P}_C = \{x \in U \mid [d(x)]^{\approx} \cap C \neq \emptyset\}$

Some simplifications are possible, however. The conditions in $\underline{K}_C$, $[e]^{\preccurlyeq} \cap \sim C = \emptyset$ (Cond. 1) and $[e]^{\succcurlyeq} \cap \sim C = \emptyset$ (Cond. 2), may be combined into $[e]^{\approx} \cap \sim C = \emptyset$. This follows from Lemma 25 and that $[e]^{\approx} = [e]^{\preccurlyeq} \cup [e]^{\succcurlyeq}$. Cond. 2 also implies $[e]^{\approx} \cap C \neq \emptyset$. Thus, the first condition in $\underline{P}_C$ may be removed.

Lemma 15. $[e]^{\succcurlyeq} \cap \sim C = \emptyset$ *implies* $[e]^{\approx} \cap C \neq \emptyset$

Proof. $[e]^{\succcurlyeq} \cap \sim C = \emptyset \xrightarrow{\text{Lem 13c}} [e]^{\approx} \cap \sim\sim C \neq \emptyset \xleftarrow{\text{Lem 11}} [e]^{\approx} \cap C \neq \emptyset$

This results in the following boundary sets.

Definition 9 (Boundary sets). *Given a set of classes $C \subseteq V_d$, the following sets constitute the upper and the lower approximations of the known and the potential objects of C:*

- $\underline{K}_C = \{x \in U \mid [d(x)]^{\approx} \cap \sim C = \emptyset\}$
- $\overline{K}_C = \{x \in U \mid [d(x)]^{\preccurlyeq} \cap \sim C = \emptyset\}$
- $\underline{P}_C = \{x \in U \mid [d(x)]^{\succcurlyeq} \cap \sim C = \emptyset\}$
- $\overline{P}_C = \{x \in U \mid [d(x)]^{\approx} \cap C \neq \emptyset\}$

8   Properties of the Boundaries

In this section, we will examine the properties of the boundary sets. The first theorem shows that the boundary sets are complementary.

Theorem 3. *Given a set of classes $C \subseteq V_d$, the following complements hold:*

a) $\underline{K}_{\sim C} = U - \overline{P}_C$

b) $\overline{K}_{\sim C} = U - \underline{P}_C$

c) $\underline{P}_{\sim C} = U - \overline{K}_C$

d) $\overline{P}_{\sim C} = U - \underline{K}_C$

Proof.

a) $\underline{K}_{\sim C} = \{x \in U \mid [d(x)]^{\approx} \cap \sim\sim C = \emptyset\} = U - \{x \in U \mid [d(x)]^{\approx} \cap \sim\sim C \neq \emptyset\}$
$= U - \{x \in U \mid [d(x)]^{\approx} \cap C \neq \emptyset\} = U - \overline{P}_C$ by Lemma 11

b) $\overline{K}_{\sim C} = \{x \in U \mid [d(x)]^{\preceq} \cap \sim\sim C = \emptyset\} = U - \{x \in U \mid [d(x)]^{\preceq} \cap \sim\sim C \neq \emptyset\}$
$= U - \{x \in U \mid [d(x)]^{\succeq} \cap \sim C = \emptyset\} = U - \underline{P}_C$ by Theorem 1

c) $\underline{P}_{\sim C} = \{x \in U \mid [d(x)]^{\succeq} \cap \sim\sim C = \emptyset\} = U - \{x \in U \mid [d(x)]^{\succeq} \cap \sim\sim C \neq \emptyset\}$
$= U - \{x \in U \mid [d(x)]^{\preceq} \cap \sim C = \emptyset\} = U - \overline{K}_C$ by Theorem 2

d) $\overline{P}_{\sim C} = \{x \in U \mid [d(x)]^{\approx} \cap \sim C \neq \emptyset\} = U - \{x \in U \mid [d(x)]^{\approx} \cap \sim C = \emptyset\}$
$= U - \underline{K}_C$

It follows from this theorem that the boundary set on $\sim\sim C$ is equal to the boundary set on C. So, applying double negation to C produces the same known and potential sets as C.

Corollary 1

a) $\underline{K}_{\sim(\sim C)} = \underline{K}_C$ c) $\underline{P}_{\sim(\sim C)} = \underline{P}_C$

b) $\overline{K}_{\sim(\sim C)} = \overline{K}_C$ d) $\overline{P}_{\sim(\sim C)} = \overline{P}_C$

Proof. Using Theorem 3b and 3c, we have for option b that $\overline{K}_{\sim(\sim C)} = U - \underline{P}_{\sim C} = U - (U - \overline{K}_C) = \overline{K}_C$. The other propositions are proven in the same manner by applying Theorem 3.

We will now examine the properties of the boundaries with respect to unions and intersections. We begin with unions. Given the discussion in Section 7, it should not be a surprise that the boundary sets, except for $\overline{P}_C$, do not maintain unions. However, we may prove a weaker property. For each boundary set, it holds that the boundary set on $C \cup D$ is a superset of the union of the individual boundary sets on C and D (where C and D are sets of classes).

Lemma 16. *Given two set of classes C and D, the following properties hold:*

a) $\underline{K}_{C \cup D} \supseteq \underline{K}_C \cup \underline{K}_D$ c) $\underline{P}_{C \cup D} \supseteq \underline{P}_C \cup \underline{P}_D$

b) $\overline{K}_{C \cup D} \supseteq \overline{K}_C \cup \overline{K}_D$ d) $\overline{P}_{C \cup D} = \overline{P}_C \cup \overline{P}_D$

Proof.

a) $\underline{K}_{C \cup D} = \{x \in U \mid [d(x)]^{\approx} \cap \sim(C \cup D) = \emptyset\}$
$= \{x \in U \mid [d(x)]^{\approx} \cap (\sim C \cap \sim D) = \emptyset\}$ (Lemma 6)
$\supseteq \{x \in U \mid [d(x)]^{\approx} \cap \sim C = \emptyset \text{ or } [d(x)]^{\approx} \cap \sim D = \emptyset\}$ (Lemma 26)
$= \underline{K}_C \cup \underline{K}_D$

b) $\overline{K}_{C \cup D} = \{x \in U \mid [d(x)]^{\preceq} \cap \sim(C \cup D) = \emptyset\}$
$= \{x \in U \mid [d(x)]^{\preceq} \cap (\sim C \cap \sim D) = \emptyset\}$ (Lemma 6)
$\supseteq \{x \in U \mid [d(x)]^{\preceq} \cap \sim C = \emptyset \text{ or } [d(x)]^{\preceq} \cap \sim D = \emptyset\}$ (Lemma 26)
$= \overline{K}_C \cup \overline{K}_D$

c) $\underline{P}_{C \cup D} = \{x \in U \mid [d(x)]^{\succcurlyeq} \cap \sim(C \cup D) = \emptyset\}$

$\qquad = \{x \in U \mid [d(x)]^{\succcurlyeq} \cap (\sim C \cap \sim D) = \emptyset\}$ $\hfill$ (Lemma 6)

$\qquad \supseteq \{x \in U \mid [d(x)]^{\succcurlyeq} \cap \sim C = \emptyset \text{ or } [d(x)]^{\succcurlyeq} \cap \sim D = \emptyset\}$ $\hfill$ (Lemma 26)

$\qquad = \underline{P}_C \cup \underline{P}_D$

d) $\overline{P}_{C \cup D} = \{x \in U \mid [d(x)]^{\approx} \cap (C \cup D) \neq \emptyset\}$

$\qquad = \{x \in U \mid [d(x)]^{\approx} \cap C \neq \emptyset \text{ or } [d(x)]^{\approx} \cap D \neq \emptyset\}$ $\hfill$ (Lemma 25)

$\qquad = \overline{P}_C \cup \overline{P}_D$

None of the boundary sets maintains intersections. Still, we can show that a boundary set on $C \cap D$ is a subset of the intersection of the boundary sets on C and D.

Lemma 17. *Given two set of classes C and D, the following properties hold:*

a) $\underline{K}_{C \cap D} \subseteq \underline{K}_C \cap \underline{K}_D$ $\qquad\qquad\qquad$ c) $\underline{P}_{C \cap D} \subseteq \underline{P}_C \cap \underline{P}_D$

b) $\overline{K}_{C \cap D} \subseteq \overline{K}_C \cap \overline{K}_D$ $\qquad\qquad\qquad$ d) $\overline{P}_{C \cap D} \subseteq \overline{P}_C \cap \overline{P}_D$

Proof.

a) This follows directly from options b and c as $\underline{K}_C$ is an intersection of $\overline{K}_C$ and $\underline{P}_C$.

b) $\overline{K}_{C \cap D} = \{x \in U \mid [d(x)]^{\preccurlyeq} \cap \sim(C \cap D) = \emptyset\}$

$\qquad \subseteq \{x \in U \mid [d(x)]^{\preccurlyeq} \cap (\sim C \cup \sim D) = \emptyset\}$ $\hfill$ (Lemmas 7 and 24)

$\qquad = \{x \in U \mid [d(x)]^{\preccurlyeq} \cap \sim C = \emptyset \text{ and } [d(x)]^{\preccurlyeq} \cap \sim D = \emptyset\}$ $\hfill$ (Lemma 25)

$\qquad = \overline{K}_C \cap \overline{K}_D$

c) $\underline{P}_{C \cap D} = \{x \in U \mid [d(x)]^{\succcurlyeq} \cap \sim(C \cap D) = \emptyset\}$

$\qquad \subseteq \{x \in U \mid [d(x)]^{\succcurlyeq} \cap (\sim C \cup \sim D) = \emptyset\}$ $\hfill$ (Lemmas 7 and 24)

$\qquad = \{x \in U \mid [d(x)]^{\succcurlyeq} \cap \sim C = \emptyset \text{ and } [d(x)]^{\succcurlyeq} \cap \sim D = \emptyset\}$ $\hfill$ (Lemma 25)

$\qquad = \underline{P}_C \cap \underline{P}_D$

d) $\overline{P}_{C \cap D} = \{x \in U \mid [d(x)]^{\approx} \cap (C \cap D) \neq \emptyset\}$

$\qquad \subseteq \{x \in U \mid [d(x)]^{\approx} \cap C \neq \emptyset \text{ and } [d(x)]^{\approx} \cap D \neq \emptyset\}$ $\hfill$ (Lemma 26)

$\qquad = \overline{P}_C \cap \overline{P}_D$

Hence, intersections are not maintained. It should be mentioned, however, that $\overline{K}_{C \cap D}$ and $\overline{K}_C \cap \overline{K}_D$ describe different concepts (the situation is similar for the other boundary sets). $C \cap D$ is the set of *classes* that C and D have in common, and $\overline{K}_{C \cap D}$ is the set of objects that belong to these classes. $\overline{K}_C \cap \overline{K}_D$, on the other hand, is the set of *objects* that C and D have in common. For example, if we consider the DAG in Figure 4 and set $C = \{e_1\}$ and $D = \{d_2\}$, then $\overline{K}_{C \cap D}$ will be empty, while $\overline{K}_C \cap \overline{K}_D$ will contain the objects that are labeled to e_1. In this case, $\overline{K}_C \cap \overline{K}_D$ seems more useful than $\overline{K}_{C \cap D}$ as it determines the common known objects of C and D. $\overline{K}_{C \cap D}$, on the other hand, ignores the relationships between the classes. It is thus a question if intersection is an interesting operator for classes. An operator that takes the relationships into account would probably be more useful. For example, we could define an operator like $C \text{ⓜ} D = \{c \in C \cup D \mid c \approx e$, for some $e \in C$ and $c \approx f$, for some $f \in D\}$. However, we will not pursue this issue any further in this paper.

9   Well-Defined DAGs

In Section 7, we had to make several adjustments to the boundary sets since they were not complementary. In particular, it was necessary to assign the objects in a class e to the known set of C if none of its subclasses were in $\sim C$ (even though e was not a subclass of any class in C). Similarly, if a class c was a lonely child of a class f, f had to be assigned to the known set of c.

It is possible, however, to avoid these complications, if we consider only single classes and assume that the DAG is well-defined such that there are no lonely children. In this case, the boundary sets may be simplified. We will consider these simplifications in this section. We begin with a definition of what it means for a DAG to be well-defined.

Definition 10 (Well-defined DAG). *A DAG is well-defined if it holds for every non-leaf class e that for each subclass of e there is another subclass of e that belong to the complement of the first subclass, i.e.,*

$$[e]^{\prec} \cap \{f\} \neq \emptyset \ \text{implies} \ [e]^{\prec} \cap \sim f \neq \emptyset$$

We may now define the simplified boundary sets for a single class.

Definition 11 (Boundary sets for a single class). *Given a (single) class c, the following sets constitute the upper and the lower approximations of the known and the potential objects of c:*

- $\underline{K}_c^* = \{x \in U \mid [d(x)]^{\succcurlyeq} \cap \{c\} \neq \emptyset \ \text{and} \ [d(x)]^{\succcurlyeq} \cap \sim\{c\} = \emptyset\}$
- $\overline{K}_c^* = \{x \in U \mid [d(x)]^{\succcurlyeq} \cap \{c\} \neq \emptyset\}$
- $\underline{P}_c^* = \{x \in U \mid [d(x)]^{\succcurlyeq} \cap \sim\{c\} = \emptyset\}$
- $\overline{P}_c^* = \{x \in U \mid [d(x)]^{\approx} \cap \{c\} \neq \emptyset\}$

It follows immediately from this definition and Definition 9 that $\underline{P}_c^* = \underline{P}_{\{c\}}$ and $\overline{P}_c^* = \overline{P}_{\{c\}}$. Hence, the single-class potential sets may be used even when the DAG is not well-defined. The problem lies with the known sets where only $\underline{K}_c^* \subseteq \underline{K}_{\{c\}}$ and $\overline{K}_c^* \subseteq \overline{K}_{\{c\}}$ hold generally. Still, if the DAG is well-defined, we may prove the following theorem. It implies that the original set-based known sets and the simplified single-class known sets must be equal for a single class.

Lemma 18. *If the DAG is well-defined, then* $[e]^{\preccurlyeq} \cap \sim\{c\} = \emptyset \Leftrightarrow [e]^{\succcurlyeq} \cap \{c\} \neq \emptyset.$

Proof.

$\Leftarrow$: Always true, since $[e]^{\succcurlyeq} \cap \{c\} \neq \emptyset \xrightarrow{\text{Lem 12d}} [e]^{\succcurlyeq} \cap \sim\sim\{c\} \neq \emptyset \xleftrightarrow{\text{Th 2}} [e]^{\preccurlyeq} \cap \sim\{c\} = \emptyset$

$\Rightarrow$: Since $[e]^{\preccurlyeq} \cap \sim c = \emptyset$, Definition 10 implies $[e]^{\preccurlyeq} \cap \{c\} = \emptyset$. Moreover,

$$[e]^{\preccurlyeq} \cap \sim\{c\} = \emptyset \xrightarrow{\text{Lem 13d}} [e]^{\approx} \cap \sim\sim\{c\} \neq \emptyset \xleftarrow{\text{Lem 11}} [e]^{\approx} \cap \{c\} \neq \emptyset$$

These two results imply $[e]^{\succcurlyeq} \cap \{c\} \neq \emptyset$ since

$$[e]^{\succcurlyeq} \cap \{c\} = ([e]^{\succcurlyeq} \cap \{c\}) \cup \emptyset = ([e]^{\succcurlyeq} \cap \{c\}) \cup ([e]^{\preccurlyeq} \cap \{c\}) = [e]^{\approx} \cap \{c\} \neq \emptyset$$

Theorem 4. *If the DAG is well-defined, then*

a) $\underline{K}_c^* = \underline{K}_{\{c\}}$
b) $\overline{K}_c^* = \overline{K}_{\{c\}}$

Proof.

a) $\underline{K}_{\{c\}} = \{x \in U \mid [d(x)]^{\approx} \cap \sim\{c\} = \emptyset\}$

$\qquad = \{x \in U \mid ([d(x)]^{\succcurlyeq} \cup [d(x)]^{\preccurlyeq}) \cap \sim\{c\} = \emptyset\}$ (By def. of $\approx$)

$\qquad = \{x \in U \mid [d(x)]^{\preccurlyeq} \cap \sim\{c\} = \emptyset \text{ and } [d(x)]^{\succcurlyeq} \cap \sim\{c\} = \emptyset\}$ (Lemma 25)

$\qquad = \{x \in U \mid [d(x)]^{\succcurlyeq} \cap \{c\} \neq \emptyset \text{ and } [d(x)]^{\succcurlyeq} \cap \sim\{c\} = \emptyset\}$ (Lemma 18)

$\qquad = \underline{K}_c^*$

b) Follows immediately from Lemma 18.

The complement of a single class is a set of classes. Thus, we need boundary sets that apply to a set of classes in order to find the known and the potential objects of the complement. Such boundary set could perhaps be found by taking the union of the single-class boundary sets, which belong to the classes in the complement. However, the previous discussion and Lemma 16 entail that a union of single-class boundary sets are only a subset of the corresponding set-based boundary set. This suggests that the complement of a single class may not be found with the single-class boundary sets. So their utility may appear limited.

However, the complement is a very special set so that the single-class boundary sets may still be useful. When a set-based boundary set is applied to a set of classes, it may add some classes that would not be included by the union of the single-class boundary set (This is the reason why a set-based boundary set is superset of the union of the corresponding single-class boundary sets). However, no classes will be added if a boundary set is applied to the complement since the complement has a special property; any class that possibly could be added by the boundary set is already in the complement. The union of single-class boundary sets is thus equal to the set-based boundary set in this case. This will be proven in Theorem 5.

This definition defines the boundary sets for the complement by using the union of single-class boundary sets.

Definition 12 (Boundary set for complement). *Let C be a set of classes. The following sets denote the the upper and the lower approximations of the known and the potential objects of $\sim C$:*

- $\underline{K}^*_{\sim C} = \bigcup_{c \in \sim C} \underline{K}^*_c$
- $\overline{K}^*_{\sim C} = \bigcup_{c \in \sim C} \overline{K}^*_c$
- $\underline{P}^*_{\sim C} = \bigcup_{c \in \sim C} \underline{P}^*_c$
- $\overline{P}^*_{\sim C} = \bigcup_{c \in \sim C} \overline{P}^*_c$

The following lemmas prove several properties that we will need in Theorem 5.

Lemma 19. $[e]^{\succcurlyeq} \cap \{c\} \neq \emptyset$ *holds for some* $c \in \sim C$ *iff* $[e]^{\preccurlyeq} \cap \sim\sim C = \emptyset$.

Proof. Lemma 25 states that $[e]^{\succcurlyeq} \cap \{c\} \neq \emptyset$ holds for some $c \in \sim C$ if and only if $[e]^{\succcurlyeq} \cap \left(\bigcup_{c \in \sim C}\{c\}\right) \neq \emptyset$. Since $\sim C = \bigcup_{c \in \sim C}\{c\}$, it follows that the condition on the left hand side is equivalent to $[e]^{\succcurlyeq} \cap \sim C \neq \emptyset$, which is again equivalent to $[e]^{\preccurlyeq} \cap \sim\sim C = \emptyset$ by Theorem 1.

Lemma 20. $[e]^{\succcurlyeq} \cap \sim\{c\} = \emptyset$ *holds for some* $c \in \sim C$ *iff* $[e]^{\succcurlyeq} \cap \sim\sim C = \emptyset$.

Proof.

$\Rightarrow$: $[e]^{\succcurlyeq} \cap \sim\{c\} = \emptyset$, for some $c \in \sim C$ $\xrightarrow{\text{Lem 26}}$ $[e]^{\succcurlyeq} \cap \left(\bigcap_{c \in \sim C} \sim\{c\}\right) = \emptyset$
$\xrightarrow{\text{Lem 5}}$ $[e]^{\succcurlyeq} \cap \sim\sim C = \emptyset$

$\Leftarrow$: $[e]^{\succcurlyeq} \cap \sim\sim C = \emptyset$ $\xleftarrow{\text{Th 2}}$ $[e]^{\preccurlyeq} \cap \sim C \neq \emptyset$ $\Leftrightarrow$ $[e]^{\preccurlyeq} \cap \bigcup_{c \in \sim C}\{c\} \neq \emptyset$
$\xrightarrow{\text{Lem 25}}$ $[e]^{\preccurlyeq} \cap \{c\} \neq \emptyset$, for some $c \in \sim C$
$\xrightarrow{\text{Lem 12c}}$ $[e]^{\preccurlyeq} \cap \sim\sim\{c\} \neq \emptyset$, for some $c \in \sim C$
$\xrightarrow{\text{Lem 13a}}$ $[e]^{\succcurlyeq} \cap \sim\{c\} = \emptyset$, for some $c \in \sim C$

Lemma 21. $[e]^{\succcurlyeq} \cap \{c\} \neq \emptyset$ *and* $[e]^{\succcurlyeq} \cap \sim\{c\} = \emptyset$ *holds for some* $c \in \sim C$ *iff* $[e]^{\approx} \cap \sim\sim C = \emptyset$

Proof.

$\Rightarrow$: $[e]^{\succcurlyeq} \cap \{c\} \neq \emptyset$ and $[e]^{\succcurlyeq} \cap \sim\{c\} = \emptyset$, for some $c \in \sim C$
$\implies$ $([e]^{\succcurlyeq} \cap \{c'\} \neq \emptyset$, for $c' \in \sim C)$ and $([e]^{\succcurlyeq} \cap \sim\{c''\} = \emptyset$, for $c'' \in \sim C)$
$\xrightarrow{\text{Lem 19 \& 20}}$ $[e]^{\preccurlyeq} \cap \sim\sim C = \emptyset$ and $[e]^{\succcurlyeq} \cap \sim\sim C = \emptyset$
$\xrightarrow{\text{Lemma 25}}$ $([e]^{\preccurlyeq} \cup [e]^{\succcurlyeq}) \cap \sim\sim C = \emptyset$ $\xrightarrow{\text{Def of } \approx}$ $[e]^{\approx} \cap \sim\sim C = \emptyset$

$\Leftarrow$: We prove first that $e \in \sim C$ when $[e]^{\approx} \cap \sim\sim C = \emptyset$.

$$[e]^{\approx} \cap \sim\sim C = \emptyset \xrightarrow{\text{Lem 11}} [e]^{\approx} \cap C = \emptyset \xrightarrow{\text{Lem 1c}} \{e\} \cap [C]^{\approx} = \emptyset$$

$$\xrightarrow{\text{Lem 8 \& 23}} \{e\} \subseteq \sim C$$

Hence, e is in $\sim C$. Moreover, e satisfies both conditions on the left hand side of the *iff* since $[e]^{\succcurlyeq} \cap \{e\} = \{e\} \neq \emptyset$ and $[e]^{\succcurlyeq} \cap \sim\{e\} \subseteq [e]^{\approx} \cap \sim\{e\} = \emptyset$. Consequently, $[e]^{\approx} \cap \sim\sim C = \emptyset$ implies that $[e]^{\succcurlyeq} \cap \{c\} \neq \emptyset$ and $[e]^{\succcurlyeq} \cap \sim\{c\} = \emptyset$ for some $c \in \sim C$.

Theorem 5

a) $\underline{K}^{*}_{\sim C} = \underline{K}_{\sim C}$

b) $\overline{K}^{*}_{\sim C} = \overline{K}_{\sim C}$

c) $\underline{P}^{*}_{\sim C} = \underline{P}_{\sim C}$

d) $\overline{P}^{*}_{\sim C} = \overline{P}_{\sim C}$

Proof.

a) $\underline{K}^{*}_{\sim C} = \bigcup_{c \in \sim C} \underline{K}^{*}_{c}$

$\quad = \{x \in U \mid [d(x)]^{\succcurlyeq} \cap \{c\} \neq \emptyset \text{ and } [d(x)]^{\succcurlyeq} \cap \sim\{c\} = \emptyset, \text{for some } c \in \sim C\}$

$\quad = \{x \in U \mid [d(x)]^{\approx} \cap \sim\sim C = \emptyset\} = \underline{K}_{\sim C}$ by Lemma 21.

b) $\overline{K}^{*}_{\sim C} = \bigcup_{c \in \sim C} \overline{K}^{*}_{c} = \{x \in U \mid [d(x)]^{\succcurlyeq} \cap \{c\} \neq \emptyset, \text{ for some } c \in \sim C\}$

$\quad = \{x \in U \mid [d(x)]^{\preccurlyeq} \cap \sim\sim C = \emptyset\} = \overline{K}_{\sim C}$ by Lemma 19.

c) $\underline{P}^{*}_{\sim C} = \bigcup_{c \in \sim C} \underline{P}^{*}_{c} = \{x \in U \mid [d(x)]^{\succcurlyeq} \cap \sim\{c\} = \emptyset, \text{ for some } c \in \sim C\}$

$\quad = \{x \in U \mid [d(x)]^{\succcurlyeq} \cap \sim\sim C = \emptyset\} = \underline{P}_{\sim C}$ by Lemma 20.

d) Follows immediately from the definitions (and Lemma 16d).

Thus, if one assumes that the DAG is well-defined and considers only single classes and their complements, one may use the simplified boundary sets instead of the set-based boundaries.

10   Set Approximations for a DAG-Decision System

The boundary sets $\underline{K}_C$, $\overline{K}_C$, $\underline{P}_C$, and $\overline{P}_C$ solve only the part of the problem that is related to the DAG. They do not consider the multiple annotation problem, nor do they consider the uncertainty in the data (due to noise), which traditionally is handled by rough set theory. Hence, the boundary sets may be inconsistent if they are considered in terms of the elementary sets induced by an indiscernibility relation. In order to take this kind of inconsistency into account, we apply the standard rough set approximations on the boundary sets:

- $\underline{B}_X(C) = \underline{B}\,X_C = \{x \in U \mid [x]_B \subseteq X_C\}$
- $\overline{B}_X(C) = \overline{B}\,X_C = \{x \in U \mid [x]_B \cap X_C \neq \emptyset\}$

where $[x]_B = \{y \in U \mid \langle x, y \rangle \in IND(B)\}$. We create two operators for each boundary set by replacing X with $\underline{K}$, $\overline{K}$, $\underline{P}$, or $\overline{P}$.

The operators are illustrated in Figure 5. However, they do not capture our intuition completely when $IND(B)$ is an indiscernibility relation on the conditional attributes. The lower approximations $\underline{B}_{\underline{K}}(C)$ and $\underline{B}_{\overline{K}}(C)$ are very conservative, and it seems that more elementary sets should belong to C. In particular, the elementary sets where at least some of the objects are known (with respect to C) and

Indiscernibility Consistent   Indiscernibility Inconsistent

DAG-Consistent DAG-Inconsistent DAG-Consistent DAG-Inconsistent

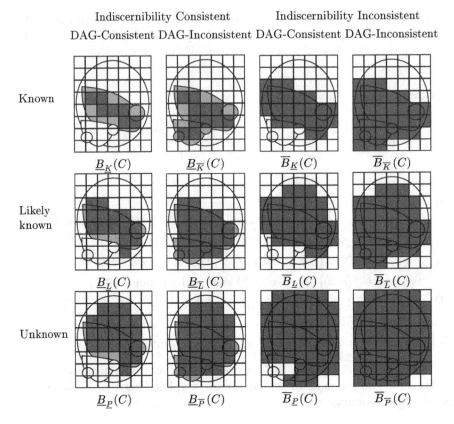

Known

$\underline{B}_K(C)$ $\underline{B}_{\overline{K}}(C)$ $\overline{B}_K(C)$ $\overline{B}_{\overline{K}}(C)$

Likely
known

$\underline{B}_L(C)$ $\underline{B}_{\overline{L}}(C)$ $\overline{B}_L(C)$ $\overline{B}_{\overline{L}}(C)$

Unknown

$\underline{B}_P(C)$ $\underline{B}_{\overline{P}}(C)$ $\overline{B}_P(C)$ $\overline{B}_{\overline{P}}(C)$

Fig. 5. *An illustration of the operators on the DAG in Figure 4 where* $C = \{c\}$. The unknown region is displayed as light gray, the known region as medium gray, and the negative region as white. The region covered by each operator is dark gray.

the remaining ones are unknown, are not included in $\underline{B}_K(C)$ and $\underline{B}_{\overline{K}}(C)$. This is unfortunate since the objects in these elementary sets probably belong to C. Some of the objects are already known to belong C. So, if we assume that the unknown objects in these sets belong to the complement classes, we will introduce new inconsistencies into the decision system. If all objects in these sets belong to C, on the other hand, no such inconsistencies will occur. Hence, the unknown objects in these elementary sets most likely belong to C since this leads to the fewest number of inconsistencies. We call these elementary sets for *C-likely-known* sets.

A key feature of the *C-likely-known* sets is that they are consistent with the potential sets, but inconsistent with the known sets. Thus, the upper approximations of the known sets and the lower approximations of the potential sets contain the *C-likely-known* sets. However, these approximations contain also other objects that should not be in a lower approximation of the known sets. $\overline{B}_K(C)$ and $\overline{B}_{\overline{K}}(C)$ cover the inconsistent elementary sets where some objects are known to belong to $\sim C$. $\underline{B}_{\overline{P}}(C)$ and $\underline{B}_P(C)$ cover completely unknown elementary sets, i.e., sets consisting of only unknown objects. The intersection of $\overline{B}_{\overline{K}}(C)$ and $\underline{B}_{\overline{P}}(C)$, on the other hand, contains only completely known and C-*likely-known* elementary sets (The situation is similar for $\overline{B}_K(C)$ and $\underline{B}_P(C)$). Hence, C-likely-known sets may be added to $\underline{B}_K(C)$ and $\overline{B}_{\overline{K}}(C)$ by using combinations of these approximations. This motivates us to define the following approximations.

Definition 13 (Approximations with C-likely-known sets)

a) $\underline{B}_L(C) = \underline{B}_P(C) \cap \overline{B}_K(C) = \{x \in U \mid [x]_B \subseteq \underline{P}_C \text{ and } [x]_B \cap \underline{K}_C \neq \emptyset\}$
b) $\overline{B}_{\overline{L}}(C) = \underline{B}_{\overline{P}}(C) \cap \overline{B}_{\overline{K}}(C) = \{x \in U \mid [x]_B \subseteq \overline{P}_C \text{ and } [x]_B \cap \overline{K}_C \neq \emptyset\}$
c) $\overline{B}_L(C) = \underline{B}_P(C) \cup \overline{B}_K(C) = \{x \in U \mid [x]_B \subseteq \underline{P}_C \text{ or } [x]_B \cap \underline{K}_C \neq \emptyset\}$
d) $\overline{B}_{\overline{L}}(C) = \underline{B}_{\overline{P}}(C) \cup \overline{B}_{\overline{K}}(C) = \{x \in U \mid [x]_B \subseteq \overline{P}_C \text{ or } [x]_B \cap \overline{K}_C \neq \emptyset\}$

Notice that $\overline{B}_L(C)$ and $\overline{B}_{\overline{L}}(C)$ contain the same elementary sets as $\overline{B}_P(C)$ and $\overline{B}_{\overline{P}}(C)$ except for the $(\sim C)$-likely-known sets, which must likely belong to the complement $\sim C$.

10.1   Some Properties of the Approximations

Each approximation has a dual approximation such that the approximation and its dual are complementary.

Definition 14 (Dual approximations). *Two approximations X and Y are duals denoted as $X \leftrightarrow Y$ if $X(C) = U - Y(\sim C)$ and $X(\sim C) = U - Y(C)$ hold.*

Corollary 2. *The approximations form the following duals.*

a) $\underline{B}_K \leftrightarrow \overline{B}_{\overline{P}}$ c) $\underline{B}_P \leftrightarrow \overline{B}_{\overline{K}}$ e) $\underline{B}_L \leftrightarrow \overline{B}_{\overline{L}}$
b) $\underline{B}_{\overline{K}} \leftrightarrow \overline{B}_P$ d) $\underline{B}_{\overline{P}} \leftrightarrow \overline{B}_K$ f) $\underline{B}_{\overline{L}} \leftrightarrow \overline{B}_L$

Proof. From rough set theory, we have that $\underline{B}(U - X) = U - \overline{B}(X)$ and $\overline{B}(U - X) = U - \underline{B}(X)$ for a set $X \subseteq U$. Case a–d follow from these properties and Theorems 3. For case e, we prove $\underline{B}_L(C) = U - \overline{B}_{\overline{L}}(\sim C)$ as follows: $\underline{B}_L(C) = \underline{B}_P(C) \cap \overline{B}_K(C) = (U - \overline{B}_{\overline{K}}(\sim C)) \cap (U - \underline{B}_{\overline{P}}(\sim C))$ (from case c and d) $= U - (\overline{B}_{\overline{K}}(\sim C) \cup \underline{B}_{\overline{P}}(\sim C)) = U - \overline{B}_{\overline{L}}(\sim C)$. The proof of $\underline{B}_L(\sim C) = U - \overline{B}_{\overline{L}}(C)$ follows directly from this property and Corollary 1. The proof of case f is similar.

The approximations are related such that some approximations are subsets of the other approximations. They may be ordered according to subset inclusion. This order is shown in Figure 6 where an arrow from one approximation to another approximation means that the former is a subset of the latter. These properties follow directly from the definitions. So no proof is given.

One may also derive properties for the approximations with regard to unions and intersections of classes. However, these follow directly from the properties of the lower and the upper approximation (see e.g., [10]) and the properties of the boundary sets that we established in Section 8. So, we will not consider them here.

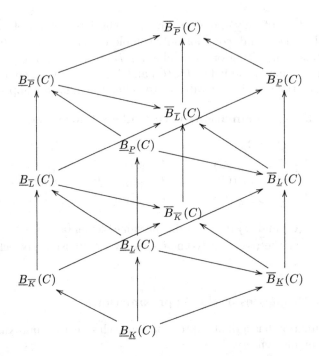

Fig. 6. *Subset inclusion of the approximations.* An arrow from A to B denotes that $A \subseteq B$.

11   Conclusion

Prediction of gene function introduces new challenges for supervised learning. The classes, which are predicted, are typically taken from an ontology. The ontology defines a *more general* ordering of the classes so that the classes may represent different detail levels of biological knowledge. Moreover, the data associated with each class are very sparse so that it is necessary to reduce the detail level in order to obtain accurate predictions.

To this end, we have presented a general rough set framework, which allow us to represent and model the structure of the DAG. Several boundary sets, which allow us to determine the known and potential objects of a class, have been introduced and their properties have been examined. In particular, it has been shown that the boundary set may be simplified if the DAG is well-defined and only single classes (and their complements) are considered. It has been demonstrated how the boundary sets may be combined with the rough set approximations, and new set of approximations have been defined to deal with the likely known set problem.

Gene function prediction introduces also another problem besides DAG structure. A gene may be labeled with more than one annotation. As we have demonstrated rough set theory lends it self well to this problem. Since a gene with multiple annotations corresponds to a inconsistent elementary set, we may apply the rough set lower and upper approximation to compute which genes belong completely and partially to a set of classes.

There are some related approaches to our framework. In our framework we have defined a partial order on the classes. The Dominance-Based Rough Set Approach (e.g. [6]) considers also an ordering of the classes. However, the ordering relation in that case is a linear ordering (or at least a total preorder). Any two classes are therefore related since the relation is total. This means that the membership of an object labeled to a class c is not unknown with respect to the subclasses of c since c does not have two or more unrelated subclasses. Hence, the unknown-property simply does not occur. Moreover, the inconsistency introduce by the DAG does not arise either. Małuszyński and Vitoria [11] present an rough set framework for Prolog with unknown objects. However, they develop a Rough Datalog language and define its semantics. They do not develop approximations for unknown objects.

It is possible to extend the framework presented here. One could utilize the objects in the superclasses to a larger extent than we have done in this paper. Instead of assuming that the class membership of these objects is unknown, we could assume that the objects of a superclass belong to the class with most similar objects and move the objects to this class. This assumption may be implemented by the means of proximity based boundary sets. These sets would generalize our boundary sets in much the same way as the variable precision rough set model (VPRS) approximations [17] generalize the standard rough set approximations. However, in this case we would use a similarity measure rather than the rough membership function (which is used in VPRS and measures degree of mismatch between an elementary sets and a set of objects). An object x labeled to a superclass of c may, for example, be assigned to the known set of c if there is some object y labeled to c or one of the subclasses of c, and the similarity is above some predetermined threshold.

In the sequel [13] we apply the framework and the approximations, which we have introduced, in a bottom-up algorithm and evaluate the performance of this algorithm.

Acknowledgment. I would like to thank Prof. Jan Komorowski and Prof. Astrid Lægreid for discussions of this work. I am also grateful to Prof. Andrzej Skowron for his suggestions, which simplified the framework.

Appendix

The lemmas presented below prove general set properties, which are used in the paper. As these are easily proven we do not present the proofs. However, the proofs can be found in [12].

Lemma 22. $A \subseteq B$ iff $(U - B) \subseteq (U - A)$

Lemma 23. $A \subseteq B$ iff $A \cap (U - B) = \emptyset$

Lemma 24. *Given, $B \subseteq C$, it is true that*

a) $A \cap C = \emptyset$ *implies* $A \cap B = \emptyset$
b) $A \cap B \neq \emptyset$ *implies* $A \cap C \neq \emptyset$

Lemma 25. *Given sets A, B, C, and $D_1, D_2, \ldots, D_n$, $n \geq 2$.*

a) $A \cap B = \emptyset$ *and* $A \cap C = \emptyset$ *iff* $A \cap (B \cup C) = \emptyset$
b) $A \cap D_i = \emptyset$, *for all* $1 \leq i \leq n$ *iff* $A \cap (\bigcup_{i=1}^{n} D_i) = \emptyset$
c) $A \cap B \neq \emptyset$ *or* $A \cap C \neq \emptyset$ *iff* $A \cap (B \cup C) \neq \emptyset$
d) $A \cap D_i \neq \emptyset$, *for some* $1 \leq i \leq n$ *iff* $A \cap (\bigcup_{i=1}^{n} D_i) \neq \emptyset$

Lemma 26. *Given sets A, B, C, and $D_1, D_2, \ldots, D_n$, $n \geq 2$.*

a) $A \cap B = \emptyset$ *or* $A \cap C = \emptyset$ *implies* $A \cap (B \cap C) = \emptyset$
b) $A \cap D_i = \emptyset$, *for some* $1 \leq i \leq n$ *implies* $A \cap (\bigcap_{i=1}^{n} D_i) = \emptyset$
c) $A \cap (B \cap C) \neq \emptyset$ *implies* $A \cap B \neq \emptyset$ *and* $A \cap C \neq \emptyset$
d) $A \cap (\bigcap_{i=1}^{n} D_i) \neq \emptyset$ *implies* $A \cap D_i = \emptyset$, *for all* $1 \leq i \leq n$

References

1. M. P. S. Brown, W. N. Grundy, D. Lin, N. Cristianini, C. W. Sugnet, T. S. Furey, M. Ares, Jr., and D. Haussler. Knowledge-based analysis of microarray gene expression data by using support vector machines. *Proceedings of the National Academy of Sciences, USA*, 97(1):262–267, 2000.
2. The Gene Ontology Consortium. Gene Ontology: Tool for the unification of biology. *Nature Genetics*, 25(1):25–29, 2000.
3. The Gene Ontology Consortium. Creating the Gene Ontology resource: Design and implementation. *Genome Research*, 11:1425–1433, 2001.
4. The Gene Ontology Consortium. The gene ontology (GO) database and informatics resource. *Nucleic Acids Research*, 32:D258–D261, 2004.
5. D. Frishman, K. Albermann, J. Hani, K. Heumann, A. Metanomski, A. Zollner, and H.-W. Mewes. Functional and structural genomics using PEDANT. *Bioinformatics*, 17(1):44–57, 2001.
6. S. Greco, B. Matarazzo, and R. Slowinski. A new rough set approach to multicriteria and multiattribute classification. In *Rough Sets and Current Trends in Computing*, volume 1424 of *Lecture Notes in Artificial Intelligence*, pages 60–67. Springer-Verlag, 1998.
7. J. W. Grzymala-Busse. Knowledge acquisition under uncertainty – a rough set approach. *Journal of Intelligent and Robotic System*, 1:3–16, 1988.
8. J. W. Grzymala-Busse. LERS – A system for learning from examples based on rough sets. In *Intelligent decision support: Handbook of Applications and Advances of Rough Sets Theory*, pages 3–18. Kluwer Academic Publishers, 1992.
9. T. R. Hvidsten, J. Komorowski, A. K. Sandvik, and A. Lægreid. Predicting gene function from gene expressions and ontologies. In *Proceedings of the Pacific Symposium on Biocomputing 6 (PSB-2001)*, pages 299–310. World Scientific Press, 2001.
10. J. Komorowski, Z. Pawlak, L. Polkowski, and A. Skowron. A rough set perspective on data and knowledge. In S K. Pal and A. Skowron, (eds), *Rough Fuzzy Hybridization*, pages 107–121. Springer-Verlag, 1999.

11. J. Małuszyński and A. Vitoria. Towards rough datalog: Embedding rough sets in prolog. In S.K. Pal, L. Polkowski, and A. Skowron (eds), *Rough-Neuro Computing: Techniques for Computing with Words*, pages 297–332, Springer-Verlag, 2004.

12. H. Midelfart. *Knowledge Discovery from cDNA Microarrays and a priori Knowledge*. PhD thesis, Department of Computer and Information Science, Norwegian University of Science and Technology (NTNU), 2003. ISBN 82-471-5617-2.

13. H. Midelfart. Supervised learning in the gene ontology — part II: A bottom-up algorithm. *Transactions on Rough Sets* IV, LNCS 3700: 98–124, 2005.

14. H. Midelfart and J. Komorowski. A rough set approach to learning in a directed acyclic graph. In James F. Peters, Andrzej Skowron, and Ning Zhong (eds.), *Proceedings of the 3th International Conference on Rough Sets and Current Trends in Computing (RSCTC-2002)*, number 2475 in Lecture Notes in Artificial Intelligence, pages 144–155. Springer-Verlag, 2002.

15. H. Midelfart, A. Lægreid, and J. Komorowski. Classification of gene expression data in an ontology. In J. Crespo, V. Maojo, and F. Martin (eds), *Proceedings of the 2nd International Symposium on Medical Data Analysis (ISMDA-2001)*, number 2199 in LNCS, pages 186–194. Springer-Verlag, 2001.

16. Z. Pawlak. *Rough Sets: Theoretical Aspects of Reasoning about Data*. Series D: System Theory, Knowledge Engineering and Problem Solving. Kluwer Academic Publishers, 1991.

17. W. Ziarko. Variable precision rough set model. *Journal of Computer and System Sciences*, 46:39–59, 1993.

Supervised Learning in the Gene Ontology
Part II: A Bottom-Up Algorithm

Herman Midelfart

Department of Biology, Norwegian University of Science and Technology,
N-7491 Trondheim, Norway
herman@bio.ntnu.no

Abstract. Prediction of gene function for expression profiles introduces
a new problem for supervised learning algorithms. The decision classes
are taken from an ontology, which defines relationships between the
classes. Supervised algorithms, on the other hand, assumes that the
classes are unrelated. Hence, we introduce a new algorithm which can
take these relationships into account. This is tested on a microarray data
set created from human fibroblast cells and on several artificial data sets.
Since standard performance measures do not apply to this problem, we
also introduce several new measures for measuring classification performance in an ontology.

1 Introduction

Gene annotations are important for interpreting biological studies such as microarray experiments as they provide information about the biological role of
genes. The Gene Ontology (GO) Consortium [4] has established several ontologies that define controlled vocabularies for describing the biological role of a gene.
These vocabularies have been applied to annotate the genome of several organisms. However, annotating the genome of an organism is a very time-consuming
task. Several attempts [3, 8, 11] have been made to predict gene function automatically, but these studies have ignored the special problems, which the ontology introduces for supervised learning.

The ontologies defined by the GO Consortium organize the classes in directed
acyclic graphs (DAG). In [13], we discussed several issues that must be handled
by a DAG-learning algorithm. These were:

1. The structure of the DAG
2. The varying detail level of the annotations
3. The scarcity of the objects with respect to the classes
4. The trade-off between the detail level and the precision of a prediction
5. The multiple annotations of an object

The framework defined in that paper solves many of these problems. It allows
us to represent the DAG structure and multiple annotations with varying level
of detail. It also introduces several approximations that allow us to determine

J.F. Peters and A. Skowron (Eds.): Transactions on Rough Sets IV, LNCS 3700, pp. 98–124, 2005.
© Springer-Verlag Berlin Heidelberg 2005

the known and potential members of a class. However, in order to develop a classifier we still have to apply the approximations in an algorithm and find a solution to other issues.

This paper presents a novel DAG learning algorithm that provides solutions to these problems. The algorithm creates a set of rules using a covering approach similar to the LEMS2 algorithm in the LERS system [6, 7]. However, it applies the new approximations instead of the standard rough set approximations in order to accommodate the DAG structure. A bottom-up pruning scheme is used to determine the best trade-off between the detail level and the precision of a prediction.

The algorithm is evaluated on several data sets. It is first tested on a micro-carray data set that was created from human fibroblast cells. It is then applied on several artificial data sets in order to obtain a better understanding of its behavior.

The DAG also constitutes a problem when a classifier is to be evaluated since standard performance measures such as accuracy or AUC do not recognize the structure of the DAG. Therefore, several new performance measures are introduced.

2   Notation and Definitions

A *DAG-decision system* is a tuple $\mathcal{A} = \langle U, A, d, \succcurlyeq \rangle$ where the universe U is a non-empty finite set of (observable) objects. A is a set of *conditional attributes* where each attribute $a \in A$ is a function $a : U \to V_a$. The *decision* attribute d is a function $d : U \to V_d$. $\succcurlyeq$ is a partial order on the classes in V_d where $p \succcurlyeq r$ denotes that p is *more general than* r $(p, r \in V_d)$. In this paper, we consider an object to be a tuple $\langle x, d \rangle$ where x is a gene in G and d is a class from V_d so that $U \subseteq G \times V_d$. The decision attribute is defined as $d(\langle x, d' \rangle) = d'$. The following indiscernibility relation partitions the universe into elementary sets corresponding to genes in G.

$$IND(G) = \{\langle x, y \rangle \in U' \times U' \mid x = \langle z, d_x \rangle \text{ and } y = \langle z, d_y \rangle\}$$

The classes $c, e \in V_d$ are *related*, denoted as $c \approx e$, iff $c \succcurlyeq e$ or $e \succcurlyeq c$. Let $C \subseteq V_d$ be a set of classes. Then the *complement* is defined as $\sim C = \{e \in V_d \mid$ for all $c \in C, e \not\approx c\}$. The classes that are *above*, *below*, or *related* to C are denoted by $[C]^{\succcurlyeq}$, $[C]^{\preccurlyeq}$, and $[C]^{\approx}$.

A DAG is *well-defined* if for every non-leaf class e and every subclass f of e there is another subclass of e that belongs to $\sim f$. If this condition holds the known and potential sets of a class c and its complement are defined as

$$\underline{K}_c^* = \{x \in U \mid [d(x)]^{\succcurlyeq} \cap \{c\} \neq \emptyset \ \& \ [d(x)]^{\succcurlyeq} \cap \sim c = \emptyset\} \quad \underline{K}_{\sim c}^* = \bigcup_{e \in \sim c} \underline{K}_e^*$$

$$\overline{K}_c^* = \{x \in U \mid [d(x)]^{\succcurlyeq} \cap \{c\} \neq \emptyset\} \quad \overline{K}_{\sim c}^* = \bigcup_{e \in \sim c} \overline{K}_e^*$$

$$\underline{P}_c^* = \{x \in U \mid [d(x)]^{\succeq} \cap \sim c = \emptyset\} \qquad \underline{P}_{\sim c}^* = \bigcup_{e \in \sim c} \underline{P}_e^*$$

$$\overline{P}_c^* = \{x \in U \mid [d(x)]^{\approx} \cap \{c\} \neq \emptyset\} \qquad \overline{P}_{\sim c}^* = \bigcup_{e \in \sim c} \overline{P}_e^*$$

A *path* from class c_0 to class c_n is a sequence of nodes $\langle c_0, c_1, \ldots, c_n \rangle$ where $c_0 \prec c_1 \prec \ldots \prec c_n$. Let $s = \langle b_0, b_1, \ldots, b_m \rangle$ and $t = \langle c_0, c_1, \ldots, c_n \rangle$ be two paths. s is a *subpath* of t, denoted as $s \sqsubseteq t$, if there is mapping such that $b_i = c_{i+j}$ for all $0 \leq i \leq m$ where $0 \leq j \leq n - m$. $Paths(a, b)$ represents the set of all paths from a to b (given $b \succ a$), i.e., $Paths(a, b) = \{\langle c_0, \ldots, c_n \rangle \mid a = c_0 \prec \ldots \prec c_n = b\}$. $Sub(c)$ denotes the immediate subclasses classes of $c \in V_d$, i.e,

$$Sub(c) = \{e \in V_d \mid c \succ e \text{ and there is a path } \langle e, c \rangle \in Paths(e, c)\}$$

The *lower* $\underline{B}$ and *upper* $\overline{B}$ approximation of a set $X \subseteq U$ are defined with respect to an indiscernibility relation $IND(B)$.

- $\underline{B}X = \{x \in U \mid [x]_B \subseteq X\}$
- $\overline{B}X = \{x \in U \mid [x]_B \cap X \neq \emptyset\}$

$[x]_B$ denotes the *elementary set* of x induced by $IND(B)$, i.e, $[x]_B = \{y \in U \mid \langle x, y \rangle \in IND(B)\}$.

A *decision rule* is denoted by $\alpha \to \beta$ and consists of an antecedent α and a conclusion β. The *antecedent* α consists of a conjunction of descriptors $\langle a_1, v_1 \rangle \wedge \cdots \wedge \langle a_n, v_n \rangle$ constructed from the conditional attributes in A. The *conclusion* consists of a single descriptor $\langle d, v \rangle$ made from the decision attribute d. $[\![r]\!]_U$ denotes the objects in U that satisfies a rule r and is defined inductively as

- $[\![\langle a, v \rangle]\!]_U = \{x \in U \mid a(x) = v\}$
- $[\![\alpha \wedge \beta]\!]_U = [\![\alpha]\!]_U \cap [\![\beta]\!]_U$
- $[\![\alpha \to \beta]\!]_U = (U - [\![\alpha]\!]_U) \cup [\![\beta]\!]_U$

A rule *covers* all of the objects that satisfy its antecedent. We denote the objects that are covered by a set RS by

$$Cov(RS) = \{x \in U \mid x \in [\![\alpha]\!]_A \text{ and } (\alpha \to \beta) \in RS\}$$

The rules, which are associated with a particular class c, are defined as

$$RS_c = \{(\alpha \to \beta) \in RS \mid \beta = \langle d, c \rangle\}$$

3   The Bottom-Up Pruning Algorithm

A main concern for DAG learning is the scarcity of the data that are available for each class. An algorithm may attempt to avoid this problem by learning rules that cover all of the known objects when it learns rules for a class c so that the objects are moved upwards. However, this means that the objects, which belong to c, will not only be covered by the rules of c, but also by the rules of any superclass of c. If c can be predicted accurately, the superclasses of c will be

Algorithm 1.1. *The main part of the bottom-up algorithm*

LearnBottomUp:

Input: A rooted DAG-decision system $\mathcal{A} = \langle U, A, d, \succcurlyeq \rangle$ with root $\top$, training accuracy γ, split ratio ρ, pruning accuracy δ, and pruning support σ.

Output: A set of rules RS.

1: Clear rule set RS $\{RS = \emptyset\}$
2: **for all** $c \in Sub(\top)$ **do** {For all immediate subclasses of $\top$}
3:　　$(RS, R) = \mathrm{RecLearnBottomUp}(c, RS, \mathcal{A}, \gamma, \rho, \delta, \sigma)$
4: **return** RS

predicted as well. The resulting classifier will thus produce a lot of redundant predictions. Hence, the algorithm must determine the most specific detail level which allows for accurate prediction and remove predictions above this level.

This problem may be solved in several ways. In [15, 14] we developed a simple voting scheme to determine the appropriate detail level and filtered out the redundant predictions. This scheme was later developed into a more robust system where the votes were based on the support of the rules [12]. However, the application of rules was not very efficient since if the objects of a class c can be predicted accurately, it is not necessary that the rules of the superclasses cover them as well. The rules of c would be sufficient. Moreover, the rules created for the most specific classes would typically cover few objects, and their predictions would not be very accurate. Hence, their contribution to the overall performance would be small so that it might actually be better to remove them all together.

A more efficient and robust strategy may be to let the learning algorithm determine the trade-off between the detail level and the precision directly rather than applying a voting system during prediction. This is the idea behind the method that will be presented in this paper. It identifies the most detailed classes that may be predicted accurately and builds the classifier from the rules of these classes.

The classes, which can be predicted accurately, can be found efficiently by examining the DAG in a bottom-up fashion. The algorithm starts with the leaf classes and moves upwards. For each class, it creates a set of rules and determines if the rules give accurate predictions. If the rules are of high quality, they are retained, and the objects covered by them are removed (such that they are not considered when rules are learned for the superclasses). If the rules are inaccurate, they are pruned, and the objects of the class are passed to the immediate superclass(es). The procedure is then repeated for the superclasses.

3.1　The Main Learning Algorithm

The details of the learning algorithm are presented in Algorithms 1.1 and 1.2. The algorithm assumes that the DAG is well-defined and has a single root (denoted by $\top$). It is executed by a call to the procedure **LearnBottomUp**, which calls the recursive procedure **RecLearnBottomUp** for each class on the top level immediately below the root. **RecLearnBottomUp** performs the main

---

Algorithm 1.2. *The recursive part of the bottom-up algorithm*

RecLearnBottomUp:

Input: a class c, a set of rules RS a rooted DAG-decision system $\mathcal{A} = \langle U, A, d, \succcurlyeq \rangle$, training accuracy γ, split ratio ρ, pruning accuracy δ, and pruning support σ.

Output: A set of rules RS and a set of uncovered objects R.

1: **if** class c has not been visited **then**
2:    Mark c as visited
3:    $R = \emptyset$
4:    **for all** $e \in Sub(c)$ **do** {For all immediate subclasses of c}
5:        $(RS, R') = \text{RecLearnBottomUp}(e, RS, \mathcal{A}, \gamma, \rho, \delta, \sigma)$
6:        $R = R \cup R'$
7:    $R_c = R \cup \left(X_c - Cov^{\prec}(RS, c) \right)$
8:    $\mathcal{P} = \overline{G}\, R_c$ and $\mathcal{N} = \underline{G}\, K^*_{\sim c}$
9:    **if** $c \in Sub(\top)$ and top level pruning is off **then**
10:       $RS = RS \cup \text{LearnRules}\,(\mathcal{P}, \mathcal{N}, A, d, c, \gamma)$
11:       $R = \emptyset$
12:   **else**
13:       $(\mathcal{P}_t, \mathcal{P}_v) = \text{SplitData}(\mathcal{P}, \rho)$ and $(\mathcal{N}_t, \mathcal{N}_v) = \text{SplitData}(\mathcal{N}, \rho)$
14:       $RS_1 = \text{LearnRules}\,(\mathcal{P}_t, \mathcal{N}_t, A, d, c, \gamma)$
15:       $RS_2 = \text{PruneRules}(RS_1, \mathcal{P}_v, \mathcal{N}_v, \delta, \sigma)$
16:       $RS = RS \cup RS_2$
17:       $R = \mathcal{P} - Cov(RS_2)$
18: **else**
19:    $R = \bigcup_{c \succcurlyeq e}(X_e - Cov^{\preccurlyeq}(RS, e) - Cov^{\succ}(RS, e))$
20: **return** (RS,R)

---

task. It traverses the DAG depth-first and considers each class in a postfix order so that rules are learned first for the subclasses and then for the class itself. As there may be several paths to a class c, the algorithm checks initially if the class has been visited already and attempts only to create rules (and visit the subclasses), if the class has not been visited before.

The algorithm attempts to create rules for each class except for the root[1]. This is done in two different ways depending on whether a class occurs at top level immediately below the root or at a more detailed level. In both cases, the objects are divided into a positive set and a negative set, and rules are learned with the subalgorithm **LearnRules**. However, the pruning algorithm **PruneRules** is not always applied to rules that are created on the top level. The reason is that the classes at this level are the most general classes that may be predicted. So the rules created for them are the most accurate that can be obtained. Moreover, the objects that are not covered at a more detailed level must at least be covered at this level. Otherwise, they will not be covered by any rule, and the classifier will not make any predictions for them. Rules are therefore not pruned at the top

---

[1] The root itself provides no information about an object since all objects belong to the root. So, if the classifier predicted the root for an object, it would just imply that the class of the object was unknown.

level by default. The rules created for the detailed classes, on the other hand, may be improved. Hence, these rules are pruned. Nevertheless, some rules at the top level may be of poor quality such that a better performance may actually be obtained by removing them. The algorithm is therefore equipped with an option that allows pruning at the top level as well.

3.2   Computation of the Positive and the Negative Set

When the rules of a class c are learned, the objects are divided into a positive set $\mathcal{P}$ containing the objects that should be covered by the rules and a negative set $\mathcal{N}$ containing the genes that should not be covered. The rules are then found by the **LearnRules** subalgorithm, which is described in the detail in Section 4.

The definitions of $\mathcal{P}$ and $\mathcal{N}$ are crucial since these sets control the kind of rules that are made by the algorithm. In our case, these sets should fulfill several requirements.

1. The rules learned for class c should not discern the objects of c from the objects of the classes related to c.
2. The scarcity of the available data for a class c should be compensated by including all known objects (with regard to c) in $\mathcal{P}$ and not only objects annotated with c.
3. A gene may have several annotations, and all of these annotations should be predicted.
4. Rules should only be learned for the objects that have not already been covered at the subclasses.

The first requirement may be fulfilled by using $\underline{K}^*_{\sim c}$ as the negative set $\mathcal{N}$ since this means that rules may cover the objects in $\overline{P}^*_c = U - \underline{K}^*_{\sim c}$. The second requirement is solved by using $\overline{K}^*_c$ for the positive $\mathcal{P}$. In order to fulfill the third requirement, we need to assign a G-elementary set (i.e., a gene) to $\mathcal{P}$ if there is an object (i.e., an annotation) in this G-elementary set that is known to belong to c. This is achieved by applying the G-upper approximation to $\overline{K}^*_c$ so that $\mathcal{P} \subseteq \overline{G}\,\overline{K}^*_c$ and a G-lower approximation on $\underline{K}^*_{\sim c}$ so that $\mathcal{N} = \underline{G}\,\underline{K}^*_{\sim c}$.

The positive set $\mathcal{P}$ is only a subset of $\overline{G}\,\overline{K}^*_c$ since algorithm should not learn new rules for the objects that have been covered at the subclasses. Hence, $\mathcal{P}$ contains only the objects in $\overline{G}\,\overline{K}^*_c$ that have not yet been covered. This set can be computed efficiently as the algorithm moves through the DAG since $\overline{K}^*_c$ can be decomposed as follows:

$$\overline{K}^*_c = \{x \in U \mid c \succcurlyeq d(x)\} = X_c \cup \left(\bigcup_{d \in Sub(c)} \overline{K}^*_d \right) = \bigcup_{c \succcurlyeq d} X_d$$

where $X_c = \{x \in U \mid d(x) = c\}$. The computation is done at several points in the algorithm. The objects that have not been covered at the subclasses are collected and added to R in line 6. The objects that belong to the class c are found in line 7. In this case, it is possible that some of these objects are covered by the rules, which have been created for the subclasses. These objects should not be covered by new rules and must be removed from X_c. This is achieved with the requirement that the objects in X_c must not be in $Cov^\prec(RS, c)$. This set contains the objects that are covered by rules of the subclass and is defined as

$$Cov^\prec(RS, c) = \{x \in U \mid x \in Cov(RS_e) \text{ and } e \prec c\}$$

The objects of $X_c - Cov^\prec(RS, c)$ and R combined into set R_c and the positive set is computed from R_c in line 8. The algorithm will then learn rules for the class and prune these rules so that it ends up with a set of rules RS_2 that have been accepted for the class. The objects in $\mathcal{P}$ that are not covered by the accepted rules in RS_2 must be passed up to the superclasses such that another attempt (to learn rules for these objects) can be made at these classes. R is therefore recomputed in line 17 such that it contains the objects in $\mathcal{P}$ that are not covered by the rules in RS_2. Unfortunately, this transfer of objects is complicated by the DAG structure since a class may be visited again if it has several parents. The contents of R are not stored since it would require a lot of memory to store this set every class in DAG (or at least for every class with multiple parents). The contents of R are therefore lost when algorithm moves upwards and must be recomputed if a class is revisited. This computation is done in line 19 where

$$Cov^\succ(RS, c, d) = \bigcap_{p \in Paths(d,c)} \{x \in U \mid x \in Cov(RS_{e_i}) \text{ and } e_i \text{ in } p\}$$

and $Cov^\preceq(RS, c)$ is defined as $Cov^\prec(RS, c)$ except that $\prec$ is replaced by $\preceq$.

Remark 1. The approximations for $\mathcal{P}$ and $\mathcal{N}$ do not consider the discernibility of the objects with regard to the conditional attributes as usual in Rough Set Theory. This discernibility is instead considered indirectly by the **LearnRules** subalgorithm, which searches for rules with accuracy above a threshold γ. A similar approach is usually taken in machine learning when flat classifiers with multiple classes are created. In this case, the positive and the negative sets for a class c are created by assigning the objects labeled with c to the positive set and the rest to the negative set (see e.g., [5, 1]), and the discernibility of the objects is handled by the covering algorithm.

Note that it is possible to apply A-approximations defined by an indiscernibility relation on the conditional attributes (i.e., $IND(A) = \{\langle x, y \rangle \in U \times U \mid a(x) = a(y), \text{ for all } a \in A\}$). We may for example create possible rules by setting $\mathcal{P} = \overline{A}\,\overline{G}\,\overline{K}_c^*$ and $\mathcal{N} = \underline{A}\,\underline{G}\,\underline{P}_{\sim c}^* \cap \overline{A}\,\underline{G}\,\underline{K}_{\sim c}^{*\,2}$. However, initial tests suggested that

---

[2] The A-lower approximation with $\sim c$-likely-known set is applied for the negative set in order to correct a problem, which occurs when some of the genes in an A-elementary set are labeled to the classes above c and some are labeled to the classes in $\sim c$, but no are labeled to c and its subclasses. In this case the, standard A-lower approximation will not assign these genes to $\mathcal{N}$ so that the rules of c may cover them. This is corrected with the lower approximation for $\sim c$-likely-known set so that the genes are assigned to $\mathcal{N}$. For a full discussion of this issues we refer the reader to [12] where the positive and negative sets with A-approximations are also used in another algorithm.

better results were obtained without A-approximations. Moreover, this simplifies the algorithm since the computation of A-approximations is complicated by splitting in line 13. Thus, A-approximations are not used.

3.3   Pruning

The pruning subsystem is responsible for removing rules that cannot be predicted accurately. It can be designed in many different ways. Some of these choices are:

- **Single rules vs. full classes**: The pruning can be made on two different levels – either on the class level or on the rule level. In the first case, we consider all rules that have been learned for a class and estimate how well they predict the class. If their performance is unsatisfactory, all of them are pruned. In the second case, each rule is tested separately and pruned if its performance is not good enough.

  The latter option has an advantage over the former since the pruning is more fine-meshed in this case. The learning task may not have the same degree of difficulty for all objects of a class c. It may be easier to learn accurate rules for some objects than for the rest. Accurate rules may thus be made for the easy objects while the rest of objects may be passed to a more general superclass and covered by rules created for these classes. With the former option, this is not possible. All of the objects must be covered either at c or at a superclass. So the classifier will either give more incorrect predictions or lose the details of objects that could be predicted to c. Note, however, that this may be the only option if another learning approach such as discriminant analysis or support vector machines is applied.

- **Validation sample**: The algorithm needs a validation sample in order to estimate the performance of the rules. The training sample can be used for this purpose. However, this may lead to overfitting as a rule that fits the training data perfectly may have a different performance on another data set. The estimated performance may thus be overly optimistic such that a rule may not be pruned even though it should.

  An alternative is to split the original training data into a training sample and a validation sample such that rules are learned from the training sample and pruned on the validation sample. Such a strategy is often used in machine learning to avoid overfitting. Unfortunately, this leaves less data for training which may be a problem especially for the most specific classes where the available data are already quite scarce.

- **Pruning criterion**: The pruning algorithm needs a criterion to determine whether a rule should be accepted or not. One possible criterion is to allow the user to specify the minimal acceptable accuracy and prune if the performance is below this value. Another is to use the rules that are learned for classes immediately below the root as a yardstick. These classes are most general classes that may be predicted and their rules should thus have the best performance. The rules that are learned for some class c can then be compared to the rules of these classes, and if the performance is worse the rules of c can be pruned.

Algorithm 1.3. Splitting of data

SplitData:

Input: A set X and splitting ratio ρ.

Output: A set T for training and a set V for validation

1: Compute the quotient set $X' = X/IND(G)$
2: Split X' into T' and V' at random such that $|T'| = \rho \cdot |X'|$ and $|V'| = (1 - \rho) \cdot |X'|$

3: $T = \{x \in X \mid [x]_G \in T'\}$ and $V = \{x \in X \mid [x]_G \in V'\}$
4: **return** (T, V)

Algorithm 1.4. Pruning of rules

PruneRules:

Input: A set of rules RS, positive set $\mathcal{P}_v$, negative set $\mathcal{N}_v$, pruning accuracy δ, and pruning support σ.

Output: A pruned rule set RS.

1: **for all** $(\alpha \rightarrow \beta) \in RS$ **do**
2:    **if** $Accuracy(\alpha, \mathcal{P}_v, \mathcal{N}_v) < \delta$ or $Support(\alpha, \mathcal{P}_v) < \sigma$ **then**
3:       $RS = RS - \{\alpha \rightarrow \beta\}$ {The rule is pruned}
4: **return** RS

In the approach presented here we have chosen to prune each rule independently. Moreover, the objects are divided into a training sample and a validation sample such that overfitting is avoided, and a rule is pruned if its accuracy is below the pruning accuracy δ, which is specified by the user. A rule is also removed if its support is below the pruning support σ.

The learning and pruning of rules are performed in lines 13-15 in Algorithm 1.2. The data in the positive set $\mathcal{P}$ and the negative set $\mathcal{N}$ are initially divided into training sets $(\mathcal{P}_t, \mathcal{N}_t)$ and validation sets $(\mathcal{P}_v, \mathcal{N}_v)$. This is done by the procedure **SplitData**, which is shown in Algorithm 1.3. This procedure splits a set X in two according to the partition induced by $IND(G)$. All objects of a G-elementary set (i.e., a gene) are therefore put either in the training set or in the validation set. This is necessary as the objects that belong to the same G-elementary set (i.e., same gene) should not occur both in the training set and the validation set. Otherwise, the estimated accuracy on the validation set would be too optimistic, and rules that should be pruned, might be retained. How the G-elementary sets are divided on the training set and validation set is controlled by the splitting ratio s. It is typically set to 2/3 so that 2/3 of the G-elementary sets end up in the training set and 1/3 in the validation set.

After the algorithm has divided the data into a training and a validation sample, it learns rules from the training sample (line 14) and prunes these rules on the validation sample (line 15). The pruning algorithm is shown in Algorithm 1.4. This procedure examines each rule in RS_1 and tests if a rule should be deleted. This situation occurs if the accuracy of the rule is below the pruning accuracy δ or the support is below the pruning support σ. The accuracy

Algorithm 1.5. Top-down search for rules of a class

LearnRulesTopDown:
Input: Positive set $\mathcal{P}$, negative set $\mathcal{N}$, conditional attributes A, decision attribute d,
  decision class c, and training accuracy γ $(0 < \gamma \leq 1)$.
Output: A set of rules RS
 1: $RS = \emptyset$
 2: **while** $\mathcal{P} \neq \emptyset$ **do**
 3:     let antecedents I and I_{best} be empty.
 4:     $B = \{\langle a, v \rangle \mid a \in A \text{ and } v \in V_a\}$
 5:     **while** $Accuracy(I, \mathcal{P}, \mathcal{N}) < \gamma$ and $B \neq \emptyset$ **do**
 6:         select $\langle a', v' \rangle \in B$ with the highest $Score(I \wedge \langle a', v' \rangle, \mathcal{P}, \mathcal{N})$
 7:         add $\langle a', v' \rangle$ to I
 8:         $B = B - \{\langle a', v \rangle \mid v \in V_{a'}\}$
 9:         **if** $Accuracy(I, \mathcal{P}, \mathcal{N}) > Accuracy(I_{best}, \mathcal{P}, \mathcal{N})$ or I_{best} is empty **then** $I_{best} = I$
10:     **for each** descriptor $\langle a, v \rangle$ in I_{best} (in the order that they were added to I) **do**
11:         let I'_{best} be I_{best} without $\langle a, v \rangle$
12:         **if** $Accuracy((I'_{best}, \mathcal{P}, \mathcal{N}) \geq \gamma$ **then** $I_{best} = I'_{best}$
13:     $\mathcal{P} = \mathcal{P} - [\![I_{best}]\!]_{\mathcal{P}}$
14:     $RS = RS \cup \{I_{best} \rightarrow \langle d, c \rangle\}$
15: **return** RS

$Accuracy(\alpha, \mathcal{P}, \mathcal{N})$ and support $Support(\alpha, \mathcal{P}_v)$ are defined as follows:

$$Accuracy(\alpha, \mathcal{P}, \mathcal{N}) = |[\![\alpha]\!]_{\mathcal{P}}| / (|[\![\alpha]\!]_{\mathcal{P}}| + |[\![\alpha]\!]_{\mathcal{N}}|)$$
$$Support(\alpha, \mathcal{P}) = |[\![\alpha]\!]_{\mathcal{P}}|$$

4   Search Algorithms

The bottom-up pruning algorithm does not learn rules on its own, but applies
a rule learning algorithm (called **LearnRules** in Algorithm 1.1) to find a set
of rules for each class. There are many different algorithms that can be used
for this task. In the experiments that we present in Section 6 we apply two
different algorithms. Both search through a hypothesis space, which consists
of conjunctions of descriptors. However, the search is conducted in different
directions. One searches the hypothesis space in a top-down fashion. The other
searches the space from the bottom and up to the root. We provide their details
in this section.

4.1   Top-Down Search

The top-down algorithm is displayed in Algorithm 1.6. It is a so-called covering
or separate-and-conquer algorithm [5]. This means that it searches for one rule
at the time. When it finds a rule that *covers* some objects of the positive set
with a certain accuracy (γ), it *separates* these objects from the rest and *conquers*
the remaining objects by repeatedly learning rules until all objects are covered.

The separate-and-conquer task is performed in the outer-while loop of Algorithm 1.6 (line 2). The loop terminates when $\mathcal{P}$ is empty. When an antecedent has been found, the objects covered by it are removed from $\mathcal{P}$ (line 13), and it is turned into a rule and added to the rule set RS (line 14).

The inner while-loop (line 5) conducts a hill-climbing search for the antecedent. Initially, the antecedent I is empty, and the set B contains all descriptors that may be added to I. In each iteration, the descriptor that has the highest score is added to I. The loop terminates when I has sufficient accuracy or no more descriptors may be added to I. The score $Score(I, \mathcal{P}, \mathcal{N})$ is defined as

$$Score(I, \mathcal{P}, \mathcal{N}) = Support(I, \mathcal{P}) \cdot Accuracy(I, \mathcal{P}, \mathcal{N})$$

Note that the selection of descriptors could also have been done according to the accuracy. However, when the support is multiplied with the accuracy, the algorithm is forced to consider both measures and not only the accuracy. In this way we ensure that the rules have a high support as well as a high accuracy.

The best antecedent is maintained in I_{best}, which is equal to I if the inner loop terminates by the first condition. When the inner-while loop terminates by the second condition, the specified accuracy cannot be obtained and I_{best} contains the most general antecedent with the best accuracy.

The inner while-loop performs a greedy search that may add redundant conditions to an antecedent. The antecedent I_{best} is therefore examined and redundant conditions are removed before a rule is created. This is done in the for-loop at line 10. The descriptors are processed in the order that they were added to the antecedent set. A descriptor is deleted from I_{best} if the accuracy without the descriptor is above training accuracy.

4.2 Bottom-Up Search

The bottom-up search algorithm is shown in Algorithm 1.6. Initially, a most specific antecedent set is created for each object in the positive set where the most specific antecedent of an object has one descriptor for each attribute in A. The algorithm tries then to merge the two most similar antecedents into a more general antecedent by dropping dissimilar descriptors. This merge operation can be described with the following function.

$$merge(I_1, I_2) = \bigwedge \{\langle a, v \rangle \mid \langle a, v \rangle \text{ in } I_1 \text{ and } \langle a, v \rangle \text{ in } I_2\}$$

The similarity of the antecedents is measured by

$$dist(I_1, I_2) = |\{a \in A \mid \langle a, v_1 \rangle \text{ in } I_1, \langle a, v_2 \rangle \text{ in } I_2, \text{ and } v_1 \neq v_2\}|$$

The generalization process is repeated as long as there are some antecedents that may be merged and the resulting antecedent has an accuracy above the training accuracy γ.

The antecedent may have redundant descriptors. So after the generalization process has terminated, each antecedent is examined and redundant descriptors are removed. This is done in almost same manner as in the top-down search.

Algorithm 1.6. Bottom-up search for rules of a class

LearnRulesBottomUp:

Input: Positive set $\mathcal{P}$, negative set $\mathcal{N}$, conditional attributes A, decision attribute d,
   decision class c, and training accuracy γ ($0 < \gamma \leq 1$).

Output: A set of rules RS.

1: $R = \{\bigwedge_{a \in A} \langle a, a(x) \rangle \mid x \in \mathcal{P}\}$ and $RS = \emptyset$
2: **while** two antecedents in R can be merged into I and $Accuracy(I, \mathcal{P}, \mathcal{N}) \geq \gamma$ **do**
3:    select the two most similar $I_1, I_2 \in R$, i.e., those with the least $dist(I_1, I_2)$
4:    create $I = merge(I_1, I_2)$
5:    remove I_1, I_2 from R and add I to R
6: **for each** $I \in R$ **do**
7:    **for each** $\langle a, v \rangle$ in I **do**
8:       let I' be I without $\langle a, v \rangle$
9:       **if** $Accuracy(I', \mathcal{P}, \mathcal{N}) \geq \gamma$ **then** $I = I'$
10:   $RS = RS \cup \{I \rightarrow \langle d, c \rangle\}$
11: **return** RS

However, the algorithm does not add descriptors to the antecedent such that the descriptors are just processed in the order that they appear in the antecedent.

5   Performance Measures

The performance of supervised learning methods is usually measured by the *accuracy* or the *area under the ROC curve* (AUC). These measures assume, however, that each object has a unique decision class and that only one prediction is made for each object. Moreover, a prediction is either correct (if it is identical to the decision class of the object) or incorrect (if it is different from the decision class). It is never partially correct.

These assumptions do not hold in our case. A gene may be annotated with several decision classes, and several predictions can be made for each gene. A prediction need not be identical to a decision class, either. It may be above or below the decision class such that it matches the class only partially.

Thus, standard performance measures of supervised learning are not applicable, and we will introduce new measures in this section.

5.1   Measuring Multiple Annotations and Predictions

Initially, we will ignore the DAG and consider only the situation where a gene has multiple annotations. In this case, we have a set of decision classes $D(x)$ and a set of predictions $\hat{D}(x)$ for each gene $x \in G$. There are two ways that a classifier may fail:

- It may not predict a class c that should be predicted (i.e., $c \in D(x)$ and $c \notin \hat{D}(x)$).
- It may predict a class c that should not be predicted (i.e., $c \notin D(x)$ and $c \in \hat{D}(x)$).

These errors can be both assessed with a metric such as

$$\frac{1}{|G|} \sum_{x \in G} \frac{|D(x) \cap \hat{D}(x)|}{|D(x) \cup \hat{D}(x)|}$$

which measures the average share of classes that $D(\cdot)$ and $\hat{D}(\cdot)$ have in common. This measure is useful for comparing the performance of different classifiers such that the best classifier can be identified. However, it is not very helpful in identifying in what way the classifier fails when it achieves less than a full score. For example, if we want to improve the performance of the classifier, we need to know what kind of mistakes it makes — does it make too few predictions such that some annotations are not predicted or does it make too many such that some predictions do not match any annotations? Unfortunately, both kinds of errors are measured in the same way by this measure. So, there is no way to tell.

One way to solve this problem is to use two measures – one for each kind of error. This strategy is taken in Information Retrieval (see e.g., [10–Chapter 8] and [2–Chapter 3]) which considers a similar problem to the multiple annotation problem. In this field they measure the *recall*, which is the ratio RA of annotations that are predicted, and *precision*, which is the ratio RP of predictions that are correct, i.e., those that correspond to annotations. We will adopt these measures here.

Definition 1 (Recall/Precision). *Let $D(x)$ be a set of decision classes for a gene $x \in G$ and assume that $\hat{D}(x)$ is a set of predictions made for x by a classifier.*

- **Recall:** $RA = \dfrac{\sum_{x \in G} |MA(x)|}{\sum_{x \in G} |D(x)|}$ *where $MA(x) = D(x) \cap \hat{D}(x)$*

- **Precision:** $RP = \dfrac{\sum_{x \in G} |MP(x)|}{\sum_{x \in G} |\hat{D}(x)|}$ *where $MP(x) = D(x) \cap \hat{D}(x)$*

5.2 Measuring the DAG

We will now consider measures for evaluating a classifier trained on a DAG. As before, we assume that $D(x)$ and $\hat{D}(x)$ denote the annotations and the predictions of gene $x \in G$.

Recall and precision may also be used in this case. However, they consider an annotation a and a prediction p to match only if a and p are equal. This is clearly a too strict requirement for the DAG as a and p may match partially if they are related. The set of matched annotations $MA(x)$ and the set of matched predictions $MP(x)$ are therefore redefined so that an annotation is considered matched if there is a related prediction, and a prediction is matched if there is a related annotation:

$$MA(x) = \{a \in D(x) \mid p \in \hat{D}(x) \text{ and } a \approx p\}$$
$$MP(x) = \{p \in \hat{D}(x) \mid a \in D(x) \text{ and } p \approx a\}$$

The definitions of recall and precision as given in Definitions 1 can then be applied on these sets instead of $MA(x) = MP(x) = D(x) \cap \hat{D}(x)$.

These measures estimate the number of the annotations and number of the predictions that are partially matched. However, they do not consider the loss of details that may occur when a prediction is more general than an annotation. In order to quantify this loss, we consider the depth of a class which is equal to the length of the path from the class to the root (i.e., the number of edges from the class to the root). There may be more than one path from a class to the root in a DAG so that a class may occur at several depths. The depth of a class is therefore defined with respect to a particular path.

Definition 2 (Depth). *Let* $t = \langle c_1, \ldots, c_n, \top \rangle$ *be a path from* $c_1 \in V_d$ *to the root* $\top$. *Then the depth of a class* c *with respect to path* t, *denoted* $Depth_t(c)$, *is* n *(i.e., the length of path* t*).*

The loss associated with a prediction p that is more general than annotation a, can be measured as the depth of p relative to the depth of a.

Definition 3 (Relative depth). *Given an annotation* a *and a prediction* p *with the associated paths* t_a *and* t_p, *the relative depth of* a *and* p *with respect to* t_a *and* t_p *is:*

$$RDepth_{t_a,t_p}(a,p) = \begin{cases} \frac{Depth_{t_p}(p)}{Depth_{t_a}(a)} & \text{if } t_a \sqsubseteq t_p \text{ or } t_p \sqsubseteq t_a \\ 0 & \text{otherwise} \end{cases}$$

where $s \sqsubseteq t$ *denotes that* s *is a subpath of* t.

This measure is illustrated in Figure 1. It will be 1 if a and p have the same depth; less than 1 if p is a superclass of a; and 0 if the a and p are unrelated. It also measures the gain in detail level that is obtained when a prediction is more specific than an annotation. In this case, $RDepth_{t_a,t_p}(a,p) > 1$.

Note that one of the paths t_a and t_p in $RDepth_{t_a,t_p}(a,p)$ must be a subpath of the other. Otherwise, the paths are not comparable. For example, if $p \not\succeq a$ and a has several immediate superclasses, there will be some paths from a to the root that include p and some that do not. The paths that do not contain p cannot be compared against the paths of p. So only paths that contain a path from p to the root as subpath should be considered.

$RDepth_{t_a,t_p}(a,p)$ depends on the paths t_a and t_p, and there may be several different relative depths associated with a and p. We need a single measure on how well p reassembles a. To this end, the maximal relative depth is used.

Definition 4 (Maximal relative depth). *Let* a *be an annotation and* p *a prediction. The maximal relative depth of* a *and* p *is defined as*

$$mrd(a,p) = \max_{\substack{t_a \in Paths(a,\top), \\ t_p \in Paths(p,\top)}} RDepth_{t_a,t_p}(a,p)$$

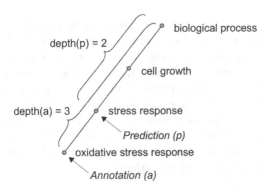

Fig. 1. *Depth and relative depth.* Assume that an object is labeled $a = oxidative$
stress response and a prediction is made for this object to $p = stress\ response$. In
this case, there are two edges from p to the root such that p has depth $Depth(p) = 2$.
Similarly, a has depth 3. The relative depth is $RDepth(a, p) = \frac{Depth(p)}{Depth(a)} = \frac{2}{3}$.

$mrd(a, p)$ measures not only the loss (when $p \not\succeq a$), but also the gain (when
$a \not\succeq p$). However, when we evaluate a classifier, we are mainly interested in how
well the classifier reproduces the actual annotations and less interested if some
details are gained. The maximal relative depth $mrd(a, p)$ is therefore restricted
such that no additional points are given if p is more detailed than a.

Definition 5 (Bounded maximal relative depth). *Let a be an annotation
and p a prediction. The bounded maximal relative depth of a and p is*

$$bmrd(a, p) = \max(mrd(a, p), 1)$$

$bmrd(a, p)$ measures only the loss in a prediction with regard to a single annotation. In order to capture the loss in all annotations and objects, we introduce
two measures that accompany recall and precision.

Definition 6 (Average recall/precision depth). *Let $x \in G$ be annotated
with the decision classes in $D(x)$, and let $\hat{D}(x)$ be a set of predictions made for
object x by a classifier. Then the average recall and precision depths are:*

- *Avg. recall depth:* $DA = \dfrac{\sum_{\substack{x \in U \\ a \in D(x)}} \max_{p \in \hat{D}(x)} bmrd(a, p)}{\sum_{x \in U} |MA(x)|}$

- *Avg. precision depth:* $DP = \dfrac{\sum_{\substack{x \in U \\ p \in \hat{D}(x)}} \max_{a \in D(x)} bmrd(a, p)}{\sum_{x \in U} |MP(x)|}$

The recall depth DA is the average $bmrd$ of the best matching prediction
for each matched annotation. This gives an indication of how well the annotations are reproduced. The precision depth DP gives the average $bmrd$ for
each prediction and its best matching annotation, indicating how well each
prediction matches.

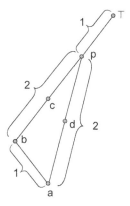

Fig. 2. *Maximal relative depth.* There are two path from a to $\top$ and one path from p to the $\top$. $Depth_{\langle a,b,c,p,\top \rangle}(p) = 4$, $Depth_{\langle a,d,p,\top \rangle}(p) = 3$, and $Depth_{\langle p,\top \rangle}(p) = 1$ The maximal relative depth is $mrd(a,p) = (Depth_{\langle p,\top \rangle}(p)/Depth_{\langle a,d,p,\top \rangle}(p)) = 1/3$.

6   Experimental Results

This section reports on several experiments with the bottom-up algorithm. The algorithm is first applied to a real life data set created with microarrays. We then examine the performance on a series of artificial data sets.

6.1   The Fibroblast Data

Our first experiment was on a data set created by Iyer et al. [9]. They studied the gene response in human fibroblast cells[3] with cDNA microarrays. In their experiments, growth factor serum was initially removed for 48 hours from a cell culture. This forced the cells into a quiescent state. Growth factor serum was then added, and samples were collected at 12 different time points (0 h, 15 min, 30 min, 1 h, 2 h, 4 h, 8 h, 12 h, 16 h, 20 h, and 24 h) and analyzed with cDNA microarrays. Each microarray contained about 8600 genes. Iyer at al. found that 517 of the genes showed substantial changes in their expression levels. These genes were clustered using hierarchical clustering, and 10 major groups were identified by inspecting the dendrogram and the heat map created by the clustering algorithm.

The data set. The data set[4] consisted of the 517 differentially expressed genes and their associated expression profiles, but contained no annotations for the genes. The genes were therefore annotated manually by using relevant information from the literature and molecular biology databases (see [11] for details).

---

[3] Fibroblasts are connective tissue cells that take part in wound healing and are capable of differentiating into specialized cells such as cartilage, bone, fat, and muscle cells. In culture, they require growth factors for proliferation (i.e., multiplication/reproduction). Growth factors are usually provided by fetal bovine serum.

[4] This is available at **http://genome-www.stanford.edu/serum/**.

The annotation classes were taken from biological process subontology (rev. 1.1152 - 25-Aug-2000) of the GO Consortium. A total of 723 annotations were found for 323 of the genes in the data set. 203 of these genes were annotated with more than one class.

The annotations referred to only 234 classes so that most of the classes in the process ontology were not used. In particular, no gene was annotated with any of the leaf classes in the ontology. Therefore, a part of the ontology was selected and only this part was used in the experiment. The part consisted of the most specific classes with annotations and their superclasses. It had a total of 313 classes where 157 were leaf classes. 113 of the classes were associated with only one gene.

The microarray measurements, which were real numbers, were discretized[5] so that they could be applied in a rule learning algorithm. This was achieved with the template approach introduced by Hvidsten et al. [8], which transforms the real numbers in a gene expression profile into templates. A template is simply an attribute value that describes a pattern over several consecutive time points.

In this experiment, we created templates that stretched over 2 to 6 time points. Thus, attributes such as 0H-15min, 0H-30min, 0H-1H, 0H-2H, and 0H-4H were constructed. Templates were then assigned to each (constructed) attribute and gene as follows: An *up* template was created if the slope between the end points in the interval was greater or equal to 0.03. A *down* template was constructed if the slope between the end points was less than or equal to −0.03. A *const* template was assigned if the absolute value of the slope between the endpoints was less than 0.03. If the interval stretched over 3 or more time points an additional requirement was set such that the slope between adjacent time points in the interval could not be below −0.02 (above 0.02) for an increasing (decreasing) template. For a constant template the absolute value of the slope the between adjacent time points had be below 0.02. If these conditions were not fulfilled no template was assigned to the gene for this attribute

Results. We applied the bottom-up ensemble method to the data set and estimated the performance with 10-fold cross-validation. The algorithm was executed with training accuracy $\gamma = 0.8$, split ratio $\rho = 0.66$, pruning accuracy $\delta = 0.8$, and pruning support $\sigma = 1$. The results are reported in Table 1. Most of the annotations in the fibroblast data were predicted, and most of the predictions were correct. However, a lot of the original detail level was lost.

The algorithm was compared to a "flat" algorithm that ignored the DAG structure. This algorithm learned possible rules for each class similarly to the LERS system [6], but used the same search algorithms as the bottom-up method. Table 1 also presents the results obtained with this algorithm. These results were quite weak. Approx. 10% of the actual annotations were predicted, and most of the details of these annotations were retained. However, about 90% of the

---

[5] Prior to discretization each expression profile was $\log_2$-transformation and normalization as follows: Let X_i $(1 \leq i \leq 12)$ be the ratios for a gene over the 12 time points. A normalized $\log_2$-ratio Y_i was then computed as: $Y_i = (\log_2 X_i)/\sqrt{\sum_{i=1}^{12}(\log_2 X_i)^2}$.

Table 1. *Results on the fibroblast data.* For each measure, the average over the 10 cross-validation folds is shown in bold, and the standard is given after the ± sign.

Method	Search	Top lev. pru.	Match		Depth	
			Recall (RA)	*Precision (RP)*	*Recall (DA)*	*Precision (DP)*
Bottom-up						
	Top-down	Yes	**0.79**±0.02	**0.72**±0.02	**0.29**±0.01	**0.32**±0.01
	Bottom-up	Yes	**0.79**±0.02	**0.73**±0.03	**0.29**±0.01	**0.32**±0.01
	Top-down	No	**0.85**±0.02	**0.63**±0.01	**0.30**±0.01	**0.34**±0.01
	Bottom-up	No	**0.84**±0.02	**0.59**±0.02	**0.29**±0.01	**0.32**±0.01
Flat						
	Top-down		**0.14**±0.01	**0.11**±0.01	**0.95**±0.01	**0.95**±0.01
	Bottom-up		**0.12**±0.01	**0.09**±0.01	**0.89**±0.03	**0.90**±0.02

predictions made by the classifier were incorrect. The results obtained with the bottom-up method were obviously better. By trading off details, it was able to reproduce many more annotations with much higher precision.

The predictions made by our algorithms had lost a lot of the details of the annotation. Most of them were made to the classes immediately below the root. We believe that this was due to the nature of the fibroblast data. The variation in the data may not be sufficient to distinguish between the more detailed classes. Iyer et al. found only 10 major clusters in the data indicating that no more than 10 classes may be distinguished. So, it seems highly unlikely that all 157 leaf classes in our annotations may be discerned. The genes participating in different processes may simply be similarly expressed in the experiment of Iyer et al., and in order to distinguish between the more detailed classes, we would require more microarray experiments. Thus, the essential variations in the fibroblast data may be best captured by the general classes at the top of the ontology.

The fibroblast data have also been classified with ROSETTA– originally, in Hvidsten et al. [8] and more recently in Lægreid et al. [11]. In both studies a subset of the classes in the ontology was selected, and the genes, which were annotated to more specific classes, were relabeled with the selected classes. In Hvidsten et al. the most specific classes, which had at least 10 genes annotated to either themselves or their subclasses, were selected. This resulted in a set of 16 classes. In Lægreid et al. 23 classes were selected manually using biological knowledge about the data.

The results obtained in these studies are summarized in Table 2. It should be mentioned that slightly different versions of the annotations were used in these studies, and that an intermediate version of these annotations was used in our experiment. There were also some differences in the definitions of the templates. This meant that only a rough comparison was possible. However, it appeared that our results had a higher precision, while more details were retained in the other studies. The precision in Hvidsten et al. was very low, and even in Lægreid et al. about 50% of the predictions were incorrect. Compared to our results,

Table 2. *Results obtained with* ROSETTA *on the fibroblast data.* These figures are only rough estimates as Hvidsten et al. only reported sensitivity and specificity for each class. Lægreid et al. reported recall and precision so that these numbers are exact. However, recall depth and precision depth were not reported in either study. These figures were estimated by comparing the original annotations to the "moved" annotations that were created by relabeling the genes with the selected classes.

Study	*No. of Classes*	*Match*		*Depth*	
		Recall (RA)	*Precision (RP)*	*Recall (DA)*	*Precision (DP)*
Hvidsten et al. [8]	16	0.67	0.33	0.71	0.71
Lægreid et al. [11]	23	0.84	0.49	0.69	0.69

it seemed that too many details were kept in these experiments. So at least some of the selected classes in these studies should have been replaced by more general classes.

A question is obviously whether our algorithm predicted too general classes such that details were lost needlessly. This issue is examined in the next section through controlled experiments with artificial data. The results that will be reported in that section showed, however, that the bottom-up method retained the detail level quite well. Hence, it is not likely that details were lost because of the algorithm.

Still, the results did not depend solely on the bottom-up method, but also on the search algorithms, which learned the rules. For example, if the rules made by these algorithms for the detailed classes were poor, they would be pruned so that the objects would be pushed upwards, and the detail level would be reduced. This may have happened since the search algorithms were fairly simple and not robust with regard to noise. They found a minimal set of rules so that at most one rule was created for each object, and for each rule they selected a minimal set of descriptors. This made them quite sensitive to noise since a rule would not be applied and no predictions would be made if an object, which otherwise matched the rule, had a distorted value for one of the descriptors in the rule. Hvidsten et al. and Lægreid et al., on the other hand, applied the genetic reduct algorithm in ROSETTA (with the object-wise option). This algorithm created several rules with different descriptors for each objects. So if an object had a distorted value for an attribute, it was likely that at least one rule did not have a descriptor for this attribute and would match the object. The genetic reduct algorithm was therefore much more robust with regard to noise.

Thus, it is possible that better results could have been obtained with the genetic reduct algorithm than with the search algorithms that were used in this study. In order to examine this hypothesis a bit further, we tried to weaken our results so that we might obtain a recall depth and a precision depth that were similar to that of Lægreid et al. This was done by changing the pruning accuracy δ. Some of the results that were achieved are reported in Table 3. However, we were unable to obtain a similar detail level as these results indicate.

Table 3. Results on the fibroblast data with relaxed settings ($\delta = 0.1$)

Method	Search	Top lev. pru.	Match		Depth	
			Recall (RA)	*Precision (RP)*	*Recall (DA)*	*Precision (DP)*
Bottom-up						
	Top-down	Yes	**0.81**±0.01	**0.56**±0.02	**0.36**±0.01	**0.39**±0.01
	Bottom-up	Yes	**0.80**±0.02	**0.54**±0.01	**0.34**±0.01	**0.37**±0.01
	Top-down	No	**0.84**±0.01	**0.49**±0.02	**0.35**±0.01	**0.38**±0.01
	Bottom-up	No	**0.83**±0.02	**0.50**±0.01	**0.35**±0.01	**0.37**±0.01

This suggests that more details could perhaps have been retained if the genetic reduct algorithm (or another less noise sensitive algorithm) had been used instead. However, it is not certain that we would have obtained a *large* increase in the detail level if this algorithm had been used. The classifier of Lægreid et al. had a low precision, and 17 of the 23 classes in their study had a depth of 2 or 3. So, if we wanted to maintain a high precision, we would have had to give up some of the details that were obtained in their study. Thus, we might still have ended up with a classifier that predicted mostly classes at the top level (i.e., at depth 1). So, even though it is possible that some more detail could have been obtained with another search algorithm, we believe that our results are reasonable given the available data. A much larger data set comprising many more measurements (created under different experimental conditions) would have been required in order to obtain more detailed predictions.

6.2 The Artificial Data

We performed several controlled experiments in order to get a better understanding of the performance of the algorithm. These experiments are reported in this section. In particular, we demonstrate the importance of avoiding discrimination between related classes, and trading off details for precision. Moreover, we examine whether the algorithms predict too general classes such that details are lost needlessly.

The data sets. Several different artificial data sets were created for these experiments. Two different DAGs – one small and one quite large – were constructed initially. The small DAG contained 13 classes where 7 classes were leaf classes. Five of these leaf classes had a depth of 3, and two had a depth of 2. The large DAG consisted of 52 classes with 30 leaf classes. Twenty of these leaf classes had a depth of 4. The rest had a depth of 3.

Objects were constructed in two steps. Model objects were first created and assigned to one or more leaf classes. Each model object had an information vector that consisted of 11 attributes where each attribute could be assigned one of the following values: *up*, *down*, and *const*. Objects were then created from the model objects. An object was produced as follows: The attribute-value pairs in the information vector of the model object were first copied to the

Table 4. Results without errors. The top-down and bottom-up search algorithms are abbreviated with T-D and B-U.

DAG	Method	Search	Match		Depth	
			Recall (RA)	Precision (RP)	Recall (DA)	Precision (DP)
Small	Bottom-up	T-D	1.000±0.000	1.000±0.000	0.989±0.005	0.989±0.005
		B-U	1.000±0.000	1.000±0.000	0.989±0.005	0.989±0.005
	Flat	T-D	1.000±0.000	1.000±0.000	1.000±0.000	0.847±0.010
		B-U	1.000±0.000	1.000±0.000	1.000±0.000	0.847±0.010
Large	Bottom-up	T-D	1.000±0.000	1.000±0.000	0.996±0.001	0.994±0.002
		B-U	1.000±0.000	1.000±0.000	0.996±0.001	0.996±0.001
	Flat	T-D	1.000±0.000	1.000±0.000	1.000±0.000	0.819±0.007
		B-U	1.000±0.000	1.000±0.000	1.000±0.000	0.819±0.007

information vector of the object. The object was then annotated by randomly generalizing the leaf class label(s) of the model object. For each leaf class label, a class was selected at random from a pool that consisted of the leaf class and its superclasses. The object was then labeled with the selected class.

Several objects were created for each model object. Fifteen instances of each model objects were created for the small DAG, 25 instances were created for the large DAG. These numbers may seem large, however, only a fraction of these objects were assigned to the leaf classes. The most detailed leaf classes in the small DAG had on average 5 instances of same model object. The deepest classes in the large DAG had 6.25. Since the performance of the classifiers was estimated with 10-fold cross-validation, these numbers were reduced by approximately 10% to 4.5 and 5.6. Instances of the same model object had to occur in both the training set and the validation set in order for the bottom-up method to learn and accept a rule. A leaf class needed at least 3 objects (when $\rho = 0.66$) such that 2 objects were placed in the training set and 1 object was put in the validation set. However, a higher number of objects was required when several model objects were annotated to the same class since the random splitting, which created the training and the validation sets, could place all instances of a model object in only one of these two sets. Hence, the available data in the leaf classes were very close to minimum required for successful learning.

Results without errors. The bottom-up algorithm was first applied to the data without introducing any errors into the datasets. The algorithm was executed with training accuracy $\gamma = 1$, split ratio $\rho = 0.66$, pruning accuracy $\delta = 1$, and pruning support $\sigma = 1$ with top level pruning[6] (These settings were used through out the artificial data experiments except for those cases where other values are explicitly given). The performance was estimated with 10-fold cross-validation and the results are shown in Table 4.

The bottom-up algorithm classified the objects almost perfectly. All predictions were correct and all annotations were matched. However, a very small loss

---

[6] Similar results were usually also obtained without top level pruning.

of details was observed. This was expected as the method requires a minimum of objects in order to accept a rule. Yet, this loss was hardly significant. By setting $\rho = 0.55$, we achieved perfect results (these are not shown). However, this depended on the seed given to the random number generator, which was used when the data were split for cross-validation. Increasing the number of instances per model object to 20 for the small DAG gave perfect results as well. Thus the method did not appear to have a tendency to predict too general classes, besides the small loss that was caused by the need to fill both the training and the validation set with objects.

The flat method was also tested on the data sets and performed quite well. The reason was that there were no distortions – in particular no irrelevant attributes – in the data set so that the algorithm had no opportunity to make mistakes. So even though it tried to discriminate between similar classes, it found rules that covered the model vectors. However, it made redundant predictions so that many of the predictions were related. This resulted in a reduced precision depth.

Results with irrelevant attributes. The data sets without errors correspond to an ideal situation where each attribute contributes to the discernibility of the classes. Such a situation will rarely occur in practice. A more realistic data set will contain noise. In particular, there will be irrelevant attributes that do not contribute to the discernibility of the classes. In order to make the data sets more realistic, we added 11 irrelevant attributes to each information vector such that each vector consisted of a total 22 attributes. The values of the irrelevant attributes were generated at random when an object was created. The same three values that were used for the relevant attributes were also assigned to these attributes.

The results from these data sets are given in Table 5. The results for the bottom-up method were very good. There were some very small errors, e.g., the bottom-up method had an RP of 0.993 and 0.995. These errors were merely not significant as reflected by the associated standard errors. They were estimation errors that were introduced by unfortunate splits made by the cross-validation procedure. This was confirmed in several cases by changing the seed of the random number generator that was used by the cross-validation procedure. In all of these cases, we were able to obtain perfect results[7].

The results obtained with the flat method were clearly much worse. The search algorithms tried to separate related classes. However, the relevant attribute could not separate between instances of the same model object that were annotated to different, but related classes. The rules were thus based on the irrelevant attributes and this resulted in very poor predictions. This clearly demonstrated that a DAG learning algorithm should not attempt to discern between related classes.

---

[7] These results are not shown as we have chosen to create of all of the results in this paper with the same seed so that they are more comparable.

Table 5. Results with irrelevant attributes

DAG	Method	Search	Match		Depth	
			Recall (RA)	Precision (RP)	Recall (DA)	Precision (DP)
Small	Bottom-up	T-D	1.000±0.000	0.993±0.007	0.989±0.006	0.989±0.006
		B-U	1.000±0.000	0.995±0.005	0.943±0.012	0.921±0.019
	Flat	T-D	0.684±0.021	0.785±0.043	0.877±0.026	0.851±0.020
		B-U	0.462±0.052	0.664±0.055	0.867±0.033	0.838±0.028
Large	Bottom-up	T-D	1.000±0.000	0.999±0.001	0.994±0.002	0.992±0.003
		B-U	1.000±0.000	0.999±0.001	0.990±0.002	0.986±0.002
	Flat	T-D	0.657±0.015	0.777±0.018	0.862±0.015	0.836±0.019
		B-U	0.444±0.011	0.532±0.015	0.815±0.022	0.809±0.018

Table 6. Results with inconsistencies. The optimal recall/precision depth was 0.846 for the small DAG and 0.901 for the large DAG.

DAG	Method	Search	Match		Depth	
			Recall (RA)	Precision (RP)	Recall (DA)	Precision (DP)
Small	Bottom-up	T-D	1.000±0.000	1.000±0.000	0.823±0.018	0.795±0.012
		B-U	1.000±0.000	1.000±0.000	0.828±0.019	0.801±0.018
	Flat	T-D	1.000±0.000	0.871±0.019	1.000±0.000	0.861±0.010
		B-U	1.000±0.000	0.871±0.019	1.000±0.000	0.861±0.010
Large	Bottom-up	T-D	1.000±0.000	1.000±0.000	0.890±0.005	0.844±0.007
		B-U	1.000±0.000	1.000±0.000	0.899±0.005	0.869±0.007
	FLat	T-D	1.000±0.000	0.937±0.004	1.000±0.000	0.832±0.003
		B-U	1.000±0.000	0.937±0.004	1.000±0.000	0.832±0.003

Results with inconsistencies. The ability of the bottom-up algorithm to trade off details for precision was also examined. This was done by adding inconsistencies to the original data sets with no errors. An inconsistency was created by assigning model objects with identical information vectors to different leaf classes. Each model object was assigned to one leaf class, and the leaf classes were selected such that they had at least one superclass in common besides the root. Objects were then created from these model objects as before, except that the class labels of these objects were not generalized. All of these objects were thus assigned to leaf classes.

Several inconsistencies were created for both DAGs. These were made with different information vectors. The classes, which were assigned to the model objects, were selected such that the depth of the common superclass varied. Given the properties of the data sets, we could compute the optimal recall/precision depth if the learning algorithm identified the common superclasses correctly. This was 0.846 for the small DAG and 0.901 for the large DAG.

These results are given in Table 6. The bottom-up method traded off details for precision so that the precision was maintained. The recall depth was very

Table 7. Results with irrelevant attributes and inconsistencies. The optimal recall/precision depth was 0.846 for the small DAG and 0.901 for large DAG.

DAG	Method	Search	σ	Top lev. pru.	Match		Depth	
					Recall (RA)	Precision (RP)	Recall (DA)	Precision (DP)
Small	Bottom-up	T-D	1	Yes	1.00±0.00	0.98±0.01	0.84±0.02	0.79±0.01
		B-U	1	Yes	1.00±0.00	0.97±0.01	0.82±0.02	0.80±0.02
		T-D	3	Yes	0.97±0.02	1.00±0.00	0.68±0.02	0.66±0.02
		B-D	3	Yes	0.97±0.02	0.99±0.01	0.70±0.02	0.67±0.02
		T-D	3	No	1.00±0.00	1.00±0.00	0.69±0.02	0.67±0.02
		B-U	3	No	1.00±0.00	0.99±0.01	0.69±0.02	0.67±0.02
	Flat	T-D			0.54±0.04	0.69±0.05	0.87±0.02	0.85±0.02
		B-U			0.40±0.03	0.53±0.03	0.88±0.03	0.87±0.03
Large	Bottom-up	T-D	1	Yes	1.00±0.00	0.96±0.00	0.90±0.01	0.85±0.01
		B-U	1	Yes	1.00±0.00	0.98±0.00	0.89±0.01	0.86±0.01
		T-D	3	Yes	1.00±0.00	1.00±0.00	0.85±0.00	0.81±0.01
		B-U	3	Yes	1.00±0.00	1.00±0.00	0.85±0.00	0.83±0.01
	Flat	T-D			0.62±0.02	0.70±0.03	0.89±0.01	0.86±0.00
		B-U			0.42±0.02	0.50±0.02	0.84±0.01	0.83±0.01

close to the optimal depth. A very small of loss of details was visible. However, this was of the same magnitude as observed previously, and it occurred most likely because this method needed a few more objects in order to learn and accept a rule.

The performance of the flat method was similar to its previous performance on the data sets without errors. However, the precision was lower because the algorithm predicted the leaf classes for the inconsistent objects and was not able to trade of details for precision.

Results with irrelevant attributes and inconsistencies. Finally, the two kinds of errors were combined in order to examine how the bottom-up algorithm handled their combined effect. 11 irrelevant attributes were added to the data sets with the inconsistencies. These attributes were assigned to the information vectors of both consistent and inconsistent objects. The inconsistent objects were therefore not inconsistent in the rough set sense of the word as they did not share the same elementary set. However, they did share the same values for the relevant attributes. These objects could therefore not be separated by any rule that was based only on the relevant attributes. A rule based on the irrelevant attributes could possibly separate them. However, it would not make any accurate predictions since the values of these attributes were generated at random. Obviously, this made it much more difficult for an algorithm to recognize that these objects should be predicted to their common superclass and not to the leaf classes.

The results are given in Table 7. The bottom-up method performed clearly much better than the flat method. However, its performance was not perfect on these data sets. It maintained the detail level quite well when the pruning support σ was set to 1, and all annotations were predicted. However, a very few predictions were incorrect. These incorrect predictions occurred because a few low quality rules slipped through the pruning system. The search algorithms created rules for the inconsistent objects with irrelevant attributes. These rules were usually pruned since they did not cover many objects in the validation set and had low accuracy. On some rare occasions, however, a rule had a high accuracy on the validation set and was accepted, and these rules created the errors. Still, the probability of such an event was very low so that the precision was only marginally reduced.

We were able to remove these errors in most cases by raising support σ to 3. However, this meant that the algorithms now needed more objects in order to learn and accept a rule. Some loss of details was therefore observed – especially for the small DAG where the objects were less abundant. This was much higher than the loss of precision that was observed when $\sigma = 1$. So, the results for $\sigma = 1$ seemed better. The precision was quite close to the optimal value. σ should thus be kept at 1 – at least when the number of objects is sparse.

7   Conclusion

In this paper we have introduced an algorithm for learning in an ontology and a framework for evaluating its performance. The experiments with the artificial data sets demonstrate that the bottom-up algorithm deals appropriately with the issues introduced by the ontology. It does not discern between related classes so that its rules have a much higher quality than the rules of a corresponding flat method. Thus, the predictions made by this algorithm are much more accurate. The algorithm is able to learn good classifiers when the annotations have a varying detail level and the data associated with each class are relatively sparse. The classes of the model objects in the artificial experiments were randomly generalized so that the annotations had a different detail level, and only a few objects were assigned to each class. The algorithms still managed to produce high quality classifiers. The algorithm seems to handle the trade-off between detail level and precision effectively. When objects are labeled with classes that cannot be predicted accurately, they identify the most specific classes that may be predicted.

It is possible that a DAG learning method may lose details that could have been retained. However, this does not appear to be a problem for the bottom-up method. It retained the detail level quite well in all of the artificial experiments, and it did not show a tendency to predict too general classes. Only a very small loss of details was witnessed in our results. This was expected as the method requires a minimum of objects in order to split the data into a training set for learning and a validation set for pruning. So if the examples are very sparse, a small loss may be observed. However, we believe that this should not be any problem in a real life situation.

The results on the fibroblast data also show that the bottom-up method performs much better than a corresponding flat method. The predictions obtained on this data set are accurate. However, they are very general such that they do not contribute much biological knowledge. Still, it seems that this is the best that can be achieved with the available data set when the top-down or the bottom-up search algorithm is used. It is possible that another search algorithm may produce more detailed predictions. However, the main problem seems to be that the fibroblast data set is insufficient for making both detailed and accurate predictions. Our algorithm should thus be applied on a much larger data set.

One possibility would be to use microarray data from several experimental studies assuming that these contain a larger variety of expression patterns such that more classes can be separated. However, it is a question whether enough microarray data can be obtained such that both accurate and detailed predictions can be made. By predicting gene function annotations from microarray data we assume that co-expressed genes are involved in the same biological process. This assumption has been contested by Shatkay et al. [16], and it is possible that the function of a gene cannot be predicted with high accuracy from gene expression data alone. Still, it is feasible to combine microarray data with other kinds of data such as DNA sequence data. One possibility would be to add information about transcription factor binding sites to a data set by creating Boolean attributes denoting the presence or absence of a particular binding site in a gene. However, the whole sequence could in principle be used. Information from the biomedical literature such as Medline abstracts may also be applied for predicting gene function.

An important feature of our method is its generality. It can be applied on many different kinds of data and is not limited to microarray data. One possibility would be to use it for text classification – a task which also involves an ontology.

Moreover, it may be combined with many different learning algorithms. The top-down and the bottom-up search algorithms that we used here were mainly chosen for their simplicity and may not be the best learning algorithms. Any rule learning algorithm can be used directly by the bottom-up method. However, the method is not limited to rule learning. It can be applied to any binary classifier. In this case, we will just prune whole classes instead of rules. More precisely, the algorithm will learn a binary classifier on the training sample when a class is visited. This binary classifier will then be tested on the validation sample and pruned if its performance is unsatisfactory.

References

1. K. M. Ali and M. J. Pazzani. HYDRA: A noise-tolerant relational concept learning algorithm. In R. Bajcsy, editor, *Proceedings of the 13th International Joint Conference on Artificial Intelligence (IJCAI-93)*, pages 1064–1071, 1993.
2. R. Baeza-Yates and B. Ribeiro-Neto. *Modern Information Retrieval*. ACM Press, 1999.
3. M. P. S. Brown, W. N. Grundy, D. Lin, N. Cristianini, C. W. Sugnet, T. S. Furey, M. Ares, Jr., and D. Haussler. Knowledge-based analysis of microarray gene expression data by using support vector machines. *Proceedings of the National Academy of Sciences, USA*, 97(1):262–267, 2000.

4. The Gene Ontology Consortium. Gene Ontology: Tool for the unification of biology. *Nature Genetics*, 25(1):25–29, 2000.

5. J. Fürnkranz. Separate-and-conquer rule learning. *Artificial Intelligence Review*, 13(1):3–54, 1999. Occurs also as Technical Report OEFAI-TR-96-25.

6. J. W. Grzymala-Busse. Knowledge acquisition under uncertainty – a rough set approach. *Journal of Intelligent and Robotic System*, 1:3–16, 1988.

7. J. W. Grzymala-Busse. LERS – A system for learning from examples based on rough sets. In *Intelligent decision support: Handbook of Applications and Advances of Rough Sets Theory*, pages 3–18. Kluwer Academic Publishers, 1992.

8. T. R. Hvidsten, J. Komorowski, A. K. Sandvik, and A. Lægreid. Predicting gene function from gene expressions and ontologies. In *Proceedings of the Pacific Symposium on Biocomputing 6 (PSB-2001)*, pages 299–310. World Scientific Press, 2001.

9. W. R. Iyer, M. B. Eisen, D. T. Ross, G. Schuler, T. Moore, J. C. F. Lee, J. M. Trent, L. M. Staudt, J. Hudson, Jr., M. S. Boguski, D. Lashkari, D. Shalon, D. Botstein, and P. O. Brown. The transcriptional program in the response of human fibroblasts to serum. *Science*, 283:83–87, 1999.

10. R. R. Korfhage. *Information Storage and Retrieval*. Wiley, 1997.

11. A. Lægreid, T. R. Hvidsten, H. Midelfart, J. Komorowski, and A. K. Sandvik. Predicting Gene Ontology biological process from temporal gene expression patterns. *Genome Research*, 13(5):965–979, 2003.

12. H. Midelfart. *Knowledge Discovery from cDNA Microarrays and a priori Knowledge*. PhD thesis, Department of Computer and Information Science, Norwegian University of Science and Technology (NTNU), 2003. ISBN 82-471-5617-2.

13. H. Midelfart. Supervised learning in the gene ontology — part I: A rough sets framework. *Transactions on Rough Sets* IV, LNCS 3700: 69–97, 2005.

14. H. Midelfart and J. Komorowski. A rough set approach to learning in a directed acyclic graph. In J. F. Peters, A. Skowron, and N. Zhong (eds.), *Proceedings of the 3th International Conference on Rough Sets and Current Trends in Computing (RSCTC-2002)*, number 2475 in Lecture Notes in Artificial Intelligence, pages 144–155. Springer-Verlag, 2002.

15. H. Midelfart, A. Lægreid, and J. Komorowski. Classification of gene expression data in an ontology. In J. Crespo, V. Maojo, and F. Martin (eds.), *Proceedings of the 2nd International Symposium on Medical Data Analysis (ISMDA-2001)*, number 2199 in LNCS, pages 186–194. Springer-Verlag, 2001.

16. H. Shatkay, S. Edwards, W. J. Wilbur, and M. Boguski. Genes, themes and microarrays: Using information retrieval for large-scale gene analysis. In *Proceedings of the 8th International Conference on Intelligent Systems for Molecular Biology (ISMB-2000)*, pages 317–328. AAAI Press, 2000.

Comparative Analysis of Deterministic and Nondeterministic Decision Tree Complexity Local Approach

Mikhail Ju. Moshkov

Institute of Computer Science, University of Silesia,
39, Będzińska St., Sosnowiec, 41-200, Poland
moshkov@us.edu.pl

Abstract. For problems over arbitrary information system we study the relationships among the complexity of a problem description, the minimal complexity of a decision tree solving this problem deterministically, and the minimal complexity of a decision tree solving this problem nondeterministically. We consider the local approach to investigation of decision trees where only attributes from a problem description are used for construction of decision trees solving this problem.

Keywords: Decision tree, rough set theory, complexity.

1 Introduction

Decision trees over finite information systems are investigated in rough set theory [12,13,17,19], test theory [4,6], theory of questionnaires [14], theory of decision tables [5], machine learning [2,16], search theory [1,20], etc. The notion of infinite information system is useful in discrete optimization [7] and computational geometry [15]. However, decision trees over infinite information systems are investigated to a lesser degree than over finite information systems. In this paper we consider arbitrary (finite and infinite) information systems.

We study problems with many-valued decisions over considered information system. Any problem is specified by a number of attributes that divide the universe into domains on which these attributes have fixed values. A nonempty finite set of decisions is attached to each domain. For a given object from the universe it is required to find a decision from the set attached to the domain containing this object.

Many problems investigated in rough set theory are reduced to problems with many-valued decisions. Let us consider a finite decision system [18]. This system is specified by a number of conditional attributes that divide the universe into domains on which these attributes have fixed values. Our aim is to find the value of a decision attribute using only values of conditional attributes. Consider an arbitrary domain. It is clear that using only values of conditional attributes we will obtain the same value of the decision attribute for all objects from the domain. If the decision attribute is constant on this domain then we can find its

J.F. Peters and A. Skowron (Eds.): Transactions on Rough Sets IV, LNCS 3700, pp. 125–143, 2005.

exact value for all objects from the domain. Otherwise, we can find exact value only for some objects. If we want to minimize the number of incorrect answers we must choose a value of the decision attribute which is true for maximal number of objects from the considered domain. It is possible that there exists more than one such value. In this case we deal with a problem with many-valued decisions.

Moreover, problems with many-valued decisions arise naturally in such areas as discrete optimization, fault diagnosis, computational geometry. For example, let $g_1, \ldots, g_m$ be functions from $\mathbb{R}^n$ to $\mathbb{R}$. For a given $\bar{a} \in \mathbb{R}^n$ it is required to find a number from the set $D(\bar{a})$ of numbers $i \in \{1, \ldots, m\}$ such that $g_i(\bar{a}) = \min\{g_1(\bar{a}), \ldots, g_m(\bar{a})\}$. If for some $\bar{a}$ the set $D(\bar{a})$ contains at least two numbers then we deal with a problem with many-valued decisions.

As algorithms for problem solving we consider decision trees which solve problems deterministically or nondeterministically. One can interpret decision trees solving a problem nondeterministically as a way for representation of arbitrary complete (applicable to any object) decision rule systems for the problem [3].

We consider various complexity measures which characterize time complexity of decision trees. One of the most known among them is the depth. The depth of a decision tree is the maximal number of nodes labeling by attributes in a path from the root to a terminal node of the tree (the depth of a decision rule system is the maximal number of conditions in the left-hand side of a rule from the system).

There are two approaches to decision tree investigation: the local approach where for a problem the decision trees are considered using only attributes from the problem description, and the global one where for problem solving all attributes from the considered information system can be used. In this paper decision trees are studied in the frameworks of the local approach.

This paper deals with comparative analysis of the three parameters of problems over arbitrary information system: the complexity of a problem description (in the case of the depth, for example, this is the number of attributes in the problem description), the minimal complexity of a decision tree solving this problem deterministically, and the minimal complexity of a decision tree solving this problem nondeterministically. Coarse classification of relations among these parameters is considered and all possible seven types of these relations are enumerated.

Note that the results of this paper were published without proofs in [8,9,10]. Similar results obtained in the frameworks of the global approach can be found in [9,10,11].

2   Basic Definitions and Results

2.1   Decision Trees

Let $\omega = \{0, 1, 2, \ldots\}$, $E_k = \{0, 1, \ldots, k-1\}$, $k \geq 2$, A be a nonempty set and F be a nonempty set of functions from A to E_k. Functions from F will be called *attributes*, and the pair $U = (A, F)$ will be called *an information system*. Denote by F^* the set of all finite words over the alphabet F including the empty word λ.

A node in a finite directed tree is called *the root* if it is the only node which has no entering edges. A tree which has such a node will be called *a finite directed tree with root*. Nodes without issuing edges will be called *terminal* nodes. Nodes which are neither the root nor terminal will be called *working* nodes. *A complete path* in a finite directed tree with root is any sequence $\xi = v_0, d_0, \ldots, v_m, d_m, v_{m+1}$ of nodes and edges of the tree such that v_0 is the root, v_{m+1} is a terminal node, and for $i = 0, \ldots, m$ the edge d_i issues from the node v_i and enters the node v_{i+1}.

A *decision tree over information system U* is a marked finite directed tree with root which has at least two nodes and possesses the following properties:

a) the root and the edges issuing from the root are not labeled;
b) each working node is labeled by an attribute from the set F;
c) each edge issuing from a working node is labeled by a number from E_k;
d) each terminal node is labeled by a number from ω.

A decision tree is called *deterministic* if it satisfies the following conditions:

a) exactly one edge issues from the root;
b) edges issuing from a working node are labeled by pairwise different numbers.

The set of decision trees over U will be denoted by $\mathrm{Tree}(U)$. Let $\Gamma \in \mathrm{Tree}(U)$. Denote by $\mathrm{At}(\Gamma)$ the set of attributes attached to working nodes of Γ. By $\mathrm{Path}(\Gamma)$ we denote the set of complete paths in Γ. Let $\xi = v_0, d_0, \ldots, v_m, d_m, v_{m+1}$ be a complete path in Γ. We denote by $\tau(\xi)$ the number attached to the node v_{m+1}. Define a word $\varphi(\xi)$ from F^* and a subset $\mathcal{A}(\xi)$ of the set A associated with ξ. If $m = 0$ then $\varphi(\xi) = \lambda$ and $\mathcal{A}(\xi) = A$. Let $m > 0$ and for $j = 1, \ldots, m$ the node v_j be labeled by the attribute f_j, and the edge d_j be labeled by the number δ_j. Then $\varphi(\xi) = f_1 \ldots f_m$ and $\mathcal{A}(\xi) = \{a : a \in A, f_1(a) = \delta_1, \ldots, f_m(a) = \delta_m\}$. Denote $\mathcal{S}(\xi) = \{f_1(x) = \delta_1, \ldots, f_m(x) = \delta_m\}$.

2.2   Problems

The set of nonempty finite subsets of the set ω will be denoted by $\mathcal{P}(\omega)$. A *problem over U* is any $(n + 1)$-tuple $z = (\nu, f_1, \ldots, f_n)$ where $n \in \omega \setminus \{0\}$, $\nu : E_k^n \to \mathcal{P}(\omega)$ and $f_1, \ldots, f_n \in F$. Denote $\mathrm{At}(z) = \{f_1, \ldots, f_n\}$. The problem z may be interpreted as a problem of searching for at least one number from the set $z(a) = \nu(f_1(a), \ldots, f_n(a))$ for a given $a \in A$. We denote by $\mathrm{Probl}(U)$ the set of problems over the information system U.

Let $z \in \mathrm{Probl}(U)$ and $\Gamma \in \mathrm{Tree}(U)$. We will say that the tree Γ solves the problem z *nondeterministically* if the following conditions hold:

a) $\bigcup_{\xi \in \mathrm{Path}(\Gamma)} \mathcal{A}(\xi) = A$.
b) For any $a \in A$ and $\xi \in \mathrm{Path}(\Gamma)$ if $a \in \mathcal{A}(\xi)$ then $\tau(\xi) \in z(a)$.

We will say that the tree Γ solves the problem z *deterministically* if Γ is a deterministic decision tree which solves z nondeterministically.

2.3   Complexity Measures

A *complexity measure over* U is any mapping $\psi : F^* \to \omega$. The complexity measure ψ will be called *limited* if it possesses the following properties:

(a) $\psi(\alpha_1 \alpha_2) \leq \psi(\alpha_1) + \psi(\alpha_2)$ for any $\alpha_1, \alpha_2 \in F^*$.

(b) $\psi(\alpha_1 \alpha_2 \alpha_3) \geq \psi(\alpha_1 \alpha_3)$ for any $\alpha_1, \alpha_2, \alpha_3 \in F^*$.

(c) For any $\alpha \in F^*$ the inequality $\psi(\alpha) \geq l(\alpha)$ holds where $l(\alpha)$ is the length of α.

We extend an arbitrary complexity measure ψ onto the set $\mathrm{Tree}(U)$ in the following way: $\psi(\Gamma) = \max\{\psi(\varphi(\xi)) : \xi \in \mathrm{Path}(\Gamma)\}$ for any $\Gamma \in \mathrm{Tree}(U)$. The value $\psi(\Gamma)$ will be called *the complexity* of the decision tree Γ.

Now we consider some examples of complexity measures. Let $w : F \to \omega \setminus \{0\}$. We define the function $\psi^w : F^* \to \omega$ in the following way: $\psi^w(\alpha) = 0$ if $\alpha = \lambda$ and $\psi^w(\alpha) = \sum_{i=1}^{m} w(f_i)$ if $\alpha = f_1 \dots f_m$. The function ψ^w is a limited complexity measure over U and it is called *a weighted depth*. If $w \equiv 1$ then the function ψ^w is called *the depth* and is denoted by h.

Let ψ be a complexity measure over U and $z = (\nu, f_1, \dots, f_n) \in \mathrm{Probl}(U)$. The value $\psi^i_U(z) = \psi(f_1 \dots f_n)$ will be called *the complexity of the problem z description*. We denote by $\psi^d_U(z)$ the minimal complexity of a decision tree $\Gamma \in \mathrm{Tree}(U)$ which solves the problem z deterministically and satisfies the condition $\mathrm{At}(\Gamma) \subseteq \mathrm{At}(z)$. We denote by $\psi^a_U(z)$ the minimal complexity of a decision tree $\Gamma \in \mathrm{Tree}(U)$ which solves the problem z nondeterministically and satisfies the condition $\mathrm{At}(\Gamma) \subseteq \mathrm{At}(z)$.

2.4   Local Types of T-Pairs

A pair (U, ψ) where U is an information system and ψ is a complexity measure over U will be called *a test-pair* (or, *t-pair*, in short). If ψ is a limited complexity measure then t-pair (U, ψ) will be called *a limited t-pair*.

Let (U, ψ) be a t-pair. We have the three parameters $\psi^i_U(z)$, $\psi^d_U(z)$ and $\psi^a_U(z)$ for any problem $z \in \mathrm{Probl}(U)$, and we will investigate the relationships between any two such parameters for problems from $\mathrm{Probl}(U)$. Let us consider, for example, the parameters $\psi^i_U(z)$ and $\psi^d_U(z)$. Let $n \in \omega$. We will study relations of the kind $\psi^i_U(z) \leq n \Rightarrow \psi^d_U(z) \leq u$ which are true for any $z \in \mathrm{Probl}(U)$. The minimal value of u is most interesting for us. This value (if exists) is equal to

$$\mathcal{U}^{di}_{U\psi}(n) = \max\{\psi^d_U(z) : z \in \mathrm{Probl}(U), \psi^i_U(z) \leq n\} \ .$$

Also we will study relations of the kind $\psi^i_U(z) \geq n \Rightarrow \psi^d_U(z) \geq l$. In this case the maximal value of l is most interesting for us. This value (if exists) is equal to

$$\mathcal{L}^{di}_{U\psi}(n) = \min\{\psi^d_U(z) : z \in \mathrm{Probl}(U), \psi^i_U(z) \geq n\} \ .$$

The two functions $\mathcal{U}^{di}_{U\psi}$ and $\mathcal{L}^{di}_{U\psi}$ describe how the behavior of the parameter $\psi^d_U(z)$ depends on the behavior of the parameter $\psi^i_U(z)$.

There are 18 similar functions for all ordered pairs of parameters $\psi^i_U(z)$, $\psi^d_U(z)$ and $\psi^a_U(z)$. These 18 functions well describe the relationships among the considered parameters. It will be very interesting to point out 18-tuples of these functions for all t-pairs as well as for all limited t-pairs. But this is a very difficult problem. In this paper instead of functions we will study types of functions. With any function we will associate its type from the set $\{\alpha, \beta, \gamma, \delta, \epsilon\}$. For example, if a function has infinite domain, and it is bounded from above then its type is equal to α. Thus, we will enumerate 18-tuples of types of functions. These tuples will be represented in tables called the local types of t-pairs.

Now we give definitions of mentioned above notions. Let $b, c \in \{i, d, a\}$. We define partial functions $\mathcal{U}_{U\psi}^{bc} : \omega \to \omega$ and $\mathcal{L}_{U\psi}^{bc} : \omega \to \omega$ by

$$\mathcal{U}_{U\psi}^{bc}(n) = \max\{\psi_U^b(z) : z \in \mathrm{Probl}(U), \psi_U^c(z) \leq n\} ,$$
$$\mathcal{L}_{U\psi}^{bc}(n) = \min\{\psi_U^b(z) : z \in \mathrm{Probl}(U), \psi_U^c(z) \geq n\}$$

for any $n \in \omega$. If the value $\mathcal{U}_{U\psi}^{bc}(n)$ is definite then it is the unimprovable upper bound on the values $\psi_U^b(z)$ for problems $z \in \mathrm{Probl}(U)$ satisfying $\psi_U^c(z) \leq n$. If the value $\mathcal{L}_{U\psi}^{bc}(n)$ is definite then it is the unimprovable lower bound on the values $\psi_U^b(z)$ for problems $z \in \mathrm{Probl}(U)$ satisfying $\psi_U^c(z) \geq n$.

Let g be a partial function from ω to ω. We denote by $\mathrm{Dom}(g)$ the domain of g. Denote $\mathrm{Dom}^+(g) = \{n : n \in \mathrm{Dom}(g), g(n) \geq n\}$ and $\mathrm{Dom}^-(g) = \{n : n \in \mathrm{Dom}(g), g(n) \leq n\}$. Now we define the value $\mathrm{typ}(g) \in \{\alpha, \beta, \gamma, \delta, \epsilon\}$ called the type of g.

- If $\mathrm{Dom}(g)$ is an infinite set and g is a bounded from above function then $\mathrm{typ}(g) = \alpha$.
- If $\mathrm{Dom}(g)$ is an infinite set, $\mathrm{Dom}^+(g)$ is a finite set, and g is an unbounded from above function then $\mathrm{typ}(g) = \beta$.
- If both sets $\mathrm{Dom}^+(g)$ and $\mathrm{Dom}^-(g)$ are infinite then $\mathrm{typ}(g) = \gamma$.
- If $\mathrm{Dom}(g)$ is an infinite set and $\mathrm{Dom}^-(g)$ is a finite set then $\mathrm{typ}(g) = \delta$.
- If $\mathrm{Dom}(g)$ is a finite set then $\mathrm{typ}(g) = \epsilon$.

We denote by $\mathrm{typ}_l(U, \psi)$ a table with three rows and three columns in which rows from top to bottom and columns from the left to the right are labeled by indices i, d, a. The pair $\mathrm{typ}(\mathcal{L}_{U\psi}^{bc})\ \mathrm{typ}(\mathcal{U}_{U\psi}^{bc})$ is on the intersection of the row with index $b \in \{i, d, a\}$ and the column with index $c \in \{i, d, a\}$. The table $\mathrm{typ}_l(U, \psi)$ will be called the local type of t-pair (U, ψ).

2.5 Basic Results

The main problem investigated in this paper is to find all local types of t-pairs as well as of limited t-pairs. The solution of this problem describes all possible (in terms of functions $\mathcal{U}_{U\psi}^{bc}$, $\mathcal{L}_{U\psi}^{bc}$ types, $b, c \in \{i, d, a\}$) relationships among the complexity of problem description, the complexity of problem solving by deterministic decision trees, and the complexity of problem solving by nondeterministic decision trees.

Now we define seven tables:

$$T_1 = \begin{array}{c|ccc} & i & d & a \\ \hline i & \epsilon\alpha & \epsilon\alpha & \epsilon\alpha \\ d & \epsilon\alpha & \epsilon\alpha & \epsilon\alpha \\ a & \epsilon\alpha & \epsilon\alpha & \epsilon\alpha \end{array} \quad T_2 = \begin{array}{c|ccc} & i & d & a \\ \hline i & \gamma\gamma & \epsilon\epsilon & \epsilon\epsilon \\ d & \alpha\alpha & \epsilon\alpha & \epsilon\alpha \\ a & \alpha\alpha & \epsilon\alpha & \epsilon\alpha \end{array} \quad T_3 = \begin{array}{c|ccc} & i & d & a \\ \hline i & \gamma\gamma & \delta\epsilon & \epsilon\epsilon \\ d & \alpha\beta & \gamma\gamma & \epsilon\epsilon \\ a & \alpha\alpha & \alpha\alpha & \epsilon\alpha \end{array} \quad T_4 = \begin{array}{c|ccc} & i & d & a \\ \hline i & \gamma\gamma & \gamma\epsilon & \epsilon\epsilon \\ d & \alpha\gamma & \gamma\gamma & \epsilon\epsilon \\ a & \alpha\alpha & \alpha\alpha & \epsilon\alpha \end{array}$$

$$T_5 = \begin{array}{c|ccc} & i & d & a \\ \hline i & \gamma\gamma & \gamma\epsilon & \gamma\epsilon \\ d & \alpha\gamma & \gamma\gamma & \gamma\gamma \\ a & \alpha\gamma & \gamma\gamma & \gamma\gamma \end{array} \quad T_6 = \begin{array}{c|ccc} & i & d & a \\ \hline i & \gamma\gamma & \gamma\epsilon & \gamma\epsilon \\ d & \alpha\gamma & \gamma\gamma & \gamma\delta \\ a & \alpha\gamma & \beta\gamma & \gamma\gamma \end{array} \quad T_7 = \begin{array}{c|ccc} & i & d & a \\ \hline i & \gamma\gamma & \gamma\epsilon & \gamma\epsilon \\ d & \alpha\gamma & \gamma\gamma & \gamma\epsilon \\ a & \alpha\gamma & \alpha\gamma & \gamma\gamma \end{array}$$

Theorem 2.1. *For any t-pair (U, ψ) the relation $\mathrm{typ}_l(U, \psi) \in \{T_1, T_2, T_3, T_4, T_5, T_6, T_7\}$ holds. For any $i \in \{1, 2, 3, 4, 5, 6, 7\}$ there exists a t-pair (U, ψ) such that $\mathrm{typ}_l(U, \psi) = T_i$.*

Theorem 2.2. *For any limited t-pair (U, ψ) the relation $\mathrm{typ}_l(U, \psi) \in \{T_2, T_3, T_5, T_6, T_7\}$ holds. For any $i \in \{2, 3, 5, 6, 7\}$ there exists a limited t-pair (U, h) such that $\mathrm{typ}_l(U, h) = T_i$.*

Example 2.1. Let $n \in \omega \setminus \{0\}$, $U(n) = (\mathbb{R}^n, L_n)$, $L_n = \{r(\sum_{i=1}^{n} a_i x_i + a_{n+1}) : a_i \in \mathbb{R}, 1 \leq i \leq n+1\}$, $r : \mathbb{R} \to \{0, 1\}$ and $r(a) = 0$ iff $a < 0$. One can prove that $\mathrm{typ}_l(U(1), h) = T_3$ and $\mathrm{typ}_l(U(n), h) = T_7$ for any $n, n \geq 2$.

Seven similar examples with full proofs are included in Sect. 4 (Lemmas 4.1 - 4.7).

3   Possible Local Upper Types of T-Pairs

Let (U, ψ) be a t-pair. We denote by $\mathrm{typ}_{lu}(U, \psi)$ a table with three rows and three columns in which rows from top to bottom and columns from the left to the right are labeled by indices i, d, a. The value $\mathrm{typ}(\mathcal{U}_{U\psi}^{bc})$ is on the intersection of the row with index $b \in \{i, d, a\}$ and the column with index $c \in \{i, d, a\}$. The table $\mathrm{typ}_{lu}(U, \psi)$ will be called *the local upper type of t-pair* (U, ψ). In this section all possible local upper types of t-pairs are enumerated.

Now we define seven tables:

$$t_1 = \begin{array}{c|ccc} & i & d & a \\ \hline i & \alpha & \alpha & \alpha \\ d & \alpha & \alpha & \alpha \\ a & \alpha & \alpha & \alpha \end{array} \quad t_2 = \begin{array}{c|ccc} & i & d & a \\ \hline i & \gamma & \epsilon & \epsilon \\ d & \alpha & \alpha & \alpha \\ a & \alpha & \alpha & \alpha \end{array} \quad t_3 = \begin{array}{c|ccc} & i & d & a \\ \hline i & \gamma & \epsilon & \epsilon \\ d & \beta & \gamma & \epsilon \\ a & \alpha & \alpha & \alpha \end{array} \quad t_4 = \begin{array}{c|ccc} & i & d & a \\ \hline i & \gamma & \epsilon & \epsilon \\ d & \gamma & \gamma & \epsilon \\ a & \alpha & \alpha & \alpha \end{array}$$

$$t_5 = \begin{array}{c|ccc} & i & d & a \\ \hline i & \gamma & \epsilon & \epsilon \\ d & \gamma & \gamma & \gamma \\ a & \gamma & \gamma & \gamma \end{array} \quad t_6 = \begin{array}{c|ccc} & i & d & a \\ \hline i & \gamma & \epsilon & \epsilon \\ d & \gamma & \gamma & \delta \\ a & \gamma & \gamma & \gamma \end{array} \quad t_7 = \begin{array}{c|ccc} & i & d & a \\ \hline i & \gamma & \epsilon & \epsilon \\ d & \gamma & \gamma & \epsilon \\ a & \gamma & \gamma & \gamma \end{array}$$

Proposition 3.1. *For any t-pair (U, ψ) the relation $\mathrm{typ}_{lu}(U, \psi) \in \{t_1, t_2, t_3, t_4, t_5, t_6, t_7\}$ holds.*

Proposition 3.2. *For any limited t-pair (U, ψ) the relation $\mathrm{typ}_{lu}(U, \psi) \in \{t_2, t_3, t_5, t_6, t_7\}$ holds.*

We divide the proofs of the propositions into a sequence of lemmas.

Lemma 3.1. *Let (U, ψ) be a t-pair and $z \in \mathrm{Probl}(U)$. Then the inequalities $\psi_U^a(z) \leq \psi_U^d(z) \leq \psi_U^i(z)$ hold.*

Proof. Let $z = (\nu, f_1, \ldots, f_n)$. It is not difficult to construct a decision tree $\Gamma_0 \in \mathrm{Tree}(U)$ which solves the problem z deterministically by sequential computation values of the attributes $f_1, \ldots, f_n$. Evidently, $\psi(\Gamma_0) = \psi_U^i(z)$ and $\mathrm{At}(\Gamma_0) \subseteq \mathrm{At}(z)$. Therefore $\psi_U^d(z) \leq \psi_U^i(z)$. If a decision tree $\Gamma \in \mathrm{Tree}(U)$ solves the problem z deterministically then the decision tree Γ solves the problem z nondeterministically. Therefore $\psi_U^a(z) \leq \psi_U^d(z)$. $\qquad \square$

Let (U, ψ) be a t-pair, $n \in \omega$ and $b, c \in \{i, d, a\}$. The notation $\mathcal{U}_{U\psi}^{bc}(n) = \infty$ means that the set $\{\psi_U^b(z) : z \in \mathrm{Probl}(U), \psi_U^c(z) \leq n\}$ is infinite. Evidently, if $\mathcal{U}_{U\psi}^{bc}(n) = \infty$ then $\mathcal{U}_{U\psi}^{bc}(n + 1) = \infty$. It is not difficult to prove the following statement.

Lemma 3.2. *Let (U, ψ) be a t-pair and $b, c \in \{i, d, a\}$. Then*

a) if there exists $n \in \omega$ such that $\mathcal{U}_{U\psi}^{bc}(n) = \infty$ then $\mathrm{typ}(\mathcal{U}_{U\psi}^{bc}) = \epsilon$;

b) if there is no $n \in \omega$ such that $\mathcal{U}_{U\psi}^{bc}(n) = \infty$ then $\mathrm{Dom}(\mathcal{U}_{U\psi}^{bc}) = \{n : n \in \omega, n \geq n_0\}$ where $n_0 = \min\{\psi_U^c(z) : z \in \mathrm{Probl}(U)\}$.

Let (U, ψ) be a t-pair and $b, c, e, f \in \{i, d, a\}$. The notation $\mathcal{U}_{U\psi}^{bc} \triangleleft \mathcal{U}_{U\psi}^{ef}$ means that for any $n \in \omega$ the following statements hold:

a) if the value $\mathcal{U}_{U\psi}^{bc}(n)$ is definite then either $\mathcal{U}_{U\psi}^{ef}(n) = \infty$ or the value $\mathcal{U}_{U\psi}^{ef}(n)$ is definite and the inequality $\mathcal{U}_{U\psi}^{bc}(n) \leq \mathcal{U}_{U\psi}^{ef}(n)$ holds;

b) if $\mathcal{U}_{U\psi}^{bc}(n) = \infty$ then $\mathcal{U}_{U\psi}^{ef}(n) = \infty$.

Let $\preceq$ be a linear order on the set $\{\alpha, \beta, \gamma, \delta, \epsilon\}$ such that $\alpha \preceq \beta \preceq \gamma \preceq \delta \preceq \epsilon$.

Lemma 3.3. *Let (U, ψ) be a t-pair. Then $\mathrm{typ}(\mathcal{U}_{U\psi}^{bi}) \preceq \mathrm{typ}(\mathcal{U}_{U\psi}^{bd}) \preceq \mathrm{typ}(\mathcal{U}_{U\psi}^{ba})$ and $\mathrm{typ}(\mathcal{U}_{U\psi}^{ab}) \preceq \mathrm{typ}(\mathcal{U}_{U\psi}^{db}) \preceq \mathrm{typ}(\mathcal{U}_{U\psi}^{ib})$ for any $b \in \{i, d, a\}$.*

Proof. From the definition of the functions $\mathcal{U}_{U\psi}^{bc}$, $b, c \in \{i, d, a\}$, and from Lemma 3.1 it follows that $\mathcal{U}_{U\psi}^{bi} \triangleleft \mathcal{U}_{U\psi}^{bd} \triangleleft \mathcal{U}_{U\psi}^{ba}$ and $\mathcal{U}_{U\psi}^{ab} \triangleleft \mathcal{U}_{U\psi}^{db} \triangleleft \mathcal{U}_{U\psi}^{ib}$ for any $b \in \{i, d, a\}$. Using these relations and Lemma 3.2 we obtain the statement of the lemma. $\qquad \square$

Lemma 3.4. *Let (U, ψ) be a t-pair and $b, c \in \{i, d, a\}$. Then*

a) $\mathrm{typ}(\mathcal{U}_{U\psi}^{bc}) = \alpha$ iff the function ψ_U^b is bounded from above on the set $\mathrm{Probl}(U)$;

b) if the function ψ_U^b is unbounded from above on $\mathrm{Probl}(U)$ then $\mathrm{typ}(\mathcal{U}_{U\psi}^{bb}) = \gamma$.

Proof. The statement a) is obvious. Let the function ψ_U^b be unbounded from above on $\mathrm{Probl}(U)$. One can show that in this case the equality $\mathcal{U}_{U\psi}^{bb}(n) = n$ holds for infinitely many $n \in \omega$. Therefore $\mathrm{typ}(\mathcal{U}_{U\psi}^{bb}) = \gamma$.   □

Corollary 3.1. *Let (U, ψ) be a t-pair and $b \in \{i, d, a\}$. Then $\mathrm{typ}(\mathcal{U}_{U\psi}^{bb}) \in \{\alpha, \gamma\}$.*

Lemma 3.5. *Let (U, ψ) be a t-pair and $\mathrm{typ}(\mathcal{U}_{U\psi}^{ii}) \neq \alpha$. Then $\mathrm{typ}(\mathcal{U}_{U\psi}^{id}) = \mathrm{typ}(\mathcal{U}_{U\psi}^{ia}) = \epsilon$.*

Proof. Using Lemma 3.4 we conclude that the function ψ_U^i is unbounded from above on $\mathrm{Probl}(U)$. Let $m \in \omega$. Then there exists a problem $z = (\nu, f_1, \ldots, f_n) \in \mathrm{Probl}(U)$ such that $\psi_U^i(z) \geq m$. Let us consider the problem $z' = (\nu', f_1, \ldots, f_n)$ where $\nu' \equiv \{0\}$. It is clear that $\psi_U^i(z') \geq m$. Let Γ be a decision tree which consists of the root, the terminal node labeling by 0 and the edge connecting these two nodes. One can show that the tree Γ solves the problem z' deterministically. Therefore $\psi_U^a(z') \leq \psi_U^d(z') \leq \psi(\Gamma) = \psi(\lambda)$. Taking into account that m is an arbitrary number from ω we obtain $\mathcal{U}_{U\psi}^{id}(\psi(\lambda)) = \infty$ and $\mathcal{U}_{U\psi}^{ia}(\psi(\lambda)) = \infty$. Using Lemma 3.2 we conclude that $\mathrm{typ}(\mathcal{U}_{U\psi}^{id}) = \mathrm{typ}(\mathcal{U}_{U\psi}^{ia}) = \epsilon$.   □

Lemma 3.6. *Let (U, ψ) be a t-pair. Then $\mathrm{typ}(\mathcal{U}_{U\psi}^{ai}) \in \{\alpha, \gamma\}$.*

Proof. Let $U = (A, F)$ and $f : A \to E_k$ for any $f \in F$. Using Lemma 3.3 and Corollary 3.1 we obtain $\mathrm{typ}(\mathcal{U}_{U\psi}^{ai}) \in \{\alpha, \beta, \gamma\}$. Assume that $\mathrm{typ}(\mathcal{U}_{U\psi}^{ai}) = \beta$. Then there exists $m \in \omega \setminus \{0\}$ such that $\mathcal{U}_{U\psi}^{ai}(n) < n$ for any $n \in \omega$, $n > m$. Let us prove by induction on n that for any problem $z \in \mathrm{Probl}(U)$ if $\psi_U^i(z) \leq n$ then $\psi_U^a(z) \leq m_0$, where $m_0 = \max\{m, \psi(\lambda)\}$. Using Lemma 3.1 we conclude that under the condition $n \leq m$ the considered statement holds. Let it hold for some n, $n \geq m$. Let us show that this statement holds for $n + 1$ too. Let $z \in \mathrm{Probl}(U)$ and $\psi_U^i(z) \leq n + 1$. Since $n + 1 > m$, we obtain $\psi_U^a(z) \leq n$. Let $\Gamma \in \mathrm{Tree}(U)$, $\mathrm{At}(\Gamma) \subseteq \mathrm{At}(z)$, $\psi(\Gamma) = \psi_U^a(z)$ and Γ solve the problem z nondeterministically. Assume that in Γ there exists a complete path ξ in which there are no working nodes. In this case a decision tree, which consists of the root, the terminal node labeling by $\tau(\xi)$ and the edge connecting these two nodes, solves the problem z nondeterministically. Therefore $\psi_U^a(z) \leq \psi(\lambda) \leq m_0$. Assume now that each complete path in the decision tree Γ contains a working node. Let $\xi \in \mathrm{Path}(\Gamma)$, $\xi = v_0, d_0, \ldots, v_p, d_p, v_{p+1}$ and for $i = 1, \ldots, p$ the node v_i be labeled by the attribute f_i, and the edge d_i be labeled by the number δ_i. Let us consider the problem $z_\xi = (\nu_\xi, f_1, \ldots, f_p)$ where $\nu_\xi(\delta_1, \ldots, \delta_p) = \{\tau(\xi)\}$ and $\nu_\xi(\bar{\sigma}) = \{\tau(\xi) + 1\}$ for any p-tuple $\bar{\sigma} \in E_k^p$ such that $\bar{\sigma} \neq (\delta_1, \ldots, \delta_p)$. It is clear that $\psi_U^i(z_\xi) \leq n$. Using the inductive hypothesis we conclude that there exists a decision tree $\Gamma_\xi \in \mathrm{Tree}(U)$ which has the following properties: Γ_ξ solves the problem z_ξ nondeterministically, $\mathrm{At}(\Gamma_\xi) \subseteq \mathrm{At}(z_\xi)$ and $\psi(\Gamma_\xi) \leq m_0$. Let $\mathcal{A}(\xi) \neq \emptyset$. We denote by $\tilde{\Gamma}_\xi$ a tree obtained from Γ_ξ by removal all nodes and edges which satisfy the following condition: there is no a complete path ξ' in Γ_ξ which contains this node or edge and for which $\tau(\xi') = \tau(\xi)$. Let $\{\xi : \xi \in \mathrm{Path}(\Gamma), \mathcal{A}(\xi) \neq \emptyset\} = \{\xi_1, \ldots, \xi_r\}$. Let us identify the roots of the trees $\tilde{\Gamma}_{\xi_1}, \dots, \tilde{\Gamma}_{\xi_r}$. We denote by G the obtained tree. It is not difficult to show that $G \in \mathrm{Tree}(U)$, $\mathrm{At}(G) \subseteq \mathrm{At}(z)$, $\psi(G) \le m_0$ and the decision tree G solves the problem z nondeterministically. Thus, the considered statement holds. Using Lemma 3.4 we conclude that $\mathrm{typ}(\mathcal{U}_{U\psi}^{ai}) = \alpha$. The obtained contradiction shows that $\mathrm{typ}(\mathcal{U}_{U\psi}^{ai}) \in \{\alpha, \gamma\}$. $\qquad\square$

Lemma 3.7. *Let (U, ψ) be a limited t-pair and $\mathrm{typ}(\mathcal{U}_{U\psi}^{ai}) = \alpha$. Then $\mathrm{typ}(\mathcal{U}_{U\psi}^{di}) \in \{\alpha, \beta\}$.*

Proof. Let $U = (A, F)$ and $f : A \to E_k$ for any $f \in F$. Using Lemma 3.4 we conclude that there exists $r \in \omega$ such that the inequality $\psi_U^a(z) \le r$ holds for any problem $z \in \mathrm{Probl}(U)$. Let $f \in F$ and $f \not\equiv const$. Let us consider a problem $z_f = (\nu, f)$ where $\nu(\delta) = \{\delta\}$ for any $\delta \in E_k$. Let $\Gamma \in \mathrm{Tree}(U)$, $\mathrm{At}(\Gamma) \subseteq \mathrm{At}(z_f)$, $\psi(\Gamma) = \psi_U^a(z_f)$ and Γ solve the problem z_f nondeterministically. Since $f \not\equiv const$, we have $f \in \mathrm{At}(\Gamma)$. Using the property (b) of the complexity measure ψ we obtain $\psi(\Gamma) \ge \psi(f)$. Consequently, $\psi(f) \le r$.

Let $f_1, \dots, f_n \in F$, $\delta_1, \dots, \delta_n \in E_k$ and the system of equations

$$\{f_1(x) = \delta_1, \dots, f_n(x) = \delta_n\} \tag{1}$$

be compatible on A. Let us consider a problem $z = (\nu, f_1, \dots, f_n)$ where $\nu : E_k^n \to \{\{0\}, \{1\}\}$ and $\nu(\bar{\sigma}) = \{1\}$ iff $\bar{\sigma} = (\delta_1, \dots, \delta_n)$. Taking into account that $\psi_U^a(z) \le r$ and the complexity measure ψ has the property (c), it is not difficult to show that there exists a subsystem of the system (1) which contains at most r equations and has the same set of solutions just as the system (1).

Let the system (1) be incompatible on A. Let us show that there exists an incompatible subsystem of the system (1) which contains at most $r+1$ equations. If the system $\{f_1(x) = \delta_1\}$ is incompatible on A then the considered statement holds. Otherwise there exists $i \in \{1, \dots, n-1\}$ such that the system

$$\{f_1(x) = \delta_1, \dots, f_i(x) = \delta_i\} \tag{2}$$

is compatible on A and the system, which is obtained from the system (2) by adding the equation $f_{i+1}(x) = \delta_{i+1}$, is incompatible on A. According to proved above, there exists a subsystem S of the system (2) which contains at most r equations and has the same set of solutions just as the system (2). By adding the equation $f_{i+1}(x) = \delta_{i+1}$ to the system S we obtain a subsystem of the system (1) which is incompatible on A and contains at most $r + 1$ equations.

Let $z_1 \in \mathrm{Probl}(U)$. It is clear that there exists a problem $z_2 \in \mathrm{Probl}(U)$ which has the following properties: $z_1(a) = z_2(a)$ for any $a \in A$, $\mathrm{At}(z_2) \subseteq \mathrm{At}(z_1)$, and $f \not\equiv const$ for any $f \in \mathrm{At}(z_2)$. Let $z_2 = (\nu, f_1, \dots, f_n)$. Now we consider the problem $z_3 = (\nu', f_1, \dots, f_n)$ from $\mathrm{Probl}(U)$ where ν' satisfies the following conditions: $|\nu'(\bar{\delta})| = 1$ for any $\bar{\delta} \in E_k^n$, and $\nu'(\bar{\delta}_1) \ne \nu'(\bar{\delta}_2)$ for any $\bar{\delta}_1, \bar{\delta}_2 \in E_k^n$, $\bar{\delta}_1 \ne \bar{\delta}_2$. It is not difficult to show that $\psi_U^d(z_1) \le \psi_U^d(z_2) \le \psi_U^d(z_3)$. For $\bar{\delta} = (\delta_1, \dots, \delta_n) \in E_k^n$ let us denote by $A(\bar{\delta})$ the set of solutions on A for the following system of equations:

$$S(\bar{\delta}) = \{f_1(x) = \delta_1, \dots, f_n(x) = \delta_n\} \ .$$

Let $\Delta(z_3) = \{\bar{\delta} : \bar{\delta} \in E_k^n, A(\bar{\delta}) \neq \emptyset\}$ and $N = |\Delta(z_3)|$. Taking into account that any compatible on A system $S(\bar{\delta})$ has a subsystem with at most r equations and with the same set of solutions just as the system $S(\bar{\delta})$, it is not difficult to show that

$$N \leq (n+1)^r \cdot k^r . \tag{3}$$

For any $i \in \{1, \ldots, n\}$ and $\delta \in E_k$ we denote by $N(i, \delta)$ the number of n-tuples from $\Delta(z_3)$ in which i-th digit is equal to δ.

It is clear that $A = \bigcup_{\bar{\delta} \in \Delta(z_3)} A(\bar{\delta})$, and for any $a_1, a_2 \in A$ the relation $z_3(a_1) \cap z_3(a_2) \neq \emptyset$ holds iff $a_1 \in A(\bar{\delta})$ and $a_2 \in A(\bar{\delta})$ for some $\bar{\delta} \in \Delta(z_3)$. Thus, when we begin to solve the problem z_3 for an element $a \in A$, we know only that the element a belongs to one of the sets $A(\bar{\delta})$, $\bar{\delta} \in \Delta(z_3)$. When the problem z_3 will be solved, we will be able to point to the concrete n-tuple $\bar{\delta} \in \Delta(z_3)$ such that $a \in A(\bar{\delta})$. If the value of the attribute f_i is computed and $f_i(a) = \delta$ then the number of sets $A(\bar{\delta})$, in which the element a can be contained, reduces from N to $N(i, \delta)$. Let $N \geq 2$. Let us show that we can choose at most $r + 1$ attributes from $\{f_1, \ldots, f_n\}$ which have the following property: the computation of these attributes, independently of obtained values, leads to reduction of the set number in twice. Let $i \in \{1, \ldots, n\}$ and σ_i be a number from E_k such that $N(i, \sigma_i) = \max\{N(i, \delta) : \delta \in E_k\}$. Evidently,

$$N(i, \delta) \leq \frac{N}{2} \tag{4}$$

for any $\delta \in E_k, \delta \neq \sigma_i$. Let $\bar{\sigma} = (\sigma_1, \ldots, \sigma_n)$. Now we consider the system of equations $S(\bar{\sigma})$. According to proved above, there exists a subsystem $\{f_{i_1}(x) = \sigma_{i_1}, \ldots, f_{i_m}(x) = \sigma_{i_m}\}$ of the system $S(\bar{\sigma})$ which has the same set of solutions just as the system $S(\bar{\sigma})$ and for which $m \leq r + 1$. Let we compute values of the attributes $f_{i_1}, \ldots, f_{i_m}$ on element $a \in A$, and let $f_{i_1}(a) = t_1, \ldots, f_{i_m}(a) = t_m$. We denote by Δ' the set $\{(\delta_1, \ldots, \delta_n) \in \Delta(z_3) : \delta_{i_1} = t_1, \ldots, \delta_{i_m} = t_m\}$. It is clear that $a \in \bigcup_{\bar{\delta} \in \Delta'} A(\bar{\delta})$. If the equality $t_j = \sigma_{i_j}$ holds for any $j \in \{1, \ldots, m\}$ then $\Delta' = \{\bar{\sigma}\}$. Let for some $j \in \{1, \ldots, m\}$ the relation $t_j \neq \sigma_{i_j}$ hold. Using (4) we obtain $|\Delta'| \leq \frac{N}{2}$. Thus, after the computation values of $m \leq r + 1$ attributes the number of the sets $A(\bar{\delta})$, in which the element a can be contained, reduced in twice. If $|\Delta'| = 1$ then the problem z_3 is solved for the element a. If $|\Delta'| > 1$ then we in the same way look for proper attributes for the set Δ', etc. Using the inequality (3) we conclude that there exists a decision tree Γ which solves the problem z_3 deterministically and for which $\mathrm{At}(\Gamma) \subseteq \mathrm{At}(z_3)$ and $h(\Gamma) \leq (r+1)\lceil \log_2 N \rceil \leq (r+1)^2 \log_2 k(n+1)$. Taking into account that $\psi(f_i) \leq r$ for any attribute $f_i \in \mathrm{At}(z_3)$ and the complexity measure ψ has the property (a), we obtain

$$\psi_U^d(z_3) \leq r(r+1)^2 \log_2 k(n+1) .$$

Consequently, $\psi_U^d(z_1) \leq r(r+1)^2 \log_2 k(n+1)$. Taking into account that the complexity measure ψ has the property (c), we obtain $\psi_U^i(z_1) \geq n$. Since z_1 is an arbitrary problem over U, we have $\mathrm{Dom}^+(\mathcal{U}_{U\psi}^{di})$ is a finite set. Therefore $\mathrm{typ}(\mathcal{U}_{U\psi}^{di}) \neq \gamma$. Using Lemma 3.3 and Corollary 3.1 we obtain $\mathrm{typ}(\mathcal{U}_{U\psi}^{di}) \in \{\alpha, \beta\}$. $\square$

Proof (of Proposition 3.1). Let (U, ψ) be a t-pair. Using Corollary 3.1 we conclude that $\text{typ}(\mathcal{U}_{U\psi}^{ii}) \in \{\alpha, \gamma\}$. Using Corollary 3.1 and Lemma 3.3 we obtain $\text{typ}(\mathcal{U}_{U\psi}^{di}) \in \{\alpha, \beta, \gamma\}$. From Lemma 3.6 it follows that $\text{typ}(\mathcal{U}_{U\psi}^{ai}) \in \{\alpha, \gamma\}$.

a) Let $\text{typ}(\mathcal{U}_{U\psi}^{ii}) = \alpha$. Using Lemmas 3.3 and 3.4 we obtain $\text{typ}_{lu}(U, \psi) = t_1$.

b) Let $\text{typ}(\mathcal{U}_{U\psi}^{ii}) = \gamma$ and $\text{typ}(\mathcal{U}_{U\psi}^{di}) = \alpha$. Using Lemmas 3.3, 3.4 and 3.5 we obtain $\text{typ}_{lu}(U, \psi) = t_2$.

c) Let $\text{typ}(\mathcal{U}_{U\psi}^{ii}) = \gamma$ and $\text{typ}(\mathcal{U}_{U\psi}^{di}) = \beta$. From Lemma 3.5 it follows that $\text{typ}(\mathcal{U}_{U\psi}^{id}) = \text{typ}(\mathcal{U}_{U\psi}^{ia}) = \epsilon$. Using Lemmas 3.3 and 3.6 we obtain $\text{typ}(\mathcal{U}_{U\psi}^{ai}) = \alpha$. From this equality and from Lemma 3.4 it follows that $\text{typ}(\mathcal{U}_{U\psi}^{ad}) = \text{typ}(\mathcal{U}_{U\psi}^{aa}) = \alpha$. Using the equality $\text{typ}(\mathcal{U}_{U\psi}^{di}) = \beta$, Lemma 3.3 and Corollary 3.1 we obtain $\text{typ}(\mathcal{U}_{U\psi}^{dd}) = \gamma$. From the equalities $\text{typ}(\mathcal{U}_{U\psi}^{dd}) = \gamma$, $\text{typ}(\mathcal{U}_{U\psi}^{aa}) = \alpha$ and from Lemmas 3.2 and 3.4 it follows that $\text{typ}(\mathcal{U}_{U\psi}^{da}) = \epsilon$. Thus, $\text{typ}_{lu}(U, \psi) = t_3$.

d) Let $\text{typ}(\mathcal{U}_{U\psi}^{ii}) = \text{typ}(\mathcal{U}_{U\psi}^{di}) = \gamma$ and $\text{typ}(\mathcal{U}_{U\psi}^{ai}) = \alpha$. Using Lemma 3.5 we obtain $\text{typ}(\mathcal{U}_{U\psi}^{id}) = \text{typ}(\mathcal{U}_{U\psi}^{ia}) = \epsilon$. From Lemma 3.4 it follows that $\text{typ}(\mathcal{U}_{U\psi}^{ad}) = \text{typ}(\mathcal{U}_{U\psi}^{aa}) = \alpha$. Using Lemma 3.3 and Corollary 3.1 we obtain $\text{typ}(\mathcal{U}_{U\psi}^{dd}) = \gamma$. From this equality, equality $\text{typ}(\mathcal{U}_{U\psi}^{aa}) = \alpha$ and from Lemmas 3.2 and 3.4 it follows that $\text{typ}(\mathcal{U}_{U\psi}^{da}) = \epsilon$. Thus, $\text{typ}_{lu}(U, \psi) = t_4$.

e) Let $\text{typ}(\mathcal{U}_{U\psi}^{ii}) = \text{typ}(\mathcal{U}_{U\psi}^{di}) = \text{typ}(\mathcal{U}_{U\psi}^{ai}) = \gamma$. Using Lemma 3.5 we conclude that $\text{typ}(\mathcal{U}_{U\psi}^{id}) = \text{typ}(\mathcal{U}_{U\psi}^{ia}) = \epsilon$. Using Lemma 3.3 and Corollary 3.1 we obtain $\text{typ}(\mathcal{U}_{U\psi}^{dd}) = \text{typ}(\mathcal{U}_{U\psi}^{ad}) = \text{typ}(\mathcal{U}_{U\psi}^{aa}) = \gamma$. Using Lemma 3.3 we obtain $\text{typ}(\mathcal{U}_{U\psi}^{da}) \in \{\gamma, \delta, \epsilon\}$. Therefore $\text{typ}_{lu}(U, \psi) \in \{t_5, t_6, t_7\}$. □

Proof (of Proposition 3.2). Let (U, ψ) be a limited t-pair. Taking into account that the complexity measure ψ has the property (c) and using Lemma 3.4 we obtain $\text{typ}(\mathcal{U}_{U\psi}^{ii}) \neq \alpha$. Therefore $\text{typ}_{lu}(U, \psi) \neq t_1$. Using Lemma 3.7 we obtain $\text{typ}_{lu}(U, \psi) \neq t_4$. From these relations and Proposition 3.1 it follows that the statement of the proposition holds. □

4   Realizable Local Upper Types of T-Pairs

In this section all realizable local upper types of t-pairs are enumerated.

Proposition 4.1. *For any $i \in \{1, 2, 3, 4, 5, 6, 7\}$ there exists a t-pair (U, ψ) such that*

$$\text{typ}_{lu}(U, \psi) = t_i \ .$$

Proposition 4.2. *For any $i \in \{2, 3, 5, 6, 7\}$ there exists a limited t-pair (U, h) such that*

$$\text{typ}_{lu}(U, h) = t_i \ .$$

We divide the proofs of the propositions into a sequence of lemmas.

Let us define a t-pair (U_1, π) as follows: $U_1 = (\omega, F_1)$ where $F_1 = \{f\}$ and $f \equiv 0$, and $\pi \equiv 0$.

Lemma 4.1. $\text{typ}_{lu}(U_1, \pi) = t_1$.

Proof. Using Lemma 3.4 we conclude that $\text{typ}(\mathcal{U}_{U_1\pi}^{ii}) = \alpha$. From this equality and from Proposition 3.1 it follows that $\text{typ}_{lu}(U_1, \pi) = t_1$. $\qquad\square$

Let us define a t-pair (U_2, h) as follows: $U_2 = (\omega, F_2)$ where $F_2 = F_1$.

Lemma 4.2. $\text{typ}_{lu}(U_2, h) = t_2$.

Proof. It is not difficult to show that $h_{U_2}^d(z) = 0$ for any problem $z \in \text{Probl}(U_2)$. It is clear that the function $h_{U_2}^i$ is unbounded from above on $\text{Probl}(U_2)$. Using Lemma 3.4 we conclude that $\text{typ}(\mathcal{U}_{U_2 h}^{ii}) = \gamma$ and $\text{typ}(\mathcal{U}_{U_2 h}^{dd}) = \alpha$. From these equalities and from Proposition 3.1 it follows that $\text{typ}_{lu}(U_2, h) = t_2$. $\qquad\square$

Let us define a t-pair (U_3, h) as follows: $U_3 = (\omega, F_3)$ where $F_3 = \{l_i : i \in \omega \setminus \{0\}\}$ and for any $i \in \omega \setminus \{0\}$, $j \in \omega$ if $j \leq i$ then $l_i(j) = 0$, and if $j > i$ then $l_i(j) = 1$.

Lemma 4.3. $\text{typ}_{lu}(U_3, h) = t_3$.

Proof. It is not difficult to show that for any compatible on ω system of equations

$$\{l_{i_1}(x) = \delta_1, \ldots, l_{i_m}(x) = \delta_m\} ,$$

where $l_{i_1}, \ldots, l_{i_m} \in F_3$ and $\delta_1, \ldots, \delta_m \in E_2$, there exists a subsystem which has the same set of solutions and which contains at most two equations. Using this fact it is not difficult to show that $h_{U_3}^a(z) \leq 2$ for any problem $z \in \text{Probl}(U_3)$. From here and from Lemma 3.4 it follows that $\text{typ}(\mathcal{U}_{U_3 h}^{ai}) = \alpha$. It is clear that (U_3, h) is a limited t-pair. Using Lemma 3.7 we conclude that $\text{typ}(\mathcal{U}_{U_3 h}^{di}) \in \{\alpha, \beta\}$. Let us show that $\text{typ}(\mathcal{U}_{U_3 h}^{di}) = \beta$. Assume the contrary: $\text{typ}(\mathcal{U}_{U_3 h}^{di}) = \alpha$. Using Lemma 3.4 we conclude that there exists a number $m \in \omega$ such that $h_{U_3}^d(z) \leq m$ for any problem $z \in \text{Probl}(U_3)$. Therefore there exists a number $n \in \omega$ that satisfies the following condition: for any problem $z \in \text{Probl}(U_3)$ there exists a decision tree $\Gamma \in \text{Tree}(U_3)$ which solves the problem z deterministically and for which $|\text{At}(\Gamma)| \leq n$ and $\text{At}(\Gamma) \subseteq \text{At}(z)$. Let us consider a problem $z' = (\nu, l_1, \ldots, l_{n+1})$ from $\text{Probl}(U_3)$ where $\nu : E_2^{n+1} \to \mathcal{P}(\omega)$ and $\nu(\bar{\delta}_1) \cap \nu(\bar{\delta}_2) = \emptyset$ for any $\bar{\delta}_1, \bar{\delta}_2 \in E_2^{n+1}$ such that $\bar{\delta}_1 \neq \bar{\delta}_2$. Let Γ be an arbitrary decision tree over U_3 which solves the problem z' deterministically and for which $\text{At}(\Gamma) \subseteq \text{At}(z')$. It is not difficult to show that $\text{At}(\Gamma) = \text{At}(z')$ and, hence, $|\text{At}(\Gamma)| \geq n + 1$. We obtained a contradiction. Therefore $\text{typ}(\mathcal{U}_{U_3 h}^{di}) = \beta$. From this equality and from Proposition 3.1 it follows that $\text{typ}_{lu}(U_3, h) = t_3$. $\qquad\square$

Let us define a t-pair (U_4, μ) as follows: $U_4 = (\omega, F_4)$ where $F_4 = F_3$, $\mu(\lambda) = 0$, $\mu(l_{i_1} \ldots l_{i_m}) = 1$ if $m = 1$ or $m = 2$ and $i_1 > i_2$, $\mu(l_{i_1} \ldots l_{i_m}) = \max\{i_1, \ldots, i_m\}$ in other cases.

Lemma 4.4. $\mathrm{typ}_{lu}(U_4, \mu) = t_4$.

Proof. By analogy with the proof of Lemma 4.3 one can show that $\mathrm{typ}(\mathcal{U}_{U_4\mu}^{ai}) = \alpha$.

Let us prove that $\mathrm{typ}(\mathcal{U}_{U_4\mu}^{di}) = \gamma$. Let $n \in \omega$ and $n \geq 4$. Now we consider a problem $z = (\nu, l_1, \ldots, l_n)$ from $\mathrm{Probl}(U_4)$ where $\nu : E_2^n \rightarrow \mathcal{P}(\omega)$ and $\nu(\bar{\delta}_1) \cap \nu(\bar{\delta}_2) = \emptyset$ for any $\bar{\delta}_1, \bar{\delta}_2 \in E_2^n$ such that $\bar{\delta}_1 \neq \bar{\delta}_2$. It is clear that $\mu_{U_4}^i(z) = n$. Let Γ be a decision tree over U_4 which solves the problem z deterministically and for which $\mathrm{At}(\Gamma) \subseteq \mathrm{At}(z)$ and $\mu(\Gamma) = \mu_{U_4}^d(z)$. It is not difficult to prove that $\mathrm{At}(\Gamma) = \mathrm{At}(z)$. Therefore $|\mathrm{At}(\Gamma)| = n \geq 4$. Using Lemma 3.1 we conclude that $\mu(\Gamma) \leq n$. Let us show that $\mu(\Gamma) = n$. Assume the contrary: $\mu(\Gamma) < n$. Since $l_n \in \mathrm{At}(\Gamma)$, there exists a complete path ξ in the decision tree Γ such that the letter l_n is in the word $\varphi(\xi)$. Since $\mu(\varphi(\xi)) < n$, we have $\varphi(\xi) = l_n$ or $\varphi(\xi) = l_n l_i$ where $i < n$. Let an edge d issue from root of the tree Γ. Then the node, which d enters, is labeled by the attribute l_n. Taking into account that $\mu(\Gamma) < n$ and using the properties of the complexity measure μ we obtain $h(\Gamma) \leq 2$. Consequently, $|\mathrm{At}(\Gamma)| \leq 3$. We obtained a contradiction. Therefore $\mu(\Gamma) = n$ and $\mu_{U_4}^d(z) = n$. Thus, $\mathcal{U}_{U_4\mu}^{di}(n) = n$. Taking into account that n is an arbitrary number from ω for which $n \geq 4$, we obtain $\mathrm{typ}(\mathcal{U}_{U_4\mu}^{di}) = \gamma$. From this equality, equality $\mathrm{typ}(\mathcal{U}_{U_4\mu}^{ai}) = \alpha$ and Proposition 3.1 it follows that $\mathrm{typ}_{lu}(U_4, \mu) = t_4$. □

Let us define a t-pair (U_5, h) as follows: $U_5 = (\omega, F_5)$ where $F_5 = \{f_i : i \in \omega \setminus \{0\}\}$ and for any $i \in \omega \setminus \{0\}$, $j \in \omega$ if $i = j$ then $f_i(j) = 1$, and if $i \neq j$ then $f_i(j) = 0$.

Lemma 4.5. $\mathrm{typ}_{lu}(U_5, h) = t_5$.

Proof. Let $z \in \mathrm{Probl}(U_5)$. We will show that $h_{U_5}^d(z) = h_{U_5}^a(z)$. Let Γ be a decision tree over U_5 which solves the problem z nondeterministically and for which $\mathrm{At}(\Gamma) \subseteq \mathrm{At}(z)$ and $h(\Gamma) = h_{U_5}^a(z)$. Let ξ be a complete path in the tree Γ such that $0 \in \mathcal{A}(\xi)$. If in the complete path ξ there are no working nodes then, as it is not difficult to show, $h_{U_5}^d(z) = h_{U_5}^a(z) = 0$. Let there be $m > 0$ working nodes in the path ξ and $\mathcal{S}(\xi) = \{f_{i_1}(x) = \delta_1, \ldots, f_{i_m}(x) = \delta_m\}$. Since $0 \in \mathcal{A}(\xi)$, we have $\delta_1 = \ldots = \delta_m = 0$. It is clear that for any $p \in \mathcal{A}(\xi)$ the relation $\tau(\xi) \in z(p)$ holds.

Let us describe a decision tree Γ_1 over U_5 which solves the problem z deterministically and for which $\mathrm{At}(\Gamma_1) \subseteq \mathrm{At}(z)$ and $h(\Gamma_1) \leq m$. For an arbitrary $p \in \omega$ the decision tree Γ_1 computes the values $f_{i_1}(p), \ldots, f_{i_m}(p)$. If $f_{i_1}(p) = \ldots = f_{i_m}(p) = 0$ then $p \in \mathcal{A}(\xi)$ and, hence, the problem z is solved since we know that $\tau(\xi) \in z(p)$. Let for some $j \in \{1, \ldots, m\}$ the equality $f_{i_j}(p) = 1$ hold. Then $p = i_j$ and, hence, the problem z is solved too since we know all the set $z(p)$.

It is clear that $h(\Gamma) \geq m$. Therefore $h(\Gamma_1) \leq h_{U_5}^a(z)$ and $h_{U_5}^a(z) \leq h_{U_5}^a(z)$. Using Lemma 3.1 we conclude that $h_{U_5}^d(z) \geq h_{U_5}^a(z)$ and, hence, $h_{U_5}^d(z) = h_{U_5}^a(z)$. From this equality it follows that there is no $n \in \omega$ for which $\mathcal{U}_{U_5h}^{da}(n) = \infty$. Using Lemma 3.2 we conclude that $\mathrm{Dom}(\mathcal{U}_{U_5h}^{da}) = \{n : n \in \omega, n \geq n_0\}$ for some $n_0 \in \omega$ and $\mathcal{U}_{U_5h}^{da}(n) \leq n$ for any $n \in \mathrm{Dom}(\mathcal{U}_{U_5h}^{da})$.

Let $n \in \mathrm{Dom}(\mathcal{U}^{da}_{U_5 h})$ and $n \geq 1$. We will show that $\mathcal{U}^{da}_{U_5 h}(n) = n$. Let $z = (\nu, f_1, \ldots, f_n)$ be a problem from $\mathrm{Probl}(U_5)$ for which $\nu(\bar{\delta}_1) \cap \nu(\bar{\delta}_2) = \emptyset$ for any $\bar{\delta}_1, \bar{\delta}_2 \in E_2^n$ such that $\bar{\delta}_1 \neq \bar{\delta}_2$. Let Γ be a decision tree over U_5 which solves the problem z nondeterministically and for which $\mathrm{At}(\Gamma) \subseteq \mathrm{At}(z)$ and $h(\Gamma) = h^a_{U_5}(z)$. Let ξ be a complete path in the tree Γ such that $0 \in \mathcal{A}(\xi)$. It is clear that in the complete path ξ there is at least one working node. Let $\mathcal{S}(\xi) = \{f_{i_1}(x) = \delta_1, \ldots, f_{i_m}(x) = \delta_m\}$. Since $0 \in \mathcal{A}(\xi)$, we have $\delta_1 = \ldots = \delta_m = 0$. It is clear that $i \notin \mathcal{A}(\xi)$ for any $i \in \{1, \ldots, n\}$. Therefore for any $i \in \{1, \ldots, n\}$ the equation $f_i(x) = 0$ is contained in the system $\mathcal{S}(\xi)$. Consequently, $h(\Gamma) \geq n$ and $h^a_{U_5}(z) \geq n$. It is clear that $h^i_{U_5}(z) = n$. Using Lemma 3.1 we obtain $h^a_{U_5}(z) = n$ and $h^d_{U_5}(z) = n$. Therefore $\mathcal{U}^{da}_{U_5 h}(n) \geq n$ and, as proved above, $\mathcal{U}^{da}_{U_5 h}(n) = n$. Taking into account that n is an arbitrary number from ω such that $n \geq \max(n_0, 1)$ we conclude that $\mathrm{Dom}^-(\mathcal{U}^{da}_{U_5 h})$ and $\mathrm{Dom}^+(\mathcal{U}^{da}_{U_5 h})$ are infinite sets and $\mathrm{typ}(\mathcal{U}^{da}_{U_5 h}) = \gamma$. Using Proposition 3.1 we obtain $\mathrm{typ}_{lu}(U_5, h) = t_5$.   □

Let us define a t-pair (U_6, h) as follows: $U_6 = (\omega, F_6)$ where $F_6 = F_5 \cup G_6$, $G_6 = \{g_{2i+1} : i \in \omega\}$ and for any $i \in \omega$, $j \in \omega$ if $j \in \{2i+1, 2i+2\}$ then $g_{2i+1}(j) = 1$, and if $j \notin \{2i+1, 2i+2\}$ then $g_{2i+1}(j) = 0$.

Lemma 4.6. $\mathrm{typ}_{lu}(U_6, h) = t_6$.

Proof. Let $z \in \mathrm{Probl}(U_6)$. We will show that $h^d_{U_6}(z) \leq h^a_{U_6}(z) + 1$. Let Γ be a decision tree over U_6 which solves the problem z nondeterministically and for which $\mathrm{At}(\Gamma) \subseteq \mathrm{At}(z)$ and $h(\Gamma) = h^a_{U_6}(z)$. Let ξ be a complete path in the tree Γ such that $0 \in \mathcal{A}(\xi)$. If in the complete path ξ there are no working nodes then, as it is not difficult to show, $h^d_{U_6}(z) = h^a_{U_6}(z) = 0$. Let there be $m > 0$ working nodes in the path ξ and $\mathcal{S}(\xi) = \{q_1(x) = \delta_1, \ldots, q_m(x) = \delta_m\}$ where $q_i \in F_6, 1 \leq i \leq m$. Since $0 \in \mathcal{A}(\xi)$, we have $\delta_1 = \ldots = \delta_m = 0$. It is clear that for any $p \in \mathcal{A}(\xi)$ the relation $\tau(\xi) \in z(p)$ holds.

Let us describe a decision tree Γ_1 over U_6 which solves the problem z deterministically and for which $\mathrm{At}(\Gamma_1) \subseteq \mathrm{At}(z)$ and $h(\Gamma_1) \leq m + 1$. For an arbitrary $p \in \omega$ the decision tree Γ_1 computes the values $q_1(p), \ldots, q_m(p)$. If $q_1(p) = \ldots = q_m(p) = 0$ then $p \in \mathcal{A}(\xi)$ and, hence, the problem z is solved since we know that $\tau(\xi) \in z(p)$. Let for some $j \in \{1, \ldots, m\}$ the equality $q_j(p) = 1$ hold. If $q_j = f_i$ then $p = i$ and, hence, the problem z is solved too since we know all the set $z(p)$. Let $q_j = g_{2i+1}$. Then $p \in \{2i+1, 2i+2\}$. Let there be no attributes f_{2i+1} and f_{2i+2} in the set $\mathrm{At}(z)$. One can show that in this case $z(2i+1) \cap z(2i+2) \neq \emptyset$ and, hence, the problem z is solved. Let at least one from the attributes f_{2i+1} and f_{2i+2} be in the set $\mathrm{At}(z)$. For example, let $f_{2i+1} \in \mathrm{At}(z)$. Then we compute the value $f_{2i+1}(p)$. If $f_{2i+1}(p) = 1$ then $p = 2i+1$, and if $f_{2i+1}(p) = 0$ then $p = 2i+2$. In these cases the problem z is solved too.

It is clear that $h(\Gamma) \geq m$. Therefore $h(\Gamma_1) \leq h^a_{U_6}(z) + 1$ and $h^d_{U_6}(z) \leq h^a_{U_6}(z) + 1$. From this inequality it follows that there is no $n \in \omega$ for which $\mathcal{U}^{da}_{U_6 h}(n) = \infty$. Using Lemma 3.2 we conclude that $\mathrm{Dom}(\mathcal{U}^{da}_{U_6 h}) = \{n : n \in \omega, n \geq n_0\}$ for some $n_0 \in \omega$.

Let $n \in \mathrm{Dom}(\mathcal{U}_{U_6 h}^{da})$ and $n \geq 1$. We will show that $\mathcal{U}_{U_6 h}^{da}(n) \geq n + 1$. Let $z = (\nu, g_1, f_1, f_2, \ldots, g_{2n-1}, f_{2n-1}, f_{2n})$ be a problem from $\mathrm{Probl}(U_6)$ such that $\nu(\bar{\delta}_1) \cap \nu(\bar{\delta}_2) = \emptyset$ for any $\bar{\delta}_1, \bar{\delta}_2 \in E_2^{3n}$, $\bar{\delta}_1 \neq \bar{\delta}_2$. It is not difficult to show that $h_{U_6}^a(z) \leq n$. Let us prove that $h_{U_6}^d(z) \geq n + 1$. Let Γ be a decision tree over U_6 which solves the problem z deterministically and for which $\mathrm{At}(\Gamma) \subseteq \mathrm{At}(z)$ and $h(\Gamma) = h_{U_6}^d(z)$. Let ξ be a complete path in the tree Γ such that $0 \in \mathcal{A}(\xi)$. It is clear that in the complete path ξ there is at least one working node. Let $\mathcal{S}(\xi) = \{q_1(x) = \delta_1, \ldots, q_m(x) = \delta_m\}$ where $q_i \in F_6, 1 \leq i \leq m$. Since $0 \in \mathcal{A}(\xi)$, we have $\delta_1 = \ldots = \delta_m = 0$. It is clear that $i \notin \mathcal{A}(\xi)$ for any $i \in \{1, \ldots, 2n\}$. Therefore the equation $g_{2i-1}(x) = 0$ or both equations $f_{2i-1}(x) = 0$ and $f_{2i}(x) = 0$ must belong to the system $\mathcal{S}(\xi)$ for any $i \in \{1, \ldots, n\}$. If the number of working nodes in the path ξ is greater than n then the considered statement holds. Let the number of working nodes in the path ξ be equal to n. Then $\mathcal{S}(\xi) = \{g_1(x) = 0, \ldots, g_{2n-1}(x) = 0\}$. Let v be the last working node in the path ξ and let v be labeled by the attribute g_{2i-1}. Let us consider the complete path ξ' in the tree Γ such that $2i - 1 \in \mathcal{A}(\xi')$. Since Γ is a deterministic decision tree, the path ξ' contains the node v. Assume that in the path ξ' there are exactly n working nodes. Then $\mathcal{S}(\xi') = \{g_{2 \cdot 1 - 1}(x) = 0, \ldots, g_{2(i-1)-1}(x) = 0, g_{2i-1}(x) = 1, g_{2(i+1)-1}(x) = 0, \ldots, g_{2n-1}(x) = 0\}$ and $\mathcal{A}(\xi') = \{2i - 1, 2i\}$, but this is impossible since $z(2i-1) \cap z(2i) = \emptyset$. Consequently, in the path ξ' there are at least $n+1$ working nodes and $h(\Gamma) \geq n + 1$. Therefore $h_{U_6}^d(z) \geq n + 1$ and $\mathcal{U}_{U_6 h}^{da}(n) \geq n + 1$. Taking into account that n is an arbitrary number from ω such that $n \geq \max(n_0, 1)$, we conclude that $\mathrm{Dom}^-(\mathcal{U}_{U_6 h}^{da})$ is a finite set and $\mathrm{Dom}^+(\mathcal{U}_{U_6 h}^{da})$ is an infinite set. Consequently, $\mathrm{typ}(\mathcal{U}_{U_6 h}^{da}) = \delta$. Using Proposition 3.1 we obtain $\mathrm{typ}_{lu}(U_6, h) = t_6$. $\square$

Let us define a t-pair (U_7, h) as follows: $U_7 = (\omega, F_7)$ where $F_7 = F_3 \cup F_5$.

Lemma 4.7. $\mathrm{typ}_{lu}(U_7, h) = t_7$.

Proof. Using Lemmas 3.2 and 4.3 we conclude that there exists $n \in \omega$ such that $\mathcal{U}_{U_3 h}^{da}(n) = \infty$. Using this equality and the relation $F_3 \subseteq F_7$ it is not difficult to show that $\mathcal{U}_{U_7 h}^{da}(n) = \infty$. From here and from Lemma 3.2 it follows that $\mathrm{typ}(\mathcal{U}_{U_7 h}^{da}) = \epsilon$.

From Lemmas 3.4 and 4.5 it follows that the function $h_{U_5}^a$ is unbounded from above on the set $\mathrm{Probl}(U_5)$. Using the relation $F_5 \subseteq F_7$ it is not difficult to show that the function $h_{U_7}^a$ is unbounded from above on the set $\mathrm{Probl}(U_7)$. Using Lemma 3.4 we obtain $\mathrm{typ}(\mathcal{U}_{U_7 h}^{aa}) = \gamma$. From the equalities $\mathrm{typ}(\mathcal{U}_{U_7 h}^{da}) = \epsilon$, $\mathrm{typ}(\mathcal{U}_{U_7 h}^{aa}) = \gamma$ and from Proposition 3.1 it follows that $\mathrm{typ}_{lu}(U_7, h) = t_7$. $\square$

Proof (of Proposition 4.1). The statement of the proposition follows from Lemmas 4.1 - 4.7. $\square$

Proof (of Proposition 4.2). The statement of the proposition follows from Lemmas 4.2, 4.3, 4.5, 4.6 and 4.7. $\square$

5   Auxiliary Statements

This section contains some auxiliary statements which will be used under the analysis of relationships between local upper types and local types of t-pairs.

Let X be a nonempty set, $f : X \to \omega$ and $g : X \to \omega$. We define partial functions $\mathcal{U}^{fg} : \omega \to \omega$ and $\mathcal{L}^{gf} : \omega \to \omega$ as follows: if $n \in \omega$ then

$$\mathcal{U}^{fg}(n) = \max\{f(x) : x \in X, g(x) \le n\} \;,$$
$$\mathcal{L}^{gf}(n) = \min\{g(x) : x \in X, f(x) \ge n\} \;.$$

The notation $\mathcal{U}^{fg}(n) = \infty$ means that $\{f(x) : x \in X, g(x) \le n\}$ is an infinite set. Evidently, if $\mathcal{U}^{fg}(n) = \infty$ then $\mathcal{U}^{fg}(n+1) = \infty$.

It is not difficult to prove

Lemma 5.1. *The following statements hold:*
a) If there exists $n \in \omega$ such that $\mathcal{U}^{fg}(n) = \infty$ then $\mathrm{typ}(\mathcal{U}^{fg}) = \epsilon$.
b) If there is no $n \in \omega$ such that $\mathcal{U}^{fg}(n) = \infty$ then $\mathrm{Dom}(\mathcal{U}^{fg}) = \{n : n \in \omega, n \ge n_g\}$ where $n_g = \min\{g(x) : x \in X\}$.
c) If the function f is bounded from above on X then $\mathrm{typ}(\mathcal{L}^{gf}) = \epsilon$.
d) If the function f is unbounded from above on X then $\mathrm{Dom}(\mathcal{L}^{gf}) = \omega$.

Lemma 5.2. *The following statements hold:*
a) $\mathrm{typ}(\mathcal{U}^{fg}) = \alpha$ iff $\mathrm{typ}(\mathcal{L}^{gf}) = \epsilon$.
b) $\mathrm{typ}(\mathcal{U}^{fg}) = \epsilon$ iff $\mathrm{typ}(\mathcal{L}^{gf}) = \alpha$.

Proof. a) Let $\mathrm{typ}(\mathcal{U}^{fg}) = \alpha$. Using Lemma 5.1 one can show that the function f is bounded from above on X and $\mathrm{typ}(\mathcal{L}^{gf}) = \epsilon$. Let $\mathrm{typ}(\mathcal{L}^{gf}) = \epsilon$. Using Lemma 5.1 one can show that the function f is bounded from above on X and $\mathrm{typ}(\mathcal{U}^{fg}) = \alpha$.

b) Let $\mathrm{typ}(\mathcal{U}^{fg}) = \epsilon$. From Lemma 5.1 it follows that the function f is unbounded from above on X and there exists $m \in \omega$ such that $\mathcal{U}^{fg}(m) = \infty$. Therefore $\mathrm{Dom}(\mathcal{L}^{gf}) = \omega$ and $\mathcal{L}^{gf}(n) \le m$ for any $n \in \omega$. Consequently, $\mathrm{typ}(\mathcal{L}^{gf}) = \alpha$. Let $\mathrm{typ}(\mathcal{L}^{gf}) = \alpha$. Using Lemma 5.1 we conclude that $\mathrm{Dom}(\mathcal{L}^{gf}) = \omega$ and there exists $m \in \omega$ such that $\mathcal{L}^{gf}(n) \le m$ for any $n \in \omega$. Therefore $\mathcal{U}^{fg}(m) = \infty$ and $\mathrm{typ}(\mathcal{U}^{fg}) = \epsilon$. $\square$

Lemma 5.3. *Let $\mathrm{typ}(\mathcal{U}^{fg}) \ne \alpha$ and $\mathrm{typ}(\mathcal{U}^{fg}) \ne \epsilon$. Then*
a) $|\mathrm{Dom}^-(\mathcal{U}^{fg})| < \infty$ iff $|\mathrm{Dom}^+(\mathcal{L}^{gf})| < \infty$.
b) $|\mathrm{Dom}^+(\mathcal{U}^{fg})| < \infty$ iff $|\mathrm{Dom}^-(\mathcal{L}^{gf})| < \infty$.

Proof. Using Lemmas 5.1 and 5.2 we obtain $\mathrm{Dom}(\mathcal{U}^{fg}) = \{n : n \in \omega, n \ge n_g\}$ and $\mathrm{Dom}(\mathcal{L}^{gf}) = \omega$ where $n_g = \min\{g(x) : x \in X\}$.

a) Let $|\mathrm{Dom}^-(\mathcal{U}^{fg})| < \infty$. Then there exists $m \in \omega$, $m \ge n_g$, such that $\mathcal{U}^{fg}(n) > n$ for any $n \in \omega$, $n \ge m$. Let $n \ge m$. Then there exists an element $x_0 \in X$ such that $g(x_0) \le n$ and $f(x_0) \ge n+1$. Therefore $\mathcal{L}^{gf}(n+1) \le n$. Consequently, $\mathcal{L}^{gf}(n) < n$ for any $n \in \omega$, $n \ge m+1$, and $|\mathrm{Dom}^+(\mathcal{L}^{gf})| < \infty$.

Let $|\mathrm{Dom}^+(\mathcal{L}^{gf})| < \infty$. Then there exists $m \in \omega$ such that $\mathcal{L}^{gf}(n) < n$ for any $n \in \omega$, $n \ge m$. Let $n \in \omega$ and $n \ge m$. Then there exists an element $x_0 \in X$

such that $g(x_0) \leq n-1$ and $f(x_0) \geq n$. Therefore $\mathcal{U}^{fg}(n-1) \geq n$. Consequently, $\mathcal{U}^{fg}(n) > n$ for any $n \in \omega, n \geq m-1$, and $|\mathrm{Dom}^-(\mathcal{U}^{fg})| < \infty$.

b) Let $|\mathrm{Dom}^+(\mathcal{U}^{fg})| < \infty$. Then there exists $m \in \omega, m \geq n_g$, such that $\mathcal{U}^{fg}(n) < n$ for any $n \in \omega, n \geq m$. Let $n \geq m$. Then for any element $x \in X$ such that $g(x) \leq n$ the inequality $f(x) < n$ holds. Therefore for any element $x \in X$ such that $f(x) \geq n$ the inequality $g(x) > n$ holds. Consequently, $\mathcal{L}^{gf}(n) > n$ and $|\mathrm{Dom}^-(\mathcal{L}^{gf})| < \infty$.

Let $|\mathrm{Dom}^-(\mathcal{L}^{gf})| < \infty$. Then there exists $m \in \omega$ such that $\mathcal{L}^{gf}(n) > n$ for any $n \in \omega, n \geq m$. Let $n \in \omega$ and $n \geq \max(m, n_g)$. Then for any element $x \in X$ such that $f(x) \geq n$ the inequality $g(x) > n$ holds. Therefore for any element $x \in X$ such that $g(x) \leq n$ the inequality $f(x) < n$ holds. Consequently, $\mathcal{U}^{fg}(n) < n$ and $|\mathrm{Dom}^+(\mathcal{U}^{fg})| < \infty$.   □

Lemma 5.4. *The following statements hold:*
a) $\mathrm{typ}(\mathcal{U}^{fg}) = \beta$ *iff* $\mathrm{typ}(\mathcal{L}^{gf}) = \delta$.
b) $\mathrm{typ}(\mathcal{U}^{fg}) = \gamma$ *iff* $\mathrm{typ}(\mathcal{L}^{gf}) = \gamma$.
c) $\mathrm{typ}(\mathcal{U}^{fg}) = \delta$ *iff* $\mathrm{typ}(\mathcal{L}^{gf}) = \beta$.

Proof. a) Let $\mathrm{typ}(\mathcal{U}^{fg}) = \beta$. Using Lemma 5.2 we obtain $\mathrm{Dom}(\mathcal{L}^{gf})$ is an infinite set. From Lemma 5.3 it follows that $|\mathrm{Dom}^-(\mathcal{L}^{gf})| < \infty$. Therefore $\mathrm{typ}(\mathcal{L}^{gf}) = \delta$. Let $\mathrm{typ}(\mathcal{L}^{gf}) = \delta$. Using Lemma 5.2 we conclude that $\mathrm{Dom}(\mathcal{U}^{fg})$ is an infinite set and the function $\mathcal{U}^{fg}$ is unbounded from above. From Lemma 5.3 it follows that $|\mathrm{Dom}^+(\mathcal{U}^{fg})| < \infty$. Therefore $\mathrm{typ}(\mathcal{U}^{fg}) = \beta$.

b) Let $\mathcal{U}^{fg} = \gamma$. Using Lemma 5.3 we obtain $\mathrm{typ}(\mathcal{L}^{gf}) = \gamma$. Let $\mathrm{typ}(\mathcal{L}^{gf}) = \gamma$. Using Lemma 5.2 we obtain $\mathrm{typ}(\mathcal{U}^{fg}) \neq \alpha$ and $\mathrm{typ}(\mathcal{U}^{fg}) \neq \epsilon$. From here and from Lemma 5.3 it follows that $\mathrm{typ}(\mathcal{U}^{fg}) = \gamma$.

c) Using Lemma 5.2 and statements a), b) of the lemma we conclude that $\mathrm{typ}(\mathcal{U}^{fg}) = \delta$ iff $\mathrm{typ}(\mathcal{L}^{gf}) = \beta$.   □

Let us define a function $\rho : \{\alpha, \beta, \gamma, \delta, \epsilon\} \rightarrow \{\alpha, \beta, \gamma, \delta, \epsilon\}$ as follows: $\rho(\alpha) = \epsilon, \rho(\beta) = \delta, \rho(\gamma) = \gamma, \rho(\delta) = \beta, \rho(\epsilon) = \alpha$.

Proposition 5.1. *Let X be a nonempty set, $f : X \rightarrow \omega, g : X \rightarrow \omega, \mathcal{U}^{fg}(n) = \max\{f(x) : x \in X, g(x) \leq n\}$ and $\mathcal{L}^{gf}(n) = \min\{g(x) : x \in X, f(x) \geq n\}$ for any $n \in \omega$. Then $\mathrm{typ}(\mathcal{L}^{gf}) = \rho(\mathrm{typ}(\mathcal{U}^{fg}))$.*

Proof. The statement of the proposition follows from Lemmas 5.2 and 5.4.   □

6   Proofs of Theorems 2.1 and 2.2

Using Proposition 5.1 we obtain the following statement.

Proposition 6.1. *Let (U, ψ) be a t-pair and $b, c \in \{i, d, a\}$. Then $\mathrm{typ}(\mathcal{L}^{cb}_{U\psi}) = \rho(\mathrm{typ}(\mathcal{U}^{bc}_{U\psi}))$.*

Proof (of Theorem 2.1). The statement of the theorem follows from Propositions 3.1, 4.1 and 6.1.   □

Proof (of Theorem 2.2). The statement of the theorem follows from Propositions 3.2, 4.2 and 6.1.   □

Acknowledgments

The author is greatly indebted to Andrzej Skowron for helpful discussions.

References

1. Ahlswede, R., Wegener, I.: Suchprobleme. B.G. Teubner, Stuttgart, 1979
2. Angluin, D.: Queries and concept learning. Machine Learning **2**(4) (1988) 319–342
3. Bazan, J., Nguyen, H. Son, Nguyen, S. Hoa, Synak, P., Wróblewski, J.: Rough set algorithms in classification problems. Rough Set Methods and Applications: New Developments in Knowledge Discovery in Information Systems (Studies in Fuzziness and Soft Computing **56**). Edited by L. Polkowski, T.Y. Lin and S. Tsumoto. Phisica-Verlag. A Springer-Verlag Company (2000) 48–88
4. Chegis, I.A., Yablonskii, S.V.: Logical methods of electric circuit control. Trudy MIAN SSSR **51** (1958) 270–360 (in Russian)
5. Humby, E.: Programs from Decision Tables. Macdonald, London and American Elsevier, New York, 1973
6. Moshkov, M.Ju.: Conditional tests. Problems of Cybernetics **40**. Edited by S.V. Yablonskii. Nauka Publishers, Moscow (1983) 131–170 (in Russian)
7. Moshkov, M.Ju.: Decision Trees. Theory and Applications. Nizhny Novgorod University Publishers, Nizhny Novgorod, 1994 (in Russian)
8. Moshkov, M.Ju.: Comparative analysis of complexity of deterministic and nondeterministic decision trees. Local Approach. Actual Problems of Modern Mathematics **1**. NII MIOO NGU Publishers, Novosibirsk (1995) 109–113 (in Russian)
9. Moshkov, M.Ju.: Two approaches to investigation of deterministic and nondeterministic decision tree complexity. Proceedings of the World Conference on the Fundamentals of AI. Paris, France (1995) 275–280
10. Moshkov, M.Ju.: Local and global approaches to comparative analysis of complexity of deterministic and nondeterministic decision trees. Actual Problems of Modern Mathematics **2**. NII MIOO NGU Publishers, Novosibirsk (1996) 110–118 (in Russian)
11. Moshkov, M.Ju.: Comparative analysis of deterministic and nondeterministic decision tree complexity. Global approach. Fundamenta Informaticae **25** (1996) 201–214
12. Pawlak, Z.: Information Systems – Theoretical Foundations. PWN, Warsaw, 1981 (in Polish)
13. Pawlak, Z.: Rough Sets – Theoretical Aspects of Reasoning about Data. Kluwer Academic Publishers, Dordrecht, Boston, London, 1991
14. Picard, C.F.: Theorie des Questionnaires. Gauthier-Villars, Paris, 1965
15. Preparata, F.P., Shamos, M.I.: Computational Geometry: An Introduction. Springer-Verlag, 1985
16. Quinlan, J.R.: Induction of decision trees. Machine Learning **1**(1) (1986) 81–106
17. Skowron, A.: Rough sets in KDD. Proceedings of the 16-th World Computer Congress (IFIP'2000). Beijing, China (2000) 1–14

18. Skowron, A., Pawlak, Z., Komorowski, J., Polkowski, L.: A rough set perspective on data and knowledge. Handbook of KDD. Edited by W. Kloesgen and J. Żytkow. Oxford University Press (2002) 134–149
19. Skowron, A., Rauszer, C.: The discernibility matrices and functions in information systems. Intelligent Decision Support. Handbook of Applications and Advances of the Rough Set Theory. Edited by R. Slowinski. Kluwer Academic Publishers, Dordrecht, Boston, London (1992) 331–362
20. Tarasova, V.P.: Opponent Strategy Method in Optimal Search Problems. Moscow University Publishers, Moscow, 1988 (in Russian)

A Fast Host-Based Intrusion Detection System Using Rough Set Theory

Sanjay Rawat[1,2], V.P. Gulati[2], and Arun K. Pujari[1]

[1] AI Lab, Dept. of Computer and Information Sciences,
University of Hyderabad, Hyderabad-500046, India
tosanjayr@gmail.com, akpcs@uohyd.ernet.in
[2] IDRBT, Castle Hills, Road No.1,
Masab Tank, Hyderabad-500057, India
vp.gulati@tcs.com

Abstract. Intrusion Detection system has become the main research focus in the area of information security. Last few years have witnessed a large variety of technique and model to provide increasingly efficient intrusion detection solutions. We advocate here that the intrusive behavior of a process is highly localized characteristics of the process. There are certain smaller episodes in a process that make the process intrusive in an otherwise normal stream. As a result it is unnecessary and most often misleading to consider the whole process in totality and to attempt to characterize its abnormal features. In the present work we establish that subsequences of reasonably small length of sequence of system calls would suffice to identify abnormality in a process. We make use of rough set theory to demonstrate this concept. Rough set theory also facilitates identifying rules for intrusion detection. The main contributions of the paper are the following- (a) It is established that very small subsequence of system call is sufficient to identify intrusive behavior with high accuracy. We demonstrate our result using DARPA'98 BSM data; (b) A rough set based system is developed that can extract rules for intrusion detection; (c) An algorithm is presented that can determine the status of a process as either normal or abnormal on-line.

Keywords: Data mining, Decision Table, Rough Set, Intrusion Detection, Anomaly, Misuse.

1 Introduction

Intrusion detection systems (IDSs) have become a major area of research and product development. They work on the premise that intrusions can be detected through examinations of various parameters such as network traffic, CPU utilization, I/O utilization, user location, and various file activities. Based on the various approaches, different types of IDS are proposed in the literature. On the basis of audit data, there are two types of IDS. The *network-based* systems collect data directly from the network that is being monitored, in the form of *packets* [29] and the *host-based* systems collect data from the host being protected [2].

J.F. Peters and A. Skowron (Eds.): Transactions on Rough Sets IV, LNCS 3700, pp. 144–161, 2005.
© Springer-Verlag Berlin Heidelberg 2005

Based on processing of data to detect attacks, IDS can also be classified into two types – *misuse-based* systems and *anomaly-based* systems. While the former keeps the signatures of known attacks in the database and compares new instances with the stored signatures to find attacks, the latter learns the normal behavior of the monitored system and then looks out for any deviation in it for signs of intrusions. It is clear that misuse based IDS cannot detect new attacks and we have to add manually any new attack signature in the list of known patterns. IDS based on anomaly detection, on the other hand, are capable of detecting new attacks as any attack is assumed to be different from normal activity. However anomaly based IDS sometimes sets false alarms because it cannot differentiate properly between deviations due to authentic user's activity and that of an intruder.

Among various IDS approaches, *signature-analysis* stores patterns of attacks as semantic descriptions [21]. The main drawback of the signature analysis technique, like all misuse-based approaches, is the need for frequent updates to keep up with the stream of new vulnerabilities/attacks discovered. *Rule-based* intrusion detection [34][20][13] assumes that intrusion attempts can be characterized by sequences of events that lead to the state of compromised-system. Such systems are characterized by their expert system properties that fire rules when audit records or system status information begin to indicate suspicious activity. The main limitations of this approach are the difficulty of extracting knowledge about attacks and the processing speed. *State transition analysis* technique describes an attack with a set of goals and transitions, and represents them as state transition diagrams [18][19][32]. The most widely used approach of anomaly-based intrusion detection is *statistical* [16][27]. User or system behavior is measured by a number of variables sampled over time and stored in a profile. The current behavior of each user is maintained in a profile. At regular intervals the current profile is merged with the stored profile. Anomalous behavior is determined by comparing the current profile with the stored profile.

Forrest *et al* [11][12] suggest that system calls trace of a process under normal execution can be taken as its normal behavior in terms of system calls, as variation in sequences of system calls is very small. On the other hand, this variation is relatively higher when compared to a sequence of system calls under abnormal execution. This variation can be attributed to the presence of one or more *alien* (thus malicious) subsequences in the abnormal process. It should be noted that not all the subsequences of an abnormal process are malicious. Thus intrusive part should be detectable as a subsequence of the whole abnormal sequence of the process.

In this paper we present a technique of discovering rules for intrusion detection. We make use of *rough set theory* for this purpose. To best of our knowledge, Lin was the first to propose the idea of applying rough sets to the problem of anomaly detection [25]. Though the paper lacks the experimental results [25], it provides some solid theoretical background. The following two theorems are important:

1. Every sequence of records in computer has a repeating sequence
2. If the audit trail is long enough, then there are repeating records

Following the argument of Forrest *et al* and in the view of above theorems, our approach is based on subsequences of system calls. We formulate the problem as a classification problem by writing the set of subsequences as a decision table. The proposed method is a combination of signature-based and anomaly based approaches. A program behavior is monitored as a sequence of system calls. These sequences are further converted into the subsequences of shorter length. These subsequences are considered as the signatures for malicious as well as normal activities. By doing so, one of the disadvantages of signature-based approach of frequently updating the signature database can be avoided. Empirical results show that the proposed system is able to detect new abnormal activities without updating the signatures. Further, these signatures are represented in the form of IF-THEN type decision rules. The advantage of representing signatures in this form is that such signatures are easy to interpret for further analysis. Rough set theory is used to induce decision rules. Rules induced by using rough set theory are very compact because before inducing rules, all the redundant features of the audit data are removed. This makes the matching of rules *faster*, thus making the system suitable for on-line detection. The proposed system is also fast in the sense that process is compared, in parts, as it starts calling system calls. So we do not have to wait until it exits.

The major contributions of the paper are:

- It is established empirically that short sequences of system calls are sufficient to detect intrusive behavior with high accuracy;
- A rough set based approach is developed that can extract decision rules for intrusion detection;
- An algorithm is presented that can classify a process as normal or abnormal on-line.

Rest of the paper is organized as follows: Section 2 gives an overview of research work on process profiling using sliding window approaches and learning rules for intrusion detection. Section 3 presents some preliminary background to understand the approach. A detailed description of the proposed scheme is given in the section 4. Section 5 covers the experimental setup and analysis of the results. Section 6 concludes the paper.

2   Related Work

Recently, process monitoring for the sign of intrusions has attracted the attention of many researchers and active research is being done in this area. In the approach, called *time-delay embedding (tide)*, initiated by Forrest *et al* [11][12], normal behavior of processes is captured because programs show a stable behavior over the period of time under normal execution. In this approach, short sequences of system calls are used to profile a process. It uses a sliding window algorithm to populate a table with the positional relationships between system calls. Forrest et al use a sliding window of size $k + 1$ to record which system calls succeed or precede each other at offsets 1 through k. This implementation is said to have a "forward" lookahead because while matching with the testing process, the current system call is used as the index to the table and anomalies are found by performing a pair wise comparison between the current system call and each system call that follows at offsets 1 through k. If the process, under consideration, has a matching system call in the table at each offset, then it can be considered normal. Otherwise it is abnormal.

Inspired by *human immune system*, tide approach is extended by Hofmeyr *et al* [17] by using a technique called *sequence time-delay embedding (stide)*. In this approach, the traces of system calls generated by a process are scanned and a database of all unique sequences of a given length, k that occurred during the trace, is built up. The database, then, is used to monitor the ongoing behavior of the processes invoked by the program, by calculating Hamming distance between two sequences. An anomaly count is defined as the number of mismatches in a temporally local region. If the count is greater than a predefined threshold, the sequence is flagged as anomalous.

A simple addition to stide, called *stide with frequency threshold (t-stide)* is proposed to test the premise that rare sequences are suspicious [37]. For each sequence in database, frequency of its occurrence in training data is also recorded. Sequences from test traces are compared to those in the database, just like stide. Rare sequences, as well as those not included in the database, are counted as mismatches. These mismatched are aggregated into locality frame count. Some better results are found if the length of the sequences is not fixed [38]. These approaches are not suitable for on-line detection as the frequency cannot be determined until after the process terminates [37]. Following the inspirations from immune system, Cabrera *et al* propose to build *Anomaly Dictionaries* as *self* for anomalous sequences [6]. These anomaly dictionaries contain short sequences of system calls spawn by processes under attacks. A string matching classifier is used to classify any new process.

A similar approach is followed by Lee *et al* [22], but they make use of a rule learner RIPPER, to form the rules for classification. The normal process is transformed into sequences of fixed length, k. Each sequence is turned into a RIPPER sample by treating all system calls, except the last in the sequence, as attributes and the last one as the target class.

All of the above approaches concentrate only on the sequences of system calls. Tandon and Chan [36] propose to consider system calls arguments and other parameters, along with the sequences of system calls. They make use of the variant of a rule learner LERAD *(Learning Rules for Anomaly Detection)*. Three variants of LERAD are proposed to generate rules under different inputs - S-LERAD for sequences of system calls only, A-LERAD for system call arguments and other key attributes and M-LERAD for argument information and sequences of system calls. A total of six system calls are used in training - first five as conditions and sixth one as decision. In A-LERAD, system calls are taken as pivotal attributes. Any value for other arguments (path, return value, error status), given a system call, which was never encountered in the value for a long time, would trigger an alarm. M-LERAD merges both S-LERAD and

A-LERAD. Each input comprises of system call, arguments, path, return value, error status and the previous five system calls.

In a recent paper [33], an efficient scheme is proposed by using *kNN classifier*. A new similarity measure is proposed to consider the frequency and ordering of system calls in a process under normal execution.

Rough set theory has well been applied to many data mining techniques like association rules, classification and clustering. The use of rough sets in the simple and improved formation of association rules is shown by Guan *et al* [15] and Delic *et al* [10]. Guan *et al* show that *maximal association rules* can be formed by applying basic rough set operations in a much simpler manner. Delic *et al* argue to reduce the computational time by applying the concept of reduct extraction directly on the produced rules, not on attribute. As a result, they propose the hybridization of *apriori algorithm* and rough set, named as *Apriori+*, to generate association rules.

With the increase of web-based transactions over internet, it is interesting and necessary to learn user's activities for better understanding and improvement of web services. Often, it is very difficult to categorize users into different clusters as boundary of user's activities is not sharp. A rough set based clustering scheme is proposed by Lingras to cluster various users, based on their access-patterns [26]. The scheme is applied on university students to cluster them into three groups– *studious, crammers* and *workers*, by using five attributes of access-patterns. The genetic algorithms are used to maximize the *prescision* value. The problem of *vague boundaries* is tackled by calculating *lower* and *upper approximations* of three groups.

An *et al* [1] investigate the idea of applying rough set to text classification, particularly web page classifications. As a standard technique, pages are described as frequencies or top n occurring words under each of the k categories. Out of these n words, many may be redundant and, therefore, can be discarded. Rough set based operation, called *positive region*, is used to calculate the importance of the attributes and thereby removing the unnecessary attributes (words). ELEM2 *rule induction* algorithm is used to learn the rules for each category. Each rule is given a score, termed as *rule quality* and based on this number, test instances are classified to different categories.

Dan Zhu *et al* [39] present a comparative study on IDS based on neural network, inductive learning and rough sets and find that, on an average, rough set based approach performs better over other techniques. But, according to their study, rough sets are not as efficient as neural network in classifying unseen objects.

Rough set theory has been applied on sequence of system calls made by a process to learn the normal behavior of process in terms of IF-THEN rules [7]. These rules are used to predict the $(k+1)^{th}$ system call in a sequence of length k. If the predicted and the actual system calls are identical, the sequence is normal otherwise abnormal. The present study takes a similar approach but instead of predicting next system call in order to classify the process, it shows that a process can directly be classified as normal or abnormal based on its subsequences of system calls.

Very recently, Lian-hua *et al* [24] apply rough set classification (RSC) technique to network-based IDS. In their approach, *decision table* consists of network connection records, provided by *KDD data set*. The approach is based on the observation that rough set *reduct* computation can be viewed as a *minimal hitting set problem*. A minimal hitting set is computed from the *multiset* which, in turn, computed from the *discernibility function*. A *hybrid genetic algorithm* is used to calculate the reducts of rough set. The results are compared with the SVM-based IDS and are found to be better for DoS and probing attacks, but inferior in case of R2L and U2R categories of attacks. The computation time for reduct calculation is also minimized by using new hybrid genetic algorithm.

3   Preliminary Background

In this section, we provide some basic definitions and notations used in our work. Later in the section we try to formulate intrusion detection as a problem in rough set theory.

3.1   Rough Set Theory

Knowledge discovery comprises of techniques from machine learning, statistics, pattern recognition, fuzzy and rough sets etc to extract knowledge or information from the huge amount of data. Rough set theory was introduced by Z. Pawlak [31] to provide a systematic mathematical framework for studying imprecise and insufficient knowledge to generate decision rules. A rough set is a set of objects that cannot be precisely characterized based on a set of available attributes. The idea of rough set consists of the approximation of a set by a pair of sets, called as *lower approximation* and *upper approximation*.

Let $S = < U, Q, V, f >$ be an information system where U- is the closed universe, a finite set of N objects $x_1, x_2, ..., x_N$; Q is a finite set of n attributes $q_1, q_2, ..., q_n$; $V = \bigcup_{q \in Q} V_q$ where V_q the domain of attribute q; and $f : U \times Q \to V$ is the total function called as information function such that $f(x, q) \in V_q$ for every $x \in U$ and $q \in Q$. A subset of attributes $A \subseteq Q$ determines as equivalence relation of the universe U, called as *indiscernibility relation* and denoted as $IND(A)$.

Definition 1: For a given $A \subseteq Q$ and $X \subseteq U$ (a concept X), the A-lower approximation $(\underline{A}X)$ and A-upper approximation $(\overline{A}X)$ of set X are defined as follows:

$$\begin{aligned} \underline{A}X &= \{x \in U : [x]_A \subseteq X\} = \bigcup\{Y \in A^* : Y \subseteq X\} \\ \overline{A}X &= \{x \in U : [x]_A \cap X \neq \phi\} = \bigcup\{Y \in A^* : Y \cap X \neq \phi\} \end{aligned} \tag{1}$$

where $[x]_A = \{y \in U : xIND(A)y\}$ and A^* is the partition of U generated by $IND(A)$ on U.

Definition 2: The accuracy of an approximation of the set X by the set of attributes A is defined as:

$$\alpha_A(X) = \frac{|\underline{A}X|}{|\overline{A}X|} \tag{2}$$

Definition 3: Given an information system $S = < U, Q, V, f >$, with condition and decision attributes $Q = A \cup D$, s.t. $A \cap D = \phi$, for a given set of condition attributes A, we can define the A-*positive region* $POS_A(D)$ in the relation $IND(D)$ as:

$$POS_A(D) = \bigcup \{ \underline{A}X : X \in D^* \} \tag{3}$$

where D^* denotes the family of equivalence classes defined by the relation $IND(D)$. $POS_A(D)$ contains all the objects in U which can be classified perfectly without error into the distinct classes defined by $IND(D)$, based only on information in relation $IND(D)$. Similarly, in general, if $A, B \subseteq Q$, then A-*positive region* of B is defined as

$$POS_A(B) = \bigcup_{X \in B^*} \underline{A}X \tag{4}$$

3.2   Decision Table

A data set is represented as a table, where each row represents an object or case. Every column represents an attribute that can be measured for each object. In supervised learning, there is an outcome of classification that is known. This a posteriori knowledge is expressed by one distinguished attribute called decision attribute. A table wherein one of the attributes is decision attribute is called a *decision table*. More precisely:

Definition 4: Given an information system $S = < U, Q, V, f >$, if Q can be expressed as condition and decision attributes i.e. $Q = A \cup D$, with $A \cap D = \phi$, then S is called a decision table (or decision system) [8].

Definition 5: A decision table is said to be *consistent* if each unique row has only one value of decision attribute. Objects from decision table can be partitioned into disjoint classes, called *concepts*, based on the decision attributes D.

Definition 6: An expression $(a = v)$, where $a \in A$ and $v \in V_q$, is called an *atomic formula* (or elementary condition) c. An elementary condition c can be interpreted as mapping:

$$c : U \rightarrow \{true, false\}$$

A conjunction C of q elementary condition is denoted by

$$C = c_1 \wedge c_2 \wedge ... \wedge c_q.$$

Definition 7: The cover of C, denoted by $[C]$, is the subset of objects (examples), which satisfy the conditions represented by C.

$$[C] = \{x \in U : C(x) = true\}$$

3.3   Decision Rules

The decision rules are logically described as

If (a conjunction of elementary conditions) **then** (decision)

In general decision rules can be considered as data patterns, which represent relationships between values of attributes in the decision table. The rule set, obtained from a consistent table, is said to be *deterministic*. Any set of rules may fall into any of the following three categories [35]:

- Minimum set of decision rules,
- Exhaustive set of decision rules,
- Satisfactory set of decision rules

First category contains the smallest number of rules sufficient to cover the set of objects belonging to one class. Second category consists of all the rules that can be induced from the table. However, time complexity for the second choice is exponential and using this approach may not be practical for larger data set [35]. For our experiment we, therefore, choose the first approach of inducing rules, as the data set used in our experiment is very large. We make use of a rough set based algorithm LEM2 for inducing rules [14], which is presented in figure 1. We use the same notations introduced in the previous section. Additionally $C(G)$ denote the set of conditions c currently considered to be added to the conjunction C. K is the concept and rule r is characterized by its conditional part R.

```
Input:K- set of objects
Output:R- set of rules
begin
    G = K;
    R = φ;
   while G ≠ φ do
      begin
         C = φ;
         C(G) = {c : [c] ∩ G ≠ φ};
         while (C = φ) or (!([C] ⊆ K)) do
            begin
               select a pair c ∈ C(G) such that |[c] ∩ G| is maximum;
               if ties, select a pair c ∈ C(G) with the smallest cardinality |[c]|;
               if further ties occur, select the first pair from the list;
               C = C ∪ {c}; G = [c] ∩ G;
               C(G) = {c : [c] ∩ G ≠ φ};
               C(G) = C(G) − C;
            end;
            for each elementary condition c ∈ C do
               if [C − c] ⊆ K then C = C − {c};
            create rule r basing the conjunction C and add it to R;
            G = K − ⋃ [R];
                   r∈R
      end;
      for each r ∈ R do
      if  ⋃  [S] = K then R = R − r
         s∈R−r
   end
```

Fig. 1. LEM2 algorithm

LEM2 algorithm follows a heuristic strategy from machine learning techniques. The strategy starts with creating a first rule by choosing sequentially the 'best' elementary conditions (conjunction of attributes values). Then, all the learning examples that matches this rules are removed from consideration. The process is repeated iteratively while some learning examples remain uncovered. The rules so obtained are capable of classifying new unseen objects [35]. In the whole process, the algorithm also discards all the dispensable attributes. Thus the number of attributes to be matched is reduced which makes the algorithm faster and hence more suitable for nearly real time detection.

4   The Proposed Approach: Rough Set and Intrusion Detection

The data mining techniques are well suited for IDS design because the aim of an intrusion detection system is to trigger alarm (present knowledge) when any intrusion occurs. Thus an IDS can be thought of a decision support system which stores huge data (host or network related) and extracts useful patterns (information about the normal and abnormal behavior) so that it can classify normal and abnormal data precisely. Forrest *et al* [12] suggested the use of small sequences of system calls, made by a process, as the profile of the process. The study done by Lee *et al* [22] also validates this observation. But if we analyze normal and abnormal processes, we find that not all parts of an abnormal process are responsible for intrusion. Thus intrusive part should be detectable as a subsequence of the whole abnormal sequence of the process. Thus one point of focus of this study is to determine the adequate length of such subsequences. Also as pointed out earlier, not all of the subsequences of an abnormal process are abnormal. Many of them will be identical to those occurred in normal process. This is the point where rough set theory can be used to derive disjoint set of subsequences.

Let P be a set of normal processes, defined as sequences of system calls. Then the lth process can be represented as

$$< p_1^l, p_2^l, \ldots, p_n^l >$$

where n is the length of the process and p_j^l is j^{th} system call. Each of these processes is transformed into the subsequences of length k. Thus for l^{th} sequence, i^{th} subsequence is given by

$$p_i^l, p_{i+1}^l \ldots, p_{i+k-1}^l$$

Each of these subsequences is labeled as normal. In case of abnormal processes, as pointed out earlier, not all of the subsequences of an abnormal process are abnormal. Many of them are identical to those occurred in normal process. Thus a subsequence corresponding to an abnormal process, matching with any of the normal subsequences, is discarded; otherwise it is labeled as abnormal. With the above formulation, we consider an intrusion detection system as an information system $S = < U, Q, V, f >$, defined in the section 3.1 with $Q = A \cup D$, $D=$ {normal, abnormal}. The number of attributes in the conditional part A equals the length of the subsequence. V consists of all the system calls appearing in all the processes used for training. U consists of all the subsequences of the chosen length that can be derived from all the processes using sliding window of the chosen length k. It should be noted that by removing duplicate subsequences, we get a consistent decision table because no subsequence can belong to normal as well as abnormal classes. To put in rough set terminology, let P and T be two normal and abnormal processes respectively given as $P = < p_1, p_2, \ldots, p_n >$ and $T = < t_1, t_2, \ldots, t_m >$ where p_i's and t_j's are system calls. Let k be the size of sliding window. Each process P and T is transformed into subsequences of length k. Then the information system S can be represented as follows: A is comprised of k attributes $A_1, A_2 \ldots, A_k$. U consists of all the subsequences of the forms $P_i = p_i, p_{i+1}, \ldots, p_{i+k-1}$ and $T_i = t_i, t_{i+1}, \ldots, t_{i+k-1}$, represented as rows of S. Let us denote normal and abnormal classes by D_{normal} and $D_{abnormal}$, which is a partition of U by the decision attribute D, denoted as D^*. As mentioned earlier that there will be many T_i's which are identical to some P_i's, but labeled as abnormal. Therefore the lower approximations of both the classes D_{normal} and $D_{abnormal}$ are calculated as $\underline{A}D_{normal}$ and $\underline{A}D_{abnormal}$. Therefore the positive region

$$POS_A(D) = \underline{A}D_{normal} \cup \underline{A}D_{abnormal}$$

contains only those subsequences that belong to either of the classes but not both. We apply LEM2 algorithm on S to form the certain rules. By doing so, we get the signature for normal and abnormal processes. Let us take an example to make it more clear. Let
$P^1 = $ <fcntl, close, close, fcntl, close, fcntl, close, open> and
$P^2 = $ <fcntl, close, fcntl, close, open, open>
be normal and abnormal processes respectively. We transform P^1 into a set of subsequences using a sliding window of length 5. We label all the 4 subsequences as normal. While calculating the subsequences of P^2, the first subsequence < fcntl, close, fcntl, close, open> matches with the last subsequence of P^1 and therefore it is discarded. The second subsequence < close, fcntl, close, open, open> is labeled as abnormal and added to the decision table. The final decision table is shown in the table 1. It can be seen in table 1 that the

Table 1. Representation of subsequences

Objects	A_1	A_2	A_3	A_4	A_5	Decision
1	fcntl	close	close	fcntl	close	normal
2	close	close	fcntl	close	fcntl	normal
3	close	fcntl	close	fcntl	close	normal
4	fcntl	close	fcntl	close	open	normal
5	close	fcntl	close	open	open	abnormal

Table 2. Representation of IF-THEN rules

1	$(A_2 = $ close$) \Rightarrow$ (Dec $=$normal)
2	$(A_1 = $ close$) \wedge (A_2 = $ fcntl$) \wedge (A_3 = $ close$) \wedge (A_4 = $ fcntl$) \Rightarrow$ (Dec $=$ normal)
3	$(A_1 = $ close$) \wedge (A_2 = $ fcntl$) \wedge (A_3 = $ close$) \wedge (A_4 = $ open$) \Rightarrow$ (Dec$=$abnormal)

Training Phase

1. Collect normal and abnormal processes
2. Calculate the subsequences of these processes of length k (number of attributes)
3. Remove duplicates
4. Construct the decision table with labels normal and abnormal
5. Calculate decision rule set DR using LEM2

Testing Phase

1. **For** each process P in the testing data **do**
2.   Convert the process into decision table of length k
3.   Classify each subsequence using the rule set DR
4.   **if** any of the subsequence is classified as abnormal **then**
5. P is abnormal
6.   **else** P is normal
7. **end do**

Fig. 2. Algorithmic representation of the proposed scheme

decision table, thus created, is consistent. We can calculate IF-THEN rules, using LEM2 algorithm, shown in table 2. The rules, induced by LEM2, are used to classify new processes. While classifying any new process, it is first transformed into a set of subsequences and each of these subsequences is classified on the basis on decision rules. If any subsequence pertaining to a process is classified as abnormal, the whole process is considered as abnormal. The algorithmic form of the proposed scheme is presented in figure 2 below.

5   Experimental Setup and Results

The scheme described in the previous section is tested upon the well-cited DARPA'98 data [9]. The whole data comprises of network level data and host level data. The host-based data is provided in two forms - *NT audit logs* and *BSM audit logs*. Process level information can be derived from BSM audit data. Therefore we use BSM audit logs from the 1998 DARPA data for training and testing of our algorithm. A detailed procedure for the extraction of process from the audit logs can be found in [33]. However, for completeness, we summarize the whole process below.

For each day of data, a separate BSM file is provided with the 'BSM List File'. All the intrusive sessions are labeled with the name of the attacks launched during the sessions. On analyzing the entire set of BSM logs (list files), we locate the five days which are free of any type of attacks. We choose the first four days for our training data and the fifth one for the testing of the normal data to determine the false positive rate. There are around 2000 normal sessions reported in the four days of data. We extract the processes occurring during these days and our training data set consists of 606 unique processes. There are 412 normal sessions on the fifth day and we extract 339 unique normal processes from these sessions. We use these 339 normal processes for the testing data. A total of 28 abnormal processes are extracted from the whole seven-week of data. Out of these 28 processes, 12 are used for training and remaining 16 are used for testing the detection rate of the approach. In case of a normal process, all of its subsequences should be normal. Therefore in order to test the false positive rate, we take normal subsequences of all the 339 processes together and define *coefficient of normal accuracy* as

$$\eta_n = \frac{N_c}{N} \tag{5}$$

Where N_c is the total number of normal subsequences correctly classified as normal and N is total normal subsequences used in testing data. In case of abnormal process, as mentioned earlier, not all the subsequences are abnormal. Therefore, we say that an abnormal process is detected is any of its subsequences is labeled as abnormal and we define *coefficient of abnormal accuracy* as

$$\eta_a = \frac{A_c}{A} \tag{6}$$

where A_c is total number of processes classified as abnormal and A is total abnormal processes used in testing data. It should be noted that coefficient of normal accuracy η_n is inversely proportional to the false positive rate i.e. higher the value of η_n, lower is the false positive rate.

We perform the experiments for different values of the length of the subsequences. For each value, a set of decision rule is calculated. Using this set of rules, the values of η_n and η_a are calculated. Table 3 shows the results of experiments for different values of length of subsequences. It should be pointed out that while doing the above experiment, no default rule is defined i.e. any subsequence, not covered by any of the rules is assigned to a special class *undefined* and such subsequences are excluded while calculating η_n. The minimum value of η_n is 0.997 i.e. in worst case the false positive rate is 0.003, which implies that per day, there are $339 \times 0.003 = 1.017$ false alarms. This type of approach is well suitable in situation where there is a second level of check to further analyze the data when we are not certain about the event and tolerance level is high in terms of attacks. But such an approach may delay in decision-making due to lack of high confidence. We, therefore, repeat our series of experiments with a default rule '*any subsequence, which is not covered by any of the rule is abnormal*'. This approach is well suited in situation wherein tolerance limit for attacks is very low. Thus no event is classified based on further analysis, but based on the rules including the default one i.e. each event is classified on-line. It may also be noted that while the former approach is a misuse-based, second

Table 3. Values of coefficients of normal and abnormal accuracy for different values of subsequence length without a default rule

Length of the subsequence	Number of the subsequences before/after removing duplicates	Number of rules	Value of η_n	Value of η_a	$\frac{(\eta_n + \eta_a)}{2}$
5	170976/2461	929	0.999	0.750	0.8745
10	167886/5968	1702	0.997	0.750	0.8735
15	164801/8441	1828	0.999	0.750	0.8745
20	161724/10525	1797	0.998	0.812	0.9050
25	158669/12299	1707	1.000	0.750	0.8750
30	155640/13939	1789	1.000	0.860	0.9300
35	152625/15398	1810	0.999	0.930	0.9645
40	149628/16641	1762	0.998	0.860	0.9290
45	146654/17666	1699	0.998	0.928	0.9630

Table 4. Values of coefficients of normal and abnormal accuracy for different values of subsequence length with a default rule

Length of the subsequence	Number of rules	Value of η_n	Value of η_a	$\frac{(\eta_n + \eta_a)}{2}$
5	929	0.966	0.875	0.9205
10	1702	0.959	0.812	0.8855
15	1828	0.954	0.812	0.8830
20	1797	0.958	0.875	0.9165
25	1707	0.960	0.875	0.9175
30	1789	0.958	0.930	0.9440
35	1810	0.959	0.930	0.9445
40	1762	0.958	1.000	0.9790
45	1699	0.959	0.928	0.9435

is a hybrid (anomaly and misuse) approach. Table 4 lists the results of the experiment. The minimum value of η_n is 0.958 i.e. in worst case the false positive rate is 0.042, which implies that per day, there are $339 \times 0.042 = 14.23$ false alarms.

If we look at figure 3, we find that for first set of experiment, the value of η_n is high (i.e. low rate of false positives (first longest bar in the figure)) but detection rate (η_a) could not reach 1.0. There is a clear distinction between the values of η_n with and without a default rule. This can be understood as we train our system only with 12 abnormal processes, which is much smaller than 606 normal processes used in training. In the second experiment when we included a default rule, we could get detection rate of 1.0 with normal accuracy as high as 0.958. The presence of the default rule has made the system anomaly-based and from this point of view, a decline in the value of η_n (rise in the rate of false positive) is anticipated as anomaly-based systems are known to have a high rate of false positive. An efficient IDS should have high values of η_n and η_a i.e. low false positive rate and high detection rate. The last column of table 3 and table 4 shows the average taken over the values of η_n and η_a. Figure 4 shows that the results obtained with a default rule (hybrid approach) outperform those obtained without a default rule (but not with a significant difference). The bold lines in the figure 4 represent the trends i.e. as we increase the length of subsequence, accuracy also increases and after a length of 35, accuracy ceases to increase. We also observe that though the accuracy increases as we increase the length of the subsequences, it is not a global pattern, particularly in the case of misuse-based approach (without a default rule). Therefore the length of the subsequence can be as short as 5 or as large as 40, we still can detect attacks without matching the whole process. The main point to observe here is that normally anomaly-based systems have good detection rate with a high false positive rate as compared to those based on misuse detection.

Our experiments show that anomaly-based IDS can have as low false positive rate as that of a misuse-based system with a very high detection rate. Also once the rules are computed, there is no further computation involved in classifying the processes. Only the subsequences of a process are matched against the rules for classifications. This makes the system very fast suitable for on-line detection.

All the above experiments are conducted using RSES and DIXER tools, developed at the University of Warsaw [5].

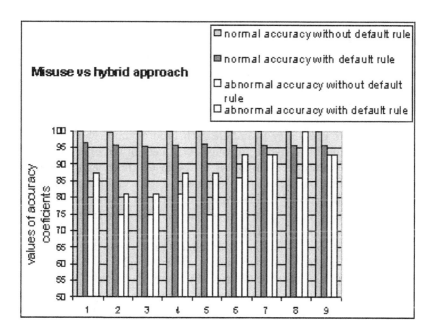

Fig. 3. Variation between values of coefficients of normal and abnormal accuracy, with and without a default rule

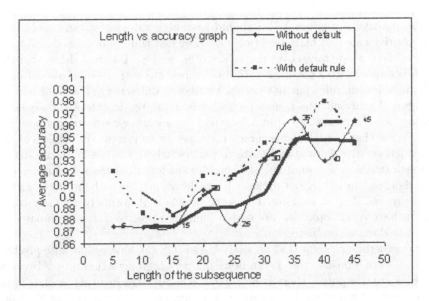

Fig. 4. Graph between average accuracy and length of subsequence, with and without a default rule

6   Conclusion

Rough set theory is applied to many areas as one of many data mining and machine learning techniques. The present paper proposes the use of rough set theory in the area of intrusion detection to make it more suitable as on-line detection system. The main motivation behind using rough set for IDS is that boundary between normal patterns and abnormal ones is not always very sharp which leads to the ambiguity in the decision of the classifiers. Rough set theory is known for its capability to handle such type of data where uncertainty and vagueness is difficult to avoid. In rough sets, most of the operations involve comparing logical operators. Therefore it is also faster in decision making. The resources used in data collection, preprocessing of data and detection of intrusion are directly proportional to the number of features under consideration for each object. Obviously, in order to have a real time response from IDS, number of features should be minimized without affecting the classification power of the system. Rough set theory is capable for of inducing rules that discard redundant attribute values. Knowledge representation is very simple and learning rate is very fast as compared to other techniques. Our study shows that it is possible to detect an attack by mare looking at some portion of the abnormal process. This reduces the dimension of the data to be processed and thus makes the subsequent computations much faster. The decision rules induced by rough set theory are easy to interpret and thus can be useful in further analyzing the events. We have tested our scheme by conducting experiments on DARPA'98 data. Empirical results, reported in the paper, justify our approach of making use of rough set for intrusion detection. As our future work, we intend to use the concept of *incremental learning* so that new rules can be learnt without retraining on whole data. We are also analyzing the IF-THEN rules to better understand the relationship among system calls to gain more insight about attacks. Our future work also includes to combine rough set method with other learning techniques, e.g. neural networks to propose a more robust IDS in terms of accuracy.

Acknowledgement

The authors are thankful to anonymous reviewers for their useful comments to improve the presentation and quality of the paper. The first author is associated with IDRBT as research fellow and thankful to IDRBT for providing financial assistance and infrastructure to carry out this work. The third author is thankful to MIT, India for its funding.

References

1. An A., Huang Y., Huang X., Cercone N.: Feature Selection with Rough Sets for Web Page Classification. In Dubois D., Grzymala-Busse J.W., Inuiguchi M., and Polkowski L. (eds), *Rough Sets and Fuzzy Sets*, Springer-Verlag (2004)
2. Bace R., Mell P.: NIST special publication on intrusion detection system. SP800-31, NIST, Gaithersburg, MD (2001)
3. Bazan J.: A Comparison of Dynamic and non-Dynamic Rough Set Methods for Extracting Laws from Decision Tables, In: Skowron A., Polkowski L.(ed.), *Rough Sets in Knowledge Discovery 1*, Physica-Verlag, Heidelberg, (1998) 321–365
4. Bazan J., Nguyen H. S., Nguyen S. H., Synak P., and Wrblewski J.: Rough set algorithms in classification problem. In: Polkowski L., Tsumoto S., Lin T.Y. (eds.), *Rough Set Methods and Applications*, Physica-Verlag, Heidelberg, (2000) 49-88.
5. Bazan J. G., Szczuka M. S., Wrblewski A.: A New Version of Rough Set Exploration System. In: Proceedings of the Third International Conference on Rough Sets and Current Trends in Computing RSCTC, Malvern, PA, Lecture Notes in Artificial Intelligence vol. 2475, Springer-Verlag (2002) 397-404
   Available at: http://logic.mimuw.edu.pl/~rses/
6. Cabrera J. B. D., Ravichandran B., Mehra R. K.: Detection and classification of intrusions and faults using sequences of system calls. In: ACM SIGMOD Record, Special Issue: Special Section on Data Mining for Intrusion Detection and treat Analysis, Vol. 30(4) (2001) 25-34
7. Cai Z., Guan X., Shao P., Peng Q., Sun G.: A Rough Set Theory Based Method for Anomaly intrusion Detection in Computer Network Systems. J Expert System 20(5) (2003) 251-259
8. Cios K., Pedrycz W., Swiniarski Roman W.: Data mining methods for Knowledge discovery. Kluwer Academic Publisher USA, (2000)
9. DARPA 1998 Data Set, MIT Lincoln Laboratory, available at: http://www.ll.mit.edu/IST/ideval/data/data_index.html
10. Delic D., Lenz Hans-J, Neiling M.: Improving the Quality of Association Rule Mining by Means of Rough Sets. In: Proceedings of the First International Workshop on Soft Methods in Probability and Statistics (SMPS'02), Warsaw (poland) (2002)

11. Forrest S., Hofmeyr S. A., Somayaji A.: Computer Immunology. Communications of the ACM, 40(10) (1997) 88-96
12. Forrest S., Hofmeyr S. A., Somayaji A., Longstaff T. A.: A Sense of Self for Unix Processes. In: Proceedings of the 1996 IEEE Symposium on Research in Security and Privacy. Los Alamitos, CA. IEEE Computer Society Press, (1996) 120-128
13. Garvey T., Lunt T. F.: Model-based Intrusion Detection. In: Proceedings of the 14th National Computer Security Conference. (1991) 372-385
14. Grzymala-Busse J. W.: A New Version of the Rule Induction System LERS. Fundamenta Informaticae, 31(1) (1997) 27-39
15. Guan J. W., Bell D. A., Liu D. Y.: The Rough Set Approach to Association Rule Mining. In: Proceedings of the Third IEEE International Conference on Data Mining (ICDM'03), (2003)
16. Helman P., Liepins G.: Statistical Foundations of Audit Trail Analysis for the Detection of Computer Misuse. IEEE Transactions on Software Engineering, 19(9) (1993) 886-901
17. Hofmeyr S. A., Forrest A., Somayaji A.: Intrusion Detection Using Sequences of System Calls. Journal of Computer Security, 6 (1998) 151-180
18. Ilgun K.: USTAT: A Real-Time Intrusion Detection System for UNIX. In: Proceedings of the 1993 IEEE Symposium on Research in Security and Privacy. (1993) 16-28
19. Ilgun K., Kemmerer R. A., Porras P. A.: State Transition Analysis: A Rule-Based Intrusion Detection Approach. IEEE Transactions on Software Engineering 21(3) (1995) 181-199
20. Kemmerer R. A.: NSTAT: A Model-based Real-time Network Intrusion Detection System. Technical Report, Number TRCS97-18, Computer Science, University of California, Santa Barbara. (1998)
21. Kumar S., Spafford E.: A Pattern-Matching Model for Intrusion Detection. In: Proceedings National Computer Security Conference, (1994) 11-21
22. Lee W., Stolfo S., Chan P.: Learning Patterns from Unix Process Execution Traces for Intrusion Detection. In: Proceedings of the AAAI97 workshop on AI methods in Fraud and risk management. AAAI Press. (1997) 50-56
23. Lee W., Stolfo Salvatore J.: Data Mining Approaches for Intrusion Detection. In: Proceedings of the 7th USENIX Security Symposium (SECURITY-98), Usenix Association, January 26-29. (1998) 79-94
24. Lian-hua Z., Guan-hua Z., Lang YU., Jie Z., Ying-cai B.: Intrusion Detection Using Rough Set Classification. Journal of Zhejiang University SCIENCE Vol. 5(9) (2004) 1076-1086
25. Lin T. Y.: Anomaly Detection: A Soft Computing Approach. In: Proceedings of the 1994 Workshop on New Security Paradigms, Little Compton, Rhode Island, United States, IEEE Computer Society Press (1994) 44-53
26. Lingras P.: Rough Set Clustering for Web Mining. In: Proceedings of the IEEE International Conference on Fuzzy Systems 2002, Honolulu, Hawaii (2002)
27. Lunt T. F.: Using Statistics to Track Intruders. In: Proceedings of the Joint Statistical Meetings of the American Statistical Association (1990)
28. Lunt T. F., Tamaru A., Gilham F., Jagannathan R., Neumann P. G., Javitz H. S., Valdes A., Garvey T. D.: A Real-Time Intrusion Detection Expert System (IDES) Technical Report, SRI Computer Science Laboratory (1992)
29. Mukherjee B., Heberlein L. T., Levitt K. N.: Network Intrusion Detection. IEEE Network. 8(3) (1994) 26-41

30. Mukkamala R., Gagnon J., Jajodia S.: Integrating Data Mining Techniques with Intrusion detection Methods. In: Research Advances in database and Information System Security: IFIPTCII, 13th working conference on Database security, July, USA, Kluwer Academic Publishers (2000)

31. Pawlak Z.: Rough sets: Theoretical aspects of reasoning about data. Kluwer Academic Publishers, Dordrecht (1991)

32. Porras P. A.: STAT – A State Transition Analysis Tool For Intrusion Detection. Technical Report, Number TRCS93-25, Computer Science. University of California, Santa Barbara (1993)

33. Rawat S., Gulati V. P., Pujari A. K.: Frequecy And Ordering Based Similarity Measure For Host Based Intrusion Detection. J Information Management and Computer Security. 12(5), Emerald Press (2004) 411-421

34. Sebring M. M., Shellhouse E., Hanna M. E., Whitehurst R. A.: Expert System in Intrusion Detection: A Case Study. In: Proceedings of the 11th National Computer Security Conference, (1988) 74-81

35. Stefanowski J.: On Rough Set Based Approaches to Induction of Decision Rules. In: Polkowski L, Skowron A (eds) *Rough Sets in Data Mining and Knowledge Discovery, vol 1.* Physica Verlag, Heidelberg. (1998) 500-529

36. Tandon G., Chan P.: Learning Rules from System Calls Arguments and Sequences for Anomaly Detection. In: ICDM Workshop on Data Mining for Computer Security (DMSEC), Melbourne, FL. (2003) 20-29

37. Warrender C., Forrest S., Pearlmutter B.: Detecting Intrusions Using System Calls: Alternative Data Modelss. In: IEEE Symposium on Security and Privacy (1999)

38. Wespi A., Dacier M., Debar H.: Intrusion Detection Using Variable-Length Audit Trail Patter. In : LNCS # 1907, RAID 2000. Toulouse, France. (2000) 110-129

39. Zhu D., Premkumar G., Zhang X., Chu Chao-Hsien: Data mining for Network Intrusion Detection: A comparison of alternative methods. J. Decision Sciences 32(4) (2001) 635-660

Incremental Learning and Evaluation of Structures of Rough Decision Tables

Wojciech Ziarko

Department of Computer Science, University of Regina,
Regina, SK, S4S 0A2, Canada

Abstract. Rough decision tables were introduced by Pawlak in the context of rough set theory. A rough decision table represents, a non-functional in general, relationship between two groups of properties of objects, referred to as condition and decision attributes, respectively. In practical applications, the rough decision tables are normally learned from data. In this process, for better coverage of the domain of interest, they can be structured into hierarchies. To achieve convergence of the learned hierarchy of rough decision tables to a stable final state, it is desirable to avoid total regeneration of the learned structure after new objects, not represented in the hierarchy, are encountered. This can be accomplished through an incremental learning process in which the structure of rough decision tables is updated, rather than regenerated, after new observations appeared. The introduction and the investigation of this incremental learning process within the framework of the rough set model is the main theme of the article. The article is also concerned with evaluation of learned decision tables and their structures by introducing the absolute gain function to measure the quality of information represented by the tables.

1   Introduction

Decision tables *learned* form data, referred here as *rough decision tables*, were originally proposed by Pawlak in the framework of rough set theory [1]. A rough decision table represents a relation, typically functional or partially functional, between a group of input values and a set of output (decision) values. As opposed to classical decision tables defined manually based on some pre-existing knowledge [13], the rough decision tables are derived from data representing observations about objects, states etc. belonging to a domain of interest referred to as *the universe*. The observations are collected in the form of property vectors accumulated from sensors, test results or by some other measurement techniques. In applications, a rough decision table obtained from data models a relationship between properties of objects. The objective is to learn the true and complete model of the relationship based on finite "training" sample collection of observations drawn from often infinite universe. The relationship of interest is a functional, or partial functional dependency between object properties.

The generalized version of rough decision tables, defined in the context of the variable precision rough set model [4],[3],[2] and called *probabilistic decision*

J.F. Peters and A. Skowron (Eds.): Transactions on Rough Sets IV, LNCS 3700, pp. 162–177, 2005.
© Springer-Verlag Berlin Heidelberg 2005

tables [5], make use of probabilistic information associated with occurrences of object properties (attribute values) or their combinations. The probabilistic decision tables are oriented towards modelling probabilistic relations among object properties while avoiding any assumptions about probabilistic independence and probability distributions of property values.

In this article, we are concerned with acquisition of functional and partial functional relations represented by rough decision tables learned from data. Relations of this kind are of interest in many applications, in particular in control-related problems where often there is no room for any uncertainty in decision making. The learning methodology presented here is however extendible to probabilistic decision tables [5]. The methodology is concerned with forming hierarchical structures of decision tables during learning, rather than using single level table. The motivation behind this approach is twofold: the reduction of the computational complexity of the learning process and the reduction of the degree of non-determinism of the learned relation [6].

The hierarchical structure of decision tables is formed by treating the rough boundary area of initial (root) table as a training sample drawn from a proper subset of the universe, to be used as a basis of the next, child layer table derivation. This basic step is then repeated recursively for the child layer table, and so on. The algorithm HDTL, originally introduced in [6], summarizes the main stages of the hierarchy derivation process. The related extension of the algorithm for deriving hierarchies of probabilistic decision tables appeared in [10].

An important aspect of any machine learning method is handling the adaptation and growth of learned knowledge structures with the arrival of new, previously unseen observations about objects of the universe of interest. In the case of decision table hierarchies, it is always possible to totally regenerate the structure of decision tables after new cases appear, but it is inefficient and may lead to the loss of previously learned model. It may also result in the lack of convergence of the learned tables to a final stable and complete state. A more desirable approach involves gradual adaption of the learned structures, to reflect new knowledge represented by new cases, in a process referred to as *incremental learning* [8],[7].

In what follows, an approach to incremental expansion of decision table hierarchy in the process of the hierarchical structure learning is presented. The presentation also includes the elementary background introduction to relevant notions of rough set theory in the context of probability theory, the presentation of the extended version of HDTL algorithm to handle multiple decisions, the introduction of incremental updating strategies for the learned hierarchy of decision tables and a comprehensive illustrative example. In addition, a new technique for evaluation of learned decision tables is introduced. The *absolute gain function* is proposed to measure the average degree of increase in quality of information, as represented by a decision table in terms of attributes and their values, relative to situation when no attributes are present. The absolute gain function is subsequently extended to evaluation of decision table hierarchies produced by the HDTL algorithm.

Most of the methods discussed in the article have been implemented as a part of an experimental control algorithm acquisition systems for vehicle control [11] and for controlling the movements of components of a simulated walking device [9]. They are also being used in experimentation with speaker-independent recognition of isolated spoken words.

2 Probabilistic View of Rough Sets

In this section we briefly review the essential assumptions and definitions of the rough set theory in the context of probability theory.

One of the prime notions is the universe of interest, a set of objects $e \in U$ about which observations are acquired by sampling sensor readings or by some other means. The existence of *probabilistic measure* P over σ-*algebra* of *measurable subsets* of U is also assumed. We will assume here that the universe is infinite in general, but that we have access to a finite sample $S \subseteq U$ expressed by accumulated observations about objects in S. The sample represents available information about the universe U. We will say that a subset $X \subseteq U$ *occurred* if $X \cap S \neq \emptyset$, where $X \cap S$ is a *set of occurrences* of X. It is assumed that all subsets X under consideration are measurable with $0 < P(X) < 1$. That is, from the probabilistic perspective, they are likely to occur but their occurrence is not certain. Regarding the notation, the notational distinction between sample $S \subseteq U$ and the universe U will be made only when there is a need to emphasize the fact that we are dealing with a sample of larger universe and an estimate of a probability rather than the actual probability. In all other situations symbol U will be used to denote both the sample and the universe.

We also assume, that observations about objects are expressed through values of a finite set of functions $C \cup D$ on U, referred to as *attributes*. The functions belonging to the set C are called *condition attributes*, whereas functions in D are referred to as *decision attributes*. We can assume, without loss of generality, that there is only one decision attribute, that is $D = \{d\}$. Each attribute a belonging to $C \cup D$ is a mapping $a : U \rightarrow V_a$, where V_a is a finite set of values called the *domain* of the attribute a. In many applications, attributes are functions obtained by discretizing values of real-valued variables representing measurements taken on objects. The set of condition attributes C defines a joint mapping denoted as $\mathbf{C} : U \rightarrow \mathbf{C}(U) \subseteq \otimes_{a \in C} V_a$, where $\otimes$ denotes Cartesian product operator of all domains of attributes in C. Both condition attributes and the decision attributes jointly define a mapping denoted as $\mathbf{C} \cup \mathbf{D} : U \rightarrow \mathbf{C} \cup \mathbf{D}(U) \subseteq \otimes_{a \in C \cup D} V_a$. The mapping $\mathbf{C} \cup \mathbf{D} : S \rightarrow \mathbf{C} \cup \mathbf{D}(S)$ can be represented by an *information table* consisting of *information vectors*, corresponding to elements of the collection $\mathbf{C} \cup \mathbf{D}(S)$. The information table summarizes all known associations between objects and attribute values, as represented by the sample $S \subseteq U$. An example information table computed from the sample S is shown in Table 1.

For each combination of condition attribute values $x \in \mathbf{C}(U)$, the set $E_x = \mathbf{C}^{-1}(x) = \{e \in U : \mathbf{C}(e) = x\}$ is called a *C-elementary set*. That is, each *C*-elementary set is a collection of objects with identical values of the attributes

Table 1. Example Information Table

U	a	b	c	d
e_1	1	1	2	**1**
e_2	1	1	2	**1**
e_3	1	0	1	**1**
e_4	1	0	1	**2**
e_5	2	2	1	**2**
e_6	2	2	1	**2**
e_7	2	0	2	**1**
e_8	1	1	2	**1**
e_9	0	2	1	**3**
e_{10}	2	2	1	**2**
e_{11}	2	2	1	**2**
e_{12}	0	2	1	**1**

belonging to the set C. Similarly, the subsets of the universe corresponding to the information vectors are called *(C,D)-elementary sets*. The elementary sets are equivalence classes of an equivalence relation, called indiscernibility relation [1], which makes equivalent any two objects having identical values of attributes in C. In general, any subset z of values of attributes $B \subseteq C \cup D$ of an information vector x corresponds to a set of objects $\mathbf{B}^{-1}(z)$ matching these values, where $\mathbf{B}^{-1}(z) = \{e \in U : \mathbf{B}(e) = z\}$.

The collection of all C-elementary sets forms a partition of the universe U, denoted as $U/C = \{E_x\}_{x \in \mathbf{C}(U)}$. Clearly, from our initial assumption and from the basic properties of the probability measure P, follows that for all $E \in U/C$ we have $0 < P(E) < 1$ and $\sum_{E \in U/C} P(E) = 1$. In practice, the partition U/C is a representation of the limits of our ability to distinguish individual objects of the universe by using values of condition attributes. The pair $\mathbb{A} = (U, U/C)$ is called the *approximation space*. Similarly to condition attributes, the decision attribute

Table 2. Information Table with Rough Region Designation

U	a	b	c	d	Region
e_1	1	1	2	**1**	POS
e_2	1	1	2	**1**	POS
e_3	1	0	1	**1**	BND
e_4	1	0	1	**2**	BND
e_5	2	2	1	**2**	POS
e_6	2	2	1	**2**	POS
e_7	2	0	2	**1**	POS
e_8	1	1	2	**1**	POS
e_9	0	2	1	**3**	BND
e_{10}	2	2	1	**2**	POS
e_{11}	2	2	1	**2**	POS
e_{12}	0	2	1	**1**	BND

induces a partition of U consisting of *decision categories* (D-elementary sets) $U/D = \{F_1, F_2, \ldots F_k\}$, $k \geq 2$, corresponding to different values of the attribute $d \in D$. That is, if the domain of d is $V_d = \{v_1^d, v_2^d, \ldots, v_k^d\}$, then $F_i = \{e \in U : d(e) = v_i^d\}$, $1 \leq i \leq k$. As with C-elementary sets, for all D-elementary sets $F \in U/D$ we have $0 < P(F) < 1$ and $\sum_{F \in U/D} P(F) = 1$

Our interest here is in the analysis of the relation between condition and decision attributes, as represented by the information table. In particular, we are interested in investigating the degree of determinism of the relation, i.e. whether it is functional, or if not, which part of it is not functional. For that purpose, the *rough approximation regions* of the decision classes F_i, $(i = 1 \ldots k)$ are defined as follows [1].

The *positive region* $POS^{C,D}(i)$ of the class F_i, $1 \leq i \leq k$ in the approximation space $\mathbb{A}$ is defined as $POS^{C,D}(i) = \cup\{E \in U/C : E \subseteq F_i\}$. The positive region $POS^{C,D}(i)$ is composed of C-elementary sets contained in the decision category F_i.

The *negative region* $NEG^{C,D}(i)$ of the class F_i is defined as $NEG^{C,D}(i) = \cup\{E \in U/C : E \subseteq (U - F_i)\}$. The negative region $POS^{C,D}(i)$ is composed of C-elementary sets contained in the complement of the decision category F_i.

The complement of the positive regions of all decision classes $BND^{C,D}(U) = U - \cup_i POS^{C,D}(i)$ is called the *boundary region* of the partition U/D in the approximation space $\mathbb{A}$. The C-elementary sets forming the boundary region are not contained in any decision category.

Finally, the *positive region* $POS^{C,D}(U)$ of the partition U/D in the approximation space $\mathbb{A}$ is defined as $POS^{C,D}(U) = \cup\{E \in U/C : \exists_i E \subseteq POS^{C,D}(i)\}$. That is, $POS^{C,D}(U) = \cup_i POS^{C,D}(i)$. The positive region of the partition U/D includes all C-elementary sets contained in a decision category.

An example of an information table with the associated designation of an approximation region for each object of the universe is shown in Table 2.

3   Rough Decision Tables

To define rough decision tables precisely, we will introduce the following notation. For any elementary set E, we will denote the by $d(E)$ the collection of associated values of decision attribute d, that is the values assigned to some objects e belonging the set E:

$$d(E) = \{v \in V_d : (\exists e \in E) d(e) = v\} \tag{1}$$

We define the *rough decision table* $DT^{C,D}$ as a mapping derived from the information table such that $\mathbf{C}(U) \rightarrow 2^{\{v_1^d, v_2^d, \ldots, v_k^d\}}$. The mapping is associating each combination of condition attribute values $x \in \mathbf{C}(U)$ with a set of possible values a decision attribute may take on objects belonging to the elementary set E_x:

$$DT_U^{C,D}(x) = d(E_x). \tag{2}$$

In tabular form, the logical disjunction symbol $\vee$ will be used to denote the possible values of a decision attribute an object may take, as illustrated in Table 3.

Table 3. Decision Table Derived from Information Table 2

a	b	c	d
1	1	2	**1**
1	0	1	**1 ∨ 2**
2	2	1	**2**
2	0	2	**1**
0	2	1	**3 ∨ 1**

Because there is only one value of the decision attribute associated with an elementary set in the positive region, the rough decision table can be expressed as a follows:

$$DT_U^{C,D}(x) = \begin{cases} d(E_x) \Leftrightarrow E_x \subseteq BND^{C,D}(U) \\ v_i^d \quad \Leftrightarrow E_x \subseteq POS^{C,D}(i) \end{cases} \tag{3}$$

In the mapping, the value $d(E_x)$ is representing the disjunction of more than one possible decisions. We say that the rough decision table is *deterministic* if the boundary region is empty. Otherwise, the rough decision table is said to be *non-deterministic*. The rough decision table is an approximate representation of the relation between condition and decision attributes. It is most useful for decision making or prediction when the relation is deterministic (functional) or largely deterministic (partially functional).

We define the *positive region decision table* $DT^{C,D}$ as a mapping associating each combination of condition attribute values $x \in \mathbf{C}(U)$ with its unique designation of the rough approximation region the respective elementary set E_x is included in:

$$\underline{DT}_U^{C,D}(x) = \begin{cases} * \Leftrightarrow E_x \subseteq BND^{C,D}(U) \\ v_i^d \Leftrightarrow E_x \subseteq POS^{C,D}(i) \end{cases} \tag{4}$$

The mapping produces one of $k + 1$ values, with the value * representing the lack of definite decision, that is the boundary region designation. Clearly, positive region decision tables are always deterministic. The positive region decision table is said to be *empty* if it is mapping onto only one symbol *, which corresponds to the absence of the positive region. The positive region decision table based on information Table 2, is demonstrated in Table 4.

Table 4. Positive Region Decision Table

a	b	c	d
1	1	2	**1**
1	0	1	*
2	2	1	**2**
2	0	2	**1**
0	2	1	*

As indicated in Section 2, in many practical applications, complete information about the universe U is unavailable. In real-life situations we normally deal with samples and probability estimates from those samples. The available information typically represents a part $S \subseteq U$ of the universe. Consequently, any decision table derived from the sample S is also incomplete in general, partially and perhaps inaccurately representing the relationships existing in the universe U. Such a decision table will be called *empirical decision table*. In general, it is not possible to assert the correctness and completeness of empirical decision tables. However, we can prove that if individual objects in S are selected at random then, with the growth of S, empirical decision tables will converge, in probabilistic sense, to true rough decision table extending over the whole universe. Consequently, empirical decision tables are subject of learning and adaptation with the arrival of new observations. An example empirical decision table derived from information table given in Table 2 is shown in Table 3.

4   Evaluation of Rough Decision Tables

Rough decision tables can be evaluated by *dependency* and *predictive gain* measures. The degree of determinism represented by the decision table can be estimated in terms of a *dependency measure*, denoted as $\gamma(DT^{C,D})$. The dependency is given by the probability of positive region of the partition U/D:

$$\gamma(DT_U^{C,D}) = P(POS^{C,D}(U)). \tag{5}$$

For the purpose of evaluation of empirical decision tables, the probability can be estimated from the sample S in usual way by:

$$\gamma(DT_S^{C,D}) = \frac{card(POS^{C,D}(S))}{card(S)}, \tag{6}$$

where $card(*)$ denotes the number of elements in the set (the set cardinality). The dependency measure computed from the sample reflects the proportion of objects in S which are associated with a unique decision category. For example, the dependency degree between condition attributes $C = \{a, b, c\}$ and the decision attributes $D = \{d\}$ of Table 2 is given by $\gamma(DTC, D) = \frac{8}{12}$.

The information gain of individual elementary set $E \in U/C$ with respect to predicting the occurrence of the target set $F \in U/D$ can be expressed by means of *local absolute gain* function defined by

$$gabs(E) = |P(F|E) - P(F)|, \tag{7}$$

where $|*|$ denotes absolute value function.

The overall *predictive gain* of a decision table, with respect to the decision category $F \in U/D$, is given by the *absolute gain function* $0 \le GABS_U^{C,D_F} < 1$ defined as

$$GABS_U^{C,D_F} = \sum_{E \in U/C} P(E)gabs(E). \tag{8}$$

Table 5. Example Summary of Elementary Sets and Cardinalities

E	a	b	c	$card(E)$	$card(F \cap E)$
E_1	1	1	2	3	3
E_2	1	0	1	2	1
E_3	2	2	1	4	0
E_4	2	0	2	1	1
E_5	0	2	1	2	1

Alternatively, the absolute gain function can be expressed as

$$GABS_U^{C,D_F} = \sum_{E \in U/C} |P(F \cap E) - P(F)P(E)|, \qquad (9)$$

in which case it can be perceived as an average measure of degree of probabilistic independence between groups of attributes C and D.

If sample $S \subseteq U$ is used to estimate the probabilities appearing in formula (9), set cardinalities can be used to estimate the value of the absolute gain function:

$$GABS_S^{C,D_F} = \frac{\sum_{E \in S/C} |card(S)card(F \cap E) - card(F)card(E)|}{card(S)^2}, \qquad (10)$$

The absolute gain function measures the average degree of change of occurrence probability of F, relative to its *prior probability* $P(F)$, as a result of occurrences of elementary sets $E \in U/C$. The prior probability represents the likelihood of occurrence of target set F in the absence of any additional information expressed in terms of attribute-value vectors. The notion of the absolute gain function stems from the idea of the *relative gain* measure introduced in [12]. More details about the gain measures are provided in [15] and [16]. A comprehensive review of related measures can be found in [14].

To illustrate the application of the absolute gain function (10), we will calculate its value for the decision category $F = \{e \in S : d(e) = 1\}$ based on the information contained in Table 2. To do it conveniently, it is useful to enumerate all elementary sets and the associated cardinalities in the Table 5. Given that $card(F) = 6$ and $card(S) = 12$, we can easily calculate from the Table 5 the absolute gain value for Table 2, which is $GABS_S^{C,D_F} = 0.333$. This represents 0.333 average increase in the certainty of prediction of $d = 1$ or $d \neq 1$, due to presence of condition attributes, in comparison to the situation when no condition attributes are available, as represented by the prior probability $P(F) = 0.5$.

The overall predictive gain of a decision table is defined as the expected value $0 \leq GABS_U^{C,D} < 1$ of absolute gain functions of its all decision categories:

$$GABS_U^{C,D} = \sum_{F \in U/D} P(F)GABS_U^{C,D_F} \qquad (11)$$

The *overall gain function* can be estimated from empirical decision tables in usual way by estimating the respective probabilities from the sample S.

5   Hierarchy Generation Algorithm HDTL

In this section, the basic procedure for forming decision table hierarchies with progressively narrower boundary regions is described.

Given information table (*a parent table*) with non-empty positive region and non-empty boundary region, one can associate with it a positive region decision table, which provides the specification of decisions to be taken for all positive region elementary sets and the definition of the boundary area. The boundary area can be next treated as a new universe, if additional condition attributes are available, to produce a "child" level positive region decision table and the definition of its boundary area. By applying this step recursively, one can define positive region decision table based on the boundary region of the parent table, and so on, as long as the boundary areas of the parent tables are non-empty and extra attributes are available to produce a non-empty positive region decision table at each level. In this way, a linear hierarchy of positive region rough decision tables can be formed, in which each non-root table is based on the boundary area of its predecessor. The final table in the hierarchy is the rough deterministic or non-deterministic decision table. The latter case occurs when no more attributes are available to obtain non-empty positive region decision table on a given level. In the process of building the hierarchy, the child decision attributes D' are simply parent decision attributes D restricted to the boundary area, that is that is, $D' = D_{|BND^{C,D}(U)}$. However, the child condition attributes C' are new attributes that can be unrelated to the parent condition attributes C. In practice, to achieve the reduction of the boundary, the condition attributes used in each non-root layer of the hierarchy should be significantly different from the ones used in the parent layer. They should not be merely restrictions of parent level attributes to the child level domain, that is, to the boundary region of the parent level. They should represent essential new information helping to categorize boundary area of a parent into classes which are impossible to achieve with parent's existing attributes.

The method for the hierarchy construction can be summarized more precisely in the form of an algorithm referred to as **HDTL-M** . The presented algorithm is a multi-valued generalization of the original **HDTL** method published in [6]. In this algorithm, U is the initial set of objects, C is the initial set of attributes and D is the decision attribute of the initial set of objects. In addition, $\mathbb{U}$, $\mathbb{C}$, $\mathbb{D}$ are variables representing current set of objects, current condition attributes and current decision attribute respectively. We will denote by α, $0 < \alpha \le 1$ the desired degree of the global dependency of the whole structure of decision tables. The degree α will be used as a termination criterion to stop the expansion of the hierarchy of decision tables once satisfactory global dependency degree Γ of the whole structure of decision tables with the decision attribute has been reached. The Γ global dependency is a generalization of the degree of dependency γ for a single table and is defined in detail in the next section.

Algorithm. HDTL-M

1. $\mathbb{U} \longleftarrow U, \mathbb{C} \longleftarrow C, \mathbb{D} \longleftarrow D$
2. **Define** *root layer information table and the associated positive region decision table* $\underline{DT}_{\mathbb{U}}^{\mathbb{C},\mathbb{D})}$
3. **Repeat**
   {
4. **Compute** *dependency of current layer* $\gamma(\underline{DT}_{\mathbb{U}}^{\mathbb{C},\mathbb{D})})$
5. **Compute** *dependency* Γ *of the current hierarchy*
6. **If** $(\gamma(\underline{DT}_{\mathbb{U}}^{\mathbb{C},\mathbb{D})}) = 0$ or $\Gamma \geq \alpha$ **Then**
   {
   Output *current layer decision table* $DT_{\mathbb{U}}^{\mathbb{C},\mathbb{D}}$
   Exit
   }
7. **Attempt** *to define new condition attributes on* $BND^{\mathbb{C},\mathbb{D}}(\mathbb{U})$
8. **If** (*new condition attributes not defined*) **Then**
   {
   Output *current layer decision table* $DT_{\mathbb{U}}^{\mathbb{C},\mathbb{D}}$
   Exit
   }
9. **Output** *current layer positive region decision table* $\underline{DT}_{\mathbb{U}}^{\mathbb{C},\mathbb{D}}$
10. **Define** *new universe, initiate new layer,* $\mathbb{U} \longleftarrow BND^{\mathbb{C},\mathbb{D}}(\mathbb{U})$
11. **Assign** *new condition attributes,* $\mathbb{C} \longleftarrow$ **new** $(\mathbb{C},\mathbb{D})$
12. **Define** *new information table and the decision table* $DT_{\mathbb{U}}^{\mathbb{C},\mathbb{D})}$
   }

The algorithm **HDTL-M** produces a sequence of connected decision tables. All tables in the sequence, with except of the last one, are positive region decision tables. The final table is a rough decision table as defined in Section 3. The computational procedure is linear in the number of elementary sets and in the number of objects.

To illustrate the algorithm **HDTL-M**, we can consider the information given in Table 2 as the initial input. After the first iteration of the algorithm, the first layer positive region decision table (Table 4) was produced and a new information table (Table 6) was defined based on the boundary area of Table 2, with new condition attributes m and n and $U' = BND^{C,D}(U)$. The corresponding second layer positive region decision table, obtained from second layer information table, is shown in Table 7.

Similarly, based on the boundary area of the second layer information table (Table 6), the final layer of the hierarchy can be constructed, assuming that there is only one new condition attribute p on this level with $U" = BND^{C',D'}(U')$. The respective information and decision tables for this layer are shown in Tables 8 and 9 respectively.

Table 6. Second Layer Information Table

U'	m	n	d	Region
e_3	x	y	1	BND
e_4	x	z	2	POS
e_9	x	y	3	BND
e_{12}	z	y	1	POS

Table 7. Second Layer Decision Table

m	n	d
x	y	*
x	z	2
z	y	1

Table 8. Final Layer Information Table

U''	p	d	Region
e_3	1	1	POS
e_9	2	3	POS

Table 9. Final Layer Decision Table

p	d
1	1
2	3

6 Evaluation of Decision Table Hierarchies

As with single layer tables, one can evaluate the total degree of determinism of the relation between condition and decision attributes as represented by the hierarchy. The total degree of dependency can be computed by recursively applying, starting from the leaf table and going up to the root table, the following formula for computing the dependency of a parent table $\gamma(DT_U^{C,D})$, if the dependency of a child table $\gamma(DT_{U'}^{C',D'})$ is given:

$$\gamma(DT_U^{C,D}) = P(POS^{C,D}(U)) + P(BND^{C,D}(U))\gamma(DT_{U'}^{C',D'}). \qquad (12)$$

The probabilities appearing in the formula (12) can be estimated from the finite sample $S \subseteq U$, leading to the following formulation applicable to empirical decision tables:

$$\gamma(DT_S^{C,D}) = \frac{card(POS^{C,D}(S)) + card(BND^{C,D}(S))\gamma(DT_{S'}^{C',D'})}{card(S)}. \qquad (13)$$

The dependency measure represents the fraction of objects that can be uniquely classified into decision categories by applying the decision tables in the hierarchy. The dependency of the whole structure of decision tables, the last dependency computed in the recursive application of formula (12) or (13), is called a *global dependency* and denoted by Γ.

For example, to compute the dependency of the structure of decision tables used in the previous section to illustrate the algorithm HDTL, the dependency of the third (final) layer table (Table 9) is computed first, which is 1. By applying formula (13), the dependency of the structure composed of third layer table (Table 9) and second layer table (Table 7) is computed next, which yields $\frac{2+2*1}{4} = 1$. Similarly, the dependency of the whole structure composed of layer 1 (Table 4) and layers 2 and 3 is computed by applying formula (13) again: $\frac{8+4*1}{12} = 1$.

In addition to evaluating the hierarchies in terms of dependency measure, they can also be evaluated in terms of absolute gain function. Let R^* represent the final partitioning of the universe U, as obtained by recursive splitting of the universe U by attribute-value vectors appearing in successive levels of the hierarchy. The absolute gain of the hierarchy of decision tables in the context of the partitioning R^* is defined as expected value of the local gains of all elementary sets present in R^*:

$$GABS_S^F = \sum_{E \in R^*} |P(F \cap E) - P(F)P(E)|. \tag{14}$$

In terms of cardinalities of elementary sets computed from the finite sample $S \subseteq U$, the formula (14) can be expressed as:

$$GABS_S^F = \frac{\sum_{E \in R^*} |card(S)card(F \cap E) - card(F)card(E)|}{card(S)^2}. \tag{15}$$

Similarly, the expected gain function can be defined for all decision categories by:

$$GABS_U = \sum_{F \in U/D} P(F)GABS_U^F. \tag{16}$$

Using set cardinalities, based on sample $S \subseteq U$, the formula (16) can be rewritten as:

$$GABS_S = \frac{\sum_{F \in S/D} card(F)GABS_S^F}{card(S)}. \tag{17}$$

The interpretation of the gain measures for hierarchies of decision tables is essentially identical to their interpretation for individual decision tables.

7 Incremental Expansion of Decision Table Hierarchies

One of the key issues in effective machine learning is handling new observations (cases). Models obtained via analysis of empirical data are typically inaccurate and incomplete. New cases may either contradict the model or can fall outside the scope of the model. A common approach to deal with situations like that is to add the new cases to the pool of data previously used to produce the model, and to regenerate the model again, taking the presence of the new cases into account. This approach however suffers from a number of problems. Firstly, the re-computation of the model may lead to the loss of what was learned previously, so that new model could not be related to the previous one, making it impossible to assess the degree of model variation or convergence to a stable state. Secondly, the total re-computation of the model is usually time consuming, particularly when large data collections are used or when the complexity of the model generation algorithms is high. The re-computation of the model can be avoided by using new cases to expand the existing model through relatively minor modifications, while preserving as much as possible of the existing structure. The main advantage of this approach is the possibility of tracking the variations of the model to evaluate its usefulness in terms of its ability to converge to the stable *fully learned* state. The process of learning by adapting the exiting model in response to occurrence of new observations is called *incremental learning*.

Not every modelling technique allows for incremental adaptation, for example it does not appear to be possible in modelling using neural nets. The hierarchies of decision tables, on the other hand, are very well suited for this kind of adaptation. Below we summarize the hierarchy adaptation strategy in response to occurrences of new observations falling into one of four possible categories. We will say that a new case is *inconsistent* if it matches the pattern of condition attributes values of an elementary set belonging to the positive region but it does not match the value of the decision attribute. Otherwise, the new case is *consistent*.

1. **New case falls into boundary region of one of the layers of the hierarchy of decision tables.**
   Action: Adjust the count of the cases belonging to the matching elementary set and the count of the universe of the affected layer; recompute the dependency of the hierarchy of decision tables.

2. **New consistent case falls into positive region of one of the layers of the hierarchy of decision tables.**
   Action: Adjust the count of the cases belonging to the matching elementary set and the count of the universe of the affected layer; recompute the dependency of the hierarchy of decision tables.

3. **New inconsistent case falls into positive region of one of the layers of the hierarchy of decision tables.**
   Action: Shift the affected elementary set E into boundary area; expand the child layer information table by the elementary set E by individually adding cases belonging to E and recursively applying the adaptation rules; recompute the dependency of the adapted hierarchy of decision tables; recompute new decision tables for all affected subordinate layers.

Table 10. First Layer Information Table After Modification

U	a	b	c	d	Region
e_1	1	1	2	1	BND
e_2	1	1	2	1	BND
e_3	1	0	1	1	BND
e_4	1	0	1	2	BND
e_5	2	2	1	2	POS
e_6	2	2	1	2	POS
e_7	2	0	2	1	POS
e_8	1	1	2	1	BND
e_9	0	2	1	3	BND
e_{10}	2	2	1	2	POS
e_{11}	2	2	1	2	POS
e_{12}	0	2	1	1	BND
e_{13}	*1*	*1*	*2*	*2*	BND

Table 11. First Layer Decision Table After Modification

a	b	c	d
1	*1*	*2*	*
1	0	1	*
2	2	1	2
2	0	2	1
0	2	1	*

4. **New case, not matching any of the elementary sets, falls into one of the layers of the hierarchy of decision tables.**
   Action: Expand the universe and the information table of the affected layer by creating a new positive region elementary set; recompute the dependency of the hierarchy of decision tables; modify decision table for the affected layer by adding an extra row.

Among the above adaptation strategies, the strategy number 3 is most complex and potentially has most significant impact on the structure of the hierarchy of decision tables due to possible propagation of change through multiple subordinate layers of the structure.

To illustrate this strategy based on the Table 2, assume that a new inconsistent case, denoted by e_{13} and represented by the information vector $(1, 1, 2, 2)$ of first layer attributes a, b, c and d, respectively, appeared. According to strategy 3, the boundary area of root information table was expanded, as shown in Table 10. The associated decision table was modified accordingly (Table 11). In addition, the universe of second layer table was expanded and the new case was classified into positive region, based on values of second layer attributes m and n. This is reflected in modified second layer information and decision tables shown in Tables 12 and 13, respectively. The final layer was not affected by the change.

Table 12. Second Layer Information Table After Modification

U'	m	n	d	Region
e_1	x	x	**1**	POS
e_2	x	x	**1**	POS
e_3	x	y	**1**	BND
e_4	x	z	**2**	POS
e_8	z	x	**1**	POS
e_9	x	y	**3**	BND
e_{12}	z	y	**1**	POS
e_{13}	y	y	*2*	POS

Table 13. Second Layer Decision Table After Modification

m	n	d
x	x	**1**
x	y	*
x	z	**3**
z	x	**1**
z	y	**1**
y	y	*2*

8 Conclusions

The focus of the article is on learning approximate models of functional, or partial functional relations from data. Learning functional or partial functional models from data seems to be significant from application perspective, as in many application areas standard mathematical modelling techniques are not applicable due to excessive problem complexity or lack of continuity. As a possible approach to dealing with this problem, the article introduced an extension HDTL-M of the original algorithm HDTL for building hierarchies of rough decision tables, absolute gain measure for evaluating the quality of the learned tables and of their structures, and several update strategies for managing the growth of the hierarchies in response to occurrences of new observations. The use of incremental learning methodology reduces the complexity of learning and allows for controlling the convergence of the learning process. It appears to be well suited for automated control applications as a technique for control algorithm acquisition from empirical data. Two promising experimental control-related applications have recently been undertaken [9], [11].

Acknowledgment. The research reported in this article has been supported in part by a research grant awarded to the author by the Natural Sciences and Engineering Research Council of Canada. The constructive and helpful remarks of anonymous referees are truly appreciated.

References

1. Pawlak, Z. Rough sets - theoretical aspects of reasoning about data. Kluwer Academic Publishers (1991).
2. Ślęzak, D. Ziarko, W. Variable precision Bayesian rough set model. Proc. of the 9th Intl. Conference on Rough Sets, Fuzzy Sets, Data Mining and Granular Computing, Chongqing, China, Lecture Notes in AI 2639, Springer Verlag, 312-315.
3. Yao, Y.Y. Wong, S.K.M. A decision theoretic framework for approximating concepts. Intl. Journal of Man-Machine Studies, vol. 37, (1992) 793-809.
4. Ziarko, W. Variable precision rough sets model. Journal of Computer and Systems Sciences, vol. 46, no. 1, (1993) 39-59.
5. Ziarko, W. Probabilistic decision tables in the variable precision rough set model. Computational Intelligence, vol. 17, no 3, (2001) 593-603.
6. Ziarko, W. Acquisition of hierarchy-structured probabilistic decision tables and rules from data. Proc. of IEEE Intl. Conference on Fuzzy Systems, Honolulu, USA, (2002) 779-784.
7. Ziarko, W. Ning, S. Machine learning through data classification and reduction. Fundamenta Informaticae, vol. 30, (1997) 371-380.
8. Marek, W. Pawlak, Z. One dimension learning. Fundamenta Informaticae, vol. 8, no. 1, (1985) 83-88.
9. Peng, Y. Application of Rough Decision Tables to Control Algorithm Acquisition. M.Sc. thesis, Computer Science Department, University of Regina, (2005).
10. Aryeetey, K. Ziarko, W. Application of variable precision rough set approach to car driver assessment. In: New Generation of Data Mining Applications, Willey-IEEE Press, (2005).
11. Shang, F. Ziarko, W. Acquisition of control algorithms. Proc. of Intl. Conference on New Trends in Intelligent Information Processing and Web Mining, Zakopane, Poland, Advances in Soft Computing, Springer Verlag, (2003) 341-350.
12. Ziarko, W. Set approximation quality measures in the variable precision rough set model. In: Soft Computing Systems, Management and Applications, IOS Press, (2001) 442-452.
13. Pollack, S. Hicks, H. Harrison, W. Decision tables: theory and practice. Willey-Interscience (1970).
14. Yao, Y, Zhong, N. An analysis of quantitative measures associated with rules. Proc. of the PAKDD'99, Lecture Notes in AI 1574, Springer Verlag, (1999) 479-488.
15. Slezak, D., Ziarko, W. Bayesian rough set model. Proc. of FDM'2002, Maebashi, Japan, 2002, 131-135.
16. Ziarko, W. Probabilistic rough sets. Proc. of RSFDGrC'05, Regina, Canada, 2005 (to appear).

A Framework for Reasoning with Rough Sets

Aida Vitória

Dept. of Science and Technology, Linköping University,
S 601 74 Norrköping, Sweden
aidvi@itn.liu.se

Abstract. Rough sets framework has two appealing aspects. First, it is a mathematical approach to deal with vague concepts. Second, rough set techniques can be used in data analysis to find patterns hidden in the data. The number of applications of rough sets to practical problems in different fields demonstrates the increasing interest in this framework and its applicability. This thesis[1] proposes a language that caters for implicit definitions of rough sets obtained by combining different regions of other rough sets. In this way, concept approximations can be derived by taking into account domain knowledge. A declarative semantics for the language is also discussed. It is then shown that programs in the proposed language can be compiled to extended logic programs under the paraconsistent stable model semantics. The equivalence between the declarative semantics of the language and the declarative semantics of the compiled programs is proved. This transformation provides the computational basis for implementing our ideas. A query language for retrieving information about the concepts represented through the defined rough sets is also discussed. Several motivating applications are described. Finally, an extension of the proposed language with numerical measures is presented. This extension is motivated by the fact that numerical measures are an important aspect in data mining applications.

1 Introduction

This thesis addresses the problem of using rough sets for knowledge representation. We propose an extension of the basic rough set formalism [1] that caters for the representation of vague concepts and reasoning about those concepts.

We present a language that allows the user to define rough sets implicitly. This contrasts with most of the currently existing systems based on rough set techniques where rough sets can only be defined explicitly by a set of examples.

We also introduce a query language to retrieve non-trivial knowledge implied by the defined rough sets.

This introductory section starts by presenting the motivation that drove us into this research (section 1.1), followed by the formulation of the concrete prob-

---

[1] This work has been partially supported by the European Commission and by the Swiss Federal Office for Education and Science within the 6th Framework Programme project REWERSE number 506779 (http://rewerse.net).

J.F. Peters and A. Skowron (Eds.): Transactions on Rough Sets IV, LNCS 3700, pp. 178–276, 2005.

lem addressed in this thesis (section 1.2), and then we highlight the main contributions of our work (section 1.3). Finally, we give an overview of the structure of this thesis (section 1.4).

1.1   Motivation

Rough sets techniques [1,2] can be used to discover new interesting data patterns (or knowledge) hidden in large tables with many lines and several columns. In more concrete terms, the main aim of rough set techniques is to synthesize descriptions of concepts from the data in the tables. The concept descriptions consist of a set of decision rules. These decision rules can be used in two different perspectives. First, rules can be used for building predictive models (i.e. classifiers) from the data. Second, rules can reveal interesting relationships in the data, i.e. each rule may represent an interesting pattern hidden in the data. These techniques have been successfully applied to many real problems in different areas like medicine [3,4], economy [5], and bioinformatics [6]. Therefore, it is not a surprise that rough set methods are enjoying an increasing popularity in the data mining field.

The rough set framework has two major appealing aspects. First, it proposes a method to handle inconsistencies due to imprecise or noisy data. Approximate concept descriptions can then be induced. This point is particularly relevant from the point of view of knowledge representation. As a consequence of using approximations, the derived decision rules describing concepts are categorized into *certain* and *possible* rules. In addition, rough set techniques have a clear mathematical foundation.

Several other important problems are also tackled in the context of rough sets:

– reducing the number of relevant attributes;
– measuring the significance of attributes;
– discovering the degree of dependency between attributes;
– generating classifiers with the possibility to predict more than one class for an object.

What we wish to emphasize here is the relevance of the rough set framework from the knowledge representation and data mining perspectives. The capability to handle vague and contradictory knowledge makes rough sets an important technique that can be incorporated in knowledge base systems. On the other hand, rough set methods can also be used to perform data exploration what makes them relevant from a data mining point of view. These two aspects account for the motivation that drove us in this research.

1.2   Problem Formulation

Most of the research in the rough sets field has been focused on the following issues: algebraic characterization and interpretation of rough sets [7]; relations of rough set theory with other theories to represent knowledge, like modal logics [8]; integration of rough sets with other techniques like inductive logic programming

[9] or fuzzy sets [10,11]; extensions to the basic rough set formalism using different types of indiscernibility relations and more general definitions of upper and lower approximations [8,12,13]; construction of software tools for data mining based on rough sets methods [14]; application of rough set techniques to real problems [3,6,15].

Most of the current rough sets techniques and software systems based on them only consider rough sets defined explicitly by concrete examples given in tabular form. The previous research mostly disregards the following problems.

- How to define rough sets in terms of other rough sets. For instance, we may wish to express that a rough set is obtained as a projection of another rough set over a subset of its attributes.
- How to incorporate domain or expert knowledge. An example of domain knowledge could be *"if a gene participates in cytoplasmic transport then it may also participate in the transport process"*. A question arises of how concept approximations can be derived by taking into account not only the examples provided explicitly by one or more tables but also the domain knowledge.

The problems described above are in the focus of this thesis. They are also addressed in [16] presenting the Computer Aided Knowledge Engineering technique supported by system *CAKE*. However, several important differences exist between this system and our framework. This issue is further discussed in section 7.

1.3   Contributions

The main contributions of this thesis are as follows.

- Definition of a language [17,18,19] that caters for implicit definitions of rough sets obtained by combining different regions of other rough sets (e.g. lower approximations, upper approximations, and boundaries). The language also allows defining rough sets in terms of explicit examples, as in most currently available systems. A declarative semantics for the language is also proposed.
- Definition of a query language [19] for retrieving information about the concepts represented through the defined rough sets.
- Definition of a computational engine for the proposed language. This engine is obtained by a translation of the proposed language to the language of extended logic programs, under the paraconsistent stable model semantics. We also prove the correctness of the proposed translation with respect to the declarative semantics of the language. In this way we establish a link between two important fields, rough set theory and paraconsistent logic programming [20].
- Several motivating applications are discussed [21]. These examples show that several useful techniques and extensions to rough sets, reported in the literature, and implemented in an "ad hoc" way can be naturally expressed in our language.

- Extension of the proposed language with numerical measures [22] is presented. This extension is motivated by the fact that numerical measures are an important aspect in data mining applications.
- A software system based on the proposed language. The system was developed by R. Andersson [23] under joint supervision of the author and J. Małuszyński.

Although the major ideas presented in this thesis have been previously published in conference and journal papers, there are some new notions that are not addressed in those previous publications. First, we present a declarative semantics for the language that caters for implicit definition of rough sets, without considering quantitative measures. Second, the declarative semantics of the query language is also formalized. Third, the correctness of the transformation of rough programs (without quantitative measures) into extended logic programs is only proved in this thesis. Finally, the correctness of the query answering algorithm presented in this work has not been previously published.

This thesis also gives a more comprehensive introduction to both rough sets and logic programming main notions than the previous publications. Hence, readers acquainted with none or just one of the fields can easily read this work.

1.4 Structure of the Thesis

The rest of this thesis is organized as follows.

- Section 2 gives an introduction to rough sets and a brief overview of how several main problems are addressed in this framework.
- Section 3 surveys some important notions of logic programming and paraconsistent stable model semantics. These topics help the reader to understand the transformation technique applied to the proposed language.
- Section 4 introduces formally a language that caters for implicit definitions of rough sets in terms of other rough sets. We present the declarative semantics of the language. We also show a transformation of programs in this language to paraconsistent logic programs. Moreover, we prove that this transformation is correct with respect to the declarative semantics of the language. In addition, a query language is also defined and an algorithm to obtain answers to the queries is discussed.
- Section 5 demonstrates the feasibility of our approach on practical applications by formulating in our language several problems, presented in the rough set literature.
- Section 6 proposes an extension of the language with numerical measures. We also give an overview of a software system, available through a Web page, based on these ideas.
- Section 7 concludes this thesis and points to several problems that deserve further research.

2   Basics of Rough Set Theory

This section provides a brief overview of rough set theory. Rough set theory was introduced by Z. Pawlak [1,24] in the early eighties as a methodology for handling uncertainty in data.

The underlying idea of rough set theory is that several objects may look similar due to the limitations in our knowledge. Hence, it is only possible to distinguish classes of objects rather than individual objects. Consequently, only approximate descriptions of concepts (sets of objects) can be constructed. Section 2.1 discusses this idea while section 2.2 presents the notion of concept approximations and formalizes the notion of rough set.

One of the important problems addressed in the rough set framework is the generation of decision rules from which classifiers are then built. Section 2.3 is devoted to this topic.

Numerical measures are another central problem in rough set theory and this is the issue addressed in section 2.4. We first survey some basic numerical measures that can be used to analyze the quality of the derived decision rules. We then discuss how to measure the degree of dependency between attributes and significance of attributes.

Classifiers obtained by the rough set techniques may predict more than one class for an object. The problem of how multiple class prediction can be handled is briefly surveyed in section 2.5.

We conclude this section by presenting, in section 2.6, an extension to the basic rough set formalism, called Variable Precision Rough Set Model, that is widely used in practical applications.

Most of the contents of this section are based on the ideas presented in the tutorials [25,26].

2.1   Rough Sets: The Main Idea

Datasets in many practical problems are presented as a single database relation or table. For instance, entries in the table may correspond to persons with sight problems and they record for each person whether he has astigmatism, the person age, whether the tear production is normal or reduced, and whether the person is currently using spectacles. Assume that all these persons have experimented the use of contact lenses. The table also records for each person whether he has experienced any major problem, related to the use of contact lenses, that led him to stop using contact lenses. A concrete example of such table is given.

Example 1. In table 1 the column headings (or attributes) have the following meaning: Ast stands for astigmatism and can have the value 0 (no astigmatism) or 1 (with astigmatism); Age can have the values 0 (not more than 20 years old), 1 (more than 20 years old but not more than 50 years old), and 2 (more than 50 years old). TearP stands for tear production and can have the value 1 (reduced tear production) or 2 (normal tear production); Spec stands for spectacles and can have the value 0 (using spectacles) or 1 (not using spectacles); and Lenses

Table 1. Table of people with sight problems.

	Ast	Age	TearP	Spec	Lenses
o_1	0	0	1	0	0
o_2	1	1	1	0	0
o_3	1	0	2	1	1
o_4	1	1	1	0	0
o_5	1	0	1	0	0
o_6	0	0	1	0	0
o_7	0	2	1	1	1
o_8	0	2	1	1	1
o_9	1	2	1	0	1
o_{10}	0	0	2	1	0
o_{11}	1	0	1	0	0
o_{12}	1	0	1	0	0
o_{13}	1	0	2	1	1
o_{14}	1	2	2	1	1
o_{15}	0	2	1	1	1
o_{16}	0	0	2	1	1
o_{17}	1	0	1	0	0
o_{18}	1	0	2	1	0

stands for contact lenses and can have the value 0 (stopped using contact lenses) or 1 (did not stop using contact lenses). □

As the example above shows, objects of a given universe U (e.g. people with sight problems) are described in terms of certain chosen attributes (e.g. tear production). An attribute a can be seen as a total function $a : U \rightarrow V_a$, where V_a is called the *value domain* of a. In this thesis, we assume that we do not have missing (unknown) attribute values. Thus, every object is associated with a tuple of attributes.

The special constant null may belong to the value domain V_a off an attribute a. If for an object $o \in U$, $a(o) = $ null, then this means that for this particular object o the value of attribute a is not defined (alternatively, attribute a could be seen as a partial function). Note that null does not denote a missing value. In the latter case a value exists but it is not known, while in our framework null means that the attribute value is not relevant. For example, if patient's age is '<2' then the value of the attribute employer should be null.

Objects of the universe are often classified as belonging or not to a predefined class determined by one of the attributes, often called *decision attribute*. Consider again the example above. We may consider two classes of persons: those who had to abandon the use of contact lenses and those who had not. Each person $o \in U$ is then classified as belonging to the former class (Lenses$(o) = 0$) or to the latter one (Lenses$(o) = 1$). Hence, Lenses is in this case the decision attribute.

The notion of decision table is formalized to capture these ideas.

Definition 1. *A decision table $\mathcal{D}$ is a triple (U, A, d), where U is a set of objects, A is a set of condition attributes, and d is a (often binary) decision attribute such that* $null \notin V_d$ *(i.e. for each object $o \in U$, $d(o)$ is defined).*

The table of example 1 can be seen as the decision table $\mathcal{L}enses = (U, A, \mathtt{Lenses})$, where $U = \{o_1, \ldots, o_{18}\}$ and $A = \{\mathtt{Ast}, \mathtt{Age}, \mathtt{TearP}, \mathtt{Spec}\}$. Moreover, using the attributes from A, persons o_3 and o_{18} are indiscernible from each other because they are represented by the same tuple of condition attributes $\langle 1, 0, 2, 1 \rangle$. However, they have different outcomes for the decision attribute: o_3 did not have to stop using contact lenses, while o_{18} did have. This points out that the information available in a decision table may be contradictory.

Definition 2. *Given a decision table $\mathcal{D} = (U, A, d)$, an object $o_i \in U$ is indiscernible from object $o_j \in U$ if and only if, for all condition attributes $a \in A$, $a(o_i) = a(o_j)$.*

Hence, a decision table $\mathcal{D} = (U, A, d)$ induces an *indiscernibility relation* R_A,

$$R_A = \{(o_i, o_j) \in U^2 \mid o_i \text{ is indiscernible from } o_j\} \ .$$

The indiscernibility relation is an equivalence relation and it induces a partition of the universe U into equivalence classes. These equivalence classes are also known in the rough set literature as *indiscernibility classes* or *elementary sets*. The pair (U, R_A) is called *approximation space* and R_A^* denotes the set of its equivalence classes.

Note that if $a(o) = null$, for some object o of an indiscernibility class, then $a(o') = null$, for every other object o' in the same equivalence class.

To simplify the presentation of several central ideas in later sections (e.g. decision rules), we introduce informally the notion of derived decision table.

Example 2. The decision table 2, $\mathcal{L}enses' = (U', A, \mathtt{Lenses'})$, is derived from table 1. Column $\mathtt{Class}$ designates an indiscernibility class of table 1. Hence, the universe U' of the derived table is composed of indiscernibility classes. The condition attributes remain the same as in table 1, i.e. $A = \{\mathtt{Ast}, \mathtt{Age}, \mathtt{TearP}, \mathtt{Spec}\}$.

Table 2. Decision table derived from Table 1

Class	Ast	Age	TearP	Spec	Lenses'
E_1	0	0	1	0	$\{0\}$
E_2	1	0	2	1	$\{0, 1\}$
E_3	1	1	1	0	$\{0\}$
E_4	0	2	1	1	$\{1\}$
E_5	1	2	1	0	$\{1\}$
E_6	1	0	1	0	$\{0\}$
E_7	1	2	2	1	$\{1\}$
E_8	0	0	2	1	$\{0, 1\}$

The values for the decision attribute **Lenses'** are now non-empty subsets of $\{0,1\}$. If an indiscernibility class E only contains objects whose outcome for the decision attribute in table 1 is 0 (1) then the decision attribute in this derived table has value $\{0\}$ ($\{1\}$). However, if some objects in an indiscernibility class E have outcome 0 for decision attribute **Lenses** while other objects belonging to E have outcome 1, then the decision attribute in this derived table has value $\{0,1\}$ (i.e. **Lenses'** $= \{0,1\}$).

There are 8 indiscernibility classes:

$$E_1 = \{o_1, o_6\},$$
$$E_2 = \{o_3, o_{13}, o_{18}\},$$
$$E_3 = \{o_2, o_4\},$$
$$E_4 = \{o_7, o_8, o_{15}\},$$
$$E_5 = \{o_9\},$$
$$E_6 = \{o_5, o_{11}, o_{12}, o_{17}\},$$
$$E_7 = \{o_{14}\},$$
$$E_8 = \{o_{10}, o_{16}\}.$$

From the second line of table 2, we can read that indiscernibility class E_2 contains some objects whose outcome for the decision attribute **Lenses** is 0 (o_{18}) while other objects in E_2 have outcome 1 (o_3 and o_{13}).

Each indiscernibility class can be described by a boolean formula. For instance, consider E_1. It can be described by

$$(\text{Ast} = 0 \wedge \text{Age} = 0 \wedge \text{TearP} = 1 \wedge \text{Spec} = 0) .$$

Alternatively, we can simply use the tuple $\langle 0, 0, 1, 0 \rangle$ to describe the same indiscernibility class. This is the approach followed in this thesis. □

As shown in the example 2, indiscernibility classes can be described by a unique tuple of $\prod_{a_i \in A} V_{a_i}{}^2$. These elementary sets are like atomic information granules that are used to define concepts, i.e. subsets of the universe. The next example illustrates this point.

Example 3. Consider the decision table 2. This decision table is associated with two concepts: those people who stopped using contact lenses and those who did not. The former concept is denoted by the subset

$$WithoutLenses = \{o_1, o_2, o_4, o_5, o_6, o_{10}, o_{11}, o_{12}, o_{17}, o_{18}\} ,$$

while the latter is denoted by the subset

$$WithLenses = \{o_3, o_7, o_8, o_9, o_{13}, o_{14}, o_{15}, o_{16}\} .$$

An important question is

---

[2] The expression $\prod_{a_i \in A} V_{a_i}$ denotes the cartesian product $V_{a_1} \times \cdots \times V_{a_k}$, where $A = \{a_1, \cdots, a_k\}$.

- When can we say that a person could be recommended the use of contact lenses, based on the attributes Ast, Age, TearP, and Spec and their values shown on the decision table 2?

To answer this question, we need to build a discriminating description of concept *WithLenses*. This description can be obtained in terms of the description of the indiscernibility classes. We remind the reader that each indiscernibility class can be described by a tuple of (condition) attribute values. For instance, $E_4 \subseteq WithLenses$. Hence, we can conclude that for people satisfying the conditions

$$(\text{Ast} = 0 \wedge \text{Age} = 2 \wedge \text{TearP} = 1 \wedge \text{Spec} = 1) ,$$

the use of contact lenses should be recommended. However, indiscernibility class E_2 seems to raise a problem in this context because E_2 is neither contained in *WithLenses* nor it is disjoint from *WithLenses*. This indicates that concept *WithLenses* cannot be defined precisely using the available information.   □

We can conclude from the example above that it is not always possible to learn precise descriptions of concepts. This is due to the fact that we may have contradictory knowledge that leads to vague concepts. It is in this context that the notion of rough set emerges naturally, since it introduces the idea of set (or concept) approximations.

2.2   Approximations and Rough Sets

Let (U, R_A) be an approximation space. The tuple describing each indiscernibility class $E \in R_A^*$ is denoted by $\overrightarrow{E^A}$. For example, in table 2, $\overrightarrow{E_1^A} = \langle 0, 0, 1, 0 \rangle$.

Definition 3. *Let $\mathcal{D} = (U, A, d)$ be a decision table and $X \subseteq U$. Rough set theory introduces two types of approximations of concept X in the approximation space (U, R_A).*

- Lower approximation *of X, denoted by $\underline{X}$,*

$$\underline{X} = \{ \overrightarrow{E^A} \mid E \in R_A^* \text{ and } E \subseteq X \} .$$

- Upper approximation *of X, denoted by $\overline{X}$,*

$$\overline{X} = \{ \overrightarrow{E^A} \mid E \in R_A^* \text{ and } E \cap X \neq \emptyset \} .$$

Let $\neg X = U \backslash X$, where U is the set of objects under consideration. The upper approximation $\overline{X}$ can informally be interpreted as a description of the objects that *possibly* belong to a given concept X. Notice that some doubt exists in this description because there may exist a tuple $t \in \overline{X} \cap \overline{\neg X}$. The lower approximation of X should informally be viewed as a description of those objects that *definitely* belong to the concept. The set $\underline{X} = \overline{X} \cap \overline{\neg X}$ is called the *boundary* and it corresponds to the conflicting cases (i.e. the doubtful ones).

Example 4. Consider once more the decision table 2 and the concepts

$$WithLenses = \{o_3, o_7, o_8, o_9, o_{13}, o_{14}, o_{15}, o_{16}\} \,,$$

$$\neg WithLenses = \{o_1, o_2, o_4, o_5, o_6, o_{10}, o_{11}, o_{12}, o_{17}, o_{18}\} \,.$$

Note that $\neg WithLenses$ represents the same concept as $WithoutLenses$, introduced in example 3. Their upper and lower approximations are given.

$$
\begin{aligned}
\overline{WithLenses} &= \{\langle 1,0,2,1 \rangle, \langle 0,2,1,1 \rangle, \langle 1,2,1,0 \rangle, \\
&\qquad \langle 1,2,2,1 \rangle, \langle 0,0,2,1 \rangle\} \,, \\
\underline{WithLenses} &= \{\langle 0,2,1,1 \rangle, \langle 1,2,1,0 \rangle, \langle 1,2,2,1 \rangle\} \,, \\
\overline{\neg WithLenses} &= \{\langle 0,0,1,0 \rangle, \langle 1,0,2,1 \rangle, \langle 1,1,1,0 \rangle, \\
&\qquad \langle 1,0,1,0 \rangle, \langle 0,0,2,1 \rangle\} \,, \\
\underline{\neg WithLenses} &= \{\langle 0,0,1,0 \rangle, \langle 1,1,1,0 \rangle, \langle 1,0,1,0 \rangle\} \,.
\end{aligned}
$$

The upper approximation of $WithLenses$ describes those persons who possibly will not have problems in using contact lenses while the lower approximation of $\neg WithLenses$ describes those person who certainly will have problems due to the use of contact lenses. Notice that the tuple $\langle 1,0,2,1 \rangle$ belongs to both $\overline{WithLenses}$ and $\overline{\neg WithLenses}$, i.e. $\langle 1,0,2,1 \rangle \in \overline{WithLenses}$. This fact indicates that for persons satisfying the condition

$$(\texttt{Ast} = 1 \wedge \texttt{Age} = 0 \wedge \texttt{TearP} = 2 \wedge \texttt{Spec} = 1)$$

there exists contradictory evidence and, therefore, it is not possible to state with certainty whether they should be recommended to use contact lenses.   □

Let (U, R_A) be an approximation space and $X, Y \subseteq U$. It has been proved that set approximations have several important properties [24]. We list some of them.

(1) $\underline{X} \subseteq X \subseteq \overline{X}$.
(2) $\overline{(X \cup Y)} = \overline{X} \cup \overline{Y}$.
(3) $\overline{(X \cap Y)} \subseteq \overline{X} \cap \overline{Y}$.
(4) $\underline{(X \cup Y)} \supseteq \underline{X} \cup \underline{Y}$.
(5) $\underline{(X \cap Y)} = \underline{X} \cap \underline{Y}$.
(6) $\overline{\neg X} = \neg \underline{X}$.
(7) $\underline{\neg X} = \neg \overline{X}$.
(8) If $X \subseteq Y$ then $\underline{X} \subseteq \underline{Y}$ and $\overline{X} \subseteq \overline{Y}$.

A concept that cannot be defined precisely is represented by a pair of sets of tuples, to be called rough set.

Definition 4. *A rough set (or rough relation) S is a pair $(\overline{S}, \neg \overline{S})$ such that $\overline{S}, \neg \overline{S} \subseteq \prod_{a_i \in A} V_{a_i}$, for some non empty set of attributes A. The rough complement of a rough set $S = (\overline{S}, \neg \overline{S})$ is the rough set $\neg S = (\neg \overline{S}, \overline{S})$.*

Example 5. Consider again example 4. Since concept *WithLenses* cannot be defined precisely in terms of the elementary sets belonging to R_A^*, where $A = \{\texttt{Ast}, \texttt{Age}, \texttt{TearP}, \texttt{Spec}\}$, this concept is then represented by the rough set *Lenses* $= (\overline{WithLenses}, \neg \overline{WithLenses})$. $\qquad\qquad\qquad\qquad\qquad\qquad\qquad\qquad\qquad\qquad\qquad$ □

We stress that there is not a unique way to define a rough set. Different definitions for the concept of rough set have been proposed in the literature [7]. For instance, a rough set could be defined as the pair $S = (\underline{S}, \overline{S})$. All these definitions formalize the idea of vague sets due to the existence of a boundary region. The reason for preferring one definition over another is related to the concrete application the definition's author has in mind. We have chosen to define a rough set as $S = (\overline{S}, \neg\overline{S})$ rather than $S = (\underline{S}, \overline{S})$ because the former definition gives information about all negative examples while the latter only indicates those negative examples in the boundary region.

The notion of rough set used in our framework differs in a number of respects from those usually presented in rough set literature. First in our framework, lower and upper approximations are sets of tuples while these approximations are usually defined as subsets of the universe U. Second, we assume that a rough set may partition the set of all possible tuples $\mathcal{W} = \prod_{a_i \in A} V_{a_i}$, where A is a set of condition attributes, into four regions. Figure 1 illustrates this point.

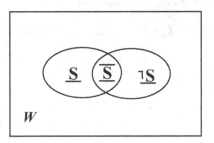

Fig. 1. The four regions generated by a rough set S

Given a rough set S, these four regions correspond to $\underline{S}$, $\neg \underline{S}$, $\overline{S}$, and the set of tuples about which there is no information, $\mathcal{W} \setminus (\underline{S} \cup \neg \underline{S} \cup \overline{S})$. Most literature considers that a rough set divides the universe U in three regions: lower approximation of a rough set S, lower approximation of $\neg S$, and boundary.

Let (U, R_A) be an approximation space, $[x]_A$ be the equivalence class of object $x \in U$ in that approximation space, and S be a rough set. Then, $\varepsilon(x, S)$ denotes whether x belongs to the concept represented by rough set S and it is defined as follows.

$$\varepsilon(x, S) = \begin{cases} \texttt{true} & \text{if } \overrightarrow{[x]_A} \in \underline{S} \\ \texttt{false} & \text{if } \overrightarrow{[x]_A} \in \neg \underline{S} \\ \top & \text{if } \overrightarrow{[x]_A} \in \overline{S} \\ \bot & \text{otherwise } . \end{cases}$$

$\varepsilon(x, S) = \top$ indicates the existence of contradictory information about object x and $\varepsilon(x, S) = \bot$ shows lack of information about x.

Let us assume, without loss of generality, that d is a binary decision attribute of a decision table $\mathcal{D} = (U, A, d)$. It is easy to see that we can associate a rough set $D = (\overline{D}, \neg\overline{D})$ with $\mathcal{D}$, where $\overline{D}$ is the set of tuples with positive outcome for the decision attribute d and $\overline{\neg D}$ is the set of tuples with negative outcome. Our definition of rough set cannot be seen as an alternative representation for a decision table. This can be explained by the fact that, for the former case, there is no information associated with each tuple of how many objects belong to the corresponding indiscernibility class or how many objects in an indiscernibility class have positive (negative) outcome for the decision attribute. From a formal point of view, this problem can be easily addressed and we will discuss it in section 2.4.

We have adopted the convention to give the same name to a decision table, to its decision attribute, and to the rough relation it defines, since all these concepts are very associated with each other. However, the appearance of the printed names is different for each case. For instance, if a decision table is called "$\mathcal{L}$enses", a name usually starting with calligraphic letter, then its decision attribute is called "**Lenses**" and it defines the rough relation "$Lenses$". We use interchangeably the expression "objects" and "individuals" to refer to the elements of the universe U under consideration.

We stress that, in this work, we only consider approximation spaces (U, R_A) where R_A is an equivalence relation. The literature also discusses its generalizations to tolerance approximation spaces [12] and similarity approximation spaces [27].

2.3 Decision Rules

In the context of supervised learning, an important task is the discovery of classification rules from the data provided in the decision tables. The decision rules not only capture patterns hidden in the data as they can also be used to classify new unseen objects.

Definition 5. *Let* $\mathcal{D} = (U, A, d)$ *be a decision table and* $\{a_1, \ldots, a_n\} \subseteq A$. *Assume also that* $\{v_{d1}, \ldots, v_{dk}\} \subseteq V_d$ *and* $v_i \in V_{a_i}$, *for* $1 \leq k$ *and* $1 \leq i \leq n$. *A decision rule is an expression of the form*

$$(a_1 = v_i) \wedge \ldots \wedge (a_n = v_n) \longrightarrow (d = v_{d1}) \vee \ldots \vee (d = v_{dk}).$$

If $k = 1$ *then the decision rule is called a* deterministc *rule. Otherwise* $(k > 1)$, *the decision rule is called* non-deterministic *rule.*

A decision rules can informally be understood as an implication. Symbols "$\wedge$", "$\vee$" can be read as "and" and "or", respectively.

Given a decision rule r, $cond(r)$ denotes the expression on the left hand side of symbol "$\longrightarrow$" and $dec(r)$ denotes the expression on the right hand side.

Example 6. Consider again table 2. Then, r_1 is a deterministic decision rule obtained from this table

$$r_1 \equiv (\mathtt{Age} = 0) \wedge (\mathtt{TearP} = 2) \wedge (\mathtt{Spec} = 1) \longrightarrow (\mathtt{Lenses'} = \{0, 1\}) \quad .$$

It states that if a person's age is not more than 20 years, and his tear production is normal, and he does not use spectacles then the outcome for the decision attribute **Lenses** is 0 or 1, i.e. the optician cannot with certainty recommend contact lenses.

The above rule corresponds to the following non-deterministic rule for decision table 1.

$$(\mathtt{Age} = 0) \wedge (\mathtt{TearP} = 2) \wedge (\mathtt{Spec} = 1) \longrightarrow (\mathtt{Lenses} = 0) \vee (\mathtt{Lenses} = 1)$$

Hence, this rule describes some of the objects belonging to the upper approximation of rough set *Lenses*.

Consider now the decision rule

$$r_2 \equiv (\mathtt{Age} = 0) \wedge (\mathtt{Spec} = 1) \longrightarrow (\mathtt{Lenses'} = \{0, 1\}) .$$

Both decision rules r_1 and r_2 above can be used to identify some of the objects belonging to $\overline{Lenses}$. However, an important difference should be pointed. The latter rule is more general than the former in the sense that it states a smaller number of conditions in $cond(r)$. Therefore, r_2 is more likely to be applied to a larger number of new objects to predict the outcome for the decision attribute. □

Let us call *decision classes* to the partitions of the universe generated by the decision attribute, i.e. elements of $R^*_{\{d\}}$. The preceding example points that rules induced from a decision table with a minimal number of conditions (i.e. attribute-value pairs, $(a = v)$) are the most useful because those rules are more general. In the context of data mining, one of the main aims of rough-set based algorithms is to either find a set of minimal rules that covers a given decision class or to compute all minimal rules for the chosen decision class.

As shown in the previous sections, the concept associated with a given decision class may not be definable in terms of the elementary sets, i.e. the concept is rough. Decision rules can then be computed either with respect to the lower approximation or upper approximation of the target concept. An important difference between these two cases is that rules generated from the lower approximation are deterministic while rules generated from the upper approximation may be non-deterministic.

We turn now to the formalization of minimal decision rules.

Let $\mathcal{D} = (U, A, d)$ be a decision table, $c_1 \equiv (a_1 = v_1) \wedge \ldots \wedge (a_n = v_n)$ and $c_2 \equiv (a_1 = v_1) \vee \ldots \vee (a_n = v_n)$ be two conditions, with $\{a_1, \ldots, a_n\} \subseteq A$ and $1 \leq n$. $Cover(c_1)$ and $Cover(c_2)$ denote the following subsets of U.

$$Cover(c_1) = \bigcap_{1 \leq i \leq n} \{o \in U \mid a_i(o) = v_i\} ,$$
$$\text{and}$$
$$Cover(c_2) = \bigcup_{1 \leq i \leq n} \{o \in U \mid a_i(o) = v_i\} .$$

A rule r *covers* all objects that match the condition on its left-hand side, denoted as $Cover(r)$, i.e. $Cover(r) = Cover(cond(r))$. This definition can be extended to a set of rules S:

$$Cover(S) = \bigcup_{r \in S} Cover(r) .$$

Let $c_1 \equiv (a_{11} = v_{11}) \wedge \ldots \wedge (a_{1n} = v_{1n})$ and $c_2 \equiv (a_{21} = v_{21}) \wedge \ldots \wedge (a_{2k} = v_{2k})$, with $n, k \geq 1$, be two conditions. We write $c_1 \preceq c_2$ to denote that every attribute-value pair $(a_{1i} = v_{1i})$ occurring in c_1, with $1 \leq i \leq n$, also occurs in c_2. The expression $c_1 \prec c_2$ means that $c_1 \preceq c_2$ and $c_1 \neq c_2$. For instance, $(\text{Age} = 2) \preceq (\text{Age} = 2) \wedge (\text{TearP} = 1)$.

Definition 6 ([26]). *Let r be a decision rule. A* value reduct *for r, denoted as $red(r)$, is a condition satisfying the following properties.*

(1) $red(r) \preceq cond(r)$.
(2) $Cover(red(r)) \subseteq Cover(dec(r))$. *i.e. value reduct preserves the inclusion relation of the set of objects covered by the rule in $Cover(dec(r))$;*
(3) *For every condition $c \prec red(r)$, $Cover(c) \not\subseteq Cover(dec(r))$, i.e. the value reduct is a minimal condition, with respect to the partial order $\prec$, satisfying properties (1) and (2).*

The definition of minimal rules is based on the notion of value reduct.

Definition 7. *Let r be a decision rule. r is a* minimal rule *if and only if $red(r) = cond(r)$.*

Given a decision rule r, a single value reduct can be computed in linear time with respect to the number of attributes that appear in $cond(r)$.

Example 7. Consider example 6 and table 2. Decision rule

$$(\text{Age} = 0) \wedge (\text{TearP} = 2) \wedge (\text{Spec} = 1) \longrightarrow (\text{Lenses'} = \{0, 1\})$$

is not minimal while

$$(\text{Age} = 0) \wedge (\text{Spec} = 1) \longrightarrow (\text{Lenses'} = \{0, 1\}) \quad \text{and}$$
$$(\text{Age} = 0) \wedge (\text{TearP} = 2) \longrightarrow (\text{Lenses'} = \{0, 1\})$$

are minimal decision rules. Note that

$$Cover((\text{Age} = 0) \wedge (\text{Spec} = 1)) = E_2 \cup E_8 = Cover(\text{Lenses'} = \{0, 1\}) . \quad \square$$

Let $\mathcal{D} = (U, A, d)$ be a decision table, $A = \{a_1, \ldots, a_n\}$, and assume that $v \in V_d$. An algorithm based on value reducts that computes a set of minimal rules covering the decision class $D_v = \{o \in U \mid d(o) = v\}$ can be easily devised. Assume that we are interested in a set of deterministic rules.

Computation of Minimal Rules [26]:

  (i) Compute $\underline{D_v}$. Let L be a set of decision rules initialized to the empty set.
 (ii) Consider the decision rule

$$r \equiv (a_1 = v_1) \wedge \ldots \wedge (a_n = v_n) \rightarrow (d = v)$$

associated with each tuple $\langle v_1, \ldots, v_n \rangle \in \underline{D_v}$.

(ii.1) Compute a value reduct for rule r, i.e. $\overline{\text{red}}(r)$.

(ii.2) If the $Cover(red(r)) \not\subseteq \bigcup_{r_i \in L} Cover(r_i)$ then

$$L = L \cup \{red(r) \rightarrow (d = v)\} .$$

(iii) Output each rule in L.

If a set of non-deterministic rules is sought instead then $\underline{D_v}$ should be replaced by $\overline{D_v}$ in the algorithm above. The algorithm can be computed in polynomial time. Its time complexity is proportional to the number of condition attributes and number of tuples in $\underline{D_v}$ $(\overline{D_v})$.

Example 8. Consider again table 2 and let us compute the minimal deterministic decision rules for $\underline{Lenses'_{\{0,1\}}}$, corresponding to the boundary region of table $\mathcal{L}enses$. We have that

$$\underline{Lenses'_{\{0,1\}}} = \{\langle 1, 0, 2, 1 \rangle, \langle 0, 0, 2, 1 \rangle\} .$$

It can be easily verified that for the decision rules associated with tuples $\langle 1,0,2,1 \rangle$ and $\langle 0, 0, 2, 1 \rangle$ there are two possible value reducts

$$(\text{Age} = 0) \wedge (\text{Spec} = 1) \quad \text{and}$$
$$(\text{Age} = 0) \wedge (\text{TearP} = 2)$$

It can be also easily verified that $(\text{Age} = 2)$ is a value reduct for each of the decision rules obtained from the tuples belonging to the set

$$\underline{Lenses'_{\{1\}}} = \{\langle 0, 2, 1, 1 \rangle, \langle 1, 2, 1, 0 \rangle, \langle 1, 2, 2, 1 \rangle\} .$$

Note that $\underline{Lenses'_{\{1\}}}$ corresponds to the lower approximation of rough set $\mathcal{L}enses$. Since

$$Cover(\text{Age} = 2) \not\subseteq Cover((\text{Age} = 0) \wedge (\text{Spec} = 1)) \quad \text{and}$$
$$Cover((\text{Age} = 0) \wedge (\text{Spec} = 1)) \not\subseteq Cover(\text{Age} = 2) ,$$

the following two minimal decision rules describe the set of objects $Lenses'_{\{0,1\}} \cup Lenses'_{\{1\}}$.

$$(\text{Age} = 0) \wedge (\text{Spec} = 1) \longrightarrow (\text{Lenses'} = \{0, 1\}) ,$$
$$(\text{Age} = 2) \longrightarrow (\text{Lenses'} = \{1\}) .$$

If we consider table 1 instead then, the deterministic decision rules above correspond to a non-deterministic rule and to a deterministic rule, respectively.

$$(\text{Age} = 0) \wedge (\text{Spec} = 1) \longrightarrow (\text{Lenses} = 0) \vee (\text{Lenses} = 1) ,$$
$$(\text{Age} = 2) \longrightarrow (\text{Lenses} = 1) .$$

These two rules cover all objects described by tuples in the upper approximation of rough set $\mathcal{L}enses$ (obtained from table 1). $\qquad\square$

In step (iii) above, L is a non-empty set of minimal rules such that $Cover(L)$ $= D_v$. Hence, these set of rules form a discriminating description of the approximated concept. However, from the point of view of knowledge discovery, it is more interesting to find all possible minimal rules than just one set of minimal rules. In contrast to the above algorithm that is polynomial, computing all possible minimal rules is NP-hard.

All minimal rules can be computed by first creating the decision-relative discernibility matrix. A Boolean expression can then be constructed from this matrix. This expression is simplified (absorption law of Boolean algebra can be applied) and prime implicants of the simplified expression are computed. Each prime implicant can finally be translated to a decision rule. More detailed descriptions of this algorithm can be obtained from [25,28]. A survey of the main algorithms for inducing decision rules using rough set theory is presented in [29,30].

2.4 Numerical Measures

The first part of this section is devoted to the discussion of several numerical measures that can be associated with a decision rule for measuring its quality. In the second part, we introduce the notion of reduct and we briefly mention an algorithm to compute decision rules based on reducts.

Measuring Quality of Decision Rules. Quality measures associated with decision rules can be used to eliminate some of the decision rules. We list below some of these quality measures [25].

Given a set S, the expression $|S|$ denotes the number of elements in S.

Definition 8 (Support). *Let r be a decision rule induced from a decision table $\mathcal{D} = (U, A, d)$. The* support *of r, denoted as $Supp(r)$, is defined as*

$$Supp(r) = |Cover(cond(r)) \cap Cover(dec(r))| \ .$$

The support of a rule represents the number of objects of the universe that match both conditions $cond(r)$ and $dec(r)$.

Definition 9 (Strength). *Let r be a decision rule induced from a decision table $\mathcal{D} = (U, A, d)$. The* strength *of r, denoted as $Strength(r)$, is defined as*

$$Strength(r) = \frac{|Supp(r)|}{|U|} \ .$$

The strength of a rule indicates the proportion of objects in the universe that match both $cond(r)$ and $dec(r)$; i.e. the percentage of objects for which the pattern expressed by the rule is true. Hence, $Strength(r)$ can be seen as an estimate of the probability $Pr(cond(r) \wedge dec(r))$.

Definition 10 (Accuracy). *Let r be a decision rule induced from a decision table $\mathcal{D} = (U, A, d)$. The* accuracy *of r, denoted as $Acc(r)$, is defined as*

$$Acc(r) = \frac{Supp(r)}{|Cover(cond(r))|} .$$

The accuracy of a rule corresponds to the conditional probability $Pr(o \in Cover(dec(r)) \mid o \in Cover(cond(r)))$. By other words, $Acc(r)$ expresses how trustworthy is the rule is drawing the conclusion $dec(r)$ for an object matching the condition on the left-hand side of the rule.

Definition 11 (Coverage). *Let r be a decision rule induced from a decision table $\mathcal{D} = (U, A, d)$. The* coverage *of r, denoted as $Cov(r)$, is defined as*

$$Cov(r) = \frac{Supp(r)}{|Cover(dec(r))|} .$$

The coverage of a rule corresponds to the conditional probability $Pr(o \in Cover(cond(r)) \mid o \in Cover(dec(r)))$. Hence, $Cov(r)$ quantifies how well the rule left-hand side, $cond(r)$, describes the set of objects covered by its right-hand side, $dec(r)$.

Reducts. The derived decision table shown in example 2 contains nearly the same information as in table 1. However in the derived table, if $\textbf{Lenses'}(E) = \{0, 1\}$, for some indiscernibility class E, then we cannot know how many objects in E have outcome 0 for decision attribute **Lenses** and how many have outcome 1. To overcome this problem, we define a family of functions. Let $\mathcal{D} = (U, A, d)$ be a decision table and $V_d = \{v_1, \ldots, v_n\}$. We have then that, for each indiscernibility class $E \in R_A^*$,

$$\lambda_{\mathcal{D}}^i(\overrightarrow{E^A}) = |\{o \in E \mid d(o) = v_i\}| ,$$

with $1 \leq i \leq n$. Moreover, $card(\overrightarrow{E^A}) = |E| = \sum_{1 \leq i \leq n} \lambda_{\mathcal{D}}^i(\overrightarrow{E^A})$, i.e. $card(\overrightarrow{E^A})$ denotes the number of objects in the indiscernibility class described by tuple $\overrightarrow{E^A}$. Function $card$ can be extended to a set of tuples T.

$$card(T) = \sum_{t \in T} card(t) .$$

Let (U, R_{B_1}) and (U, R_{B_2}) be two approximation spaces. We now define those objects of the universe for which knowing the values of attributes B_1 is sufficient for determining the values of attributes B_2, denoted as $Pos_{B_1}(B_2)$.

$$Pos_{B_1}(B_2) = \bigcup_{X \in R_{B_2}^*} \underline{X} ,$$

where $\underline{X}$ is the lower approximation of X in the approximation space (U, R_{B_1}).

An interesting numerical measure associated with a decision table is the degree of functional dependency [25,26] between two subsets of attributes of the table.

Definition 12. *Let $\mathcal{D} = (U, A, d)$ be a decision table and $B_1, B_2 \subseteq A \cup \{d\}$. The degree of functional dependency in the relationship between attribute sets B_1 and B_2, denoted as $\kappa(B_1, B_2)$, is defined as*

$$\kappa(B_1, B_2) = \frac{card(Pos_{B_1}(B_2))}{|U|} .$$

Function $\kappa(B_1, B_2)$ can be understood as the proportion of objects of the universe U for which knowing the values of attributes B_1 is enough to determine the values of attributes B_2. Obviously, $0 \leq \kappa(B_1, B_2) \leq 1$. If $\kappa(B_1, B_2) = 1$ then functional dependency $B_1 \rightarrow B_2$ exists in the table and it can be easily shown that $R_{B_1} \subseteq R_{B_2}$ (i.e. the partition generated by attributes B_1 is finer than the partition generated by B_2). If $\kappa(B_1, B_2) = 0$ then no values of attributes B_2 can be determined by values of attributes B_1. If $0 < \kappa(B_1, B_2) < 1$ then the values of attributes B_2 can be determined by values of attributes B_1, only for some objects (but not all).

Example 9. Consider decision tables of example 1 and 2. For the first table,

$$\kappa(\{\textsf{Ast}, \textsf{Age}, \textsf{TearP}, \textsf{Spec}\}, \{\textsf{Lenses}\}) \simeq 0.72$$

indicating that the decision attribute is not functionally determined by the condition attributes. For the second table $\kappa(\{\textsf{Ast}, \textsf{Age}, \textsf{TearP}, \textsf{Spec}\}, \{\textsf{Lenses'}\}) = 1$. Hence, we can conclude that the functional dependency

$$\{\textsf{Ast}, \textsf{Age}, \textsf{TearP}, \textsf{Spec}\} \rightarrow \{\textsf{Lenses'}\}$$

holds for this table. □

If we consider decision table 2, we can conclude that

$$\kappa(\{\textsf{Age}, \textsf{TearP}\}, \{\textsf{Lenses'}\}) = 1 .$$

Hence, we only need condition attributes $\textsf{Age}$ and $\textsf{TearP}$ in order to be able to determine the decision class of an object, i.e. to determine $\textsf{Lenses'}(E_i)$ ($1 \leq i \leq 8$). This makes possible savings in the amount of information that needs to be represented and it may also lead to a table where regularities are more easy to find. Relative reducts formalize this idea.

Definition 13. *Let A and B be two sets of attributes. A relative reduct of A with respect to $\kappa(A, B)$, denoted as $red(A, B)$, is a subset of A having the following properties.*

(1) $\kappa(A, B) = \kappa(red(A, B), B)$.
(2) *For all $a \in red(A, B)$,*

$$\kappa(red(A, B) \setminus \{a\}, B) \neq \kappa(A, B) ,$$

i.e. $red(A, B)$ is a minimal subset of A satisfying (1).

From a practical point of view, we are often interested in discovering relative reducts with respect to $\kappa(A, \{d\})$, for a decision table $\mathcal{D} = (U, A, d)$.

A single relative reduct can be computed in linear time. However, computing all relative reducts or a minimal reduct is NP-hard. Most of the algorithms for determining all reducts are based on a decision-relative discernibility matrix [25,28].

In section 2.3, we have presented an algorithm to compute a set S of (minimal) decision rules forming a discriminating description for a given decision class D_v. This algorithm is based on a covering approach since $\overline{D_v} = Cover(S)$ or $D_v = Cover(S)$. A set of non-deterministic rules is obtained in the former case, while only deterministic rules are computed for the latter. However, rules can be computed in a different way: find first relative reducts and then create decision rules by overlaying the reducts over objects in the table. We illustrate this second approach with the next example.

Example 10. Consider decision table 2. A relative reduct of {Ast, Age, TearP, Spec}, with respect to $\kappa(\{$Ast, Age, TearP, Spec$\}$, Lenses'), is {Age, TearP}. The following decision rules would then be obtained

$$(\texttt{Age} = 0) \wedge (\texttt{TearP} = 1) \longrightarrow (\texttt{Lenses} = 0) \, ,$$
$$(\texttt{Age} = 0) \wedge (\texttt{TearP} = 2) \longrightarrow (\texttt{Lenses} = 0) \vee (\texttt{Lenses} = 1) \, ,$$
$$(\texttt{Age} = 1) \wedge (\texttt{TearP} = 1) \longrightarrow (\texttt{Lenses} = 0) \, ,$$
$$(\texttt{Age} = 2) \wedge (\texttt{TearP} = 1) \longrightarrow (\texttt{Lenses} = 1) \, ,$$
$$(\texttt{Age} = 2) \wedge (\texttt{TearP} = 2) \longrightarrow (\texttt{Lenses} = 1) \, .$$

Notice that the above decision rules are not minimal. For instance, (Age = 2) is a value reduct of the last two rules above.   □

It is also possible to use approximations of a relative reduct $red(A, B)$, with respect to $\kappa(A, B)$. These approximations are subsets of A that "almost" preserve the same capability as the attribute set A in determining the values of attributes B. An advantage of using approximations of reducts is that decision rules synthesized from them are less sensitive to noise in the data and, therefore, the quality of classification of new objects tends to increase. We briefly mention two approaches: dynamic reducts [31] and s-reducts [28].

Let $\mathcal{D} = (U, A, d)$ be a decison table and $red(A, B)$ be a relative reduct, with respect to $\kappa(A, B)$, where $B \subseteq A \cup \{d\}$. A *dynamic reduct* is a subset of A appearing "sufficiently often" as a relative reduct in random sample subtables obtained from $\mathcal{D}$. Notice that "sufficiently often" should be understood as a parameter that is tuned according to the data in the table.

We can measure how the dropping of a number of attributes from attribute set A_1 changes the coefficient $\kappa(A_1, B)$. Let $A_2 \subseteq A_1$.

$$\alpha_{(A_1, B)}(A_2) = 1 - \frac{\kappa(A_2, B)}{\kappa(A_1, B)} \, .$$

It is worth noting that if A_2 is the minimal subset of A_1 such that $\kappa(A_1, B) = \kappa(A_2, B)$ (i.e. A_2 is a relative reduct of A_1 with respect to $\kappa(A_1, B)$) then

$\alpha_{(A_1,B)}(A_2) = 0$. The attribute set A_2 is an *s-reduct* if $\alpha_{(A_1,B)}(A_2)$ is not larger than a given threshold called error level. This error level should be tuned for the table being considered.

2.5   Prediction

In sections 2.3 and 2.4, we have introduced decision rules and gave an idea about how they can be computed. These decision rules can then be used to make predictions for unseen objects. For instance, the decision rules obtained from one relative reduct form a classifier. Classifiers with better prediction capabilities are usually obtained by combining rules obtained from several reducts.

Classifiers may be non-deterministic because

- the new object matches non-deterministic rules, or
- the object matches several (deterministic) rules that lead to different decisions.

An obvious question is how to solve the problem of conflicting decisions. This issue can be addressed by *voting* [25]. We describe below this strategy.

Without loss generality, we assume that a classifier only contains deterministic decision rules. A non-deterministic rule

$$(a_1 = v_1) \wedge \ldots \wedge (a_n = v_n) \rightarrow (d = v_d) \vee (d = v_{d'})$$

can always be replaced by two deterministic rules,

$$(a_1 = v_1) \wedge \ldots \wedge (a_n = v_n) \rightarrow (d = v_d) \text{ and}$$
$$(a_1 = v_1) \wedge \ldots \wedge (a_n = v_n) \rightarrow (d = v_{d'}) .$$

Let $\mathcal{C}$ be a classifier, i.e. a set of (deterministic) decision rules. Moreover, $Rul_{\mathcal{C}}(o)$ denotes the set of decision rules of $\mathcal{C}$ that match object o and it is formally defined as

$$Rul_{\mathcal{C}}(o) = \{r \in \mathcal{C} \mid o \in Cover(r)\} .$$

Voting Algorithm:

(1) If $Rul_{\mathcal{C}}(o) = \emptyset$ then classification is not possible. Otherwise, proceed with step (2).
(2) For each possible decision $v \in V_d$ compute the number of votes casted by each rule.

$$votes(v) = \sum_{r \in Rul_{\mathcal{C}}(o)} \nu_v(r) ,$$

where

$$\nu_v(r) = \begin{cases} 0 & \text{if } dec(r) \neq v \\ Supp(r) & \text{otherwise} \end{cases}$$

(3) Compute $\delta = \sum_{v' \in V_d} votes(v')$, i.e. the total number of casted votes.

(4) For each possible decision $v \in V_d$ compute the certainty factor

$$certainty(o, v) = \frac{votes(v)}{\delta} \ .$$

(5) Output the decision with the largest certainty.

The voting algorithm described above can be modified in a number of ways. For instance, the number of votes casted by each rule fired can be based on some other measured instead of rules's support.

2.6   The Variable Precision Rough Set Model

In section 2.2, we introduced the notions of concept approximations, lower and upper approximations. These ideas have been further generalized by W. Ziarko, see [13], who introduced the *variable precision rough set model* (VPRSM).

We start by discussing informally the VPRSM. Consider two parameters l and u, called *precision control parameters*, such that $0 \leq l < u \leq 1$. Generalization of lower (upper) approximation and boundary region of a rough relation can be obtained as follows. The lower approximation of a concept X ($\neg X$) is obtained from those indiscernibility classes E such that its degree of overlapping with the set X ($\neg X$) is larger or equal than u $(1 - l)$. Those indiscernibility classes E such that their degree of overlapping with X is between l and u remain in the boundary region. This technique can also be seen as a way to "thin" the boundary region and it has the advantage of making concept approximations less sensitive to possible noise contained in the data.

To formalize this idea, we need to introduce a function assigning to each indiscernibility class E a measure of the degree of overlapping of set X with E. This function corresponds to the conditional probability

$$Pr(o \in X \mid o \in E) = \frac{|(X \cap E)|}{|E|} \ .$$

Let (U, R_A) be an approximation space. Concept approximations can then be defined as

$$\underline{X} = \{\overrightarrow{E^A} \mid E \in R_A^* \text{ and } Pr(o \in X \mid o \in E) \geq u\} \ ,$$
$$\underline{\neg X} = \{\overrightarrow{E^A} \mid E \in R_A^* \text{ and } Pr(o \in X \mid o \in E) \geq (1 - l)\} \ ,$$
$$\overline{X} = \{\overrightarrow{E^A} \mid E \in R_A^* \text{ and } u < Pr(o \in X \mid o \in E) < l\} \ .$$

It is worth to note that if $u = 1$ and $l = 0$ then the above definitions of lower (upper) approximations and boundary are equivalent to the ones presented in section 2.2.

Let $X \subseteq U$ and $Pr(X) = \frac{|X|}{|U|}$. To obtain some gain in the predictive capability, it is required that $u > Pr(X)$ and $l < Pr(X)$. Requiring that $u > Pr(X)$ $(l < Pr(X))$ will enable us to predict that an object $o \in X$ ($o \in \neg X$) more accurately than random guess.

3   Logic Programming Framework

This section surveys the logic programming concepts needed in the sequel and it is self-contained. It gives to the reader the essential notions to understand the compilation technique discussed in sections 4 and 6.

We start by introducing definite logic programs, in section 3.1, and then present in section 3.2 a more general class of logic programs, called extended logic programs. We also discuss the declarative semantics of extended logic programs and queries.

3.1   The Main Idea: Definite Logic Programs

Logic programming [32,33,34,35] is a computational formalism that uses logic (e.g. first-order logic) to express knowledge and inference to manipulate the knowledge in order to be able to extract new knowledge.

In this work, the syntax of a logic program makes use of three disjoint alphabets: an alphabet of variable symbols Var, an alphabet of constant symbols $Const$, and an alphabet of predicate symbols $Pred$. Moreover, the set of symbols $\{\neg, not\} \not\subseteq Pred$. A *term* t is either a constant symbol or a variable, i.e $t \in Var \cup Const$. To distinguish between constants and variable symbols, we follow the usual convention: variables start with upper case letter (e.g. $X, Dist \in Var$), while using names beginning with lower case letters for constants (e.g. $small, c \in Const$). An *atom* is an expression of the form $p(t_1, \ldots, t_n)$, where p is an n-ary predicate symbol ($p \in Pred$) and each $t_1, \ldots, t_n$ is a term. We write p/n, with $n \geq 0$, to express that p is an n-ary predicate symbol. An atom with zero arguments is simply written as p .

Intuitively, predicates denote n-ary relations and atoms can be seen as statements saying that an n-ary tuple belongs to an n-ary relation. For instance, the atom office(xana, spetsen, 7) expresses that xana's office is on 7th floor of building spetsen, i.e. the tuple $\langle$xana, spetsen, 7$\rangle$ belongs to the relation denoted by predicate office.

In the logic programming framework, knowledge is represented through clauses.

Definition 14. *A definite clause is an expression of the form*

$$H \; \text{:-} \; A_1, \ldots, A_n. \; ,$$

where H and each A_i ($0 \leq i \leq n$) is an atom.

The left side of a (definite) clause (with respect to :-) is called the *head* and the right side is designated as *body* of the clause. A *fact* is a clause with empty body (i.e. $n = 0$), succinctly represented by H. . When no confusion arises, we will refer to "clauses" instead of "definite clauses".

Clauses can informally be understood as implications: if every atom in the body is true then the head must also be true. Therefore, the comma symbol "," is interpreted as conjunction.

Example 11. We give an example of a definite clause and a fact.

(1) `fly(tom) :- bird(tom).` *"If tom is a bird then tom flies.*
(2) `bird(tom).`   *"Tom is a bird."*

☐

The order by which atoms appear in the body of a clause is irrelevant. Thus, both clauses `fly(tom) :- bird(tom), healthy(tom).` and `fly(tom) :- healthy(tom), bird(tom).` have the same meaning.

If no variables occur in the atoms of a clause (atom), then the clause (atom) is *ground*. For instance, clause (1) of the example above is ground.

Definition 15. *Let $X_1, \ldots, X_n$ be variables occurring in some atom $q(t_1, \ldots, t_m)$, with $1 \leq n \leq m$. A grounding substitution θ is a set of bindings*

$$\theta = \{X_1/c_1, \ldots, X_n/c_n\}$$

(including the empty set) of variables X_i $(1 \leq i \leq n)$ to constants $c_i \in Const$.

A substitution $\theta = \{X_1/c_1, \ldots, X_n/c_n\}$ can be applied to a clause C (atom A), written as $C\theta$ ($A\theta$), and it represents the clause (atom) obtained from C (A) by substituting each variable X_i for c_i $(1 \leq i \leq n)$. For instance, if $\theta = \{X_1/c_1, X_2/c_2\}$ then $p(X_1, X_2)\theta$ is the atom $p(c_1, c_2)$. We can also say that variable X_1 (X_2) is *instantiated* with constant c_1 (c_2).

A *ground instance of a clause* C is obtained by applying a grounding substitution θ to clause C and $C\theta$ is ground. A non-ground clause stands for all its ground instances. Therefore, variables are implicitly universally quantified.

Example 12. We give an example of a non-ground definite clause.

(3) `fly(X) :- bird(X).` *"All birds fly".*

More formally, clause (3) represents the following implication

$$\forall X (\texttt{bird}(X) \Rightarrow \texttt{flies}(X)) .$$

Clause (1), in example 11, can be obtained by applying substitution $\{X/tom\}$ to clause (3).

☐

Definition 16. *A* definite logic program *is a set of definite clauses.*

Given a definite logic program $\mathcal{P}$, *ground*($\mathcal{P}$) represents the set of all ground instances of any clause $C \in \mathcal{P}$. This notation will also be used for sequence of atoms $A_1, \ldots, A_n$ $(n \geq 1)$, i.e. *ground*($A_1, \ldots, A_n$).

Example 13. Consider the following definite logic program.

$\mathcal{P} = \{\texttt{fly(X) :- bird(X).}$,
        `bird(piu).` ,
        `bird(tom).`}.

Note that for this logic program, $Const = \{\texttt{piu}, \texttt{tom}\}$. The non-ground clause `fly(X) :- bird(X).` stands for the two ground clauses below, corresponding to its ground instances.

    fly(piu) :- bird(piu).   fly(tom) :- bird(tom).

The atom `fly(piu)` is a ground atom, whereas `fly(X)` is not. We can also say that the atom `fly(piu)` is a ground instance of `fly(X)` (i.e. `fly(piu)` $\in$ $ground(\texttt{fly(X)})$).   □

An interesting aspect of the logic programming framework is the possibility to query the knowledge encoded in logic programs. By querying a logic program one may retrieve interesting non-trivial knowledge. For instance, given the program of example 13, we may ask whether `tom` flies or which birds fly. In the former case, we expect to obtain a positive answer. In the latter case, we should get as answer `tom` and `piu`. This point is particularly relevant for the operational semantic (i.e. implementation) of the query languages proposed in later sections.

We discuss in more detail queries in next section. However, we first introduce a more general class of logic programs that includes definite logic programs.

3.2   Extended Logic Programs

Extended Logic Programming (ELP) is the target language of the transformations discussed in section 4, and we resort only to the disjunctive free fragment of the languages described in [36,37], generalizing Answer Set Semantics [38] to the paraconsistent case.

In contrast to ELP, it is not possible to represent negative information in a definite logic program. The main distinctive feature of ELP is that it allows to express two forms of negation, explicit and default, allowing both open-world and closed-world reasoning. Explicit negation describes negative evidence, e.g. *"Tom does not fly."*, while default negation allows reasoning with lack of information, e.g. *"There is no evidence that tom flies."*.

The language in which extended logic programs are expressed is also based in an alphabet of variable, constant, and predicate symbols, i.e. $Var \cup Const \cup Pred$. Let At denote the set of all atoms built with alphabet symbols. An *objective literal* L is either an atom $A \in At$ or its explicit negation $\neg A$. The set of all objective literals is $OLit = At \cup \neg At$, where $\neg At = \{\neg A : A \in At\}$. The default negation of a literal L is represented by $not\ L$, also called default negated literal. A *literal* is either an objective literal L or its default negation $not\ L$, and the set of all literals is

$$Lit = OLit\ \cup\ not\ OLit = \{A, \neg A, not\ A, not\ \neg A : A \in At\}\ .$$

Intuitively, an objective literal represents a (positive or negative) evidence, while the default negated literal represents lack of (respectively, positive or negative) evidence. This makes it possible, for example, to represent differently the information that a flight departed without delay obtained from the flight control, from lack of the delay announcement.

Similar to definite logic programs, knowledge is encoded as sets of clauses in ELP. However, clauses of an extended logic program may include explicit and default negation.

Definition 17. *A* clause *is an expression*

$$L_0 :- L_1, \ldots, L_m, not\ L_{m+1}, \ldots, not\ L_n.$$

where each L_i is an objective literal and $0 \le m \le n$.

An *integrity constraint* has the form

$$:- L_1, \ldots, L_m, not\ L_{m+1}, \ldots, not\ L_n. \ ,$$

with $n \ge m \ge 1$ and it can be seen as a clause with the head being the atom *false* (or $\perp$) representing falsehood. For instance, the integrity constraint

$$:- \mathtt{human}(X), \mathtt{male}(X), \mathtt{female}(X).$$

expresses that no human can be male and female, simultaneously.

Definition 18. *An* extended logic program *is a finite set of clauses and integrity constraints.*

The notions of ground atom previously introduced can be easily extended to literals and sequences of literals. Moreover, the definitions of ground program and substitution presented for definite logic programs apply also to extended logic programs.

Example 14. Assume that the following clauses belong to the extended logic program $\mathcal{P}$.

"*Someone is guilty if he is guilty*"
(1) `guilty(X) :- guilty(X).`

"*Someone is innocent if we cannot prove he is guilty.*"
(2) `innocent(X) :- not guilty(X).`

"*Someone is not guilty if he is innocent.*"
(3) `¬guilty(X) :- innocent(X).`

"*A person is a male if we cannot prove he is a female.*"
(4) `male(X) :- person(X), not female(X).`

"*A person is a female if we cannot prove she is a male.*"
(5) `female(X) :- person(X), not male(X).`

"*A person cannot be guilty and non-guilty, simultaneously.*"
(6) `:- person(X), guilty(X), ¬guilty(X).`

"*Tommy is a person.*"
(7) `person(tommy)`

Clauses (4) and (5) together express the idea that a person must be either a male or a female.   □

Declarative Semantics of Extended Logic Programs The declarative semantics of a program captures its meaning. The declarative semantics of (extended) logic programs is based on the notion of interpretation. An interpretation is simply a subset of the ground objective literals, also known as the *extended Herbrand base*.

Definition 19. *An* interpretation $\mathcal{I}$ *of an extended logic program $\mathcal{P}$ is any subset of ground$(OLit) = ground(At) \cup \neg ground(At)$.*

As usual, an interpretation settles the set of *true* literals. If $L \in \mathcal{I}$ then the objective literal L has the truth value *true*, and if $L \notin \mathcal{I}$ then the objective literal L is *false*. Clearly, if an objective literal L is *false* then *not* L is *true*.

An interpretation induces the following consequence relation:

$$\mathcal{I} \models L \qquad \text{if an only if } L \in \mathcal{I},$$
$$\mathcal{I} \models \text{not } L \qquad \text{if and only if } L \notin \mathcal{I},$$
$$\mathcal{I} \models L_1, \ldots, L_n \text{ if and only if } \mathcal{I} \models L_1 \text{ and } \ldots \text{ and } \mathcal{I} \models L_n,$$

where L is an arbitrary ground objective literal and each L_i $(1 \leq i \leq n)$ is an arbitrary ground literal.

Example 15. Consider the extended logic program $\mathcal{P}$ of example 14. A possible interpretation for $\mathcal{P}$ is

$$\mathcal{I} = \{\texttt{innocent(tommy)}, \neg\texttt{guilty(tommy)}, \texttt{male(tommy)}, \texttt{person(tommy)}\}.$$

In this interpretation the literals $\texttt{guilty(tommy)}$, $\texttt{female(tommy)}$, *not* $\texttt{male(tommy)}$ are *false*, i.e. $\mathcal{I} \not\models \texttt{guilty(tommy)}$, $\mathcal{I} \not\models \texttt{female(tommy)}$, $\mathcal{I} \not\models not \texttt{male(tommy)}$. Obviously,

$$\mathcal{I} \models \neg\texttt{guilty(tommy)}, not \texttt{female(tommy)} . \qquad \square$$

An interpretation $\mathcal{I}$ *satisfies* a program clause if the corresponding implication holds in $\mathcal{I}$, and *satisfies* an integrity constraint if at least one literal in its body is *false*. Next definition formalizes this idea.

Definition 20. *A model $\mathcal{M_P}$ of an extended logic program $\mathcal{P}$ is any interpretation that satisfies every clause and integrity constraint of ground$(\mathcal{P})$, i.e. $(0 \leq m \leq n)$*

1. *For every $L_0 :\text{-} L_1, \ldots, L_m, not\ L_{m+1}, \ldots, not\ L_n \in ground(\mathcal{P})$, if $\mathcal{I} \models L_1, \ldots, L_m, not\ L_{m+1}, \ldots, not\ L_n$ then $\mathcal{I} \models L_0$.*
2. *For every $:\text{-} L_1, \ldots, L_m, not\ L_{m+1}, \ldots, not\ L_n \in ground(\mathcal{P})$, then $\mathcal{I} \not\models L_1, \ldots, L_m, not\ L_{m+1}, \ldots, not\ L_n$.*

For instance, the interpretation $\mathcal{I}$ in example 15 is also a model of the program. Intuitively, an integrity constraint discards all model candidates that make every literal in its body *true*.

An extended logic program may have zero, one, or more models. Moreover, set inclusion is a partial order for the set of models of an extended logic program. Since we want to consider only the models such that each objective literal can be justified by some evidence in the program, only (some of) the minimal models are of interest. To capture formally this intuition, we need to recall an important property of definite logic programs (theorem 1) and introduce the notion of reduct [3] of an extended logic program [39].

Note that an extended logic program may have several minimal models while a definite logic program has always a least model (unique minimal model).

Theorem 1 ([32]). *Let $\mathcal{P}$ be a definite logic program. Then, $\mathcal{P}$ has a least model.*

Definition 21 ([39]). *Let $\mathcal{P}$ be an extended logic program and $\mathcal{I}$ an interpretation. The reduct of $\mathcal{P}$ with respect to $\mathcal{I}$ is the definite logic program $\psi_{\mathcal{I}}(\mathcal{P})$ such that $L_0 \text{:-} L_1, \ldots, L_m. \in \psi_{\mathcal{I}}(\mathcal{P})$ if and only if there is a program clause of the form $L_0 \text{:-} L_1, \ldots, L_m, not\ L_{m+1}, \ldots, not\ L_n. \in ground(\mathcal{P})$ such that $\mathcal{I} \models not\ L_{m+1}, \ldots, not\ L_n$, where $0 \leq n \leq m$.*

It should be stressed that $\psi_{\mathcal{I}}(\mathcal{P})$ is always a definite logic program. Hence by theorem 1, it must have a least model.

Definition 22. *Let $\mathcal{P}$ be an extended logic program. An interpretation $\mathcal{I}$ is a paraconsistent stable model of $\mathcal{P}$ if and only $\mathcal{I}$ is the least model of $\psi_{\mathcal{I}}(\mathcal{P})$ and $\mathcal{I}$ satisfies all integrity constraints of $\mathcal{P}$.*

The semantics of extended logic programs is captured by those minimal models that are also paraconsistent stable models. The semantics is paraconsistent because a piece of information and its explicit negation can simultaneously hold. Note that an extended logic program may have no paraconsistent stable models, although it can have several minimal models. Intuitively, these programs are meaningless. The paraconsistent stable model semantics coincides with the stable model semantics [39,40,41] whenever explicit negated literals do not occur in the program.

Example 16. Consider once more the extended logic program $\mathcal{P}$ of example 14 and assume that $\mathcal{P}_1 = \mathcal{P} \cup \{\text{:- female(X).}\}$. Although,

$$\mathcal{I}_1 = \{\text{innocent(tommy)}, \neg\text{guilty(tommy)},$$
$$\text{female(tommy)}, \text{person(tommy)}\}$$

is a model of $\mathcal{P}$, it cannot be a model of $\mathcal{P}_1$ because the new integrity constraint rejects those interpretations containing any ground instance of literal female(X). The following interpretations

$$\mathcal{M}_1 = \{\text{guilty(tommy)}, \text{male(tommy)}, \text{person(tommy)}\} \text{ and}$$
$$\mathcal{M}_2 = \{\text{innocent(tommy)}, \neg\text{guilty(tommy)},$$
$$\text{male(tommy)}, \text{person(tommy)}\}$$

---

[3] Reduct of a logic program and reduct of a decision table (commonly used in the rough set framework) are two independent notions.

are minimal models of $\mathcal{P}_1$. However, only $\mathcal{M}_2$ is a paraconsistent stable model, whereas $\mathcal{M}_1$ is not.

Intuitively, the reason for $\mathcal{M}_1$ not being a paraconsistent stable model is that `guilty(tommy)` is a justification to itself (see clause (1) of example 14). However, we can justify having `innocent(tommy)` in $\mathcal{M}_2$ because there is no evidence that `guilty(tommy)` is *true* (i.e. *not* `guilty(tommy)` is *true*) and then, by clause (2), we must have that `innocent(tommy)` is *true*.

Formally, taking definition 22, the reduct of $\mathcal{P}_1$ with respect to $\mathcal{M}_1$ is the definite logic program $\varphi_{\mathcal{M}_1}(\mathcal{P}_1)$ consisting of the following clauses.

```
guilty(tommy) :- guilty(tommy).
¬guilty(tommy) :- innocent(tommy).
male(tommy) :- person(tommy).
person(tommy).
```

The least model of $\varphi_{\mathcal{M}_1}(\mathcal{P}_1)$ is $\mathcal{M} = \{$`male(tommy)`, `person(tommy)`$\}$. Since $\mathcal{M}_1 \neq \mathcal{M}$, we conclude that $\mathcal{M}_1$ is not a paraconsistent stable model. It can be easily checked that $\mathcal{M}_2$ is in this case a paraconsistent stable model. □

We introduce now the notion of a ground literal l to be implied by an extended logic program $\mathcal{P}$, denoted as $\mathcal{P} \models l$.

Definition 23. *Let $\mathcal{P}$ be an extended logic program and L be a ground literal. There is a paraconsistent stable model $\mathcal{M}$ of $\mathcal{P}$ such that $\mathcal{M} \models L$ if and only if $\mathcal{P} \models L$.*

We conclude this section with a final remark. The semantics presented in this section is non-monotonic because by adding a new statement to a program the set of literals implied by the program may decrease. For instance consider again the program of example 14. If the fact `guilty(tommy).` would be added to $\mathcal{P}$ then $\mathcal{P} \not\models$`innocent(tommy)`.

Queries As we mentioned in the end of the section 3.1, one of the main aims of the ELP framework is to extract information from extended logic programs by querying them. Let us introduce the notion of query.

Definition 24. *A query is a pair $(L_1, \ldots, L_n , \mathcal{P})$, with $n \geq 1$, where $\mathcal{P}$ is an extended logic program and each L_i is a literal.*

We need now to define the notion of answer.

Definition 25. *Let $(\mathcal{Q}, \mathcal{P})$ be a query. An* answer *to the query is the set of ground substitutions*

$$\{\theta \mid \mathcal{Q}\theta \in ground(\mathcal{Q}) \text{ and } \mathcal{M} \models \mathcal{Q}\theta\} ,$$

for some paraconsistent stable model $\mathcal{M}$ of $\mathcal{P}$.

Example 17. Consider the extended logic program $\mathcal{P}$ of example 14 and the queries

(innocent(X), $\mathcal{P}$)   "Who is innocent?"
(guilty(tommy), $\mathcal{P}$) "Is Tommy guilty?"

The answer to the first query is $\{\{X/\text{tommy}\}\}$, since innocent(tommy) belongs to a paraconsistent stable model of $\mathcal{P}$ (see $\mathcal{M}_2$ in example 16). However, the answer to the second question is $\emptyset$ because the literal guilty(tommy) does not belong to any paraconsistent stable model of $\mathcal{P}$.   □

Finally, we refer that *Smodels* [42,43] and *dlv* [44,45] are currently available systems for computing stable models of programs (often with tens of thousands of clauses). Both systems can also handle integrity constraints, and can be used in practice to determine paraconsistent stable models of extended logic programs. Moreover, any standard *Prolog* system [46] can be used to compute answers to queries $(\mathcal{Q}, \mathcal{P})$, when $\mathcal{P}$ is a definite logic program.

4   A Language for Defining Rough Relations

This section presents a new language [19,21] for defining and querying rough relations, based on logic programming.

The main intuitive idea underlying this language is as follows. A rough relation S divides the universe in four regions: those examples that definitely belong to the concept represented by S (to be denoted $\underline{S}$); those examples that definitely do not belong to the concept (to be denoted $\neg\underline{S}$), those examples for which there is contradictory evidence (to be denoted $\overline{S}$); and those examples for which there is not any information of whether they belong to the concept (i.e. the remaining part of the universe not contained in $\underline{S} \cup \neg\underline{S} \cup \overline{S}$). Using clauses we can then combine regions of different rough relations to define implicitly a new rough relation. The language introduced in this section does not take into account quantitative measures. This extension is discussed in section 6.

The declarative semantics of the language (discussed in section 4.2) associates a rough set S with each predicate symbol s of the language. Hence, we give indirectly a four-valued interpretation to each predicate: if tuple $t \in \underline{S}$ then $s(t)$ has the logic value true; if tuple $t \in \neg\underline{S}$ then $s(t)$ has the logic value false; if tuple $t \in \overline{S}$ then $s(t)$ has the logic value $\top$ (denoting contradictory evidence); otherwise, $s(t)$ has the logic value $\bot$ (denoting lack of information). Four-valued logics have been studied by other authors, of which the most well-known is Belnap's four-valued logic [47,48]. However, statements in our language make explicit reference to one of the three regions of a rough set (lower approximation, upper approximation, or boundary region) or the remaining part of the world not belonging to any of these three regions. In other words, the statements contain explicit tests of whether a tuple t belongs to one of those four regions (as shown in the next section, predicate symbols can only occur in rough or testing literals). Since each of those four regions is a crisp set, we do not need to resort to four-valued logical operations (e.g. disjunction). We use instead two-valued logic. The main advantage of using two-valued semantics is that there are several systems [46,43] readily available that can be used for making computations and answering queries. This aspect is particularly relevant from the point of view of implementing a system that can answer queries about knowledge bases encoded in our language. Formulating a language with a four-valued semantics that caters for representing and reasoning with rough concepts is out of the scope of this thesis, although this could be an issue for future work.

Section 4.1 introduces the language and its declarative semantics is then formalized in section 4.2. A transformation of the proposed language into the language of extended logic programs and a proof of its correctness is presented in section 4.3. Finally, section 4.4 puts forward a query language to extract information from rough relations defined in a program.

4.1 The Syntax

The language we are going to introduce uses three disjoint alphabets: an alphabet of variable symbols Var, an alphabet of constant symbols $Const$, and an alphabet of predicate symbols $Pred$. The notions of term and atom are similar to the ones introduced for logic programs. A *term* t is any symbol belonging to $Var \cup Const$. We follow the usual convention that variables start with upper case letter (e.g. $X, Dist \in Var$) and constants begin with lower case (e.g. $small, c \in Const$). Moreover, an n-ary predicate p is often denoted $p/n \in Pred$ and the set of symbols $\{\neg, not\} \not\subseteq Pred$. An *atom* A is an expression of the form $p(t_1, \ldots, t_n)$, where p is an n-ary predicate symbol and each $t_1, \ldots, t_n$ is a term.

Given a predicate p/n of arity $n > 0$, formulas of the form $\underline{l}(t_1, \ldots, t_n)$, $\bar{l}(t_1, \ldots, t_n)$, or $\underline{\bar{l}}(t_1, \ldots, t_n)$, where l is either p or $\neg p$, are called *rough literals*. Moreover, the expression $p?(t_1, \ldots, t_n)$ represents a *testing literal*.

Definition 26. *A rough clause is any expression of the form*

$$H :\text{-} B_1, \ldots, B_n, T_1, \ldots, T_m.$$

where H and every B_i ($0 \le i \le n$) is a rough literal, and each T_j ($0 \le j \le m$) is a testing literal such that all variables occurring in a testing literal also occur in some B_i.

Rough clauses with an empty body (i.e. $n = 0$ and $m = 0$) are called *rough facts*.

The order by which rough (testing) literals occur in a rough clause is irrelevant. Thus, both rough clauses

$$\bar{p}(X, Y) :\text{-} \underline{q_1}(X, Y), \neg q_2(X, Y), r_1?(X), r_2?(Y).$$
$$\bar{p}(X, Y) :\text{-} \underline{\neg q_2}(X, Y), \underline{q_1}(X, Y), r_2?(X), r_1?(Y).$$

have the same meaning.

We can now define the notion of rough program.

Definition 27. *A rough program $\mathcal{P}$ is a finite set of rough clauses.*

Intuitively, each predicate p denotes a rough relation P and we use rough literals to represent evidence about tuples. The lower (upper) approximation of P is represented by $\underline{p}(t_1, \ldots, t_n)$ ($\overline{p}(t_1, \ldots, t_n)$). The boundary of P is denoted by $\overline{p}(t_1, \ldots, t_n)$. Obviously, the rough literals $\overline{p}(t_1, \ldots, t_n)$ and $\overline{\neg p}(t_1, \ldots, t_n)$ have the same meaning. For instance, the rough facts[4]

$$\overline{\text{recommendLenses}(\text{young}, \text{myope}, \text{yes})}.$$

and

$$\overline{\neg\text{recommendLenses}(\text{young}, \text{myope}, \text{yes})}.$$

express the information that the tuple $\langle\text{young}, \text{myope}, \text{yes}\rangle$ belongs to both $\overline{RecommendLenses}$ and to $\overline{\neg RecommendLenses}$ (thus, to the boundary of $RecommendLenses$). Informally, the first fact states that young myope people with astigmatism should use contact lenses while the second asserts exactly the opposite (perhaps, because different opticians have different opinions for these customers). The rough fact

$$\underline{\text{recommendLenses}(\text{young}, \text{myope}, \text{no})}.$$

states that the tuple $\langle\text{young}, \text{myope}, \text{no}\rangle$ is a positive example of rough relation $RecommendLenses$ but cannot be a negative example of it (i.e. $\langle\text{young}, \text{myope}, \text{no}\rangle \in \underline{RecommendLenses}$). Notice that rough literals of the form $\overline{\neg p}(t_1, \ldots, t_n)$ or $\neg p(t_1, \ldots, t_n)$ express negative evidence.

A testing literal $p?(t_1, \ldots, t_n)$ asserts that there is no information whether tuple $\langle t_1, \ldots, t_n \rangle$ describes a positive and/or a negative example of the concept represented by rough relation P.

A decision table $\mathcal{D} = (U, A, d)$ can be easily represented in our language, if quantitative measures are ignored. A row $\langle c_1, \ldots, c_n \rangle$ of $\mathcal{D}$ corresponding to a positive (negative) example, where each $c_i \in V_{a_i}$ is the value of a condition attribute $a_i \in A$, is represented as the fact $\overline{d}(c_1, \ldots, c_n)$ ($\overline{\neg d}(c_1, \ldots, c_n)$). An important aspect to bear in mind is that the proposed language does not represent the individuals in U. Rough relations are represented as sets of tuples of attribute values, not as sets of individuals.

We stress that condition attributes are not referred by their name (e.g. Age) in the rough literals. They are instead identified by their position in the argument list of the rough literal. For instance, we use the convention that the first argument of the predicate recommendLenses represents the condition attribute Age. The condition attribute associated with a term t_i in a rough literal

$$\overline{q}(t_1, \ldots, t_n) \text{ or } \overline{\neg q}(t_1, \ldots, t_n) \text{ or}$$
$$\underline{q}(t_1, \ldots, t_n) \text{ or } \neg q(t_1, \ldots, t_n) \text{ or}$$
$$\overline{q}(t_1, \ldots, t_n) \text{ or } \overline{\neg q}(t_1, \ldots, t_n),$$

is represented as $att_Q(i)$. Each term t_i can only represent values belonging to $V_{att_Q(i)}$.

---

[4] The third condition attribute refers to whether the person suffers from astigmatism.

A rough clause represents an implication, as in the context of logic programs. The use of variables in a rough literal of a rough clause indicate that the underlying implication is valid for each possible value of the corresponding condition attribute. Since rough clauses allow lower and upper approximations of a relation as well as boundaries to occur both in the body and in the head of a clause, it is possible to define separately each of the regions (i.e. lower and upper approximations and boundary) of a rough relation in terms of regions of other rough relations. For instance, we can state that the boundary of a rough relation Q is contained in the lower approximation of another rough relation P. If predicates $q/3$ and $p/3$ denote the rough relations Q and P, respectively, then the rough clause

$$\underline{p}(X_1, X_2, X_3) :- \overline{\underline{q}}(X_1, X_2, X_3).$$

captures such information.

Given a rough relation P with n attributes, an n-ary tuple t is *undefined* with respect to P if and only if t is neither a positive nor a negative example of the relation, i.e.

$$\langle t_1, \ldots, t_n \rangle \notin \overline{P} \text{ and } \langle t_1, \ldots, t_n \rangle \notin \overline{\neg P},$$

where each $t_i \in V_{att_P(i)}$. We can test in the body of a rough clause whether a tuple $\langle t_1, \ldots, t_n \rangle$ is undefined with respect to P, by using the testing literal $p?(t_1, \ldots, t_n)$.

The following rough clause

$$\underline{p}(X_1, X_2, X_3) :- \overline{q}(X_1, X_2, X_3), r?(X_1, X_2, X_3).$$

asserts that if a tuple $t \in \overline{Q}$ and and t is undefined with respect to R (i.e. there is no information whether t describes a positive or a negative example of the concept represented by rough relation R) then t also belongs to $\underline{P}$.

The following two examples motivate the potential usefulness of our language. More examples are presented in the next section.

Example 18. A relation $Train$ has two arguments (condition attributes) representing time and location, respectively. Two (or more) sensors automatically detect presence/absence of an approaching train at a crossing, producing facts like `train(12:50,montijo).` automatically added to the knowledge base. A malfunction of a sensor may result in the contradictory fact `¬train(12:50, montijo).` being added, too. Crossing is allowed if for sure no train approaches. This can be described by the following clause involving lower approximation in the body.

$$\overline{\text{cross}}(\text{X,Y}) \ :- \ \neg\text{train}(\text{X,Y}). \qquad\qquad \square$$

Example 19. Statistical data on purchases of certain product during a calendar year is organized as a decision table with the following 3 condition attributes defining groups of customers:

`Area` - zip code of the area where the customer lives
`Income` - customer's income interval
`Age` - customer's age interval

Notice that the decision table may define a rough relation: a `young` customer living in `Norrköping` and having `medium` income may be considered `inactive` (perhaps, because he has not bought any product item during last year) while another `young` customer, also living in `Norrköping` and with `medium` income is classified as active.

The marketing department uses the activity tables `act1` and `act2` from two consecutive years to identify the groups of *growing activity* (`ga`). The tables are represented as rough facts in our language. The activity of a group may be defined: (1) as *definitely growing*, if the group was possibly inactive in year 1 and definitely active in year 2; (2) as *definitely non growing*, if its activity changed from possibly active to definitely inactive; and (3) as a *boundary*, if the activity was boundary in both years. This can be described by the following rough clauses.

(1) $\underline{\mathtt{ga}}$(Area, Inc, Age) :- $\overline{\neg\mathtt{act1}}$(Area, Inc, Age),
$\underline{\mathtt{act2}}$(Area, Inc, Age).

(2) $\neg\underline{\mathtt{ga}}$(Area, Inc, Age) :- $\overline{\mathtt{act1}}$(Area, Inc, Age),
$\underline{\neg\mathtt{act2}}$(Area, Inc, Age).

(3) $\overline{\underline{\mathtt{ga}}}$(Area, Inc, Age) :- $\overline{\mathtt{act1}}$(Area, Inc, Age),
$\overline{\mathtt{act2}}$(Area, Inc, Age).   □

The language described in this section extends substantially the language presented in [17,18]. The language discussed in this previous work only allows the use of upper approximations in the definition of new rough relations.

4.2   The Declarative Semantics

The main idea underlying the proposed language is that each predicate symbol occurring in a rough program denotes a rough relation. We formalize this idea in this section.

Definition 28. *Let $\mathcal{P}$ be a rough program. A rough interpretation $\mathcal{I}$ of $\mathcal{P}$ is a function mapping each predicate symbol q/n occurring in P into a rough relation $Q^{\mathcal{I}} = (\overline{Q^{\mathcal{I}}}, \overline{\neg Q^{\mathcal{I}}})$ such that $\overline{Q^{\mathcal{I}}}, \overline{\neg Q^{\mathcal{I}}} \subseteq \prod_{1 \leq i \leq n} V_{att_Q(i)}.$*

If no variables occur in a rough (testing) literal then the rough (testing) literal is called a *ground rough (testing) literal*. A rough clause is ground, if all rough and testing literals occurring in it are also ground.

The notion of a rough literal L being true in a rough interpretation $\mathcal{I}$, denoted as $\mathcal{I} \models L$, is defined by statements (1) − (8) below.

(1) $\mathcal{I} \models \overline{q}(c_1, \ldots, c_n) \Leftrightarrow \langle c_1, \ldots, c_n \rangle \in \overline{Q^{\mathcal{I}}}.$
(2) $\mathcal{I} \models \underline{q}(c_1, \ldots, c_n) \Leftrightarrow (\langle c_1, \ldots, c_n \rangle \in \overline{Q^{\mathcal{I}}} \text{ and } \langle c_1, \ldots, c_n \rangle \notin \overline{\neg Q^{\mathcal{I}}}).$

(3) $\mathcal{I} \models \overline{q}(c_1, \ldots, c_n) \Leftrightarrow (\langle c_1, \ldots, c_n \rangle \in \overline{Q^{\mathcal{I}}}$ and $\langle c_1, \ldots, c_n \rangle \in \overline{\neg Q^{\mathcal{I}}})$.

(4) $\mathcal{I} \models \overline{\neg q}(c_1, \ldots, c_n) \Leftrightarrow \langle c_1, \ldots, c_n \rangle \in \overline{\neg Q^{\mathcal{I}}}$.

(5) $\mathcal{I} \models \neg q(c_1, \ldots, c_n) \Leftrightarrow (\langle c_1, \ldots, c_n \rangle \in \overline{\neg Q^{\mathcal{I}}}$ and $\langle c_1, \ldots, c_n \rangle \notin \overline{Q^{\mathcal{I}}})$.

(6) $\mathcal{I} \models \overline{\neg q}(c_1, \ldots, c_n) \Leftrightarrow (\langle c_1, \ldots, c_n \rangle \in \overline{Q^{\mathcal{I}}}$ and $\langle c_1, \ldots, c_n \rangle \in \overline{\neg Q^{\mathcal{I}}})$.

(7) $\mathcal{I} \models q?(c_1, \ldots, c_n) \Leftrightarrow (\langle c_1, \ldots, c_n \rangle \notin \overline{Q^{\mathcal{I}}}$ and $\langle c_1, \ldots, c_n \rangle \notin \overline{\neg Q^{\mathcal{I}}})$.

(8) $\mathcal{I} \models B_1, \cdots, B_n, T_1, \cdots, T_m \Leftrightarrow (\mathcal{I} \models B_1, \cdots, \mathcal{I} \models B_n,$
$\mathcal{I} \models T_1, \cdots, \mathcal{I} \models T_m)$, where each B_i $(0 \le i \le n)$ is a ground rough literal and each T_j $(0 \le j \le m)$ is a ground testing literal.

Note that for any rough interpretation $\mathcal{I}$, it is possible that

$$\mathcal{I} \models \overline{q}(t_1, \ldots, t_n), \overline{\neg q}(t_1, \ldots, t_n) ,$$

for two rough literals $\overline{q}(t_1, \ldots, t_n)$ and $\overline{\neg q}(t_1, \ldots, t_n)$. However, if we consider lower approximations instead then we must have

$$\mathcal{I} \not\models \underline{q}(t_1, \ldots, t_n), \underline{\neg q}(t_1, \ldots, t_n) .$$

Consider a rough literal $l(t_1, \ldots, t_n)$, with $1 \le n$ and l is either $\overline{q}$, or $\underline{q}$, or $\overline{q}$, or $\overline{\neg q}$, or $\underline{\neg q}$, or $\overline{\neg q}$, for some predicate q/n. Recall that each term t_i is associated with a condition attribute of rough relation Q whose value domain is $V_{att_Q(i)}$.

Definition 29. *Let* $X_{1j}, \ldots, X_{nk}$ *$(1 \le n$ and $1 \le j \le k)$ be variables occurring in some rough literal, where the second index indicates the variable's position in the argument list of the literal. Assume also that q is the predicate symbol occurring in the rough literal. A grounding substitution θ is a set of bindings* $\{X_{1j}/c_1, \ldots, X_{nk}/c_n\}$ *(including the empty set) of variables X_{im} $(1 \le i \le n$ and $j \le m \le k)$ to constants $c_i \in V_{att_Q(m)}$.*

The notion of ground rough program is similar to the notion of ground (extended) logic program, presented previously. A *ground instance of a rough clause* C, denoted by $C\theta$, is obtained by applying a grounding substitution θ to rough clause C and $C\theta$ is ground. Given a rough program $\mathcal{P}$, $ground(\mathcal{P})$ represents the set of all ground instances of any rough clause $C \in \mathcal{P}$.

A rough interpretation $\mathcal{I}$ of a rough program $\mathcal{P}$ *satisfies* a rough clause

$$H \text{:- } B_1, \ldots, B_n, T_1, \ldots, T_m. \in ground(\mathcal{P})$$

if and only if

- if $\mathcal{I} \models B_1, \ldots, B_n, T_1, \ldots, T_m$ then $\mathcal{I} \models H$.

Definition 30. *A* model *of a rough program $\mathcal{P}$ is a rough interpretation that satisfies each rough clause of $ground(\mathcal{P})$.*

A rough program may have several models or no models at all. For instance, the rough program

$$\mathcal{P} = \{\underline{p}(a) \text{:- } \overline{q}(b). \, , \overline{\neg p}(a). \, , \underline{q}(b).\}$$

has no models. Example 20 below shows a rough program with more than one model.

We turn now to the definition of a partial order between the models of a rough program.

Definition 31. *Let $\mathcal{M}_1$ and $\mathcal{M}_2$ be two models of a rough program $\mathcal{P}$. $\mathcal{M}_1 \preceq \mathcal{M}_2$ if and only if*

$$\overline{Q^{\mathcal{M}_1}} \subseteq \overline{Q^{\mathcal{M}_2}} \text{ and } \overline{\neg Q^{\mathcal{M}_1}} \subseteq \overline{\neg Q^{\mathcal{M}_2}} ,$$

for every predicate symbol q/n occurring in $\mathcal{P}$.

Based on the partial order $\preceq$ defined above, we introduce the notion of minimal models.

Definition 32. *A model $\mathcal{M}$ is a* minimal model *of a rough program $\mathcal{P}$ if and only if $\mathcal{M}' \not\preceq \mathcal{M}$, for every other model $\mathcal{M}'$ of $\mathcal{P}$.*

We give next an example illustrating a situation where a rough program has more than one model.

Example 20. Consider that we want to represent some expert knowledge saying that

> *"If someone possibly has an* infection *but his* temperature *is* normal, *then he either might suffer from* diseaseA *or from* diseaseB *(but never from both)."*

Moreover, two decision tables are given. Based on the existence of certain symptoms and results of some clinical tests, several experts decide independently whether a patient has diseaseA or diseaseB. Symptoms and clinical test results form the condition attributes. For instance, a condition attribute is temperature that can have the values low, normal, or high.

The expert knowledge can be represented as follows. Assume that predicates diseaseA and diseaseB have arity three (i.e. the corresponding decision tables have three condition attributes).

$\overline{\text{diseaseA}}$(infect, normal, Z) :- $\neg$diseaseB(infect, normal, Z).

$\overline{\text{diseaseB}}$(infect, normal, Z) :- $\neg$diseaseA(infect, normal, Z).

Note that rough relations *DiseaseA* and *DiseaseB*, denoted by predicates diseaseA and diseaseB respectively, are partially defined explicitly by decision tables. Some other tuples belonging to these relations are obtained from the clauses above. In reality experts may decide in different ways whether a person may have a certain disease.

Consider the rough program $\mathcal{P}$ consisting of the two clauses above and including also the following facts obtained from the decision tables.

$\overline{\neg\text{diseaseA}}$(infect, normal, c). $\overline{\neg\text{diseaseB}}$(infect, normal, c).

Program $\mathcal{P}$ has (at least) two models, $\mathcal{M}_1$ and $\mathcal{M}_2$, reflecting two possible situations according to the available knowledge.

- $\langle \text{infect}, \text{normal}, c \rangle \in \overline{DiseaseA^{\mathcal{M}_1}}$ and
  $\langle \text{infect}, \text{normal}, c \rangle \in \overline{\neg DiseaseB^{\mathcal{M}_1}}$.
- $\langle \text{infect}, \text{normal}, c \rangle \in \overline{DiseaseB^{\mathcal{M}_2}}$ and
  $\langle \text{infect}, \text{normal}, c \rangle \in \overline{\neg DiseaseA^{\mathcal{M}_2}}$.

In practice it may be desirable to find preferred models or at least discard some models seen as not relevant. This issue has been studied in the context of logic programming and the proposed techniques may be applicable here. In the context of rough sets, we could address this problem by extending to clauses the quantitative measures associated with decision tables. In the example above, model $\mathcal{M}_1$ is obtained by applying the first clause above, while model $\mathcal{M}_2$ is obtained by applying the second clause. If the tuple $\langle \text{infect}, \text{normal}, c \rangle$ appears as a negative example many more times in the decision table for diseaseB than in the decision table for diseaseA then we may decide to discard the second model. □

Let $\mathcal{P}$ be a rough program and q be a relation symbol occurring in $\mathcal{P}$. We are obviously not interested in any model $\mathcal{M}$ of $\mathcal{P}$ that makes more tuples to belong to $\overline{Q^{\mathcal{M}}}$ or to $\overline{\neg Q^{\mathcal{M}}}$ than what is needed to satisfy the rough clauses of $\mathcal{P}$. Thus, minimal models (as defined in 32) seem good candidates to express the meaning of a rough program. However, not all minimal models may properly capture the semantics of a rough program. Next example tries to illustrate this point.

Example 21. Consider the following rough program

$$\mathcal{P} = \{\overline{r}(c) \colon\!\text{-} \ \underline{\neg r}(c). \, , \overline{\neg r}(c).\} \ .$$

A minimal model $\mathcal{M}$ of $\mathcal{P}$ maps predicate r into the rough relation $R^{\mathcal{M}} = (\{\langle c \rangle\}, \{\langle c \rangle\})$ i.e. $\langle c \rangle$ belongs to the boundary of rough relation $R^{\mathcal{M}}$. However, no information encoded in the rough clauses of $\mathcal{P}$ leads to the conclusion that $\langle c \rangle \in \overline{R^{\mathcal{M}}}$. In order to be able to conclude that $\langle c \rangle \in \overline{R^{\mathcal{M}}}$, it would be needed that $\mathcal{M} \models \underline{\neg r}(c)$. By rough clause

$$\overline{r}(c) \colon\!\text{-} \ \underline{\neg r}(c). \tag{1}$$

we could then conclude that $\langle c \rangle \in \overline{R^{\mathcal{M}}}$. However, $\mathcal{M} \not\models \underline{\neg r}(c)$. Hence, it seems reasonable to reject model $\mathcal{M}$. Note that a rough interpretation $\mathcal{I}$ such that $R^{\mathcal{I}} = (\emptyset, \{\langle c \rangle\})$ is not a model of $\mathcal{P}$ because it does not satisfy rough clause (1). Thus, $\mathcal{P}$ seems to bear a contradiction. Rough clause (1) informally states that if there is evidence that $\langle c \rangle$ is only a negative example of the concept represented by rough relation R (i.e. there is no evidence that $\langle c \rangle$ is a positive example of the concept) then we can conclude that $\langle c \rangle$ is also a positive example of that concept. □

Let $\mathcal{I}$ be a rough interpretation of a rough program $\mathcal{P}$. In order to be able to define the declarative semantics of a rough program, we need to introduce the following function, $\Psi_{\mathcal{I}}$, transforming $\mathcal{P}$ into a ground rough program $\Psi_{\mathcal{I}}(\mathcal{P})$ such that neither lower approximations nor testing literals occur in the body of any of its rough clauses. This transformation can be informally described as follows. Assume that C is a rough clause of $ground(\mathcal{P})$. For every ground rough literal $\underline{q}(t_1, \ldots, t_n)$ referring to a lower approximation and occurring in the body of C, if $\mathcal{I} \not\models \overline{\neg q}(t_1, \ldots, t_n)$ then $\underline{q}(t_1, \ldots, t_n)$ in the body of C is replaced by $\overline{q}(t_1, \ldots, t_n)$. Moreover, if a ground testing literal occurring in the body of C is true in $\mathcal{I}$, then it is removed from the body of the clause. The underlying idea behind this transformation is that if $\mathcal{I}$ is a model of $\mathcal{P}$ then it should also be a model of the transformed rough program. We give a simple example of this transformation.

Example 22. Consider the following ground rough clause

$$\underline{p}(a, b) :- \underline{q}(a, b), r?(b, c). \in ground(\mathcal{P})$$

and an interpretation $\mathcal{I}$ of $\mathcal{P}$ such that $\overline{Q}^{\mathcal{I}} = \overline{\neg Q}^{\mathcal{I}} = \overline{R}^{\mathcal{I}} = \overline{\neg R}^{\mathcal{I}} = \emptyset$. Since $\mathcal{I} \not\models \overline{\neg q}(a, b)$ and $\mathcal{I} \models r?(b, c)$, we have that

$$\underline{p}(a, b) :- \overline{q}(a, b). \in \Psi_{\mathcal{I}}(\mathcal{P}) .$$ □

We present below the formal definition of function Ψ. This definition extends the notion of reduct proposed in [39] to rough programs.

Consider that $\neg\neg q \equiv q$, for any predicate symbol q.

Definition 33. *Let $\mathcal{P}$ be a rough program and $\mathcal{I}$ be a rough interpretation of $\mathcal{P}$. Assume also that each l_j $(1 \leq j \leq k)$ in the expression below is either q_i or $\neg q_i$, for some predicate q_i. Then $\Psi_{\mathcal{I}}(\mathcal{P})$ maps $\mathcal{P}$ into a ground rough program satisfying the following condition ($n, i, k \geq 0$ and $m_1, \ldots, m_k \geq 0$):*

$$H :- B_1, \ldots, B_n, \overline{l_1}(t_{11}, \ldots t_{1m_1}), \ldots, \overline{l_k}(t_{k1}, \ldots t_{km_k}). \in \Psi_{\mathcal{I}}(\mathcal{P})$$

if and only if there is a rough clause

$$\begin{aligned} H :- &B_1, \ldots, B_n, \\ &\underline{l_1}(t_{11}, \ldots t_{1m_1}), \ldots, \underline{l_k}(t_{k1}, \ldots t_{km_k}), \\ &T_1, \ldots, T_i. \in ground(\mathcal{P}) \end{aligned}$$

such that

$$\begin{aligned} &\mathcal{I} \not\models \overline{\neg l_1}(t_{11}, \ldots t_{1m_1}), \ldots, \overline{\neg l_k}(t_{k1}, \ldots t_{km_k}) \text{ and} \\ &\mathcal{I} \models T_1, \ldots, T_i , \end{aligned}$$

where each B_j $(0 \leq j \leq n)$ is a rough literal not referring to a lower approximation and each T_l $(0 \leq l \leq i)$ is a testing literal.

An important property of a rough program $\Psi_{\mathcal{I}}(\mathcal{P})$ is that it either has a least model (a unique minimal model) or no model at all.

Lemma 1. *Let $\mathcal{P}$ be a rough program and $\mathcal{I}$ be a rough interpretation of $\mathcal{P}$. If $\Psi_{\mathcal{I}}(\mathcal{P})$ has a model then it has a least model, with respect to partial relation $\preceq$.*

Proof. Let us assume that $\Psi_{\mathcal{I}}(\mathcal{P})$ has a model. To prove that $\Psi_{\mathcal{I}}(\mathcal{P})$ has a least model, we show that any rough interpretation $\mathcal{M}$ defined as below is also a model of $\Psi_{\mathcal{I}}(\mathcal{P})$. Let V be any non-empty set of models of $\Psi_{\mathcal{I}}(\mathcal{P})$.

$$Q^{\mathcal{M}} = (\bigcap_{\mathcal{M}' \in V} \overline{Q^{\mathcal{M}'}}, \bigcap_{\mathcal{M}' \in V} \overline{\neg Q^{\mathcal{M}'}}),$$

for each rough relation Q.

Assume that $H :\!- B. \in \Psi_{\mathcal{I}}(\mathcal{P})$ and that $\mathcal{M} \models B$. Then, $W \models B$, for every $W \in V$. The key to understanding this point is that the body B can only contain rough literals referring to upper approximations and to boundaries. Moreover, upper approximations and boundaries can be seen as monotonic operators.

We conclude that $W \models H$, for every model $W \in V$, and consequently, $\mathcal{M} \models H$.

It is obvious that $\mathcal{M} \preceq W$, for all $W \in V$. □

The meaning of a rough program is captured by those minimal models that satisfy the condition described in the following definition.

Definition 34. *Let $\mathcal{P}$ be a rough program and $min(\mathcal{P})$ be the set of minimal models of $\mathcal{P}$. The semantics of $\mathcal{P}$, denoted as $sem(\mathcal{P})$, is defined as*

$$sem(\mathcal{P}) = \{\mathcal{M} \in min(\mathcal{P}) \mid \mathcal{M} \text{ is the least model of } \Psi_{\mathcal{M}}(\mathcal{P})\} .$$

Example 23. Consider a rough program $\mathcal{P}$ containing only the rough clauses and facts of example 20. Then, $sem(\mathcal{P}) = \{\mathcal{M}_1, \mathcal{M}_2\}$.

$$DiseaseA^{\mathcal{M}_1} = (\{\langle \texttt{infect, normal, c}\rangle\}, \{\langle \texttt{infect, normal, c}\rangle\}),$$
$$DiseaseB^{\mathcal{M}_1} = (\emptyset, \{\langle \texttt{infect, normal, c}\rangle\}),$$

$$DiseaseA^{\mathcal{M}_2} = (\emptyset, \{\langle \texttt{infect, normal, c}\rangle\})$$
$$DiseaseB^{\mathcal{M}_2} = (\{\langle \texttt{infect, normal, c}\rangle\}, \{\langle \texttt{infect, normal, c}\rangle\}).$$

Note that $\Psi_{\mathcal{M}_1}(\mathcal{P})$ has only the following rough clauses

$\overline{\texttt{diseaseA}}(\texttt{infect, normal, c}) :\!- \overline{\neg\texttt{diseaseB}}(\texttt{infect, normal, c}).$

$\overline{\neg\texttt{diseaseA}}(\texttt{infect, normal, c}).$

$\overline{\neg\texttt{diseaseB}}(\texttt{infect, normal, c}).$

It is easy to see that $\mathcal{M}_1$ ($\mathcal{M}_2$) is the least model of $\Psi_{\mathcal{M}_1}(\mathcal{P})$ ($\Psi_{\mathcal{M}_2}(\mathcal{P})$). □

Different minimal models $\mathcal{M} \in sem(\mathcal{P})$ can be informally understood as different alternative scenarios implied by the knowledge encoded in $\mathcal{P}$.

Example 24. Consider again the rough program presented in example 21,

$$\mathcal{P} = \{\overline{r}(c) :\!- \overline{\neg r}(c). , \overline{\neg r}(c).\} ,$$

and the minimal model $\mathcal{M}$ such that $R^{\mathcal{M}} = (\{\langle c\rangle\}, \{\langle c\rangle\})$. Note that $\Psi_{\mathcal{M}}(\mathcal{P}) = \{\overline{\neg r}(c).\}$. Obviously, $\mathcal{M}$ is not the least model of $\Psi_{\mathcal{M}}(\mathcal{P})$. Hence, $\mathcal{M} \notin sem(\mathcal{P})$.

We can easily see that $sem(\mathcal{P}) = \emptyset$. □

Finally, we introduce the notion of a rough (testing) literal l to be implied by a rough program $\mathcal{P}$, denoted as $\mathcal{P} \models l$.

Definition 35. *Let $\mathcal{P}$ be a rough program and l be a rough or testing literal. There is a model $\mathcal{M} \in sem(\mathcal{P})$ such that $\mathcal{M} \models l$ if and only if $\mathcal{P} \models l$.*

4.3   Computing the Semantics of Rough Programs

In the previous section, we define the declarative semantics of a rough program $\mathcal{P}$ as a subset of its minimal models. An obvious question is how such models, belonging to $sem(\mathcal{P})$, can be computed. This section addresses this problem.

Each rough program is compiled to an extended logic program. As we show in section 4.3, each paraconsistent stable model, of the extended logic program obtained by compiling a rough program $\mathcal{P}$, is isomorphic to a model belonging to $sem(\mathcal{P})$, and vice-versa. The operational semantics of extended logic programs is well studied [42,34] and there are several systems, like *dlv* [44,45] and *Smodels* [49,43], that can be used to compute paraconsistent stable models of extended logic programs.

Some rough programs have at most one model. Absence of recursion is a sufficient condition for a rough program to have either a least model or no models. These rough programs can be compiled to a non-recursive extended logic program. Rough programs can be queried (queries are discussed in section 4.4). Queries to a rough program are also transformed into queries to the compiled program. Given a non-recursive extended logic program, any standard *Prolog* system [46] can be used to determine whether this program has a paraconsistent stable model and answer queries. Hence for non-recursive rough programs, we could implement a system based on our ideas in *Prolog*. An easy way to verify whether a rough program is recursive consists in checking whether the ground compiled extended logic program is recursive.

We give below some examples of recursive and non-recursive rough programs.

Example 25. Consider the following rough programs.

$$\mathcal{P}_1 = \{\underline{q}(a,b) \mathbin{:-} \overline{r}(a,b). \, , \overline{r}(a,b) \mathbin{:-} \overline{q}(a,b).\} \, ,$$
$$\mathcal{P}_2 = \{\underline{q}(a,b) \mathbin{:-} \overline{r}(a,b). \, , \overline{r}(a,b) \mathbin{:-} \overline{q}(c,b).\} \, ,$$
$$\mathcal{P}_3 = \{\underline{q}(a,b) \mathbin{:-} \neg\overline{q}(a,b).\} \, .$$

Rough program $\mathcal{P}_1$ is recursive while rough programs $\mathcal{P}_2$ and $\mathcal{P}_3$ are not.   □

Generally speaking, the application problems discussed in rough set literature that can be formulated in our language do not seem to require recursive rough programs and, therefore, they are not compiled to recursive extended logic programs. Some of these applications are discussed in the next section. Moreover, we have also implemented in *Prolog* a system that is able to reason about rough relations defined in a non-recursive rough program. This system is the topic of section 6.

Compiling Rough Programs into Extended Logic Programs. In this section, we discuss in detail how rough programs can be transformed into extended logic programs.

The transformation to be presented has the following property. A model $\mathcal{M}$ of a rough program $\mathcal{P}$ belongs to $sem(\mathcal{P})$ if and only if there is a paraconsistent stable model $\mathcal{M}'$ of the transformed program $\mathcal{P}'$ such that each predicate symbol q/n occurring in $\mathcal{P}$ denotes the rough relation

$$Q^{\mathcal{M}} = (\{\langle c_1, \ldots, c_n \rangle \mid q(c_1, \ldots, c_n) \in \mathcal{M}'\}, \{\langle c_1, \ldots, c_n \rangle \mid \neg q(c_1, \ldots, c_n) \in \mathcal{M}'\}) \ .$$

In section 4.3, we prove that this property is in fact guaranteed by the proposed compilation.

The intuition underlying the compilation procedure is as follows. Assume that P and Q are the rough relations denoted by predicates p and q occurring in a rough program, respectively. Then, the literal $p(t_1, \ldots, t_n)$ states that the tuple $\langle t_1, \ldots, t_n \rangle$ belongs to P and the literal $\neg p(t_1, \ldots, t_n)$ indicates that tuple $\langle t_1, \ldots, t_n \rangle$ is not in P. (i.e. belongs to $\neg P$). The default negated literal $not\ p(t_1, \ldots, t_n)$ $(not\ \neg p(t_1, \ldots, t_n))$ states that there is no evidence that the tuple $\langle t_1, \ldots, t_n \rangle$ is a positive (negative) example of P. Now the notions of approximations and boundary reflected by rough literals can be equivalently expressed by conjunctions of literals of extended logic programs, as formalized by the following transformation τ_2. This transformation can be used to compile rough literals in the bodies of rough clauses.

$$\tau_2(p?(t_1, \ldots, t_n)) = not\ p(t_1, \ldots, t_n), not\ \neg p(t_1, \ldots, t_n) \ ,$$
$$\tau_2(\underline{p}(t_1, \ldots, t_n)) = p(t_1, \ldots, t_n), not\ \neg p(t_1, \ldots, t_n) \ ,$$
$$\tau_2(\neg \underline{p}(t_1, \ldots, t_n)) = \neg p(t_1, \ldots, t_n), not\ p(t_1, \ldots, t_n) \ ,$$
$$\tau_2(\overline{p}(t_1, \ldots, t_n)) = p(t_1, \ldots, t_n) \ ,$$
$$\tau_2(\overline{\neg p}(t_1, \ldots, t_n)) = \neg p(t_1, \ldots, t_n) \ ,$$
$$\tau_2(\underline{\overline{p}}(t_1, \ldots, t_n)) = p(t_1, \ldots, t_n), \neg p(t_1, \ldots, t_n) \ ,$$
$$\tau_2(\overline{\neg p}(t_1, \ldots, t_n)) = \tau_2(\underline{\overline{p}}(t_1, \ldots, t_n)) \ ,$$
$$\tau_2((B_1, \ldots, B_n)) = \tau_2(B_1), \ldots, \tau_2(B_n) \ .$$

The translation above is not directly applicable to the heads of rough clauses, since the heads in the target programs can contain neither conjunctions of literals nor default negated literals. In order to address this problem, rough clauses in the source program are compiled into a clause and an integrity constraint of the target program, as described below. For example, consider the rough clause

$$\underline{p}(X_1, X_2, X_3) :\!- \ \overline{q}(X_1, X_2, X_n).$$

stating that the boundary of Q (rough relation denoted by predicate $q/3$) is contained in the lower approximation of P (rough relation denoted by predicate $p/3$). Any element in the boundary of Q should be also considered a positive example of P but it should be excluded that those tuples are examples of $\neg P$. Moreover, a tuple t belongs to the boundary of Q if and only if it represents both positive and negative evidence of it. Thus,

$$p(X_1, X_2, X_3) :\!- \ q(X_1, X_2, X_3), \neg q(X_1, X_2, X_3). \tag{1}$$

and

$$:- \neg p(X_1, X_2, X_3), q(X_1, X_2, X_3), \neg q(X_1, X_2, X_3). \qquad (2)$$

capture the same information as the rough clause above. Clause (1) states that tuples belonging to both Q and $\neg Q$ also belong to P, while the integrity constraint (2) does not allow those tuples to belong to $\neg P$.

The discussion above gives a motivation for the formalization of the translation of rough clauses into clauses of an extended logic program. This formalization is defined as the following function τ_1 which refers to the above defined function τ_2. Note that $\neg p$ in an extended logic program should essentially be viewed as a new predicate symbol representing explicit negation.

$$\tau_1(\underline{p}(t_1, \ldots, t_n) :- B.) = \{p(t_1, \ldots, t_n) :- \tau_2(B). , :- \neg p(t_1, \ldots, t_n), \tau_2(B).\} ,$$

$$\tau_1(\overline{p}(t_1, \ldots, t_n) :- B.) = \{p(t_1, \ldots, t_n) :- \tau_2(B).\} ,$$

$$\tau_1(\underline{\neg p}(t_1, \ldots, t_n) :- B.) = \{\neg p(t_1, \ldots, t_n) :- \tau_2(B). , :- p(t_1, \ldots, t_n), \tau_2(B).\} ,$$

$$\tau_1(\overline{\neg p}(t_1, \ldots, t_n) :- B.) = \{\neg p(t_1, \ldots, t_n) :- \tau_2(B).\} ,$$

$$\tau_1(\underline{\overline{p}}(t_1, \ldots, t_n) :- B.) = \{\neg p(t_1, \ldots, t_n) :- \tau_2(B). , p(t_1, \ldots, t_n) :- \tau_2(B).\} ,$$

$$\tau_1(\overline{\underline{\neg p}}(t_1, \ldots, t_n) :- B.) = \tau_1(\underline{\overline{p}}(t_1, \ldots, t_n) :- B.) .$$

A rough program $\mathcal{P}$ will be transformed into an extended logic program by compiling each rough clause. Thus, $\tau_1(\mathcal{P}) = \bigcup_{C \in \mathcal{P}} \tau_1(C)$.

Next example illustrates the proposed encoding of rough programs.

Example 26. Assume that we have two similar decision tables

$$\mathcal{D}eathmi_1 = (U_1, \{\texttt{Age}, \texttt{Hypert}, \texttt{Scanabn}\}, \texttt{Deathmi}_1) ,$$
$$\mathcal{D}eathmi_2 = (U_2, \{\texttt{Age}, \texttt{Hypert}, \texttt{Scanabn}\}, \texttt{Deathmi}_2) ,$$

referring to different periods of time (e.g. year 1 and year 2, respectively). These tables record for several patients their age group (`Age`), whether they have hypertension (`Hypert`), and the result of a medical test to the heart (`Scanabn`). The decision attribute indicates whether the patient had a major heart problem during the follow up period. Both tables are represented as a set of facts in our language.

Our aim is to monitor changes in the boundary region from one period of time to another. For instance, this can give us an idea whether there are groups of patients for who the risk of having a serious cardiac problem has increased, decreased, or remained stable from the first period of time to the second period. Thus, if a tuple t describing an indiscernibility class belongs to the boundary of table $\mathcal{D}eathmi_1$ and the same tuple belongs the lower approximation of the rough relation represented by table $\mathcal{D}eathmi_2$ then, we may interpret this fact as an increase of risk for those patients having the symptoms and test results indicated by t.

These ideas can be expressed by the following rough clauses defining a new rough relation, denoted by predicate `risk`, in such a way that $\underline{Risk}$, $\neg Risk$, and $\overline{Risk}$ correspond to an increase, decrease, and stability of the risk of a cardiac event, respectively.

(1)$\underline{\mathrm{risk}}$(Age, Hypert, Scanabn) :- $\overline{\mathrm{deathmi}_1}$(Age, Hypert, Scanabn),
deathmi$_2$(Age, Hypert, Scanabn).

(2)$\underline{\neg\mathrm{risk}}$(Age, Hypert, Scanabn):- $\overline{\mathrm{deathmi}_1}$(Age, Hypert, Scanabn),
¬deathmi$_2$(Age, Hypert, Scanabn).

(3)$\underline{\overline{\mathrm{risk}}}$(Age, Hypert, Scanabn) :- $\overline{\mathrm{deathmi}_1}$(Age, Hypert, Scanabn),
$$$\overline{\mathrm{deathmi}_2}$(Age, Hypert, Scanabn).

Next, we show the result of compiling (i.e. applying function τ_1 to) each rough clause above.

– Compilation of rough clause (1).

risk(Age, Hypert, Scanabn) :- deathmi$_1$(Age, Hypert, Scanabn),
¬deathmi$_1$(Age, Hypert, Scanabn),
deathmi$_2$(Age, Hypert, Scanabn),
not ¬deathmi$_2$(Age, Hypert, Scanabn).

:- ¬risk(Age, Hypert, Scanabn), deathmi$_1$(Age, Hypert, Scanabn),
¬deathmi$_1$(Age, Hypert, Scanabn), deathmi$_2$(Age, Hypert, Scanabn),
not ¬deathmi$_2$(Age, Hypert, Scanabn).

– Compilation of rough clause (2).

¬risk(Age, Hypert, Scanabn) :- deathmi$_1$(Age, Hypert, Scanabn),
¬deathmi$_1$(Age, Hypert, Scanabn),
¬deathmi$_2$(Age, Hypert, Scanabn),
not deathmi$_2$(Age, Hypert, Scanabn).

:- risk(Age, Hypert, Scanabn), deathmi$_1$(Age, Hypert, Scanabn),
¬deathmi$_1$(Age, Hypert, Scanabn), ¬deathmi$_2$(Age, Hypert, Scanabn),
not deathmi$_2$(Age, Hypert, Scanabn).

– Compilation of rough clause (3).

risk(Age, Hypert, Scanabn) :- deathmi$_1$(Age, Hypert, Scanabn),
¬deathmi$_1$(Age, Hypert, Scanabn),
deathmi$_2$(Age, Hypert, Scanabn),
¬deathmi$_2$(Age, Hypert, Scanabn).

¬risk(Age, Hypert, Scanabn) :- deathmi$_1$(Age, Hypert, Scanabn),
¬deathmi$_1$(Age, Hypert, Scanabn),
deathmi$_2$(Age, Hypert, Scanabn),
¬deathmi$_2$(Age, Hypert, Scanabn).

$\square$

Recall that, for each rough interpretation $\mathcal{I}$ of a rough program $\mathcal{P}$, a predicate q occurring in $\mathcal{P}$ may denote a different rough relation, represented as $Q^{\mathcal{I}}$. Consequently, the denotation of a predicate is always with respect to a rough interpretation.

Correctness of the Compilation Procedure. In order to be able to prove that the compilation function τ_1 is correct, we first show that each model of a rough program $\mathcal{P}$ corresponds to a model of $\tau_1(\mathcal{P})$, and vice-versa.

We start by defining a bijective function that maps each model of a rough program into a model of an extended logic program.

Definition 36. *Let $\mathcal{I}$ be a rough interpretation of a rough program. Then, $\varphi(\mathcal{I})$ is a interpretation of an extended logic program defined as follows.*

(i) If $\langle c_1, \ldots, c_n \rangle \in \overline{Q^{\mathcal{I}}}$ then $q(c_1, \ldots, c_n) \in \varphi(\mathcal{I})$.
(ii) If $\langle c_1, \ldots, c_n \rangle \in \neg \overline{Q^{\mathcal{I}}}$ then $\neg q(c_1, \ldots, c_n) \in \varphi(\mathcal{I})$.
(iii) $\varphi(\mathcal{I})$ is the smallest set (with respect to set inclusion) satisfying both conditions(i) and (ii).

Lemma 2. *Let $\mathcal{I}$ be a rough interpretation of a rough program. Then, $\varphi^{-1}(\varphi(\mathcal{I})) = \mathcal{I}$.*

Proof. Note that φ is a bijection, i.e. it is a surjection and an injection.   □

Lemma 3. *Let $\mathcal{I}_1$ and $\mathcal{I}_2$ be two rough interpretations of a rough program. Then, $\mathcal{I}_1 \preceq \mathcal{I}_2$ if and only if $\varphi(\mathcal{I}_1) \subseteq \varphi(\mathcal{I}_2)$.*

Proof. The statement $\mathcal{I}_1 \preceq \mathcal{I}_2 \Leftrightarrow \varphi(\mathcal{I}_1) \subseteq \varphi(\mathcal{I}_2)$ can be easily proved by taking into account the definition of function φ and the definition of partial relation $\preceq$.   □

Lemma 4. *Let $B_1, \ldots, B_n$ $(n \geq 1)$ be rough or testing literals. Assume also that $\mathcal{I}$ is a rough interpretation.*

- *$\mathcal{I} \models B_1, \ldots, B_n$ if and only if $\varphi(\mathcal{I}) \models \tau_2(B_1, \ldots, B_n)$.*

Proof. The prove can be simply done by structural induction. We start with the base case.

(i.1) First, we show that if $\mathcal{I} \models B$ then $\varphi(\mathcal{I}) \models \tau_2(B)$. Assume that $\mathcal{I} \models B$ and let us consider the different possibilities for B.
- $B \equiv \overline{q}(c_1, \ldots, c_n)$. Thus, $\langle c_1, \ldots, c_n \rangle \in \overline{Q^{\mathcal{I}}}$ and, by definition 36, $q(c_1, \ldots, c_n) \in \varphi(\mathcal{I})$. Since $\tau_2(\overline{q}(c_1, \ldots, c_n)) = q(c_1, \ldots, c_n)$, we conclude that $\varphi(\mathcal{I}) \models \tau_2(B)$.
- $B \equiv \neg\overline{q}(c_1, \ldots, c_n)$. The argument is similar to the previous case.
- $B \equiv \underline{q}(c_1, \ldots, c_n)$. Thus, $\langle c_1, \ldots, c_n \rangle \in \underline{Q^{\mathcal{I}}}$ and $\langle c_1, \ldots, c_n \rangle \notin \neg \overline{Q^{\mathcal{I}}}$. By definition 36, $q(c_1, \ldots, c_n) \in \varphi(\mathcal{I})$ and $\neg q(c_1, \ldots, c_n) \notin \varphi(\mathcal{I})$. We have then that
$$\varphi(\mathcal{I}) \models q(c_1, \ldots, c_n), not \; \neg q(c_1, \ldots, c_n) \; .$$
Since
$$\tau_2(\underline{q}(c_1, \ldots, c_n)) = q(c_1, \ldots, c_n), not \; \neg q(c_1, \ldots, c_n) \; ,$$
we conclude that $\varphi(\mathcal{I}) \models \tau_2(B)$.

- $B \equiv \neg q(c_1, \ldots, c_n)$. The argument is similar to the previous case.
- $B \equiv \overline{q}(c_1, \ldots, c_n)$. Thus, $\langle c_1, \ldots, c_n \rangle \in \overline{Q^{\mathcal{I}}}$ and $\langle c_1, \ldots, c_n \rangle \in \overline{\neg Q^{\mathcal{I}}}$. By definition 36, $q(c_1, \ldots, c_n) \in \varphi(\mathcal{I})$ and $\neg q(c_1, \ldots, c_n) \in \varphi(\mathcal{I})$. We have then that

$$\varphi(\mathcal{I}) \models q(c_1, \ldots, c_n), \neg q(c_1, \ldots, c_n) \ .$$

  Since

$$\tau_2(\overline{q}(c_1, \ldots, c_n)) = q(c_1, \ldots, c_n), \neg q(c_1, \ldots, c_n) \ ,$$

  we conclude that $\varphi(\mathcal{I}) \models \tau_2(B)$.
- $B \equiv \overline{\neg q}(c_1, \ldots, c_n)$. The argument of the previous case applies to this case, too.
- $B \equiv q?(c_1, \ldots, c_n)$. Thus, $\langle c_1, \ldots, c_n \rangle \notin \overline{Q^{\mathcal{I}}}$ and $\langle c_1, \ldots, c_n \rangle \notin \overline{\neg Q^{\mathcal{I}}}$. By definition 36, $q(c_1, \ldots, c_n) \notin \varphi(\mathcal{I})$ and $\neg q(c_1, \ldots, c_n) \notin \varphi(\mathcal{I})$. We have then that

$$\varphi(\mathcal{I}) \models not\ q(c_1, \ldots, c_n), not\ \neg q(c_1, \ldots, c_n) \ .$$

  Since $\tau_2(q(c_1, \ldots, c_n)) = not\ q(c_1, \ldots, c_n), not\ \neg q(c_1, \ldots, c_n)$, we conclude that $\varphi(\mathcal{I}) \models \tau_2(B)$.

(i.2) Second, we show that if $\varphi(\mathcal{I}) \models \tau_2(B)$ then $\mathcal{I} \models B$. Assume that $\varphi(\mathcal{I}) \models \tau_2(B)$ and let us consider the different possibilities for B.
- $B = \overline{q}(c_1, \ldots, c_n)$. Since $\tau_2(\overline{q}(c_1, \ldots, c_n)) = q(c_1, \ldots, c_n)$ and $\varphi(\mathcal{I}) \models q(c_1, \ldots, c_n)$, we conclude that $q(c_1, \ldots, c_n) \in \varphi(\mathcal{I})$. By definition 36, we have that $\langle c_1, \ldots, c_n \rangle \in \overline{Q^{\mathcal{I}}}$. Hence, $\mathcal{I} \models B$.
- $B = \overline{\neg q}(c_1, \ldots, c_n)$. The argument is similar to the previous case.
- $B = \underline{q}(c_1, \ldots, c_n)$. Since

$$\tau_2(\overline{q}(c_1, \ldots, c_n)) = q(c_1, \ldots, c_n), not\ \neg q(c_1, \ldots, c_n)$$

  and $\varphi(\mathcal{I}) \models q(c_1, \ldots, c_n), not\ \neg q(c_1, \ldots, c_n)$, we conclude that $q(c_1, \ldots, c_n) \in \varphi(\mathcal{I})$ and $\neg q(c_1, \ldots, c_n) \notin \varphi(\mathcal{I})$. By definition 36, we have that $\langle c_1, \ldots, c_n \rangle \in \overline{Q^{\mathcal{I}}}$ and $\langle c_1, \ldots, c_n \rangle \notin \overline{\neg Q^{\mathcal{I}}}$. Hence, $\mathcal{I} \models B$.
- $B = \underline{\neg q}(c_1, \ldots, c_n)$. The argument is similar to the previous case.
- $B \equiv \overline{q}(c_1, \ldots, c_n)$. Then by definition of τ_2, we have that $\varphi(\mathcal{I}) \models q(c_1, \ldots, c_n), \neg q(c_1, \ldots, c_n)$. Moreover by definition 36, we also have that $\langle c_1, \ldots, c_n \rangle \in \overline{Q^{\mathcal{I}}}$ and $\langle c_1, \ldots, c_n \rangle \in \overline{\neg Q^{\mathcal{I}}}$. We can then conclude that $\mathcal{I} \models B$.
- $B \equiv \overline{\neg q}(c_1, \ldots, c_n)$. This case is equal to the previous one.
- $B \equiv q?(c_1, \ldots, c_n)$. Then by definition of τ_2, we have that $\varphi(\mathcal{I}) \not\models q(c_1, \ldots, c_n)$ and $\varphi(\mathcal{I}) \not\models \neg q(c_1, \ldots, c_n)$. Moreover by definition 36, we also have that $\langle c_1, \ldots, c_n \rangle \notin \overline{Q^{\mathcal{I}}}$ and $\langle c_1, \ldots, c_n \rangle \notin \overline{\neg Q^{\mathcal{I}}}$. We can then conclude that $\mathcal{I} \models B$.

We proceed now to the inductive step.

(ii.1) First, we prove that if $\mathcal{I} \models B_1, \ldots, B_n$ then $\varphi(\mathcal{I}) \models \tau_2(B_1, \ldots, B_n)$. Note that the base case ($n = 1$) has been proved in (i.1). If $\mathcal{I} \models B_1, \ldots, B_n$ then $\mathcal{I} \models B_1, \ldots, \mathcal{I} \models B_n$. Consequently, by inductive hypothesis, $\varphi(\mathcal{I}) \models \tau_2(B_1), \ldots, \varphi(\mathcal{I}) \models \tau_2(B_n)$. The same is to say that $\varphi(\mathcal{I}) \models \tau_2(B_1), \ldots, \tau_2(B_n)$. By definition of τ_2, $\varphi(\mathcal{I}) \models \tau_2(B_1, \ldots, B_n)$.

(ii.2) Second, we show that if $\varphi(\mathcal{I}) \models \tau_2(B_1, \ldots, B_n)$ then $\mathcal{I} \models B_1, \ldots, B_n$. Note that the base case ($n = 1$) has been proved in (i.2). If $\varphi(\mathcal{I}) \models \tau_2(B_1, \ldots, B_n)$ then, by definition of τ_2, $\varphi(\mathcal{I}) \models \tau_2(B_1), \ldots, \tau_2(B_n)$. This is the same as $\varphi(\mathcal{I}) \models \tau_2(B_1), \ldots, \varphi(\mathcal{I}) \models \tau_2(B_n)$. By inductive hypothesis, $\mathcal{I} \models B_1, \ldots,$ $\mathcal{I} \models B_n$ and, consequently, $\mathcal{I} \models B_1, \ldots, B_n$. □

Lemma 5. *Let $\mathcal{P}$ be a rough program and $\mathcal{P}' = \tau_1(\mathcal{P})$. If $\mathcal{M}_{\mathcal{P}'}$ is a model of $\mathcal{P}'$ then $\varphi^{-1}(\mathcal{M}_{\mathcal{P}'})$ is a model of $\mathcal{P}$.*

Proof. Assume that $\mathcal{M}_{\mathcal{P}} = \varphi^{-1}(\mathcal{M}_{\mathcal{P}'})$ and let us prove that $\mathcal{M}_{\mathcal{P}}$ is a model of $\mathcal{P}$. Hence, we need to show that $\mathcal{M}_{\mathcal{P}}$ satisfies each rough clause $H :\text{-} B.$ $\in$ $ground(\mathcal{P})$. If $\mathcal{M}_{\mathcal{P}} \not\models B$ then $\mathcal{M}_{\mathcal{P}}$ trivially satisfies $H :\text{-} B.$. Otherwise, let us assume that $\mathcal{M}_{\mathcal{P}} \models B$. The head H of the rough clause can be one of the rough literals:

(i) $H \equiv \underline{q}(c_1, \ldots, c_n)$. Then, the compilation function τ_1 ensures that $q(c_1, \ldots, c_n) :\text{-} \tau_2(B).$ $\in \mathcal{P}'$. By lemma 4, $\mathcal{M}_{\mathcal{P}'} \models \tau_2(B)$. Consequently, $\mathcal{M}_{\mathcal{P}'} \models q(c_1, \ldots, c_n)$ because $\mathcal{M}_{\mathcal{P}'}$ is a model of $\mathcal{P}'$. By definition of function φ, we can conclude that $\langle c_1, \ldots, c_n \rangle \in \underline{Q^{\mathcal{M}_{\mathcal{P}}}}$. Hence, $\mathcal{M}_{\mathcal{P}} \models \underline{q}(c_1, \ldots, c_n)$.

(ii) $H \equiv \overline{\neg q}(c_1, \ldots, c_n)$. The argument is similar to case (i).

(iii) $H \equiv q(c_1, \ldots, c_n)$. Then, the compilation function τ_1 ensures that

$$\{q(c_1, \ldots, c_n) :\text{-} \tau_2(B). , \quad :\text{-} \neg q(c_1, \ldots, c_n), \tau_2(B).\} \subseteq \mathcal{P}' .$$

By lemma 4, $\mathcal{M}_{\mathcal{P}'} \models \tau_2(B)$. Consequently, $\mathcal{M}_{\mathcal{P}'} \models q(c_1, \ldots, c_n)$ and $\mathcal{M}_{\mathcal{P}'} \not\models \neg q(c_1, \ldots, c_n)$ because $\mathcal{M}_{\mathcal{P}'}$ is a model of $\mathcal{P}'$. By definition of function φ, we can conclude that $\langle c_1, \ldots, c_n \rangle \in \overline{Q^{\mathcal{M}_{\mathcal{P}}}}$ but $\langle c_1, \ldots, c_n \rangle \notin \overline{\neg Q^{\mathcal{M}_{\mathcal{P}}}}$. Hence, $\mathcal{M}_{\mathcal{P}} \models q(c_1, \ldots, c_n)$.

(iv) $H \equiv \neg q(c_1, \ldots, c_n)$. The argument is similar to case (iii).

(v) $H \equiv \overline{q}(c_1, \ldots, c_n)$. Then, the compilation function τ_1 ensures that

$$\{q(c_1, \ldots, c_n) :\text{-} \tau_2(B). , \quad \neg q(c_1, \ldots, c_n) :\text{-} \tau_2(B).\} \subseteq \mathcal{P}' .$$

By lemma 4, $\mathcal{M}_{\mathcal{P}'} \models \tau_2(B)$. Consequently, $\mathcal{M}_{\mathcal{P}'} \models q(c_1, \ldots, c_n)$ and $\mathcal{M}_{\mathcal{P}'} \models \neg q(c_1, \ldots, c_n)$ because $\mathcal{M}_{\mathcal{P}'}$ is a model of $\mathcal{P}'$. By definition of function φ, we can conclude that $\langle c_1, \ldots, c_n \rangle \in \overline{Q^{\mathcal{M}_{\mathcal{P}}}}$ and $\langle c_1, \ldots, c_n \rangle \in \overline{\neg Q^{\mathcal{M}_{\mathcal{P}}}}$. Hence, $\mathcal{M}_{\mathcal{P}} \models \overline{q}(c_1, \ldots, c_n)$.

(vi) $H \equiv \overline{\neg q}(c_1, \ldots, c_n)$. This case is equivalent to case (v). □

Lemma 6. *Let $\mathcal{P}$ be a rough program and $\mathcal{P}' = \tau_1(\mathcal{P})$. If $\mathcal{M}_{\mathcal{P}}$ is a model of $\mathcal{P}$ then $\varphi(\mathcal{M}_{\mathcal{P}})$ is a model of $\mathcal{P}'$.*

Proof. Assume that $\mathcal{M}_{\mathcal{P}}$ is a model of $\mathcal{P}$. Hence, we need to show that $\varphi(\mathcal{M}_{\mathcal{P}})$ satisfies each clause $H :\text{-} B_1.$ $\in ground(\mathcal{P}')$ and each integrity constraint $:\text{-} B_2.$ $\in ground(\mathcal{P}')$. Note that clauses and integrity constraints belonging to $ground(\mathcal{P}')$ can only have some particular forms determined by the compilation function τ_1. Let us then consider each possible case of function τ_1.

(i) Assume that the rough clause $\overline{q}(c_1, \ldots, c_n)\text{:- } B. \in \mathcal{P}$. Since $\mathcal{M}_\mathcal{P}$ satisfies $\overline{q}(c_1, \ldots, c_n)\text{:- } B.$, we have either that $\mathcal{M}_\mathcal{P} \not\models B$ or $\mathcal{M}_\mathcal{P} \models B, \overline{q}(c_1, \ldots, c_n)$. If $\mathcal{M}_\mathcal{P} \not\models B$ then, by lemma 4, $\varphi(\mathcal{M}_\mathcal{P}) \not\models \tau_2(B)$, and consequently, $\varphi(\mathcal{M}_\mathcal{P})$ trivially satisfies the clause in $\tau_1(\overline{q}(c_1, \ldots, c_n)\text{:- } B.) \subseteq \mathcal{P}'$. Otherwise, $\mathcal{M}_\mathcal{P} \models B, \overline{q}(c_1, \ldots, c_n)$ and, by lemma 4, $\varphi(\mathcal{M}_\mathcal{P}) \models \tau_2(B)$. Moreover by definition 36, $q(c_1, \ldots, c_n) \in \varphi(\mathcal{M}_\mathcal{P})$. Hence, $\varphi(\mathcal{M}_\mathcal{P})$ satisfies the clause in $\tau_1(\overline{q}(c_1, \ldots, c_n)\text{:- } B.)$.

(ii) Assume that the rough clause $\neg\overline{q}(c_1, \ldots, c_n)\text{:- } B. \in \mathcal{P}$. This case can be justified in a way similar to the previous one.

(iii) Assume that the rough clause $\underline{q}(c_1, \ldots, c_n)\text{:- } B. \in \mathcal{P}$. Since $\mathcal{M}_\mathcal{P}$ satisfies $\underline{q}(c_1, \ldots, c_n)\text{:- } B.$, we have either that $\mathcal{M}_\mathcal{P} \not\models B$ or $\mathcal{M}_\mathcal{P} \models B, \underline{q}(c_1, \ldots, c_n)$. If $\mathcal{M}_\mathcal{P} \not\models B$ then, by lemma 4, $\varphi(\mathcal{M}_\mathcal{P}) \not\models \tau_2(B)$, and consequently, $\varphi(\mathcal{M}_\mathcal{P})$ satisfies the clause and the integrity constraint in $\tau_1(\underline{q}(c_1, \ldots, c_n)\text{:- } B.) \subseteq \mathcal{P}'$. Otherwise, $\mathcal{M}_\mathcal{P} \models B$ and $\mathcal{M}_\mathcal{P} \models \underline{q}(c_1, \ldots, c_n)$. By lemma 4, $\varphi(\mathcal{M}_\mathcal{P}) \models \tau_2(B)$ and by definition of $\models$, $\langle c_1, \ldots, c_n \rangle \in \underline{Q^{\mathcal{M}_\mathcal{P}}}$ and $\langle c_1, \ldots, c_n \rangle \notin \overline{\neg Q^{\mathcal{M}_\mathcal{P}}}$. By definition 36, we have that $q(c_1, \ldots, c_n) \in \varphi(\mathcal{M}_\mathcal{P})$ but $\neg q(c_1, \ldots, c_n) \notin \varphi(\mathcal{M}_\mathcal{P})$. Hence, $\varphi(\mathcal{M}_\mathcal{P})$ satisfies the clause $q(c_1, \ldots, c_n)\text{:- } \tau_2(B).$ and the integrity constraint $\text{:-}\neg q(c_1, \ldots, c_n), \tau_2(B).$ obtained by compiling the rough clause.

(iv) Assume that the rough clause $\neg\underline{q}(c_1, \ldots, c_n)\text{:- } B. \in \mathcal{P}$. An argument similar to the previous case can be also used here.

(v) Assume that the rough clause $\widetilde{q}(c_1, \ldots, c_n)\text{:- } B. \in \mathcal{P}$. Since $\mathcal{M}_\mathcal{P}$ satisfies $\widetilde{q}(c_1, \ldots, c_n)\text{:- } B.$, we have either that $\mathcal{M}_\mathcal{P} \not\models B$ or $\mathcal{M}_\mathcal{P} \models B, \widetilde{q}(c_1, \ldots, c_n)$. If $\mathcal{M}_\mathcal{P} \not\models B$ then, by lemma 4, $\varphi(\mathcal{M}_\mathcal{P}) \not\models \tau_2(B)$, and consequently, $\varphi(\mathcal{M}_\mathcal{P})$ satisfies both clauses in $\tau_1(\widetilde{q}(c_1, \ldots, c_n)\text{:- } B.) \subseteq \mathcal{P}'$. Otherwise, $\mathcal{M}_\mathcal{P} \models B$ and $\mathcal{M}_\mathcal{P} \models \widetilde{q}(c_1, \ldots, c_n)$. By lemma 4, $\varphi(\mathcal{M}_\mathcal{P}) \models \tau_2(B)$ and by definition of $\models$, $\langle c_1, \ldots, c_n \rangle \in \overline{Q^{\mathcal{M}_\mathcal{P}}}$ and $\langle c_1, \ldots, c_n \rangle \in \overline{\neg Q^{\mathcal{M}_\mathcal{P}}}$. By definition 36, we have that $q(c_1, \ldots, c_n) \in \varphi(\mathcal{M}_\mathcal{P})$ and $\neg q(c_1, \ldots, c_n) \in \varphi(\mathcal{M}_\mathcal{P})$. Thus, $\varphi(\mathcal{M}_\mathcal{P})$ satisfies both clauses $q(c_1, \ldots, c_n)\text{:- } \tau_2(B).$ and $\neg q(c_1, \ldots, c_n)\text{:- } \tau_2(B).$ obtained by compiling the rough clause.

(vi) Assume that the rough clause $\widetilde{\neg q}(c_1, \ldots, c_n)\text{:- } B. \in \mathcal{P}$. This case is equivalent to the previous one. $\qquad\Box$

Lemma 7. *Let $\mathcal{P}$ be a rough program. $\mathcal{M}$ is the least model (with respect to $\preceq$) of $\mathcal{P}$ if and only if $\varphi(\mathcal{M})$ is the least model (with respect to $\subseteq$) of $\tau_1(\mathcal{P})$.*

Proof. This lemma is direct consequence of lemmas 3, 5, and 6. $\qquad\Box$

Lemma 8. *Let $\mathcal{P}$ be a rough program and $\mathcal{M}$ be one of its models. Then,*

$$\psi_{\varphi(\mathcal{M})}(\tau_1(\mathcal{P})) = \{H \text{:- } B. \in \tau_1(\Psi_\mathcal{M}(\mathcal{P}))\}.$$

Proof. To simplify the presentation of this proof, we represent (rough) predicate argument tuples as $\overrightarrow{t}$, i.e. $q(t_1, \ldots, t_n)$ is represented as $q(\overrightarrow{t})$.

(i) We assume that

$$q(\overrightarrow{t}) \text{ :- } B. \in \tau_1(\Psi_\mathcal{M}(\mathcal{P})).$$

and then show that $q(\overrightarrow{t})$:- $B. \in \psi_{\varphi(\mathcal{M})}(\tau_1(\mathcal{P}))$.

From the hypotheses, it follows that one of the rough clauses below belongs to $\Psi_{\mathcal{M}}(\mathcal{P})$.

$$\overline{q}(\overrightarrow{t}) \text{ :- } B'.$$
$$q(\overrightarrow{t}) \text{ :- } B'.$$
$$\overline{q}(\overrightarrow{t}) \text{ :- } B'. \text{ ,}$$

where B' is such that $\tau_2(B') = B$.

Assume that $\overline{q}(\overrightarrow{t})$:- $B'. \in \Psi_{\mathcal{M}}(\mathcal{P})$ (the other two cases follow a similar reasoning). By definition of $\Psi_{\mathcal{M}}$, there is a rough clause

$$\overline{q}(\overrightarrow{t}) \text{ :- } B_1, \underline{r_1(\overrightarrow{t_1})}, \dots, \underline{r_k(\overrightarrow{t_k})}, s_1?(\overrightarrow{t_1'}), \dots, s_m?(\overrightarrow{t_m'}). \in ground(\mathcal{P}) ,$$

with $k, m \geq 0$,

$$\mathcal{M} \not\models \overline{\neg r_1(\overrightarrow{t_1})}, \dots, \overline{\neg r_k(\overrightarrow{t_k})} \tag{1}$$

$$\mathcal{M} \models s_1?(\overrightarrow{t_1'}), \dots, r_k?(\overrightarrow{t_k'}) \tag{2}$$

and $B' \equiv B_1, \overline{r_1(\overrightarrow{t_1})}, \dots, \overline{r_k(\overrightarrow{t_k})}$. Consequently,

$$q(\overrightarrow{t}) \text{ :- } \tau_2(B_1), r_1(\overrightarrow{t_1}), \ not \ \neg r_1(\overrightarrow{t_1}) \dots, r_k(\overrightarrow{t_k}), \ not \ r_k(\overrightarrow{t_1 k}),$$
$$not \ s_1(\overrightarrow{t_1'}), \ not \ \neg s_1(\overrightarrow{t_1'}) \dots, \ not \ \neg s_m(\overrightarrow{t_m'}), \neg s_m(\overrightarrow{t_m'}).$$

belongs to $\tau_1(ground(\mathcal{P}))$. Moreover, we conclude from (1) that

$$\varphi(\mathcal{M}) \models \ not \ \neg r_1(\overrightarrow{t_1}), \dots, \ not \ \neg r_k(\overrightarrow{t_k})$$

and (2) implies that

$$\varphi(\mathcal{M}) \models \ not \ s_1(\overrightarrow{t_1'}), \ not \ \neg s_1(\overrightarrow{t_1'}) \dots, \ not \ s_m(\overrightarrow{t_m'}), \ not \ \neg s_m(\overrightarrow{t_m'}) .$$

But then,

$$q(\overrightarrow{t}) \text{ :- } \tau_2(B_1), r_1(\overrightarrow{t_1}), \dots, r_k(\overrightarrow{t_k}). \in \psi_{\varphi(\mathcal{M})}(\tau_1(\mathcal{P})) .$$

Note that $\psi_{\varphi(\mathcal{M})}(\tau_1(ground(\mathcal{P}))) = \psi_{\varphi(\mathcal{M})}(\tau_1(\mathcal{P}))$.

Since $(\tau_2(B_1), r_1(\overrightarrow{t_1}), \dots, r_k(\overrightarrow{t_k})) = \tau_2(B') = B$, we can conclude that

$$q(\overrightarrow{t}) \text{ :- } B. \in \psi_{\varphi(\mathcal{M})}(\tau_1(\mathcal{P})) .$$

(ii) We now assume

$$q(\overrightarrow{t}):\text{- } B. \in \psi_{\varphi(\mathcal{M})}(\tau_1(\mathcal{P}))$$

and then show that $q(\overrightarrow{t}):\text{- } B. \in \tau_1(\Psi_{\mathcal{M}}(\mathcal{P}))$.

From the hypotheses, it follows that there is a clause

$$q(\overrightarrow{t}) \text{ :- } B, \ not \ \neg r_1(\overrightarrow{t_1}), \dots, \ not \ \neg r_k(\overrightarrow{t_k}),$$
$$not \ s_1(\overrightarrow{t_1'}), \ not \ \neg s_1(\overrightarrow{t_1'}), \dots, \ not \ s_m(\overrightarrow{t_m'}), \ not \ \neg s_m(\overrightarrow{t_m'}).$$

belonging to $\tau_1(ground(\mathcal{P}))$, with $m, k \geq 0$, and

$$\varphi(\mathcal{M}) \models not\ \neg r_1(\overrightarrow{t_1}), \ldots,\ not\ \neg r_k(\overrightarrow{t_k}) \tag{3}$$

$$\varphi(\mathcal{M}) \models not\ s_1(\overrightarrow{t'_1}),\ not\ \neg s_1(\overrightarrow{t'_1}), \ldots,$$
$$not\ s_m(\overrightarrow{t'_m}),\ not\ \neg s_m(\overrightarrow{t'_m}) \tag{4}$$

Moreover, $B \equiv B', r_1(\overrightarrow{t_1}), \ldots, r_k(\overrightarrow{t_k})$. One of the following rough clauses has then to belong to $ground(\mathcal{P})$.

$$\overline{q}(\overrightarrow{t})\ :-\ B_1, \underline{r_1(\overrightarrow{t_1})}, \ldots, \underline{r_k(\overrightarrow{t_k})}, s_1?(\overrightarrow{t'_1}), \ldots, s_m?(\overrightarrow{t'_m}).$$
$$q(\overrightarrow{t})\ :-\ B_1, \underline{r_1(\overrightarrow{t_1})}, \ldots, \underline{r_k(\overrightarrow{t_k})}, s_1?(\overrightarrow{t'_1}), \ldots, s_m?(\overrightarrow{t'_m}).$$
$$\underline{q}(\overrightarrow{t})\ :-\ B_1, \underline{r_1(\overrightarrow{t_1})}, \ldots, \underline{r_k(\overrightarrow{t_k})}, s_1?(\overrightarrow{t'_1}), \ldots, s_m?(\overrightarrow{t'_m}).\ ,$$

where $\tau_2(B_1) = B'$.
Assume that

$$\overline{q}(\overrightarrow{t})\ :-\ B_1, \underline{r_1(\overrightarrow{t_1})}, \ldots, \underline{r_k(\overrightarrow{t_k})}, s_1?(\overrightarrow{t'_1}), \ldots, s_m?(\overrightarrow{t'_m}).\ \in ground(\mathcal{P})$$

(the other two cases follow a similar reasoning). We can conclude from (3) that

$$\mathcal{M} \not\models \underline{\neg r_1(\overrightarrow{t_1})}, \ldots, \underline{\neg r_k(\overrightarrow{t_k})}$$

and (4) implies that

$$\mathcal{M} \models s_1?(\overrightarrow{t'_1}), \ldots, s_m?(\overrightarrow{t'_m})\ .$$

We have then that $\overline{q}(\overrightarrow{t})\ :-\ B_1, \overline{r_1(\overrightarrow{t_1})}, \ldots, \overline{r_k(\overrightarrow{t_k})}. \in \Psi_{\mathcal{M}}(\mathcal{P})$. Consequently,

$$q(\overrightarrow{t})\ :-\ B', r_1(\overrightarrow{t_1}), \ldots, r_k(\overrightarrow{t_k}). \in \tau_1(\Psi_{\mathcal{M}}(\mathcal{P}))\ .$$

Hence, $q(\overrightarrow{t})\ :-\ B. \in \tau_1(\Psi_{\mathcal{M}}(\mathcal{P}))$.

Lemma 9. *Let $\mathcal{P}$ be a rough program. If $\mathcal{M}$ is the least model of $\Psi_{\mathcal{M}}(\mathcal{P})$ then $\varphi(\mathcal{M})$ is the least model of $\{H :- B. \in \tau_1(\Psi_{\mathcal{M}}(\mathcal{P}))\}$.*

Proof. Let $\mathcal{P}' = \{H :- B. \in \tau_1(\Psi_{\mathcal{M}}(\mathcal{P}))\}$. If $\mathcal{M}$ is the least model of $\Psi_{\mathcal{M}}(\mathcal{P})$ then, by lemma 7, $\varphi(\mathcal{M})$ is the least model of $\tau_1(\Psi_{\mathcal{M}}(\mathcal{P}))$ and, therefore, it is a model of $\mathcal{P}'$.

Note that default negated literals cannot occur in any clause or integrity constraint of $\tau_1(\Psi_{\mathcal{M}}(\mathcal{P}))$. The reason is that neither lower approximations nor testing literals occur in the body of any rough clause in $\Psi_{\mathcal{M}}(\mathcal{P})$. In addition, $\tau_1(\Psi_{\mathcal{M}}(\mathcal{P}))$ is definite logic program with integrity constraints.

Assume that $\varphi(\mathcal{M})$ is not the least model of $\mathcal{P}'$. Then, there is a model (e.g. the least model) $\mathcal{M}'$ of $\mathcal{P}'$ such that $\mathcal{M}' \prec \varphi(\mathcal{M})$. Therefore, there is one atom $q(t_1, \ldots, t_n)$ (or $\neg q(t_1, \ldots, t_n)$) such that $q(t_1, \ldots, t_n) \in \varphi(\mathcal{M})$ but $q(t_1, \ldots, t_n) \notin \mathcal{M}'$. Only the occurrence of a default negated literal, e.g. *not* $q(t_1, \ldots, t_n)$, in an integrity constraint belonging to $\tau_1(\Psi_{\mathcal{M}}(\mathcal{P}))$ could force the entrance of an atom in the least model of $\mathcal{P}'$ in order to be also able to satisfy the integrity constraints. However, as we pointed out above, this cannot be the case. $\qquad \square$

Lemma 10. *Let $\mathcal{P}$ be a rough program and $\mathcal{I}$ be a rough interpretation of $\mathcal{P}$. If $\varphi(\mathcal{I})$ satisfies each integrity constraint in $\tau_1(\mathcal{P})$ then $\varphi(\mathcal{I})$ satisfies each integrity constraint in $\tau_1(\Psi_\mathcal{I}(\mathcal{P}))$.*

Proof. To simplify the presentation of this proof, we represent predicate argument tuples as $\overrightarrow{t}$, i.e. $q(t_1, \ldots, t_n)$ is represented as $q(\overrightarrow{t})$.

Assume that $\varphi(\mathcal{I})$ satisfies each integrity constraint in $\tau_1(\mathcal{P})$. Taking into account the definition of τ_1, we conclude that integrity constraints originate from compilation of rough clauses with a lower approximation in their head. Hence, suppose that

$$\underline{q(\overrightarrow{t})} :- B_1, \underline{r_1(\overrightarrow{t_1})}, \ldots, \underline{r_k(\overrightarrow{t_k})}, s_1?(\overrightarrow{t_1'}), \ldots, s_m?(\overrightarrow{t_m'}). \in \mathcal{P} \, ,$$

where $k, m \geq 0$ and no lower approximations or testing literals occur in B_1. Let i_c be an integrity constraint defined as follows.

$$i_c = :- \neg q(\overrightarrow{t}), \tau_2(B_1), r_1(\overrightarrow{t_1}), \; not \; \neg r_1(\overrightarrow{t_1}), \ldots, r_k(\overrightarrow{t_k}), \; not \; \neg r_k(\overrightarrow{t_k}),$$
$$not \; s_1(\overrightarrow{t_1'}), \; not \; \neg s_1(\overrightarrow{t_1'}), \ldots, not \; s_m(\overrightarrow{t_m'}), \; not \; \neg s_m(\overrightarrow{t_m'}).$$

Then, $i_c \in \tau_1(\mathcal{P})$ and $\varphi(\mathcal{I}) \models i_c$.

Assume also that

(a) $\mathcal{I} \not\models \overrightarrow{\neg r_1}(\overrightarrow{t_1}), \ldots, \overrightarrow{\neg r_k}(\overrightarrow{t_k})$ and
(b) $\mathcal{I} \models s_1?(\overrightarrow{t_1'}), \ldots, s_m?(\overrightarrow{t_m'})$.

From (a) and (b) and by definition of $\Psi_\mathcal{I}$,

$$\underline{q(\overrightarrow{t})} :- B_1, \overline{r_1}(\overrightarrow{t_1}), \ldots, \overline{r_k}(\overrightarrow{t_k}). \in \Psi_\mathcal{M}(\mathcal{P}) \, .$$

Thus, $i_c' \in \tau_1(\Psi_\mathcal{I}(\mathcal{P}))$, with

$$i_c' = \neg q(\overrightarrow{t}), \tau_2(B_1), r_1(\overrightarrow{t_1}), \ldots, r_k(\overrightarrow{t_k}). \in \tau_1(\Psi_\mathcal{I}(\mathcal{P})) \, .$$

Moreover,

  – from (a), we have that $\varphi(\mathcal{I}) \models \; not \; \neg r_1(\overrightarrow{t_1}), \ldots \; not \; \neg r_k(\overrightarrow{t_k})$ and
  – from (b), we have that

$$\varphi(\mathcal{I}) \models \; not \; s_1(\overrightarrow{t_1'}), \; not \; \neg s_1(\overrightarrow{t_1'}), \ldots, not \; s_m(\overrightarrow{t_m'}), \; not \; \neg s_m(\overrightarrow{t_m'}) \, .$$

Since $\varphi(\mathcal{I}) \models i_c$, we can conclude that $\varphi(\mathcal{I}) \models i_c'$.   □

Theorem 2. *Let $\mathcal{P}$ be a rough program and $\mathcal{P}' = \tau_1(\mathcal{P})$. Then, $\mathcal{M} \in sem(\mathcal{P})$ if and only if $\varphi(\mathcal{M})$ is a paraconsistent stable model of $\mathcal{P}'$.*

Proof. (i) First, we prove that if $\mathcal{M} \in sem(\mathcal{P})$ then $\varphi(\mathcal{M})$ is a paraconsistent stable model of $\mathcal{P}'$.

Assume that $\mathcal{M} \in sem(\mathcal{P})$. Then, $\mathcal{M}$ is the least model (with respect to $\preceq$) of $\Psi_\mathcal{M}(\mathcal{P})$ and, by lemma 9, $\varphi(\mathcal{M})$ is the least (with respect to $\subseteq$) model of $\{H \; :- \; B. \; \in \tau_1(\Psi_{\mathcal{M}}(\mathcal{P}))\}$. Consequently, by lemma 8, $\varphi(\mathcal{M})$ is the least model of $\psi_{\varphi(\mathcal{M})}(\tau_1(\mathcal{P}))$. Moreover, $\varphi(\mathcal{M})$ satisfies each integrity constraint belonging to $\tau_1(\mathcal{P})$ because $\mathcal{M}$ is a model of $\mathcal{P}$ and, by lemma 6, $\varphi(\mathcal{M})$ is a model of $\tau_1(\mathcal{P})$. We can then conclude that $\varphi(\mathcal{M})$ is a paraconsistent stable model of $\tau_1(\mathcal{P})$.

(ii) Second, we show that if $\varphi(\mathcal{M})$ is a paraconsistent stable model of $\mathcal{P}'$ then $\mathcal{M} \in sem(\mathcal{P})$. Assume that $\varphi(\mathcal{M})$ is a paraconsistent stable model of $\mathcal{P}'$. Then,

(a) $\varphi(\mathcal{M})$ satisfies all integrity constraints belonging to $\tau_1(\mathcal{P})$ and
(b) $\varphi(\mathcal{M})$ is the least model of $\psi_{\varphi(\mathcal{M})}(\tau_1(\mathcal{P}))$.

From (a) and lemma 10, we conclude that $\varphi(\mathcal{M})$ satisfies all integrity constraints in $\tau_1(\Psi_{\mathcal{M}}(\mathcal{P}))$. From (b) and lemma 8, we have that $\varphi(\mathcal{M})$ is the least model of $\{H \; :- \; B. \; \in \tau_1(\Psi_{\mathcal{M}}(\mathcal{P}))\}$. These two facts lead us to the conclusion that $\varphi(\mathcal{M})$ has to be the least model of $\tau_1(\Psi_{\mathcal{M}}(\mathcal{P}))\}$. Then, by lemma 7, $\mathcal{M}$ is the least model of $\Psi_{\mathcal{M}}(\mathcal{P})$ and, therefore, $\mathcal{M} \in sem(\mathcal{P})$.   □

Theorem 3 (Correctness). *Let $\mathcal{P}$ be a rough program and $\mathcal{P}' = \tau_1(\mathcal{P})$ and l be a rough or testing literal. Then, $\mathcal{P} \models l$ if and only if $\mathcal{P}' \models \tau_2(l)$.*

Proof. This theorem is a direct consequence of theorem 2, of lemma 4, and of definitions 23 and 35.

4.4   Queries

This section proposes a query language for querying rough programs. This can be achieved by adapting existing systems based on the stable model semantics [44,43]. Here, we only present queries and their expected answers. Since there might exist more than one model for a rough program $\mathcal{P}$, answers are computed with respect to one model of $sem(\mathcal{P})$. If a rough program has a unique model, which may often be the case[5], the answers will refer to this model.

Definition 37. *A rough query is a pair $(\mathcal{Q}, \mathcal{P})$, where $\mathcal{P}$ is a rough program and $\mathcal{Q}$ is defined by the following abstract syntax rules*

$$\mathcal{Q}_1 \longrightarrow A? \mid A?, \mathcal{Q}_1 \, .$$
$$\mathcal{Q}_2 \longrightarrow L_1 \mid L_1, \mathcal{Q}_2 \mid \mathcal{Q}_2, \mathcal{Q}_1 \, .$$
$$\mathcal{Q}_3 \longrightarrow L_1 \subseteq L_2 \mid L_1 \subseteq L_2, \mathcal{Q}_3 \, .$$
$$\mathcal{Q} \longrightarrow \mathcal{Q}_1 \mid \mathcal{Q}_2 \mid \mathcal{Q}_3 \, .$$

where $A?$ is a testing literal and each L_i $(i = 1, 2)$ is a rough literal. Moreover, a rough query is well-formed *if the following conditions are satisfied.*

(i) Any testing literal $A?$ is ground (i.e. it does not contain any variables) or its variables occur also in some rough literal of the query.

---

[5] For instance, any rough program whose rough clauses do not contain lower approximations or testing literals in their bodies either has a least model or no model at all.

(ii) For an expression of the form $L_1 \subseteq L_2$ occurring in the query, any variable occurring in L_1 should also occur in L_2, and vice-versa.

For example, rough queries $(p(X1, X2), q?(X2, X3), \mathcal{P})$ and $(p(X1, X2, X3) \subseteq \overline{q}(X1, X2), \mathcal{P})$ are not well-formed because the former does not satisfy condition (i) and the latter violates condition (ii). In what follows, we always assume that rough queries are well-formed.

Consider the rough query $(\overline{q}(c_1, c_2), \mathcal{P})$. Before presenting the notion of answer, we explain informally what is being queried and the corresponding answer. With that query we want to know whether the tuple $\langle c_1, c_2 \rangle$ belongs to the boundary region of the rough relation denoted by q, in some model of $\mathcal{P}$ belonging to $sem(\mathcal{P})$. If the atom occurring in the query is not ground then, as answer, we may obtain a list of examples valid in a certain model. For example, the query $(\overline{q}(X, Y), \mathcal{P})$ requests a list of pairs that belong to $\overline{Q^{\mathcal{M}}}$, for some some model $\mathcal{M} \in sem(\mathcal{P})$.

We formalize now the notion of answer to a rough query.

Definition 38. *Let $(\mathcal{Q}, \mathcal{P})$ be a rough query.*

(i) If $\mathcal{Q}$ is of the form $\mathcal{Q}_1$ or $\mathcal{Q}_2$ (see definition 37) then an answer to the rough query is the set of ground substitutions

$$\{\theta \mid \mathcal{Q}\theta \in ground(\mathcal{Q}) \text{ and } \mathcal{M} \models \mathcal{Q}\theta\} ,$$

for some model $\mathcal{M} \in sem(\mathcal{P})$.

(ii) If $\mathcal{Q}$ is of the form $L_{11} \subseteq L_{21}, \ldots, L_{1n} \subseteq L_{2n}$, where each L_{1i} and L_{2i} ($1 \leq i \leq n$) are rough literals, then the answer to the rough query is
(ii.1) **yes** *, if there is a model $\mathcal{M} \in sem(\mathcal{P})$ such that*

$$\mathcal{M} \models L_{11}\theta \Rightarrow \mathcal{M} \models L_{21}\theta ,$$

$$\vdots$$

$$\mathcal{M} \models L_{1n}\theta \Rightarrow \mathcal{M} \models L_{2n}\theta ,$$

for every ground substitution θ;
(ii.2) **no** *, otherwise.*

Note that $\{\emptyset\}$ is a possible answer to a rough query. This answer can be obtained in case (i) when the query is ground and it should essentially be viewed as an affirmative answer. This contrasts with the empty set answer that should be interpreted as a negative answer. For instance, if the answer to the rough query $(\overline{q}(c_1, c_2), \mathcal{P})$ is $\emptyset$ then this means that $\langle c_1, c_2 \rangle$ does not belong to the upper approximation of rough relation Q (whatever is the model $\mathcal{M} \in sem(\mathcal{P})$ that is considered).

The notion of answer to a rough query introduced above is declarative. Hence, we need to discuss how such answers can be computed. A rough query $(\mathcal{Q}, \mathcal{P})$, where $\mathcal{Q}$ is of the form $\mathcal{Q}_1$ or $\mathcal{Q}_2$ (see definition 37) is translated into a query to the extended logic program $\tau_1(\mathcal{P})$

$$(\tau_2(\mathcal{Q}), \tau_1(\mathcal{P})) ,$$

and each set of substitutions obtained as answer to this query is also an answer to the rough query.

We now discuss how a query of the form $(L_1 \subseteq L_2, \mathcal{P})$, where L_1 and L_2 are rough literals, could be answered. The idea is to translate it to a set of integrity constraints that are added to the compiled program $(\tau_1(\mathcal{P}))$. Hence, a new extended logic program $\mathcal{P}'$ is obtained in this way. Then, the answer to the query is **yes** (i.e. the test succeeds) if $\mathcal{P}'$ has at least one paraconsistent stable model. Otherwise, the answer is **no** (i.e. the test fails). Thus, we reduce the answering problem for this kind of queries to the problem of checking the existence of paraconsistent stable models of an extended logic program where certain properties, expressed by the integrity constraints, hold.

Given an objective literal L, we assume that $\neg\neg L$ and L have the same meaning. Moreover, consider the rough query $(L_1 \subseteq L_2, \mathcal{P})$, where L_1 and L_2 are rough literals. We define a function τ_3 that transforms these queries into an extended logic program with integrity constraints, for each possible case of L_2 (i.e. $\overline{L}, \underline{L}, \overline{\underline{L}}$).

$$\tau_3((L_1 \subseteq \overline{L}, \mathcal{P})) = \tau_1(\mathcal{P}) \cup \{ :- \ \tau_2(L_1), not \ L. \} \ ,$$
$$\tau_3((L_1 \subseteq \underline{L}, \mathcal{P})) = \tau_1(\mathcal{P}) \cup \{ :- \ \tau_2(L_1), not \ L. \ , \ :- \ \tau_2(L_1), \neg L. \} \ ,$$
$$\tau_3((L_1 \subseteq \overline{\underline{L}}, \mathcal{P})) = \tau_1(\mathcal{P}) \cup \{ :- \ \tau_2(L_1), not \ L. \ , \ :- \ \tau_2(L_1), not \ \neg L. \} \ .$$

It is trivial to extend function τ_3 for compiling queries of the form $(\mathcal{Q}_3, \mathcal{P})$. Assume that $\mathcal{P}$ is a rough program, L_{1i} and L_{2i} are rough literals, with $1 \leq i \leq n$, then

$$\tau_3((L_{11} \subseteq L_{21}, \cdots, L_{1n} \subseteq L_{2n}, \mathcal{P})) = \bigcup_{1 \leq i \leq n} \tau_3((L_{1i} \subseteq L_{2i}, \mathcal{P})) \ .$$

Thus, given a rough program $\mathcal{P}$, we have that the answer to the query $(\mathcal{Q}_3, \mathcal{P})$ is **yes**, if the extended logic program $\tau_3((\mathcal{Q}_3, \mathcal{P}))$ has a paraconsistent stable model. Otherwise, the answer is **no**.

Example 27. Consider again the rough program

$$\mathcal{P} = \{\overline{\texttt{diseaseA}}(\texttt{infect, normal, Z}) :- \quad \neg\texttt{diseaseB}(\texttt{infect, normal, Z}). ,$$
$$\overline{\texttt{diseaseB}}(\texttt{infect, normal, Z}) :- \quad \neg\texttt{diseaseA}(\texttt{infect, normal, Z}). ,$$
$$\neg\texttt{diseaseA}(\texttt{infect, normal, Z}). ,$$
$$\neg\texttt{diseaseB}(\texttt{infect, normal, Z}).\} \ .$$

and the queries

(i) $(\overline{\texttt{diseaseA}}(X, Y, Z), \mathcal{P})$,
(ii) $(\overline{\texttt{diseaseA}}(X, Y, Z) \subseteq \neg\texttt{diseaseB}(X, Y, Z), \mathcal{P})$.

Recall that there are two models that belong to $sem(\mathcal{P})$, see example 23. The answer to query (i) is the set of substitutions

$$\{\{X/\texttt{infect}, Y/\texttt{normal}, Z/\texttt{c}\}\}$$

because in one of the models, $\mathcal{M}_1$, $\langle \texttt{infect}, \texttt{normal}, \texttt{c} \rangle \in \overline{DiseaseA}^{\mathcal{M}_1}$.

Note that

$$\tau_3(\overline{\texttt{diseaseA}}(X, Y, Z) \subseteq \underline{\texttt{diseaseB}}(X, Y, Z),\ \mathcal{P})\ =\ \tau_1(\mathcal{P})\ \cup$$
$$\{:-\ \tau_2(\overline{\texttt{diseaseA}}(X, Y, Z)),\ not\ diseaseB(X, Y, Z).\ ,$$
$$:-\ \tau_2(\overline{\texttt{diseaseA}}(X, Y, Z)),\ \neg diseaseB(X, Y, Z).\}$$

Since $\mathcal{M}_1$ is a paraconsistent stable model of this extended logic program, obtained after compiling rough query (ii), we conclude that the answer to

$$(\overline{\texttt{diseaseA}}(X, Y, Z) \subseteq \underline{\neg\texttt{diseaseB}}(X, Y, Z),\ \mathcal{P})$$

is yes.   □

The query language proposed here is slightly more general than the one presented in [19], since now we allow for testing arbitrary inclusions between lower and upper approximations. For instance in [19], we could not test whether the lower approximation of one rough relation R_1, denoted by predicate r_1/n, was included in the upper approximation of another rough relation R_2, denoted by another predicate r_2/n. With the rough query language discussed in this section, this can be achieved through the query $(\underline{r_1}(X_1, \ldots, X_n) \subseteq \overline{r_2}(X_1, \ldots, X_n),\ \mathcal{P})$.

In some applications it is necessary to check rough inclusion or rough equality of given rough relations. We recall the notions of rough inclusion and rough equality [24].

Definition 39. *Rough relation Q_1 is* roughly included *in rough relation Q_2, denoted as $Q_1 \sqsubseteq Q_2$, if and only if $\underline{Q_1} \subseteq \underline{Q_2}$ and $\overline{Q_1} \subseteq \overline{Q_2}$.*

Definition 40. *The rough sets Q_1 and Q_2 are* roughly equal, *denoted as $Q_1 \approx Q_2$, if and only if $\overline{Q_1} = \overline{Q_2}$ and $\underline{Q_1} = \underline{Q_2}$.*

Given a rough program $\mathcal{P}$ and two predicates q_1/n and q_2/n denoting rough relations Q_1 and Q_2, respectively, we can easily test whether $Q_1 \sqsubseteq Q_2$ or $Q_1 \approx Q_2$. The rough query

$$(\overline{q_1}(X_1, \ldots, X_n) \subseteq \overline{q_2}(X_1, \ldots, X_n), \underline{q_1}(X_1, \ldots, X_n) \subseteq \underline{q_2}(X_1, \ldots, X_n),\ \mathcal{P})$$

tests for rough inclusion. Rough equality can be tested through the rough query

$$(\overline{q_1}(X_1, \ldots, X_n) \subseteq \overline{q_2}(X_1, \ldots, X_n),$$
$$\overline{q_2}(X_1, \ldots, X_n) \subseteq \overline{q_1}(X_1, \ldots, X_n),$$
$$\underline{q_1}(X_1, \ldots, X_n) \subseteq \underline{q_2}(X_1, \ldots, X_n),$$
$$\underline{q_2}(X_1, \ldots, X_n) \subseteq \underline{q_1}(X_1, \ldots, X_n),\ \mathcal{P}).$$

Finally, we show the equivalence between the proposed technique to compute answers of a rough query and its declarative semantics. To that end, we need to prove the following two lemmas.

Lemma 11. *Let $(\mathcal{Q}, \mathcal{P})$ be a rough query.*

(i) *Assume that $\mathcal{Q}$ is of the form $\mathcal{Q}_1$ or $\mathcal{Q}_2$ (see definition 37). If θ is a ground substitution belonging to an answer of $(\tau_2(\mathcal{Q}), \tau_1(\mathcal{P}))$ then θ also belongs to an answer of $(\mathcal{Q}, \mathcal{P})$.*

(ii) *Assume that $\mathcal{Q}$ is of the form $\mathcal{Q}_3$. If the extended logic program $\tau_3((\mathcal{Q}, \mathcal{P}))$ has a paraconsistent stable model then the answer to the rough query $(\mathcal{Q}, \mathcal{P})$ is* yes. *Otherwise, the answer is* no.

Proof. (i) Assume that θ is a ground substitution belonging to an answer of $(\tau_2(\mathcal{Q}), \tau_1(\mathcal{P}))$. Hence by definition 23, we have that

$$\tau_1(\mathcal{P}) \models \tau_2(\mathcal{Q})\theta .$$

It easy to see that $\tau_2(\mathcal{Q})\theta = \tau_2(\mathcal{Q}\theta)$. By theorem 3, we have then that $\mathcal{P} \models \mathcal{Q}\theta$.

(ii) Assume that the extended logic program $\tau_3((\mathcal{Q}, \mathcal{P}))$ has a paraconsistent stable model. Let $\mathcal{Q} \equiv L_{11} \subseteq L_{12}, \ldots, L_{n1} \subseteq L_{n2}$.

(ii.1) If some $L_{i1} \subseteq L_{i2} \equiv L_{i1} \subseteq \overline{L}$ $(1 \leq i \leq n)$ then for all paraconsistent stable models $\mathcal{M}$ of $\tau_3((\mathcal{Q}, \mathcal{P}))$ and ground substitutions θ we have that

$$\mathcal{M} \not\models \tau_2(L_{i1})\theta, \ not\ L\theta . \tag{1}$$

Consider a paraconsistent stable model $\mathcal{M}$ of $\tau_3((\mathcal{Q}, \mathcal{P}))$ and a ground substitution θ such that $\mathcal{M} \models \tau_2(L_{i1})\theta$. By (1), we have then that $\mathcal{M} \not\models not\ L\theta$. Therefore, $L\theta \in \mathcal{M}$ and, consequently, $\mathcal{M} \models L\theta$.
If $\mathcal{M} \models \tau_2(L_{i1}\theta), L\theta$ then, by lemma 4 and definition of function τ_2, $\varphi^{-1}(\mathcal{M}) \models L_{i1}\theta, \overline{L}\theta$. Since $\mathcal{M}$ is a paraconsistent stable model of τ_3 $((\mathcal{Q}, \mathcal{P}))$, $\mathcal{M}$ is also a paraconsistent stable model of $\tau_1(\mathcal{P})$. Hence by lemma 2, $\varphi^{-1}(\mathcal{M}) \in sem(\mathcal{P})$.
We can then conclude that there is a model $\mathcal{M}' \in sem(\mathcal{P})$ such that, if $\mathcal{M}' \models L_{i1}\theta$ then $\mathcal{M}' \models \overline{L}\theta$, for any ground substitution θ.

(ii.2) If some $L_{i1} \subseteq L_{i2} \equiv L_{i1} \subseteq \underline{L}$ $(1 \leq i \leq n)$ then for all paraconsistent stable models $\mathcal{M}$ of $\tau_3((\mathcal{Q}, \mathcal{P}))$ and ground substitutions θ we have that

$$\mathcal{M} \not\models \tau_2(L_{i1})\theta, \ not\ L\theta . \tag{2}$$

and

$$\mathcal{M} \not\models \tau_2(L_{i1})\theta, \neg L\theta . \tag{3}$$

Consider a paraconsistent stable model $\mathcal{M}$ of $\tau_3((\mathcal{Q}, \mathcal{P}))$ and a ground substitution θ such that $\mathcal{M} \models \tau_2(L_{i1})\theta$. By (2) and (3), we have then that $\mathcal{M} \models L\theta$ and $\mathcal{M} \models not\ \neg L\theta$, i.e. $\mathcal{M} \models (L, \ not\ \neg L)\theta \Leftrightarrow \mathcal{M} \models \tau_2(\underline{L}\theta)$.
If $\mathcal{M} \models \tau_2(L_{i1}\theta)$ and $\mathcal{M} \models \tau_2(\underline{L}\theta)$ then, by lemma 4, $\varphi^{-1}(\mathcal{M}) \models L_{i1}\theta$ and $\varphi^{-1}(\mathcal{M}) \models \underline{L}\theta$. Moreover, as it was shown in the previous case, $\varphi^{-1}(\mathcal{M}) \in sem(\mathcal{P})$.
We can then conclude that there is a model $\mathcal{M}' \in sem(\mathcal{P})$ such that, if $\mathcal{M}' \models L_{i1}\theta$ then $\mathcal{M}' \models \underline{L}\theta$, for any ground substitution θ.

(ii.3) If some $L_{i1} \subseteq L_{i2} \equiv L_{i1} \subseteq \overline{L}$ $(1 \leq i \leq n)$ then for all paraconsistent stable models $\mathcal{M}$ of $\tau_3((\mathcal{Q}, \mathcal{P}))$ and ground substitutions θ we have that

$$\mathcal{M} \not\models \tau_2(L_{i1})\theta, \; not \; L\theta \; . \tag{4}$$

and

$$\mathcal{M} \not\models \tau_2(L_{i1})\theta, \; not \; \neg L\theta \; . \tag{5}$$

Consider a paraconsistent stable model $\mathcal{M}$ of $\tau_3((\mathcal{Q}, \mathcal{P}))$ and a ground substitution θ such that $\mathcal{M} \models \tau_2(L_{i1})\theta$. By (4) and (5), we have then that $\mathcal{M} \models L\theta$ and $\mathcal{M} \models \neg L\theta$, i.e. $\mathcal{M} \models (L, \neg L)\theta \Leftrightarrow \mathcal{M} \models \tau_2(\overline{L}\theta)$. If $\mathcal{M} \models \tau_2(L_{i1}\theta)$ and $\mathcal{M} \models \tau_2(\overline{L}\theta)$ then, by lemma 4, $\varphi^{-1}(\mathcal{M}) \models L_{i1}\theta$ and $\varphi^{-1}(\mathcal{M}) \models \overline{L}\theta$. Moreover, as it was shown previously, $\varphi^{-1}(\mathcal{M}) \in sem(\mathcal{P})$.

We can then conclude that there is a model $\mathcal{M}' \in sem(\mathcal{P})$ such that, if $\mathcal{M}' \models L_{i1}\theta$ then $\mathcal{M}' \models \overline{L}\theta$, for any ground substitution θ. $\qquad\square$

Lemma 12. *Let $(\mathcal{Q}, \mathcal{P})$ be a rough query.*

(i) *Assume that $\mathcal{Q}$ is of the form $\mathcal{Q}_1$ or $\mathcal{Q}_2$ (see definition 37). If θ is a ground substitution belonging to an answer of $(\mathcal{Q}, \mathcal{P})$ then θ also belongs to an answer of $(\tau_2(\mathcal{Q}), \tau_1(\mathcal{P}))$.*

(ii) *Assume that $\mathcal{Q}$ is of the form $\mathcal{Q}_3$. If the answer to the rough query $(\mathcal{Q}, \mathcal{P})$ is* **yes** *then the extended logic program $\tau_3((\mathcal{Q}, \mathcal{P}))$ has a paraconsistent stable model.*

Proof. (i) The statement above is direct consequence of theorem 3 and of definition of answer of a query to an extended logic program.

(ii) Let $\mathcal{M} \in sem(\mathcal{P})$. Assume also that $\mathcal{Q} \equiv L_1 \subseteq L_2$ and that the answer to the rough query $(L_1 \subseteq L_2, \mathcal{P})$ is **yes**. Thus by definition of answer to a rough query,

$$\mathcal{M} \models L_1\theta \Rightarrow \mathcal{M} \models L_2\theta \; , \tag{6}$$

for every ground substitution θ.

If $\mathcal{M} \in sem(\mathcal{P})$ then, by theorem 2, $\varphi(\mathcal{M})$ is a paraconsistent stable model of $\tau_1(\mathcal{P})$. Moreover, from (6) and by lemma 4,

$$\varphi(\mathcal{M}) \models \tau_2(L_1)\theta \Rightarrow \varphi(\mathcal{M}) \models \tau_2(L_2)\theta \tag{7}$$

We need now to show that $\tau_3((L_1 \subseteq L_2, \mathcal{P}))$ has a paraconsistent stable model. We consider each possible case for the rough literal L_2.

(ii.1) Let $L_2 \equiv \overline{L}$. Recall that the integrity constraint

$$:- \; \tau_2(L_1), \; not \; L.$$

is added in this case to $\tau_1(\mathcal{P})$. From (7), we have that

$$\varphi(\mathcal{M}) \models \; :- \; \tau_2(L_1), \; not \; L. \; .$$

Since $\varphi(\mathcal{M})$ is a paraconsistent stable model of $\tau_1(\mathcal{P})$, we can conclude that $\varphi(\mathcal{M})$ is a paraconsistent stable model of $\tau_3((L_1 \subseteq L_2, \mathcal{P}))$.

(ii.2) Let $L_2 \equiv L$. Recall that the set of integrity constraints

$$\{:- \ \tau_2(L_1), \ not \ L. \ , \ :- \ \tau_2(L_1), \neg L.\}$$

is added in this case to $\tau_1(\mathcal{P})$. By reasoning in a way similar to case (i), we can conclude that $\varphi(\mathcal{M})$ is a paraconsistent stable model of $\tau_3((L_1 \subseteq L_2, \mathcal{P}))$.

(ii.3) Let $L_2 \equiv \overline{L}$. Recall that the set of integrity constraints

$$\{:- \ \tau_2(L_1), \ not \ L. \ , \ :- \ \tau_2(L_1), \ not \ \neg L.\}$$

is added in this case to $\tau_1(\mathcal{P})$. By reasoning in a way similar to case (i), we can conclude that $\varphi(\mathcal{M})$ is a paraconsistent stable model of $\tau_3((L_1 \subseteq L_2, \mathcal{P}))$.

This proof generalizes easily to the case $\mathcal{Q} \equiv L_{11} \subseteq L_{12}, \ldots, L_{n1} \subseteq L_{n2}$, with $1 < n$. □

Theorem 4. *Let $(\mathcal{Q}, \mathcal{P})$ be a rough query.*

(i) *Assume that $\mathcal{Q}$ is of the form $\mathcal{Q}_1$ or $\mathcal{Q}_2$ (see definition 37). A ground substitution θ belongs to the answer of $(\tau_2(\mathcal{Q}), \tau_1(\mathcal{P}))$ if and only if θ also belongs to the answer of $(\mathcal{Q}, \mathcal{P})$.*

(ii) *Assume that $\mathcal{Q}$ is of the form $\mathcal{Q}_3$. The extended logic program $\tau_3((\mathcal{Q}, \mathcal{P}))$ has a paraconsistent stable model if and only if the answer to the rough query $(\mathcal{Q}, \mathcal{P})$ is* yes.

Proof. This theorem is a direct consequence of both lemmas 11 and 12.   □

The theorem above shows that the problem of answering rough queries reduces to one of the two problems: to compute answer substitutions for a query to the compiled program; or to test whether the compiled program has a paraconsistent stable model. This provides a foundation for implementation of the language. Since the compilation procedure is polynomial with respect to the size of a rough program, the efficiency of the algorithm to answer rough queries is mainly determined by the system (e.g. *Prolog, dlv, Smodels*) used to compute answers to the queries for the compiled program. Howevere, deciding the existence of a (paraconsistent) stable model for an extended logic program is a NP-complete problem [50].

5   Application Examples

This section presents several examples [21] that highlight the applicability of the language previously discussed.

We have chosen three different relevant problems reported in the rough set literature and show how these problems can be encoded in the proposed language. In contrast to the specific-purpose solutions usually presented, our language offers a general framework where the solution to different types of problems can be declaratively expressed. Moreover, another particularly important aspect illustrated is the integration of rough sets with domain knowledge.

We start by presenting, in section 5.1, a technique to reduce the boundary region of a rough relation. Then in section 5.2, we show how to monitor changes in the boundary region of a rough relation, when some condition attributes are eliminated. Finally, section 5.3 illustrates the integration of expert knowledge through the use of default rules encoded in our language.

5.1   Hierarchy-Structured Decision Tables

A technique to reduce, and eventually eliminate, the boundary region of a rough relation R (or decision table) is introduced in [51], in the context of the variable precision rough set model [13]. This is a relevant issue because any object belonging to the boundary cannot be classified with certainty as belonging to R or $\neg R$. If we interpret a rough relation as a classifier then a large boundary might imply that the classifier is of little value.

One way to cope with the above problem is, for instance, to add more condition attributes to the table. Alternatively, if some attributes have been subject to discretization then, we could increase the precision of the existing attributes by providing more cut-points (i.e. the number of attribute values would increase). However, the disadvantage of these ideas is the rapid growth in the number of decision rules, each of them with a smaller domain coverage, i.e. $cov(r)$ tends to decrease for each decision rule r.

The main idea described in [51] is to associate only with the boundary examples a new layer of decision tables. For instance, more cut-points could be introduced for discretization of attribute values of objects in the boundary region. This would lead to the thinning of the boundary region, in may cases. Note that this "refining" process is only applied to that part of the table corresponding to the boundary region, instead of considering the whole decision table. This idea can be concretized in two ways: by building a hierarchical tree structure of decision tables or by creating a hierarchical linear structure of tables. These techniques are described below.

Each indiscernibility class contained in the boundary region can be treated as new independent universe of objects by itself and a new decision table is associated with each class, forming a new layer of decision tables. The attributes of the decision tables in a new layer have to be "more precise" in the sense they split each indiscernibility class (of the previous layer) into several sub-equivalence classes. This process can be applied recursively yielding a *hierarchical tree structure of decision tables*.

Next example shows how a tree-structured hierarchy of decision tables could be easily encoded in our language. We remind the reader that the value of an attribute a for an object o is null, i.e. $a(o) = \texttt{null}$, if the valued of a is not defined for that particular object o.

Example 28. Consider the decision table shown in table 3.

Each line of the decision table above can be encoded in our language as a fact. For instance, the first lines would be represented as

Table 3. Table of patients with heart problems

	Age	Hypert	Scanabn	Deathmi
o_1	< 70	no	no	no
o_2	> 70	no	no	no
o_3	> 70	yes	yes	yes
o_4	> 70	yes	no	yes
o_5	> 70	yes	no	no
o_6	> 70	yes	no	no
o_7	< 70	yes	yes	no
o_8	< 70	yes	yes	yes

```
¬deathmi(<70, no, no).
deathmi(>70, yes, yes).
deathmi(>70, yes, no).
    ⋮
```

We stress that expressions like "<70", used as arguments of a predicate, should be understood as constants.

It is easy to see that the indiscernibility classes

$$E_1 = \{o_4, o_5, o_6\} ,$$
$$E_2 = \{o_7, o_8\}$$

are in the boundary region.

In order to reduce the boundary region, a different set of condition attributes can be considered for some of these indiscernibility classes (in the boundary area), i.e. the new set of attributes considered for one class may be different from the set of attributes considered for another indiscernibility class in the boundary. The new combination of attributes may have been defined by experts in the field of application. It could also be the case that for some other indiscernibility classes the same attributes as in the original table have been considered, but increased discretization precision of the condition attributes has been applied to each of these classes, possibly with different cut-points for each of them. Let us illustrate these ideas with the table above.

Suppose that experts decided to consider a different set of attributes for patients belonging to E_1: instead of the age, it was considered whether the patient was a smoker. For patients in E_2 only different discretization for the Age attribute was applied.

As the reader can see from tables 4 and 5, the boundary region has been reduced to one indiscernibility class with two patients only, $E_3 = \{o_4, o_6\}$.

The decision tables 4 and 5 can be represented by the following facts. Note that each decision table is recorded under a different predicate name (deathmi, deathmi$_1$, and deathmi$_2$).

```
deathmi₁(yes, no, yes).            ¬deathmi₂(<40, yes, yes).
¬deathmi₁(yes, no, no).            deathmi₂(>40, yes, yes).
¬deathmi₁(yes, no, yes).
```

Table 4. Decison table associated with class E_1

	Hypert	Scanabn	Smoke	Deathmi
o_4	yes	no	yes	yes
o_5	yes	no	no	no
o_6	yes	no	yes	no

Table 5. Decison table associated with class E_2

	Age	Hypert	Scanabn	Deathmi
o_7	< 40	yes	yes	no
o_8	> 40	yes	yes	yes

Putting together all the above decision tables, we create a new rough relation shown in table 6. It corresponds to the initial decision table $\mathcal{D}$eathmi with a reduced boundary, by integrating tables 4 and 5. Using rough clauses, the rough relation corresponding to this table can be easily encoded. Predicate `deathmi`$_3$ denotes this rough relation.

Table 6. Decision table obtained by integrating tables 4 and 5 with table 3

	Age	Hypert	Scanabn	Smoke	Deathmi
o_1	< 70	no	no	null	no
o_2	> 70	no	no	null	no
o_3	> 70	yes	yes	null	yes
o_4	null	yes	no	yes	yes
o_5	null	yes	no	no	no
o_6	null	yes	no	yes	no
o_7	< 40	yes	yes	null	no
o_8	> 40	yes	yes	null	yes

(1) $\overline{\text{deathmi}_3}$(Age,Hypert,Scanabn,null) :- $\underline{\text{deathmi}}$(Age,Hypert,Scanabn).
(2) $\neg\text{deathmi}_3$(Age,Hypert,Scanabn,null) :- $\neg\text{deathmi}$(Age,Hypert,Scanabn).
(3) $\overline{\text{deathmi}_3}$(null,Hypert,Scanabn,Smoke) :- $\overline{\text{deathmi}_1}$(Hypert,Scanabn,Smoke).
(4) $\neg\text{deathmi}_3$(null,Hypert,Scanabn,Smoke) :- $\neg\text{deathmi}_1$(Hypert,Scanabn,Smoke).
(5) $\overline{\text{deathmi}_3}$(Age,Hypert,Scanabn,null) :- $\overline{\text{deathmi}_2}$(Age,Hypert,Scanabn).
(6) $\neg\text{deathmi}_3$(Age,Hypert,Scanabn,null) :- $\neg\text{deathmi}_2$(Age,Hypert,Scanabn).

As we can see in this example, it is possible that some indiscernibility classes in the boundary region have been split into new classes (for instance, $E_1 = \{o_4, o_5, o_6\}$ was split into $E_{11} = \{o_4, o_6\}$ and $E_{12} = \{o_5\}$) and that some of them are still in the boundary region ($E_{11} = \{o_4, o_6\}$). Then, the same idea could be applied once more generating another layer of decision tables.   □

A slightly different method (also proposed in [51]) for reducing the boundary region is obtained by treating the whole subset of the universe corresponding to the boundary as a new domain by itself. Thus, a new decision table is associated with this subset of the universe forming a new layer. However, in this case each new layer has one table only and, consequently, we get a *hierarchical linear structure of decision tables*.

The following example shows how a linear-structured hierarchy of decision tables could be encoded in our language.

Example 29. Consider the decision table 3 of example 28. The whole boundary region is treated as new domain by itself and we associate with it a new decision table. In this case the hypertension attribute was replaced by the sex of the patient and one more condition attribute indicating whether the patient is a smoker was considered. The aim is that when considering these new set of attributes, the boundary region will be reduced (or even eliminated).

Table 7. Decison table associated with the boundary region of table 3

	Age	Scanabn	Sex	Smoke	**Deathmi**
o_4	> 70	no	M	yes	yes
o_5	> 70	no	F	no	no
o_6	> 70	no	M	no	no
o_7	< 70	yes	M	yes	no
o_8	< 70	yes	M	yes	yes

Looking at the table 7, we see that the boundary has been reduced to two objects $E_3 = \{o_7, o_8\}$. Experts decided to try a different discretization for the age attribute, with the aim to eventually eliminate the boundary region. Thus, another decision table (see table 8) was associated with the boundary of table 7.

Table 8. Decison table associated with the boundary of table 7

	Age	Scanabn	Sex	Smoke	**Deathmi**
o_7	< 40	yes	M	yes	no
o_8	> 40	yes	M	yes	yes

Decision tables 7 and 8 are represented in our language as a set of facts under predicates deathmi$_1$ and deathmi$_2$, respectively.

deathmi$_1$(>70, no, M, yes).   ¬deathmi$_2$(<40, yes, M, yes).
¬deathmi$_1$(>70, no, F, no).   deathmi$_2$(>40, yes, M, yes).
¬deathmi$_1$(>70, no, M, no).   ¬deathmi$_1$(<70, yes, M, yes).
deathmi$_1$(<70, yes, M, yes).

The relation between the decision tables 3, 7, and 8, a hierarchical linear structure, can be encoded using rough clauses. All these decision tables are related to the same rough set (relation) that we designate by predicate deathmi$_3$.

```
(1) deathmi₃(Age,Hypert,Scanabn,null,null) :-  deathmi(Age,Hypert,Scanabn).
(2) ¬deathmi₃(Age,Hypert,Scanabn,null,null) :-  ¬deathmi(Age,Hypert,Scanabn).
(3) deathmi₃(Age,null,Scanabn,Sex,Smoke) :-  ¬deathmi₁(Age,Scanabn,Sex,Smoke).
(4) ¬deathmi₃(Age,null,Scanabn,Sex,Smoke) :-  ¬deathmi₁(Age,Scanabn,Sex,Smoke).
(5) deathmi₃(Age,null,Scanabn,Sex,Smoke) :-  deathmi₂(Age,Scanabn,Sex,Smoke).
(6) ¬deathmi₃(Age,null,Scanabn,Sex,Smoke) :-  ¬deathmi₂(Age,Scanabn,Sex,Smoke).
```

When considering a graphical interface, possibly showing the hierarchical structure of the tables, the clauses above could be generated automatically. Moreover, the graphical interface could also hide those predicate arguments corresponding to attributes with null value.

5.2   Avoiding Expensive Tests

In many practical applications, the attribute values correspond to the outcome of a certain test applied to objects of the universe (e.g. a medical test performed on patients). Thus, we may intuitively associate with each attribute a cost, corresponding to the cost of the test that must be performed. Obviously, some attributes may be more expensive than others. For instance, a medical test may be considered expensive because it requires the use of expensive equipment, or because it may cause a lot of discomfort to the patient, or because in general the underlying procedure is expensive.

Given an object o of the universe U and, based on the values of a set of attributes A, a certain decision d is taken (e.g. whether a patient suffers from a certain disease). This information, for each object, can be thought as recorded in the form of a decision table $\mathcal{D}_A = (U, A, d)$. Assume that a set of attributes $B \subset A$ has been identified as being expensive and, therefore, desirable to avoid. We would like to identify those objects o for which the knowledge about attributes B is absolutely necessary for making a decision, i.e. for determining $d(o)$.

The problem described has been studied in [3] and following the approach suggested there, identification of the above mentioned set of objects requires monitoring changes in the boundary of D_A (the rough relation defined by decision table $\mathcal{D}_A$) when considering only the set of attributes $A \setminus B$ (i.e. removing expensive tests B). Next, we summarize the main idea described in [3].

Given two sets A and B, the expression $A \setminus B$ denotes set difference.

Let $\mathcal{D}_A = (U, A, d)$ be a decision table, with a binary decision attribute, and $[t]$ denote the set of objects belonging to the indiscernibility class described by tuple t. If β is a set of tuples then the set of objects described by the tuples belonging to β is given by

$$obj(\beta) = \bigcup_{t \in \beta} [t] \ .$$

Assume also that $\mathcal{D}_{A \setminus B} = (U, A \setminus B, d)$ corresponds to the decision table $\mathcal{D}_A$ without attributes B (i.e. it is a projection of table $\mathcal{D}_A$). When considering a subset $A \setminus B$ of attributes, we have that

$$R_A \subseteq R_{A \setminus B} \ ,$$

where R_A and $R_{A \setminus B}$ are the indiscernibility relations induced by decision tables $\mathcal{D}_A$ and $\mathcal{D}_{A \setminus B}$, respectively. Intuitively, this means that when considering the set of attributes $A \setminus B$, several indiscernibility classes may be merged into one single class. Thus, when considering less attributes, the approximation space may be formed by a smaller number of larger indiscernibility classes. Consequently, the number of objects belonging to $obj(\underline{D_A})$ tends to decrease while the number of objects in $obj(\overline{D_A})$ tends to increase. Hence, it can also be easily concluded that

$$obj(\underline{D_{A \setminus B}}) \subseteq obj(\underline{D_A}) \quad \text{and} \quad obj(\neg\underline{D_{A \setminus B}}) \subseteq obj(\neg\underline{D_A}) \ .$$

When considering the reduced set of attributes $A \setminus B$, the number of objects in the boundary region of D_A also increases, since some indiscernibility classes previously belonging to $\neg\underline{D}_A$ or belonging to $\underline{D}_A$ have now migrated to the new boundary. Intuitively, characterization of these indiscernibility classes defines the set of objects for which knowledge about attributes B is crucial for making a decision. For all other objects, knowledge about B will not change the region, $\neg D_A$, D_A or $\overline{D_A}$, where they already belong. The set of migrating objects can then be defined as

$$Migrate(A, B, D) = (obj(\overline{D_{A \setminus B}}) \cap obj(\underline{D_A}))$$
$$\cup$$
$$(obj(\overline{D_{A \setminus B}}) \cap obj(\neg\underline{D_A})). \tag{1}$$

Although the definition above looks different from the one used in [3], they are both equivalent. However, the formulation presented here is more suitable in the context of our framework, as the reader will see soon.

Obviously, the set on non-migrating objects is defined as

$$\neg Migrate(A, B, D) = obj(\underline{D_{A \setminus B}}) \cup obj(\neg\underline{D_{A \setminus B}}) \cup obj(\overline{D_A}). \tag{2}$$

It is important to note that the set of (non)migrating objects is rough, if only attributes $A \setminus B$ are considered. Otherwise, if all attributes A are considered then the set is crisp. The example 30 illustrates this point.

The migration set enables us to find those objects that require the results of the tests associated with attributes B to be known in order to be able to make a decision. Thus, we aim at finding a description of this set of objects using attributes $A \setminus B$. This description can then be applied to new (unseen) objects to decide whether tests B should be performed.

In practice, for all objects falling in the upper approximation of the migrate set (i.e. conforming to the description of the upper approximation), tests associated with attributes B could be requested. However, if the upper approximation gets very large when attributes B are removed, then not that much is gained. This points to the need of associating some numerical measures with the upper and lower approximations giving some information about the number of objects they might contain. This issue is discussed further in section 6.

Both expressions above, the set of migrating and non-migrating objects, can be translated to a set of rough clauses. These rough clauses permit the user to discover a set of tuples describing those objects belonging to $Migrate(A, B, D)$ ($\neg Migrate(A, B, D)$). The next example illustrates this application.

Table 9. Decision table of patients with heart problems

Age	Test A_1	Test A_2	Deathmi
> 70	b_1	c_1	no
> 70	b_1	c_2	yes
> 70	b_1	c_2	no
$> 40 < 70$	b_2	c_3	yes
$> 40 < 70$	b_2	c_4	no
< 40	b_3	c_5	yes
< 40	b_3	c_5	no
< 40	b_3	c_5	no
< 40	b_4	c_3	yes

Example 30. Consider the decision table $\mathcal{D}eathmi = (U, \{\text{Age}, \text{Test } A_1, \text{Test } A_2\},$ Deathmi), where U is a set of patients with heart problems. Assume that the condition attributes A_1 and A_2 represent two medical tests. Moreover, test A_2 is usually considered as being expensive, and therefore, desirable to avoid.

From table 9, it is easy to see that

(i) $\{\langle \text{>70, b1, c1}\rangle, \langle \text{>40 <70, b2, c4}\rangle\} \subseteq \neg Deathmi$;
(ii) $\{\langle \text{>40 <70, b2, c3}\rangle, \langle \text{<40, b4, c3}\rangle\} \subseteq \underline{Deathmi}$;
(iii) $\{\langle \text{>70, b1, c2}\rangle, \langle \text{<40, b3, c5}\rangle\} \subseteq \overline{Deathmi}$;

The table is encoded as facts in our language.

```
¬deathmi(>70, b1, c1).        deathmi(>70, b1, c2).
¬deathmi(>70, b1, c2).        deathmi(>40 <70, b2, c3).
¬deathmi(>40 <70, b2, c4).    deathmi(<40, b3, c5).
¬deathmi(<40, b3, c5).        deathmi(<40, b4, c3).
```

Moreover, the following clauses monitor the impact in the boundary region of not considering test A_2. Basically, these clauses translate the set of migrating and non-migrating patients represented by formulas (1) and (2) above. The predicate d denotes the rough relation D corresponding to the projection in the first two attributes of $\mathcal{D}eathmi$ (table 9).

(1) $\overline{d}$(Age,Test_A1) :− $\overline{deathmi}$(Age,Test_A1,Test_A2).

(2) $\overline{\neg d}$(Age,Test_A1) :− $\overline{\neg deathmi}$(Age,Test_A1,Test_A2).

(3) $\overline{migrate}$(Age,Test_A1) :− $\underline{d}$(Age,Test_A1),
                          deathmi(Age,Test_A1,Test_A2).

(4) $\overline{migrate}$(Age,Test_A1) :− $\underline{d}$(Age,Test_A1),
                          ¬deathmi(Age,Test_A1,Test_A2).

(5) $\overline{\neg migrate}$(Age,Test_A1) :− $\underline{\neg d}$(Age,Test_A1).

(6) $\overline{\neg migrate}$(Age,Test_A1) :− $\underline{d}$(Age,Test_A1).

(7) $\overline{\neg migrate}$(Age,Test_A1) :− $\overline{deathmi}$(Age,Test_A1,Test_A2).

Thus, by clauses (1) and (2), we have that $\{\langle$>70, b1$\rangle$, $\langle$>40 <70, b2$\rangle$, $\langle$<40, b3$\rangle\} \subseteq \overline{D}$. By clause (4), $\langle$>70, b1$\rangle \in \overline{Migrate}$ and it corresponds to a (class of) patient(s) that migrated from the lower approximation of rough set $\neg \underline{Deathmi}$. But by clause (7) and taking into account (iii), $\langle$>70, b1$\rangle \in \overline{\neg Migrate}$. Thus, $\langle$>70, b1$\rangle$ is in the boundary of relation migrate, consequently, showing that the set of migrating patients is rough in this case.

By clause (3) or (4), $\langle$>40 <70, b2$\rangle \in \overline{Migrate}$ and it corresponds to the merging of two indiscernibility classes, one originating from $\underline{Deathmi}$ and the other from $\neg \underline{Deathmi}$.

By clause (6), $\langle$<40, b4$\rangle \in \overline{\neg Migrate}$. This indiscernibility class remains in the lower approximation, even after dropping attribute A_2. Thus, nothing is gained in performing the expensive test for these patients.

By clause (7), $\langle$<40, b3$\rangle \in \overline{\neg Migrate}$ and it corresponds to a non-migrating (class of) patient(s) that remained in the boundary after dropping the attribute corresponding to the 3rd argument of deathmi, i.e. the expensive medical test.

Let $\mathcal{P}$ be the rough program obtained from the set of facts encoding decision table 9 together with the rough clauses $(1) - (7)$. The query $(\overline{\texttt{migrate}}(\texttt{Age},\texttt{A1}),\mathcal{P})$ requests a description of all patients that may migrate when the expensive test is dropped. The answer is the set of substitutions θ,

$$\theta = \{\{Age/\text{>40 <70},\ A1/b2\}, \{Age/\text{>70},\ A1/b1\}\}\,,$$

indicating that the tuples $\langle$>40 <70, b2$\rangle$ and $\langle$>70, b1$\rangle$ belong to the upper approximation of rough relation denoted by migrate (i.e. $\overline{Migrate}$). We may then conclude that for a new patient whose age is between 40 and 70 and who obtained the result b2 for the test A_1, it is advisable to perform the medical test A_2. We may also ask

- *"For which patients older than 70 years it is worth to perform test A_2?"*

This can be translated to the rough query $(\overline{\texttt{migrate}}(\text{>70},\texttt{A1}),\ \mathcal{P})$. As answer we get the singleton $\{\{A1/b1\}\}$ stating that only patients with outcome b1 for test A_1 should be submitted to test A_2.

Another relevant question is

- *"Which patients may not be submitted to test A_2?"*

It can be represented by the rough query $(\overline{\neg \texttt{migrate}}(\texttt{Age},\texttt{A1}),\ \mathcal{P})$. As answer we get the set of substitutions θ'

$$\theta' = \{\{Age/\text{<40},\ A1/b3\}, \{Age/\text{<40},\ A1/b4\}, \{Age/\text{>70},\ A1/b1\}\}\,.$$

This answer can be interpreted as stating that if a patient conforms to the case

$$((\texttt{Age} < 40) \wedge (\texttt{Test } A_1 = b3)) \vee$$
$$((\texttt{Age} < 40) \wedge (\texttt{Test } A_1 = b4)) \vee$$
$$((\texttt{Age} > 70) \wedge (\texttt{Test } A_1 = b1))$$

then test A_2 may be rather irrelevant. $\square$

In [3], the migration set is defined in a different way.

$$Migrate(A, B, D) = obj(\overline{D_{A \setminus B}}) \cap (U \setminus obj(\overline{D_A})) \,. \tag{3}$$

Note that U is the set of all objects represented in the decision tables $\mathcal{D}_A$ and $\mathcal{D}_{A \setminus B}$. Thus,

$$U \setminus obj(\overline{D_A}) = obj(\underline{D_A}) \cup obj(\neg D_A) \,.$$

This shows that expressions (1) and (3) above are equivalent. Based on (3), a new decision table $\mathcal{D}_1 = (U, A \setminus B, d_1)$ is defined in [3]:

$$d_1(o) = \begin{cases} \textbf{yes} & o \in Migrate(A, B, D) \\ \textbf{no} & \textbf{otherwise} \end{cases}$$

A specific program then takes as input the decision table $\mathcal{D}_A = (U, A, d)$ and creates as output table $\mathcal{D}_1 = (U, A \setminus B, d_1)$, as defined above.

What we wish to emphasize here is that the language we propose is a general framework to create new rough relations and to describe them declaratively in terms of other rough relations. This contrasts with the way the problem was tackled in [3], since there a specific program to create a specific rough relation had to be built.

5.3 Representing Default Knowledge

In this section, we show through a couple of examples that we can also easily express default knowledge in our language and, as in system $CAKE$ [16], define priorities between defaults.

Intuitively, default knowledge corresponds to conclusions assumed to be true in general (we may also call it common sense knowledge), even if we do not have a direct evidence of their truth. For example, we assume that

– *"If someone is driving a car then he has a driving licence."*

However, this does not always have to be true. We may have information that invalidates this conclusion by default (e.g. the person is less than 18 years old).

Representation of default knowledge has been addressed by Reiter who has proposed default logic [52]. The following example shows how *normal default rules* of the default logic can be encoded in our formalism. A formal comparison of default logic with our formalism is out of the scope of this thesis.

Example 31. Consider table 10

Distance $= (U, \{\texttt{Dif}, \texttt{Road Conditions}, \texttt{Physical Distance}\}, \texttt{Distance})$.
This table takes a set of traffic situations U characterized by

– the difference between the actual speed of a vehicle and the speed limit at the road where the vehicle circulates (attribute $\texttt{Dif}$);
– road conditions (dry, wet, snow, or ice); and
– the distance between the vehicle and the one in front of it.

Table 10. Decision table classifying vehicle distances

Dif	Road Conditions	Physical Distance	Distance
10	dry	9	medium
10	ice	15	medium
10	ice	12	small
30	wet	30	large
−10	snow	9	medium
−10	snow	9	small

This data could have been acquired from a number of different sources. For instance, road conditions could have been obtained by sensors, a camera records traffic images, while speed limit in roads are obtained from a database. Then, one or more experts in traffic safety, decide for each situation whether the distance between vehicles is large, medium, or small. It is easy to accept that this classification depends on the attributes mentioned above. It may also happen that, given the same traffic situation, different experts classify differently the distance (i.e. one expert might say that the distance is small and another consider it as medium).

Note that the decision attribute Distance is not binary in this case, since it may assume the values small, medium, or large. However, it is easy to see this table as three decision tables defining the (rough) concepts of small, medium, and large distance. Moreover, this idea can be easily expressed in our language as rough clauses $(1) - (9)$ together with the facts under predicate p (see below). The whole table is encoded as a set of positive facts.

$\overline{p}$(10, dry, 9, medium). $\overline{p}$(10, ice, 15, medium).
$\overline{p}$(10, ice, 12, small). $\overline{p}$(30, wet, 30, large).
$\overline{p}$(-10, snow, 9, medium). $\overline{p}$(-10, snow, 9, small).

(1) $\overline{\text{large}}$(X1,X2,X3):- $\overline{p}$(X1,X2,X3,large).
(2) $\overline{\neg\text{large}}$(X1,X2,X3):- $\overline{\text{medium}}$(X1,X2,X3).
(3) $\overline{\neg\text{large}}$(X1,X2,X3):- $\overline{\text{small}}$(X1,X2,X3).

(4) $\overline{\text{medium}}$(X1,X2,X3):- $\overline{p}$(X1,X2,X3,medium).
(5) $\overline{\neg\text{medium}}$(X1,X2,X3) :- $\overline{\text{small}}$(X1,X2,X3).
(6) $\overline{\neg\text{medium}}$(X1,X2,X3) :- $\overline{\text{large}}$(X1,X2,X3).

(7) $\overline{\text{small}}$(X1,X2,X3):- $\overline{p}$(X1,X2,X3,small).
(8) $\overline{\neg\text{small}}$(X1,X2,X3):- $\overline{\text{medium}}$(X1,X2,X3).
(9) $\overline{\neg\text{small}}$(X1,X2,X3):- $\overline{\text{large}}$(X1,X2,X3).

Another decision table, see table 11,

$$\mathcal{D}\text{anger} = (U, \{\text{Dif}, \text{Road Conditions}, \text{Distance}\}, \text{Danger}),$$

shows whether a number of traffic situations has been classified as dangerous by an expert. As usual, this table is represented as a set of facts.

Table 11. Decision table classifying the danger of several traffic situations

Dif	Road Conditions	Distance	Danger
10	dry	large	no
20	ice	small	yes
0	wet	medium	no
10	wet	medium	yes
10	wet	medium	no
−10	snow	medium	yes

$\overline{\neg\text{danger}}$(10, dry, large). $\overline{\text{danger}}$(20, ice, small).
$\overline{\neg\text{danger}}$(0, wet, medium). $\overline{\text{danger}}$(10, wet, medium).
$\overline{\neg\text{danger}}$(10, wet, medium). $\overline{\text{danger}}$(-10, snow, medium).

We also would like to add some common sense knowledge to the set of facts above. For instance, consider the following statement expressing that usually people assume by default that small distances between two vehicles yield to a dangerous situation.

— *"If the distance between two vehicles is small, then we may assume that the situation is dangerous (unless there is evidence to the contrary)."*

This statement could be expressed by the following (normal) default rule, where the variables should be understood as universally quantified.

$$\frac{\text{small}(X1, X2, X3) : \text{danger}(X1, X2, \text{small})}{\text{danger}(X1, X2, \text{small})} . \tag{1}$$

The default rule (1) can be formally read as follows. If in a certain situation $\text{small}(x1, x2, x3)$ holds (i.e. it can be proved) and $\text{danger}(x1, x2, \text{small})$ is consistent with the current knowledge, then (by default) we assume that $\text{danger}(x1, x2, \text{small})$ holds, too. Note that $\text{danger}(x1, x2, \text{small})$ is consistent with the current knowledge, if we have no evidence that $\neg\text{danger}(x1, x2, \text{small})$ is true. Thus, no contradiction with the available knowledge arises by the fact that $\text{danger}(x1, x2, \text{small})$ is assumed.

Moreover, consider that we also want to express the next common sense (default) idea.

— *"If the distance between two vehicles is not small, then we may assume that the situation is not dangerous (unless it can be proved otherwise)."*

This statement could be expressed in default logic by default rules (2) and (3).

$$\frac{\text{medium}(X1, X2, X3) : \neg\text{danger}(X1, X2, \text{medium})}{\neg\text{danger}(X1, X2, \text{medium})} , \tag{2}$$

$$\frac{\text{large}(X1, X2, X3) : \neg\text{danger}(X1, X2, \text{large})}{\neg\text{danger}(X1, X2, \text{large})} . \tag{3}$$

Our next step is to show that the above default rules can be expressed in the proposed language.

(10) $\overline{\text{danger}_1}$(X1,X2,small):- $\overline{\text{small}}$(X1,X2,X3), danger?(X1,X2,small).

(11) $\overline{\neg\text{danger}_1}$(X1,X2,medium):- $\overline{\text{medium}}$(X1,X2,X3), danger?(X1,X2,medium).

(12) $\overline{\neg\text{danger}_1}$(X1,X2,large):- $\overline{\text{large}}$(X1,X2,X3), danger?(X1,X2,large).

Rough clauses $(10) - (12)$ express the default rules $(1)-(3)$, respectively. For instance, consider rough clause (10). The testing literal

$$\text{danger?(X1,X2,small)}$$

in its body allows to test whether a tuple $\langle c_1, c_2, \text{small} \rangle$ is undefined, i.e. $\langle c_1, c_2, \text{small} \rangle \notin \overline{Danger}$ and $\langle c_1, c_2, \text{small} \rangle \notin \overline{\neg Danger}$. If $\langle c_1, c_2, \text{small} \rangle \in \overline{Danger}$ then rough clause (10) is not applicable because no new information would be obtained. If $\langle c_1, c_2, \text{small} \rangle \in \overline{\neg Danger}$ then rough clause (10) is not applicable because its application would lead to a conclusion that would not be consistent with the available knowledge.

Finally, we put together the knowledge coming from the table $Danger$ with the default knowledge. To achieve this we define a new rough relation (clauses $(13)-(16)$) and use a new predicate name (danger_2).

(13) $\overline{\text{danger}_2}$(X1,X2,X3):- $\overline{\text{danger}}$(X1,X2,X3).

(14) $\overline{\text{danger}_2}$(X1,X2,X3):- $\overline{\text{danger}_1}$(X1,X2,X3).

(15) $\overline{\neg\text{danger}_2}$(X1,X2,X3):- $\overline{\neg\text{danger}}$(X1,X2,X3).

(16) $\overline{\neg\text{danger}_2}$(X1,X2,X3):- $\overline{\neg\text{danger}_1}$(X1,X2,X3).

From the fifth row of the first table we see that medium(-10,snow,9) holds. But, the tuple $\langle$-10,snow,medium$\rangle$ corresponds to a dangerous traffic situation (see last line of the second table). Thus, rough clause (11) (encoding default rule (2)) cannot be applied because danger(-10,snow,medium) holds and consequently $\langle$-10,snow,medium$\rangle$ is not undefined with respect to rough relation $Danger$.

Note that small(-10,snow,9) holds (last line of table $Distance$) but the tuple $\langle$-10,snow,small$\rangle$ does not exist in the table $Danger$ (i.e. thus, it does not correspond to a traffic situation known as non-dangerous). Hence, from rough clause (10) (corresponding to default rule (1)), we conclude danger (-10,snow,small) .   □

Next example, illustrates the use of priorities between default rules.

Example 32. Consider once more the (decision) tables of the previous example and default rules (1) and (2). Moreover, we also assume that

– *"If distance between vehicles is medium and the road conditions are icy, then we may conclude that the traffic situation is dangerous (unless it can be proved otherwise)."*

Using a default rule, we could express this statement as

$$\frac{\texttt{medium}(X1,\texttt{ice},X3):\texttt{danger}(X1,\texttt{ice},\texttt{medium})}{\texttt{danger}(X1,\texttt{ice},\texttt{medium})}. \tag{4}$$

Informally, looking at the second row of the first table, we conclude that `medium(10,ice,15)` holds. Moreover, from the second table we can conclude that the tuple ⟨10,ice,medium⟩ is not considered as a dangerous (or non-dangerous) situation (actually, there is no such tuple in the second table). Thus, default rule (2) can be applied and we conclude that ¬`danger(10,ice,medium)` also holds. Similarly, default rule (4) can be applied, to conclude that `danger`(10,ice,medium) holds, too.

Hence, from this example, we conclude that by applying different default rules, we may obtain contradictory information. Although this may be acceptable in some situations (it is a case belonging to the boundary region), in other situations we may wish to express priorities between several applicable defaults. For instance, if both default rules (2) and (4) are applicable, then we may give priority to (4) and block application of default (2) for safety reasons. To achieve this idea we first define a new rough relation expressing the default rules (1) and (4).

(10) $\overline{\texttt{danger}_1}$`(X1,X2,small):- ` $\overline{\texttt{small}}$`(X1,X2,X3), danger?(X1,X2,small).`

(11) $\overline{\texttt{danger}_1}$`(X1,ice,medium):- ` $\overline{\texttt{medium}}$`(X1,ice,X3), danger?(X1,ice,medium).`

We put then together the knowledge coming from the table $\mathcal{D}$anger with the default knowledge. The last rough clause, (15), encodes default rule (2) and gives it lower priority than rough clause (11) (encoding default rule (4)).

(12) $\overline{\texttt{danger}_2}$`(X1,X2,X3):- ` $\overline{\texttt{danger}}$`(X1,X2,X3).`

(13) $\overline{\neg\texttt{danger}_2}$`(X1,X2,X3):- ` $\overline{\neg\texttt{danger}}$`(X1,X2,X3).`

(14) $\overline{\texttt{danger}_2}$`(X1,X2,X3):- ` $\overline{\texttt{danger}_1}$`(X1,X2,X3).`

(15) $\overline{\neg\texttt{danger}_2}$`(X1,X2,medium):- ` $\overline{\texttt{medium}}$`(X1,X2,X3),`
$\qquad\qquad\qquad$`danger?(X1,X2,medium), danger`$_1$`?(X1,X2,medium).`

6   The Rough Knowledge Base System

We present the principles of a system, called *Rough Knowledge Base System* (*RKBS*). The system is available through a Web page

`http://www.ida.liu.se/rkbs` .

It can reason about rough relations defined in a rough program and answer queries. The implementation was done by R. Andersson as a master thesis [23,53]. The ideas on which this implementation is based were already previously explored in a prototype written by A. Vitória and C. V. Damásio.

The language supported by *RKBS* (to encode rough programs) extends that of section 4 by associating quantitative measures with each tuple of a rough relation [22]. However, this system may not be able to compute answers for all queries to a recursive rough program.

We believe that several interesting data mining applications using rough sets could be encoded in our language. For instance, the example discussed in section 5.2 illustrates an application of our language to a problem in the data mining field. Note that quantitative measures are an important aspect in data mining applications while recursion does not seem to be required by many of them. For this reason, we have extended our language to support the former feature while the latter is not yet supported.

A distinction must also be made between the compilation technique presented in section 4 and the compilation of rough programs with quantitative measures. The latter generates extended logic programs which require aggregate functions[6], while these functions are not needed by the former. *RKBS* compiles rough programs with quantitative measures to standard *Prolog* [33,46] programs. This opens for the use of *Prolog* built-in predicates and structured terms like lists. Note that each extended logic program, obtained by compiling a non-recursive rough program encoded in the language discussed in section 4, corresponds also to a *Prolog* program.

Another reason for considering only non-recursive rough programs is that the semantics of aggregates and stable models has been an open problem under investigation [54,55]. We avoid this problem because each non-recursive rough program with quantitative measures can be compiled to a logic program with at most one paraconsistent stable model.

The user interface of the system has been implemented in *Java* [56]. We have chosen *XSB Prolog* system [57] to write the compiler and to reason with the compiled programs for the following reasons. First, *XSB Prolog* supports *definite clause grammars*. This fact simplifies the writing of a parser and compiler for a rough program. Second, *XSB* provides ISO-predicates such as `setof/3` and `findall/3`. These predicates can be used to implement aggregate functions, like `sum` and `count`. A third reason is that *XSB* allows the use of a technique called *tabling* when computing answers to queries for logic programs. Due to this fact, it is possible to obtain answers to queries for a large class of recursive logic programs, while more traditional *Prolog* systems based on SLDNF-resolution [32] would simply loop forever. This class corresponds to non-floundering logic programs that enjoy the bounded term-depth property [58,59,60,61]. An important well-known subclass of this class of programs is *Datalog*. This aspect opens the future possibility of easily extending our system to applications that require a limited use of recursive rough programs. Fourth, version 2.6 of *XSB* has the *XASP* package that provides an efficient interface to *Smodels* [42,49,43] from *XSB* system. Note that *XSB* cannot be used to compute stable models, while *SModels* can do it. Thus, this connection between both systems could make possible to extend *RKBS* to support any recursive rough program. Finally, *XSB Prolog* is free and well-documented software.

---

[6] `sum` and `count` from SQL are examples of aggregate functions.

Since we are presenting work in progress, we have restricted ourselves to the implementation issues of the language supported by *RKBS*. Hence, we do not formally present the declarative semantics for this language, as we did for the language introduced in chaper 4. A fuller description of the declarative semantics of rough programs with quantitative measures will be submitted for later publication.

As for the language discussed in section 4, predicates of a rough program supported by *RKBS* denote rough sets. However, the notion of rough set has been extended to account for quantitative measures. In section 6.1, we extend the notion of rough set and review some quantitative measures. We then present in section 6.2 the language supported by our system. Section 6.3 is devoted to the compilation of (non-recursive) rough programs with quantitative measures. The query language of the system is discussed in section 6.4. Finally, we describe some examples in section 6.5 .

6.1   Rough Sets Revisited

This section presents an extension of the rough set notion discussed previously in section 2 that explicitly takes into account quantitative measures. We then review some quantitative measures associated with rough sets in the context of our framework.

Recall that the set of values associated with an attribute a is denoted as V_a.

Definition 41. *Given a set of attributes* $A = \{a_1, \ldots, a_n\}$, *a rough set (or rough relation)* S *is a pair of sets* $(\overline{S}, \overline{\neg S})$ *satisfying conditions (i) and (ii).*

(i) The elements of sets $\overline{S}$ *and* $\overline{\neg S}$ *are expressions of the form*

$$\langle t_1, \ldots, t_n \rangle : k \ ,$$

where $\langle t_1, \ldots, t_n \rangle \in \prod_{a_i \in A} V_{a_i}$ *and* k *is an integer larger than zero.*
(ii) The following implications are true.

$$\langle t_1, \ldots, t_n \rangle : k \in \overline{S} \ \Rightarrow \forall k' \neq k (\langle t_1, \ldots, t_n \rangle : k' \notin \overline{S}) \ ,$$
$$\langle t_1, \ldots, t_n \rangle : k \in \overline{\neg S} \Rightarrow \forall k' \neq k (\langle t_1, \ldots, t_n \rangle : k' \notin \overline{\neg S}) \ .$$

The rough complement *of a rough set* $S = (\overline{S}, \overline{\neg S})$ *is the rough set* $\neg S = (\overline{\neg S}, \overline{S})$.

For simplicity, we denote by t a general tuple $\langle t_1, \ldots, t_n \rangle$ and by $[t]$ the indiscernibility class described by tuple t. Moreover, we may also write $t \in \overline{S}$ ($t \in \underline{S}$ or $t \in \overline{S}$ or $t \in \overline{\neg S}$ or $t \in \underline{\neg S}$ or $\overline{\neg S}$), if the associated quantitative measure k is irrelevant.

Intuitively, an element $t : k \in \overline{S}$ ($t : k \in \overline{\neg S}$) indicates that the indiscernibility class described by the tuple t belongs to the upper approximation of the rough set S ($\neg S$) and that this class contains $k > 0$ individuals that are positive examples of the concept described by S ($\neg S$). *Lower approximation* of rough set S, represented $\underline{S}$, is then defined as

$$\underline{S} = \{t : k_1 \in \overline{S} \mid \forall k_2 > 0 \, (t : k_2 \notin \overline{\neg S})\}$$

and the *boundary region*, represented $\overline{\underline{S}}$, is defined as

$$\underline{S} = \{t : k_1 : k_2 \mid \exists k_1, k_2 > 0 \, (t : k_1 \in \overline{S} \text{ and } t : k_2 \in \overline{\neg S})\} \, .$$

A rough set $D = (\overline{D}, \overline{\neg D})$, as defined above, can be seen as an alternative representation of a decision table $\mathcal{D} = (U, A, \mathsf{d})$. An expression $t : k_1 \in \overline{D}$ corresponds to $k_1 > 0$ lines t of the table with positive outcome for the decision attribute, while $t : k_2 \in \overline{\neg D}$ corresponds to $k_2 > 0$ lines t with negative outcome for the decision attribute. The fact that we consider only binary decision attributes is not a restriction in practice, as shown in section 5 (see example 31).

Recall that in our work a rough set is not defined in terms of individuals of the universe, but instead in terms of the tuples that describe each indiscernibility class to which the individuals belong.

Quantitative Measures. Let a tuple t be the description of an indiscernibility class $[t]$ of a decision table $\mathcal{D} = (U, A, \mathsf{d})$. Assume also that $|d|$ $(|\neg d|)$ is the number of individuals (or lines of the table) that have positive (negative) outcome for the decision attribute d. Thus, $|d| + |\neg d|$ is the number of objects (lines) in the corresponding table. The following quantitative measures are then defined.

- The *support* of $d(t)$, denoted $\mathsf{supp}(d(t))$, corresponds to the number of individuals in the indiscernibility class $[t]$ that are positive examples. Thus, if $t : k \in \overline{D}$ then $\mathsf{supp}(d(t)) = k$.
- The *strength* of $d(t)$, denoted $\mathsf{strength}(d(t))$, indicates how often individuals in the indiscernibility class $[t]$ have positive outcome for the decision attribute d. Thus, if $t : k \in \overline{D}$ then $\mathsf{strength}(d(t)) = \frac{k}{|d| + |\neg d|}$.
- The *accuracy* of $d(t)$, denoted $\mathsf{acc}(d(t))$, corresponds to the conditional probability $Pr(d(i) = \mathsf{yes} \mid i \in [t])$. By other words, $\mathsf{acc}(d(t))$ expresses how trustworthy the indiscernibility class described by t is in drawing the conclusion that the outcome for the decision attribute d is positive. Thus, if $t : k_1 \in \overline{D}$ and $t : k_2 \in \overline{\neg D}$ then, $\mathsf{acc}(d(t)) = \frac{k_1}{k_1 + k_2}$.
- The *coverage* of $d(t)$, denoted $\mathsf{cov}(d(t))$, corresponds to the conditional probability $Pr(i \in [t] \mid d(i) = \mathsf{yes})$. By other words, $\mathsf{cov}(d(t))$ expresses how well the indiscernibility class $[t]$ describes the positive decision class. Thus, if $t : k \in \overline{D}$ then $\mathsf{cov}(d(t)) = \frac{k}{|d|}$.

Obviously, the same measures can also be defined for $\neg d(t)$. For instance, if $t : k_1 \in \overline{D}$ and $t : k_2 \in \overline{\neg D}$ then $\mathsf{acc}(\neg d(t)) = \frac{k_2}{k_1 + k_2}$. Moreover, $\mathsf{supp}(d(t)) + \mathsf{supp}(\neg d(t)) = |[t]|$ and $\mathsf{acc}(d(t)) + \mathsf{acc}(\neg d(t)) = 1$.

In section 2.4, we introduce the notions of support, strength, accuracy, and coverage for decision rules. Let $\langle t_1, \ldots, t_n \rangle$ be a tuple describing an indiscernibility class of a rough relation D such that $A_i = att_D(i)$, with $1 \leq i \leq n$. If we interpret any statement

$$\langle t_1, \ldots, t_n \rangle \in \overline{D} \quad \text{or}$$
$$\langle t_1, \ldots, t_n \rangle \in \overline{\neg D}$$

as a decision rule

$$(A_1 = t_1) \wedge \ldots \wedge (A_n = t_n) \longrightarrow (d = \mathsf{yes}) \text{ or}$$
$$(A_1 = t_1) \wedge \ldots \wedge (A_n = t_n) \longrightarrow (d = \mathsf{no}),$$

respectively, then we can easily see that the quantitative measures presented above correspond to the ones discussed in section 2.4.

6.2   A Language with Numerical Measures

In this section, we extend the language presented in section 4 with quantitative measures. We add now to the language the capability to keep track of the number of individuals that belong to each indiscernibility class of a rough relation. Moreover, quantitative measures such us support, strength, accuracy, and coverage, discussed in section 6.1, can be used to define a rough relation. We start by an informal introduction of the language.

A *rough program* is a set of *rough facts* and *rough clauses*. Rough facts encode rough relations defined explicitly by a decision table, while rough clauses are used to define implicitly new rough relations obtained by combining different regions (e.g. lower approximation, upper approximation, and boundary) of other rough relations. For instance,

$$\overline{r}(c_1, c_2, c_3) : 5. \quad \text{and}$$
$$\overline{\neg r}(c_1, c_2, c_3) : 8.$$

are two rough facts. The first says that the indiscernibility class described by the tuple of attribute values $\langle c_1, c_2, c_3 \rangle$ has 5 individuals. Moreover, these individuals are positive examples of the concept represented by the rough relation denoted by r, designated as R. The second rough fact states that the same indiscernibility class has 8 individuals that are negative examples of R (or positive examples of $\neg R$). Next, we give an example of a rough clause in the extended language.

$$\overline{p}(X_1, X_2) \ \text{:-}[\alpha, F] \ \underline{q}(X_1, X_2), \overline{\neg r}(X_1, X_2).$$

The expression to the right of $\text{:-}[\alpha, F]$ (i.e. $\underline{q}(X_1, X_2), \overline{\neg r}(X_1, X_2)$) is called the *body* and the expression to the left (i.e. $\overline{p}(X_1, X_2)$) is called the *head* of the rough clause. Moreover, α should be a rational number between 0 and 1 and F should be an associative and commutative binary function. (e.g. the minimum). If quantitative measures are ignored, the rough clause above can informally be interpreted in a way similar to the interpretation of rough clauses discussed in section 4. The expression $\underline{q}(X_1, X_2)$ can be seen as representing an indiscernibility class belonging to the lower approximation of rough relation Q, denoted by q. Note that X_1 and X_2 are variables that can be thought as representing any attribute value. Hence, the body of the rule above captures those indiscernibility classes $[\langle c_1, c_2 \rangle]$ that are in the intersection of the lower approximation of the rough relation Q with the upper approximation of rough relation $\neg R$. Moreover, the rough rule above expresses that each of these indiscernibility classes belongs to the upper approximation of rough relation P, denoted by p.

Let us now intuitively explain how quantitative information is handled. Assume that there are two indiscernibility classes described by tuple $\langle c_1, c_2 \rangle$: one indiscernibility class is part of $\underline{Q}$ and the other indiscernibility class belongs to $\overline{\neg R}$. Function F is then used to combine $\mathbf{supp}(\underline{q}(c_1, c_2))$ with the $\mathbf{supp}(\overline{\neg r}(c_1, c_2))$.

Hence, the rough clause above states that, given a tuple $\langle c_1, c_2 \rangle$ describing an indiscernibility class, if

$$\langle c_1, c_2 \rangle : k_2 \in \underline{Q} \text{ and } \langle c_1, c_2 \rangle : k_3 \in \overline{\neg R}$$

then $\langle c_1, c_2 \rangle : k_1 \in \overline{P}$, where

$$\mathtt{supp}(p(c_1, c_2)) = k_1 \geq \lfloor \alpha \times F(k_2, k_3) \rfloor .$$

Note that the support k_1 should be computed by taking into account all clauses of a rough program, as shown in example 33. This is the reason for writing $k_1 \geq \lfloor \alpha \times F(k_2, k_3) \rfloor$ instead of writing $k_1 = \lfloor \alpha \times F(k_2, k_3) \rfloor$.

In contrast with the language presented in section 4, the head of a rough clause cannot refer to the boundary region of a rough relation, i.e. an expression as $\overline{p}(X_1, X_2)$ could not be the head of a rough clause. However, this is not a real restriction as shown in example 33. If rough literals referring to the boundary region were allowed in the head of a rough clause then we could not know how many individuals computed from the body would be positive examples and how many would be negative examples. This is the motivation behind this restriction. Moreover, no testing literals (e.g. $l?(t)$) can occur in the body of a rough clause, although this feature could be easily added.

We argue now on the usefulness of having user parameterized rough clauses $H \colon \mathtt{-}[\alpha, F]\ B.\ \mathrm{,}$ where α and F are the parameters. An example illustrating the importance of parameter α is when the user wants to decrease his trust on certain data. For example, assume that the user strongly doubts of the reliability of the information carried by 20% of the examples belonging to any indiscernibility class only with positive examples of Q and for which the second attribute has value c. A rough clause like

$$q_1(X, c) \colon \mathtt{-}[0.8, -]\ q(X, c).$$

could be used to express such doubt. The new predicate q_1 denotes the same rough relation as q except that any indiscernibility class described by a tuple $\langle t_1, c \rangle \in Q_1$ has only 80% of the individuals in the corresponding indiscernibility class $\langle t_1, c \rangle \in Q$.

We also think that the way the support information, obtained from the expressions in the body, should be combined strongly depends on the application. For instance, if the user wants to represent the join of two decision tables then parameter F should correspond to the product function. But, if he wants to define a new rough relation R that captures those indiscernibility classes belonging to the same region of two rough relations P and Q (i.e. having the same description) then, it might make more sense to use the minimum function. The rough clauses below could together express this idea.

$$
\begin{aligned}
\overline{r}(X_1, X_2) &\colon \mathtt{-}[1, \min]\ \overline{p}(X_1, X_2), \overline{q}(X_1, X_2). \\
\underline{r}(X_1, X_2) &\colon \mathtt{-}[1, \min]\ \underline{p}(X_1, X_2), \underline{q}(X_1, X_2). \\
\overline{\neg r}(X_1, X_2) &\colon \mathtt{-}[1, \min]\ \overline{\neg p}(X_1, X_2), \overline{\neg q}(X_1, X_2). \\
\underline{\neg r}(X_1, X_2) &\colon \mathtt{-}[1, \min]\ \underline{\neg p}(X_1, X_2), \underline{\neg q}(X_1, X_2).
\end{aligned}
$$

Thus, if $t : k_1 \in \underline{P}$ and $t : k_2 \in Q$ then one may conclude that, for both relations, there is an indiscernibility class described by tuple t in the lower approximation and $\mathrm{supp}(r(t)) \geq \min(k_1, k_2)$.

Example 33. We give an example of a rough program $\mathcal{P}$ and discuss informally its meaning.

$$
\begin{aligned}
\mathcal{P} = \{ &\overline{p}(X_1, X_2) \ \text{:-}[1, \min] \ \underline{q}(X_1, X_2), \overline{\neg r}(X_1, X_2). \,, \\
&\overline{p}(X, c) \ \text{:-}[1, -] \ \overline{q_1}(X, c). \,, \\
&\overline{\neg p}(X, c) \ \text{:-}[1, -] \ \overline{\neg q_1}(X, c). \,, \\
&\overline{q}(a, c) : 2. \,, \\
&\overline{r}(a, c) : 3. \,, \overline{\neg r}(a, c) : 4. \,, \\
&\overline{q_1}(a, c) : 3. \,, \overline{\neg q_1}(a, c) : 7. \}
\end{aligned}
$$

The body of the first rough clause represents the intersection of the lower approximation of the rough relation Q, denoted by q, with the boundary of rough relation $\neg R$, denoted by $\neg r$. From the facts of $\mathcal{P}$, we get that $\langle a, c \rangle : 2 \in \overline{Q}$ and $\langle a, c \rangle : 4 : 3 \in \overline{\neg R}$. Hence, from the first rough clause can be concluded that $\mathrm{supp}(p(a, c)) \geq 1 \times \min(2, 4)$ ($\mathrm{supp}(q(a, c)) = 2$ and $\mathrm{supp}(\neg r(a, c)) = 4$).

The second and third rough clauses together state that if an indiscernibility class $[t]$ belongs to the boundary of rough relation $\neg Q_1$ and its second attribute has value c then, $[t]$ also belongs to the boundary of P. Moreover, the same number of positive and negative examples in $[t]$ should be inherited by rough relation P (i.e. $\mathrm{supp}(q_1(a, c)) = 3$ individuals should be considered as representing positive examples of P, while $\mathrm{supp}(\neg q_1(a, c)) = 7$ individuals should be considered as representing negative examples of P). Since the body of each of the rough clauses has only one expression, the choice of function F is irrelevant. This can be represented by the use of '$-$' instead of some concrete function.

Putting all together, it can be concluded from $\mathcal{P}$ that $\mathrm{supp}(p(a, c)) = \min (2, 4) + 3 = 5$ and $\mathrm{supp}(\neg p(a, c)) = 7$.

As this example shows, information concerning an indiscernibility class may be obtained independently from different rough clauses. For instance, information related with indiscernibility class $\langle a, c \rangle \in \overline{P}$ is obtained from the three rough clauses of $\mathcal{P}$. Note that $\mathrm{supp}(p(a, c))$ is computed by summing the support obtained from different rough clauses.   □

Another important aspect of the language is the possibility of using *quantitative measure expressions* in the body of a rough clause. For example,

$$
\begin{aligned}
&\mathrm{acc}(p(c_1, c_2)) > \mathrm{acc}(\neg q(c_1, c_2)) \ \text{and} \\
&\mathrm{supp}(\neg p(c_1, c_2)) > 7
\end{aligned}
$$

are quantitative measure expressions. The first quantitative measure expression states that the accuracy of the indiscernibility class described by $\langle c_1, c_2 \rangle$ of rough relation P is larger than the accuracy of the indiscernibility class $[\langle c_1, c_2 \rangle]$ of rough relation $\neg Q$. The second states that indiscernibility class $[\langle c_1, c_2 \rangle]$ has more than 7 individuals with negative outcome for the decision attribute p.

The next step is to define formally the language supported by *RKBS*, to encode rough programs. To this end, we first define the notion of quantitative measure expression.

Definition 42. *Assume that m stands either for* supp, *or* strength, *or* acc, *or* cov *and that relOp is one of the relation symbols* $=, <, \leq, >, \geq, \neq$. *A quantitative measure expression is a formula of the form*

$$m(l(t_1, \ldots, t_n)) \quad relOp \quad k \qquad\qquad or$$
$$m(l_1(t_1, \ldots, t_n)) \quad relOp \quad m(l_2(t_1, \ldots, t_n)) \,,$$

where

- l, l_1, *and* l_2 *are either* p *or* $\neg p$, *for some predicate symbol* p; *and*
- $n \geq 0$; *and*
- k *is a positive rational number.*

Note that not all quantitative measure expressions are meaningful. For example, $\mathtt{acc}(q(a, b)) > \mathtt{supp}(r(a, b))$ is meaningless because it does not make sense to compare accuracy with support.

Definition 43. *A rough clause in RKBS is any expression of the form*

$$H \quad :\text{-}[\alpha, F] \quad B_1, \ldots, B_i, M_1, \ldots, M_k. \,,$$

where

- H *is either* $\bar{l}(t_1, \ldots, t_n)$ *or* $\underline{l}(t_1, \ldots, t_n)$, *with* $n \geq 0$ *and* l *being either* p *or* $\neg p$, *for some predicate symbol* p; *and*
- α *is a rational number such that* $0 < \alpha \leq 1$; *and*
- F *is a commutative and associative binary function; and*
- *each* B_j *is a rough literal, with* $1 \leq j \leq i$; *and*
- *each* M_j *is a quantitative measure expression such that all variables occurring in* M_j *also occur in some rough literal in the body of the rough clause, with* $1 \leq j \leq k$; *and*
- $i, k \geq 0$.

As expected, a rough fact in *RKBS* is a rough clause with empty body (i.e. $i = k = 0$) and a rough program supported by *RKBS* is a finite set of non-recursive rough clauses (as defined in 43).

Each predicate occurring in a rough program, supported by *RKBS*, denotes a rough relation as defined in section 6.1 (see definition 41).

6.3   Compilation of *RKBS* Programs

Our system transforms a rough program into a *Prolog* program, where the special predicates `bagof/3` and `findall/3` occur. These predicates are used for the following purposes.

- To count the number of individuals in each indiscernibility class of a rough relation that are positive (negative) examples of the underlying concept.
- To count the number of individuals in the universe of a rough relation D that have positive (negative) outcome for the decision attribute d (i.e. to compute $|d|$ and $|\neg d|$).

We first informally introduce those special predicates. We then discuss the details about the compilation of (non-recursive) rough programs with quantitative measures.

All Solutions Predicates in *Prolog*. Any standard *Prolog* system [46] has two built-in predicates, `bagof/3` and `findall/3`, to collect together all solutions to a problem. We explain their meaning through a couple of examples.

Example 34. Consider the following (definite) logic program containing a number of facts about employees and their salaries.

$$\mathcal{P} = \{\texttt{salary(peter, 100).,}$$
$$\texttt{salary(terry, 150).,}$$
$$\texttt{salary(john, 100).,}$$
$$\texttt{salary(susan, 150).,}$$
$$\texttt{who(X, L) :- bagof(Y, salary(Y,X), L).}$$

For instance, the fact `salary(peter, 100).` states that "Peter's salary is 100.".

The query

$$(\texttt{who(150, L)}, \mathcal{P})$$

requests a list of all employees who earn 150. The answer is the set of substitutions

$$\{\{[\,\texttt{terry},\texttt{susan}\,]/L\}\}.$$

Note that a list of items is represented between square brackets, '[' and ']'. Thus, [terry, susan] represent the list with constants `terry` and `susan`.

The query

$$(\texttt{who(X, L)}, \mathcal{P})$$

requests all salaries together with a list of people who earn each salary. The answer is the set of substitutions

$$\{\{100/X, [\,\texttt{peter},\texttt{jonh}\,]/L\}, \{150/X, [\,\texttt{terry},\texttt{susan}\,]/L\}\}. \qquad \square$$

The atom `bagof(Template, Goal, List)` can be interpreted as the "List of all instances of `Template` such that `Goal` is satisfied". Note that `List` may contain duplicates, if the same instance of `Template` can be proved in several ways.

Next example illustrates the use of predicate `findall`.

Example 35. Consider the following (definite) logic program.

$\mathcal{P} = \{$salary(peter, 100).,
       salary(terry, 150).,
       salary(john, 100).,
       salary(susan, 150).,
       allSalaries(L) :- findall(X, salary(Y,X), L).,
       total(T) :- findall(X, salary(Y,X), L), sum(L, T).$\}$

Assume that predicate sum(L, T) has the following meaning: "T is the sum of all numbers in list L". For example, the atom sum([2, 5, 4], 11) is true. This predicate can be easily defined in *Prolog*.

The query

$$(\text{allSalaries(L)}, \mathcal{P})$$

requests a list of all salaries paid to the employees. The answer is the set of substitutions

$$\{\{[\, 100, 150, 100, 150\,]/L\}\}\,.$$

The query

$$(\text{total(T)}, \mathcal{P})$$

requests the total amount spent in salaries. The answer is the set of substitutions $\{\{500/T\}\}$. □

An empty list (i.e. a list with no elements) is represented as "[]". Moreover, the atom sum([], 0) is always true.

The main difference between

bagof(Template, Goal, List)   and
findall(Template, Goal, List)

is that the former may produce as answer a set with several substitutions, if there are variables that occur in Goal but do not appear in Template, while the latter produces only singleton answers. A query

$$(\text{bagof(Template, Goal, List)}, \mathcal{P})$$

produces a substitution for each possible instantiation of the variables of Goal that do not appear in Template.

The Compilation. For each indiscernibility class t of a rough relation R, we need now to compute the support measures, i.e. $\mathsf{supp}(r(t))$ and $\mathsf{supp}(\neg r(t))$. This point is complicated by the following. By applying one rough clause of the program, we may conclude that a certain number $k_1 > 0$ of individuals belong to $[t]$ and that they are positive examples. It may also be the case that by applying another rough clause of the same program, we conclude that other $k_2 > 0$ individuals belong to the same indiscernibility class and that they are also positive examples. Hence to compute the number of individuals belonging to $[t]$ that are positive (negative) examples of a rough relation, we may need to consider different rough clauses. To this end, the special atoms introduced in the previous section are used in the transformed program.

We start by giving an overview of the compilation procedure. Figure 2 shows the different functions that are called during compilation of a rough program in *RKBS*.

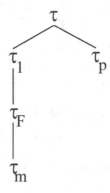

Fig. 2. Compilation procedure of *RKBS*

Function τ compiles a rough program $\mathcal{P}$ to a program $\tau(\mathcal{P})$ corresponding to a standard *Prolog* program. It calls two other functions, τ_1 and τ_p. Function τ_1 compiles each rough clause to a set of clauses and integrity constraints. τ_p generates a set of clauses for each predicate symbol q occurring in $\mathcal{P}$ that gather support information for each indiscernibility class of rough relation Q. The compilation $\tau_1(C)$ of a rough clause $C \in \mathcal{P}$ originates a call to function τ_F to compile the body of C. If the body of C contains quantitative measure expressions then τ_F calls function τ_m to compile each quantitative measure. In the reminder of this section, we discuss in detail each of the functions mentioned above.

Without loss of generality, we assume that each rough relation defined in a rough program $\mathcal{P}$ has a different name. Thus, no two equal predicate symbols with different arity may occur in $\mathcal{P}$.

To simplify the presentation of the compilation procedure, we use some notation shortcuts.

- The compiled programs may contain clauses with explicit disjunction, represented as ';', in the body of a clause. These clauses can easily be re-written as clauses without disjunction in the body. For example, the clause

$$p(X)\colon\!- q(a, X), r(X) \ ; \ q(b, X).$$

can be replace by the following two clauses

$$p(X)\colon\!- q(a, X), r(X).$$
$$p(X)\colon\!- q(b, X).$$

- If we are not interested in the constant value with which a variable X gets instantiated then, we use '$-$' instead of X. For instance, if the value with which the first argument of q is instantiated is irrelevant then we write $q(-, Y)$.
- We may write $q(\overline{X})$ instead of $q(X_1, \ldots, X_n)$, for a predicate q/n.
- Given a pair $u = \langle c_1, c_2 \rangle$, we write $u.1st$ to denote c_1 and $u.2nd$ to denote c_2.

It should be stressed that the compiled program $\tau(\mathcal{P})$ may contain integrity constraints, given a rough program $\mathcal{P}$. Integrity constraints are not allowed in standard *Prolog* programs. However, each logic program $\tau(\mathcal{P})$ can be easily transformed into a standard *Prolog* program. Assume that no predicate symbol named `false` occurs in $\tau(\mathcal{P})$. Each integrity constraint `:- B.` $\in \tau(\mathcal{P})$ can be replaced by a clause `false :- B.`. If the atom `false` belongs to the least model of the *Prolog* program obtained this way then we can conclude that some integrity constraint cannot be satisfied.

Compiling a rough program $\mathcal{P}$ implies compilation of each rough clause and rough fact. For each predicate symbol p occurring in $\mathcal{P}$, p^*, p^π, $\neg p$, $\neg p^*$, and $\neg p^\pi$ should be seen as new predicate symbols not occurring in $\mathcal{P}$. Note that $\neg p$, $\neg p^*$, and $\neg p^\pi$ represent explicit negation.

Each predicate p in the compiled program has an extra argument that carries information about the support. For example, both atoms $p(c_1, c_2, k)$ and $p^*(c_1, c_2, k)$ ($\neg p(c_1, c_2, k)$ and $\neg p^*(c_1, c_2, k,)$) indicate that the indiscernibility class $[\langle c_1, c_2 \rangle]$ belongs to $\overline{P}$ ($\neg \overline{P}$). However, the former states that the **supp** $(p(c_1, c_2))$ (**supp**$(\neg p(c_1, c_2))$) is exactly k, while the latter says that **supp**$(p(c_1, c_2))$ (**supp** $(\neg p(c_1, c_2))$) is at least k. An atom $p^\pi(c_1, c_2)$ ($\neg p^\pi(c_1, c_2)$) indicates that the indiscernibility class $[\langle c_1, c_2 \rangle]$ belongs to the upper approximation of P ($\neg P$) but it does not keep any information about the support.

In the reminder of this section we assume that l ($l_1, l_2, \cdots$) is either p or $\neg p$ and that $\neg\neg p$ is equivalent to p, for some predicate symbol p.

We introduce first a function τ_m to compile quantitative measures in the body of a rough clause. For example, the compilation of a quantitative measure such as $\tau_m(\text{supp}(q(t)))$ returns a pair u, where

- $u.1st$ is the body of a clause, and
- $u.2nd$ is a variable that will be instantiated with the support of atom $q(t)$.

If t describes an indiscernibility class of rough relation Q then literal $q(t, K)$ expresses that the $\text{supp}(q(t)) = K$. The default negated literal `not` $q^\pi(t)$ is true, if t describes an indiscernibility class that does not belong to $\overline{Q}$. The conjunction of atoms

$$\text{findall}(K_1, q(\overline{X}, K_1), L_1), \text{sum}(L_1, K')$$

imposes that $K' = |q|$, i.e. K' represents the number of individuals in the universe of rough relation Q that have positive outcome for the decision attribute. Hence, the following conjunction of atoms expresses that K is the total number of individuals in universe of Q.

$$\texttt{findall}(K_1, q(\overline{X}, K_1), L_1),$$
$$\texttt{findall}(K_2, \neg q(\overline{X}, K_2), L_2),$$
$$\texttt{sum}(L_1, K_1'), \texttt{sum}(L_2, K_2'), K = K_1' + K_2'$$

Function τ_m that compiles quantitative measures is formally presented below.

$$\tau_m(\texttt{supp}(l(t))) = \langle \texttt{code}, K \rangle$$
$$\texttt{code} = l(t, K) \ ;$$
$$\texttt{not} \ l^\pi(\bar{t}), K = 0$$

$$\tau_m(\texttt{strength}(l(\bar{t}))) = \langle \texttt{code}, K \rangle$$
$$\texttt{code} = l(t, K_0),$$
$$\texttt{findall}(K_1, l(\overline{X}, K_1), L_1),$$
$$\texttt{findall}(K_2, \neg l(\overline{X}, K_2), L_2),$$
$$\texttt{sum}(L_1, K_1'), \texttt{sum}(L_2, K_2'), K = \frac{K_0}{K_1' + K_2'} \ ;$$
$$\texttt{not} \ l^\pi(t), K = 0$$

$$\tau_m(\texttt{acc}(l(t))) = \langle \texttt{code}, K \rangle$$
$$\texttt{code} = l(t, K_1), \neg l(t, K_2), K = \frac{K_1}{K_1 + K_2} \ ;$$
$$l(t, K_1), \texttt{not} \ \neg l^\pi(t), K = 1 \ ;$$
$$\texttt{not} \ l^\pi(t), K = 0$$

$$\tau_m(\texttt{cov}(l(\bar{t}))) = \langle \texttt{code}, K \rangle$$
$$\texttt{code} = l(t, K_1)$$
$$\texttt{findall}(K_2, l(\overline{X}, K_2), L), \texttt{sum}(L, K_3), K = \frac{K_1}{K_3} \ ;$$
$$\texttt{not} \ l^\pi(t), K = 0 \ .$$

Let us informally describe the first two cases above.

- The compilation of $\texttt{supp}(l(t))$ considers two possible cases. If atom $l(t, K)$ is true then $\texttt{supp}(l(t)) = K$. If t does not describe an indiscernibility class in the upper approximation, i.e. $\texttt{not} \ l^\pi(t)$ is true, then $\texttt{supp}(l(t)) = K = 0$.
- The compilation of $\texttt{strength}(l(t))$ considers also two cases. If t describes an indiscernibility class in the upper approximation then, atom $l(t, K_0)$ is true, $\texttt{supp}(l(t)) = K_0$, and the number of individuals in the universe is given by $K_1' + K_2'$. Otherwise, the default negated literal $\texttt{not} \ l^\pi(t)$ is true and $\texttt{supp}(l(t)) = \texttt{strength}(l(t)) = 0$.

The next step is to show how to compile the body of a rough clause (with quantitative measures). Let $\mathbf{I}^+$ denote the set of positive integers including zero and F be a commutative and associative binary function such that $F : \mathbf{I}^+ \to \mathbf{I}^+$. This function indicates how the support of the atoms $p(t)$ or $\neg p(t)$, occurring in rough literals of the body, should be combined to compute the support of the atom corresponding to the head. Given the body B of a rough clause, $\tau_F(B)$ returns a pair such that $(\tau_F(B)).1st$ is the body of a clause of a logic program and $(\tau_F(B)).2nd$ represents a variable. Assume that m is either $\texttt{supp}, \texttt{strength},$

acc, or cov. E_1 and E_2 represent quantitative measure expressions. B_1 and B_2 stand for rough literals.

$$\tau_F(m(l(\bar{t}))\ relOp\ k) = \langle \text{code}\ ,\ -\rangle$$
$$\text{code} = \tau_m(m(l(\bar{t}))).1st,$$
$$\tau_m(m(l(\bar{t}))).2nd\ relOp\ k$$

$$\tau_F(m(l_1(\overline{t_1}))\ relOp\ m(l_2(\overline{t_2}))) = \langle \text{code}\ ,\ -\rangle$$
$$\text{code} = \tau_m(m(l_1(\overline{t_1}))).1st, \tau_m(m(l_2(\overline{t_2}))).1st,$$
$$\tau_m(m(l_1(\overline{t_1}))).2nd$$
$$relOp$$
$$\tau_m(m(l_2(\overline{t_2}))).2nd$$

$$\tau_F((E_1, E_2)) = \langle \text{code}\ ,\ -\rangle$$
$$\text{code} = (\tau_F(E_1)).1st, (\tau_F(E_2)).1st$$

$$\tau_F(\underline{p}(\bar{t})) = \langle \text{code}\ ,\ K\rangle$$
$$\text{code} = p(\bar{t}, K), \text{not}\ \neg p^\pi(\bar{t})$$

$$\tau_F(\neg \underline{p}(\bar{t})) = \langle \text{code}\ ,\ K\rangle$$
$$\text{code} = \neg p(\bar{t}, K), \text{not}\ p^\pi(\bar{t})$$

$$\tau_F(\overline{p}(\bar{t})) = \langle \text{code}\ ,\ K\rangle$$
$$\text{code} = p(\bar{t}, K)$$

$$\tau_F(\overline{\neg p}(\bar{t})) = \langle \text{code}\ ,\ K\rangle$$
$$\text{code} = \neg p(\bar{t}, K)$$

$$\tau_F(\underline{\overline{p}}(\bar{t})) = \langle \text{code}\ ,\ K\rangle$$
$$\text{code} = p(\bar{t}, K), \neg p^\pi(\bar{t})$$

$$\tau_F(\underline{\overline{\neg p}}(\bar{t})) = \langle \text{code}\ ,\ K\rangle$$
$$\text{code} = \neg p(\bar{t}, K), p^\pi(\bar{t})$$

$$\tau_F((B_1, B_2)) = \langle \text{code}\ ,\ K\rangle$$
$$\text{code} = (\tau_F(B_1)).1st, (\tau_F(B_2)).1st,$$
$$K = F((\tau_F(B_1)).2nd, (\tau_F(B_2)).2nd)$$

$$\tau_F((B_1, E_1)) = \langle \text{code}\ ,\ K\rangle$$
$$\text{code} = (\tau_F(B_1)).1st, \tau_m(E_1).1st,$$
$$K = (\tau_F(B_1)).2nd\ .$$

The reader can confirm that the underlying idea to compile the body of a rough clause is similar to what was discussed in section 4.

We need to define a function that compiles a rough clause into a set of clauses and integrity constraints. This function is presented below and it is based on ideas similar to the ones discussed in section 4.

$$\tau_1(\underline{p}(\bar{t}):\text{-}[\alpha, F]\ B.) = \{p^*(\bar{t}, K):\text{-}\ (\tau_F(B)).1st,$$
$$K = \lfloor \alpha \times (\tau_F(B)).2nd \rfloor.,$$
$$:\text{-}\ \neg p^\pi(\bar{t}), (\tau_F(B)).1st.\} \ ,$$

$$\tau_1(\overline{p}(\bar{t}):\text{-}[\alpha, F]\ B.) = \{p^*(\bar{t}, K):\text{-}\ (\tau_F(B)).1st,$$
$$K = \lfloor \alpha \times (\tau_F(B)).2nd \rfloor.\} \ ,$$

$$\tau_1(\neg \underline{p}(\bar{t}):\text{-}[\alpha, F]\ B.) = \{\neg p^*(\bar{t}, K):\text{-}\ (\tau_F(B)).1st,$$
$$K = \lfloor \alpha \times (\tau_F(B)).2nd \rfloor.,$$
$$:\text{-}\ p^\pi(\bar{t}), (\tau_F(B)).1st.\} \ ,$$

$$\tau_1(\overline{\neg p}(\bar{t}):\text{-}[\alpha, F]\ B.) = \{\neg p^*(\bar{t}, K):\text{-}\ (\tau_F(B)).1st,$$
$$K = \lfloor \alpha \times (\tau_F(B)).2nd \rfloor.\} \ .$$

The compilation of a rough clause might generate clauses with repeated literals. Thus, it is convenient to eliminate repeated literals from the bodies of compiled clauses.

Let $[t]$ be an indiscernibility class of a rough relation R. Application of different rough clauses may lead to the conclusion that a certain number of individuals belonging to $[t]$ are positive examples of a rough relation R ($\neg R$). Hence, it is needed to gather support information for each indiscernibility class of a rough relation. Function τ_p formalizes this idea. For each predicate symbol r occurring in the rough program, the following set of clauses is generated.

$$\tau_p(r) = \{r(\overline{X}, K):\text{-}\ \texttt{bagof}(K', r^*(\overline{X}, K'), L), \texttt{sum}(L, K).,$$
$$\neg r(\overline{X}, K):\text{-}\ \texttt{bagof}(K', \neg r^*(\overline{X}, K'), L), \texttt{sum}(L, K).,$$
$$r^\pi(\overline{X}):\text{-}\ r(\overline{X}, \_).,$$
$$\neg r^\pi(\overline{X}):\text{-}\ \neg r(\overline{X}, \_).\} \ .$$

Functions τ_1 and τ_p can also be applied to a rough program $\mathcal{P}$. Let $\beta_\mathcal{P}$ be the set of all predicate symbols occurring in $\mathcal{P}$.

$$\tau_1(\mathcal{P}) = \bigcup_{C \in \mathcal{P}} \tau_1(C) \ ,$$
$$\tau_p(\mathcal{P}) = \bigcup_{p \in \beta_\mathcal{P}} \tau_p(p).$$

Compilation of a rough program $\mathcal{P}$ is obtained by applying function τ to $\mathcal{P}$,

$$\tau(\mathcal{P}) = \tau_1(\mathcal{P}) \cup \tau_p(\mathcal{P}) \ .$$

The example below illustrates the compilation of a simple rough program in RKBS.

Example 36. Consider again the program

$$\mathcal{P} = \{\overline{p}(X_1, X_2)\ :\text{-}[1, -]\ \overline{q}(X_1, X_2), \texttt{acc}(q(X_1, X_2)) > 0.85.,$$
$$\overline{p}(X, c)\ :\text{-}[1, -]\ \overline{q_1}(X, c).,$$
$$\neg \overline{p}(X, c)\ :\text{-}[1, -]\ \overline{\neg q_1}(X, c).,$$
$$\overline{q}(a, c) : 2.,$$
$$\overline{q_1}(a, c) : 3.,\overline{\neg q_1}(a, c) : 7.\}$$

– **Compilation of the first rough clause**

$$\tau_1(\overline{p}(X_1, X_2) \ \text{:-}\, [1, -] \ \overline{q}(X_1, X_2), \text{acc}(q(X_1, X_2)) > 0.85.)$$

adds the following clauses to $\tau(\mathcal{P})$.

$$p^*(X_1, X_2, K_1)\text{:-}\ q(X_1, X_2, K_1), \neg q(X_1, X_2, K_2),$$
$$K_3 = \tfrac{K_1}{K_1 + K_2}, K_3 > 0.85.$$

$$p^*(X_1, X_2, K_1)\text{:-}\ q(X_1, X_2, K_1), not \, \neg q^\pi(X_1, X_2),$$
$$K_2 = 1, K_2 > 0.85.$$

$$p^*(X_1, X_2, K_1)\text{:-}\ q(X_1, X_2, K_1), not \, q^\pi(X_1, X_2),$$
$$K_2 = 0, K_2 > 0.85.$$

– **Compilation of the second rough clause**

$$\tau_1(\overline{p}(X, c) \ \text{:-}\, [1, -] \ \overline{q_1}(X, c).)$$

adds the following clause to $\tau(\mathcal{P})$.

$$p^*(X, c, K)\text{:-}\ q_1(X, c, K), \neg q_1^\pi(X, c).$$

– **Compilation of the third rough clause**

$$\tau_1(\overline{\neg p}(X, c) \ \text{:-}\, [1, -] \ \overline{\neg q_1}(X, c).)$$

adds the following clause to $\tau(\mathcal{P})$.

$$\neg p^*(X, c, K)\text{:-}\ \neg q_1(X, c, K), q_1^\pi(X, c).$$

– **Compilation of the rough facts** adds the following facts to $\tau(\mathcal{P})$.

$$q^*(a, c, 2).$$

$$q_1^*(a, c, 3).$$

$$\neg q_1^*(a, c, 7).$$

Finally, function τ_p is called for each predicate symbol.

$$
\begin{aligned}
\tau_p(p) \ = \ \{ &p(X_1, X_2, K)\text{:-}\ \text{bagof}(K', p^*(X_1, X_2, K'), L), \text{sum}(L, K).\,, \\
&\neg p(X_1, X_2, K)\text{:-}\ \text{bagof}(K', \neg p^*(X_1, X_2, K'), L), \text{sum}(L, K).\,, \\
&p^\pi(X_1, X_2)\text{:-}\ p(X_1, X_2, -).\,, \\
&\neg p^\pi(X_1, X_2)\text{:-}\ \neg p(X_1, X_2, -).\} \ .
\end{aligned}
$$

$$
\begin{aligned}
\tau_p(q) \ = \ \{ &q(X_1, X_2, K)\text{:-}\ \text{bagof}(K', q^*(X_1, X_2, K'), L), \text{sum}(L, K).\,, \\
&\neg q(X_1, X_2, K)\text{:-}\ \text{bagof}(K', \neg q^*(X_1, X_2, K'), L), \text{sum}(L, K).\,, \\
&q^\pi(X_1, X_2)\text{:-}\ q(X_1, X_2, -).\,, \\
&\neg q^\pi(X_1, X_2)\text{:-}\ \neg q(X_1, X_2, -).\} \ .
\end{aligned}
$$

$$
\begin{aligned}
\tau_p(q_1) \ = \ \{ &q_1(X_1, X_2, K)\text{:-}\ \text{bagof}(K', q_1^*(X_1, X_2, K'), L), \text{sum}(L, K).\,, \\
&\neg q_1 X_1, X_2, K)\text{:-}\ \text{bagof}(K', \neg q_1^*(X_1, X_2, K'), L), \text{sum}(L, K).\,, \\
&q_1^\pi(X_1, X_2)\text{:-}\ q_1(X_1, X_2, -).\,, \\
&\neg q_1^\pi(X_1, X_2)\text{:-}\ \neg q_1(X_1, X_2, -).\} \ . \hfill \square
\end{aligned}
$$

6.4   The Query Language of *RKBS*

Our system allows rough programs to be queried. Quantitative measure expressions may also appear in queries.

A rough query is a pair consisting of the query itself and a rough program from which an answer should be retrieved. We start by formally defining the query language of the system.

Definition 44. *A rough query in RKBS is a pair* $(\mathcal{Q}, \mathcal{P})$, *where* $\mathcal{P}$ *is a rough program supported by RKBS and* $\mathcal{Q}$ *is defined by the following abstract syntax rules*

$$
\begin{aligned}
RelOp &\longrightarrow \; ==|\neq|>|\geq|<|\leq \; . \\
\mathcal{Q}_1 &\longrightarrow K = M \mid K = M, \mathcal{Q}_1 \mid K = M, K \;\; RelOp \;\; k, \mathcal{Q}_1 \mid \\
&\qquad K_1 = M, K_2 = M, K_1 \;\; RelOp \;\; K_2, \mathcal{Q}_1 \, . \\
\mathcal{Q}_2 &\longrightarrow \; classify(A) \, . \\
\mathcal{Q}_3 &\longrightarrow L \mid L, \mathcal{Q}_3 \mid \mathcal{Q}_1 \, . \\
\mathcal{Q} &\longrightarrow \mathcal{Q}_3 \mid \mathcal{Q}_2 \, .
\end{aligned}
$$

where M *is a quantitative measure,* K, K_1, K_2 *are variables,* k *is a rational number,* L *is a rough literal, and* A *is an objective literal. A rough query is well-formed, if all variables occurring in a quantitative measure expression also occur in some other rough literal of the query.*

In contrast to the query language presented in section 4, it is not possible to test whether a region of a rough relation (e.g. lower approximation) is a subset of another region of some other rough relation. Thus, rough inclusion and rough equality cannot be tested, either. On the other side, the system supports queries of the form $(\texttt{classify}(d(t)) , \mathcal{P})$.

We describe informally the meaning of some (well-formed) rough queries. Consider the rough query

$$(\texttt{classify}(d(c_1, X, c_3)) , \mathcal{P}) \; .$$

Each tuple $t = \langle c_1, c_2, c_3 \rangle$ describing an indiscernibility class of a rough relation D can be seen as a decision rule. Assume that rough relation D corresponds (implicitly) to the decision table $\mathcal{D} = (U, \{a_1, a_2, a_3\}, \texttt{d})$. If $t \in \overline{D}$ then it induces the decision rule $(a_1 = c_1) \wedge (a_2 = c_2) \wedge (a_3 = c_3) \rightarrow (\texttt{d} = \texttt{yes})$. If $t \in \neg\overline{D}$ then it induces the decision rule $(a_1 = c_1) \wedge (a_2 = c_2) \wedge (a_3 = c_3) \rightarrow (\texttt{d} = \texttt{no})$. The query above requests a prediction for the decision class to which a new individual i described by $(a_1 = c_1 \wedge a_3 = c_3)$ may belong. To answer this query the strategy described in section 2.5 is followed. The answer to the rough query is either the pair $(\texttt{d} = \texttt{yes}, C_F)$, or $(\texttt{d} = \texttt{no}, C_F)$, or $(\texttt{d} = \texttt{unknown}, 0)$, where C_F is the certainty factor of the prediction. The last case corresponds to the situation where no decision rule is fired.

Consider another rough query

$$(\underline{p}(X_1, X_2), K_1 = \texttt{supp}(p(X_1, X_2)), K_2 = \texttt{supp}(\neg p(X_1, X_2)) , \mathcal{P}) \; .$$

This rough query requests the description of all indiscernibility classes in the boundary region of P with indication, for each indiscernibility class, of how many individuals of that class are positive examples and how many individuals are negative examples. Hence, the substitution

$$\{a/X_1, b/X_2, K_1/5, K_2/7\}$$

could be an answer stating that $\langle a, b \rangle \in \overline{P}, \mathsf{supp}(p(a, b)) = 5$, and $\mathsf{supp}(\neg p(a, b)) = 7$.

The rough query

$$(\overline{p}(X_1, b), K = \mathsf{acc}(p(X_1, b)), K > 0.6 \; , \; \mathcal{P})$$

requests a description of all indiscernibility classes in the upper approximation of P such that the second attribute has value b and their corresponding accuracy is larger than 0.6.

The system answers a rough query for a rough program $\mathcal{P}$ by compiling it to one or more *Prolog* queries to the *Prolog* program corresponding to $\tau(\mathcal{P})$. The compilation of rough queries is based on ideas similar to the compilation functions τ_m and τ_F, presented in section 6.3. Answers to (rough) queries are sets of substitutions, like for the query language introduced in section 4. We give an example showing how a rough query can be answered by querying the compiled program.

Example 37. Consider the rough query

$$(\overline{\underline{p}}(X_1, b), K = \mathsf{supp}(r(X_1, b)), K < 10 \; , \; \mathcal{P})$$

The following two queries are generated for the compiled program.

$$(p^\pi(X_1, b), \neg p^\pi(X_1, b), r(X_1, b, K), K < 10 \; , \; \tau(\mathcal{P})) \qquad \text{and}$$
$$(p^\pi(X_1, b), \neg p^\pi(X_1, b), \mathsf{not} \; r^\pi(X_1, b), K = 0, K < 10 \; , \; \tau(\mathcal{P})) \; .$$

The union of the sets representing the answers to the queries above is the answer to the initial rough query. □

As the example above shows, a rough query may be answered by issuing more than one query to the compiled program. The compilation of quantitative measures (see definition of function τ_m) involves more than one possible case. This explains why more than one query might be needed. Assume that variable X_1 is instantiated with a constant c. In the example above, the first query, obtained by compiling the rough query, corresponds to the case where the indiscernibility class described by $\langle c, b \rangle \in \overline{R}$, i.e. $\mathsf{supp}(r(c, b)) > 0$. The second query corresponds to the case where $\langle c, b \rangle \notin \overline{R}$, i.e. $\mathsf{supp}(r(c, b)) = 0$.

6.5   Application Examples

Variable Precision Rough Relations. We show below how quantitative measure expressions in the body of rough clauses can be used to build generalized rough approximations of a relation, in the spirit of the variable precision rough set model [13], described in section 2.6. This aspect illustrates an important application of our language, since the VPRSM is the rough set model often used in practical applications.

Example 38. Let q denote a rough relation Q (possibly obtained directly from a decision table and encoded as a set of rough facts in our language) , l and u be two precision control parameters such that $l < u$.

We define then a new rough relation Q_1 such that

- if $\mathtt{acc}(q(t)) \geq u$ then $t \in Q_1$;
- if $\mathtt{acc}(q(t)) \leq l$ then $t \in \neg Q_1$;
- If $l < \mathtt{acc}(q(t)) < u$ then $t \in \overline{Q_1}$.

$$\overline{q_1}(X_1, X_2) \ :\text{-}[1,-] \ \underline{q}(X_1, X_2).$$

$$\overline{\neg q_1}(X_1, X_2) \ :\text{-}[1,-] \ \neg q(X_1, X_2).$$

%Any indiscernibility class t in the boundary s.t.
% $acc(q(t)) \geq u$ is considered to be in Q
$$\underline{q_1}(X_1, X_2) \ :\text{-}[1,\mathtt{sum}] \ \overline{q}(X_1, X_2), \overline{\neg q}(X_1, X_2), \mathtt{acc}(q(X_1, X_2)) \geq u.$$

% Any indiscernibility class t in the boundary s.t.
% $acc(q(t)) \leq l$ is considered to be in $\neg Q$
$$\underline{\neg q_1}(X_1, X_2) \ :\text{-}[1,\mathtt{sum}] \ \overline{q}(X_1, X_2), \overline{\neg q}(X_1, X_2), \mathtt{acc}(q(X_1, X_2)) \leq l.$$

% Any indiscernibility class t in the boundary s.t.
% $l < acc(q(t)) < u$ remains in the boundary
$$\overline{q_1}(X_1, X_2) \ :\text{-}[1,-] \ \overline{q}(X_1, X_2), \mathtt{acc}(q(X_1, X_2)) > l, \mathtt{acc}(q(X_1, X_2)) < u.$$

$$\overline{\neg q_1}(X_1, X_2) \ :\text{-}[1,-] \ \overline{\neg q}(X_1, X_2), \mathtt{acc}(q(X_1, X_2)) > l, \mathtt{acc}(q(X_1, X_2)) < u.$$

Note that the use of $\overline{q}(X_1, X_2), \overline{\neg q}(X_1, X_2)$ in the body of the third and fourth rough clauses captures those indiscernibility classes $[t]$ in the boundary of Q. Moreover, it is worth to note the use of function $\mathtt{sum}$ to combine $\mathtt{supp}(q(t))$ with $\mathtt{supp}(\neg q(t))$, since $\mathtt{supp}(q(t)) + \mathtt{supp}(\neg q(t))$ gives the total number of individuals in the indiscernibility class $[t]$.   □

The rough program above shows that our framework caters for extending the VPRSM to implicitly defined rough relations.

Avoiding Expensive Tests Revisited. In section 5.2, we describe a possible technique to identify those individuals for who expensive tests, corresponding to some of the condition attributes, can (cannot) be avoided. We show now the same problem formulated in the language supported by our system. In addition, we illustrate how quantitative measure expressions can be used to retrieve relevant information.

Example 39. Consider the decision table $\mathcal{D}eathmi = (U, \{\mathtt{Age}, \mathtt{A_1}, \mathtt{A_2}\}, \mathtt{Deathmi})$, where U is a set of patients with heart problems. This decision table is encoded as a set of rough facts shown below. Assume that the conditional attributes A_1 and A_2 represent two medical tests. Moreover, test A_2 is usually considered as being expensive, and therefore, desirable to avoid.

```
¬deathmi(>70, b1, c1): 2.
¬deathmi(>70, b1, c2): 2.
¬deathmi(>40 <70, b2, c4): 2.
¬deathmi(<40, b3, c5): 2.

deathmi(>70, b1, c2): 3.
deathmi(>40 <70, b2, c3): 4.
deathmi(<40, b3, c5): 3.
deathmi(<40, b4, c3): 8.
```

The following rough clauses monitor the impact in the boundary region of not considering test A_2.

(1) $\overline{d}$(Age, A1) :-[1,_] $\overline{deathmi}$(Age, A1, A2).

(2) $\overline{\neg d}$(Age, A1) :-[1,_] $\overline{\neg deathmi}$(Age, A1, A2).

(3) $\overline{migrate}$(Age, A1) :-[1,min] $\underline{d}$(Age, A1), $\underline{deathmi}$(Age, A1, A2).

(4) $\overline{migrate}$(Age, A1) :- [1,min] $\underline{d}$(Age, A1), $\neg deathmi$(Age, A1, A2).

(5) $\overline{\neg migrate}$(Age, A1) :-[1,_] $\underline{\neg d}$(Age, A1).

(6) $\overline{\neg migrate}$(Age, A1) :-[1,_] $\underline{d}$(Age, A1).

(7) $\overline{\neg migrate}$(Age, A1) :-[1,sum] $\overline{deathmi}$(Age, A1, A2), $\overline{\neg deathmi}$(Age, A1, A2).

Predicate `migrate` denotes the rough relation

$$Migrate = (\{\langle >70, b1\rangle : 2, \langle >40 <70, b2\rangle : 6\},$$
$$\{\langle >70, b1\rangle : 5, \langle <40, b3\rangle : 5, \langle <40, b4\rangle : 8\}) .$$

We show some useful queries and their answers. Assume that rough program $\mathcal{P}$ contains all rough facts above and the rough clauses $(1) - (7)$.

- *"For which patients it may be useful to request the expensive test A_2? And what is the expected gain if only those patients undergo test A_2?"*

This request can be formulated by the rough query $\mathcal{Q}_1$

$$\mathcal{Q}_1 = (\overline{migrate}(Age, A1),$$
$$K_1 = \text{strength}(\overline{migrate}(Age, A1)),$$
$$K_2 = \text{strength}(\overline{\neg migrate}(Age, A1)) , \mathcal{P}) .$$

The answer to this rough query is the set

$$\{\{>70/Age, b1/A1, 0.0769/K_1, 0.1923/K_2\} ,$$
$$\{>40 <70/Age, b2/A1, 0.2308/K_1, 0/K_2\}\} .$$

This answer indicates that

* ⋆ for patients who are more than 70 years old and have got result **b1** in the test **A1**, or
* ⋆ for patients who are between 40 and 70 years and have got result **b2** in the test **A1**,

it may be advisable to perform additionally test **A2**. Moreover, if only the patients suggested by this answer undergo the expensive test then, we may expect to avoid the test for about 50% of the patients. Notice that if

$$\sum_{t \in \overline{Migrate}} (\texttt{strength}(\texttt{migrate}(t)) + \texttt{strength}(\neg\texttt{migrate}(t)))$$

would get too close to one then, this would indicate that not that much would be gained by not requesting test **A2** for all patients.

– *"Make a prediction of whether individuals with result* **b1** *for test* A_1 *need to be submitted to test* A_2*."*

This query could be formulated as follows

$$\mathcal{Q}_2 = (\texttt{classify}(\texttt{migrate}(Age, b1)) , \mathcal{P}) .$$

The answer is
$$(\texttt{migrate} = \texttt{no} , 0.7193)$$

Thus, the prediction of the system is that the expensive test is not needed and the confidence factor on this prediction is 0.7 .   □

The example discussed in this section has been tested in *RKBS*. The system has also been tested with the same problem but more realistic data was used: the `deathmi` table contained 418 patients and 12 condition attributes. Figure 3 shows the interface of the *RKBS*. On the top part there is a text area where the user can enter the rough clauses and rough facts of a rough program. In this case, the text area displays some of the rough clauses belonging to the rough program presented in the example above. It is also possible to load a rough program from a local file or from a Web page.

Figure 4 displays the first rough query of the example above. On the bottom part, it is shown the answer to the rough query in table format. Rough queries can be directly entered in the text area or can be constructed with the help of several menus. The second rough query and its answer is shown in figure 5.

7   Conclusions and Future Work

This thesis has introduced a language to define intensionally rough relations. An extension of the language supporting basic quantitative measures has also been considered.

This section is intended to summarize the work presented in this thesis (section 7.1) and point out possible directions for future research (section 7.2).

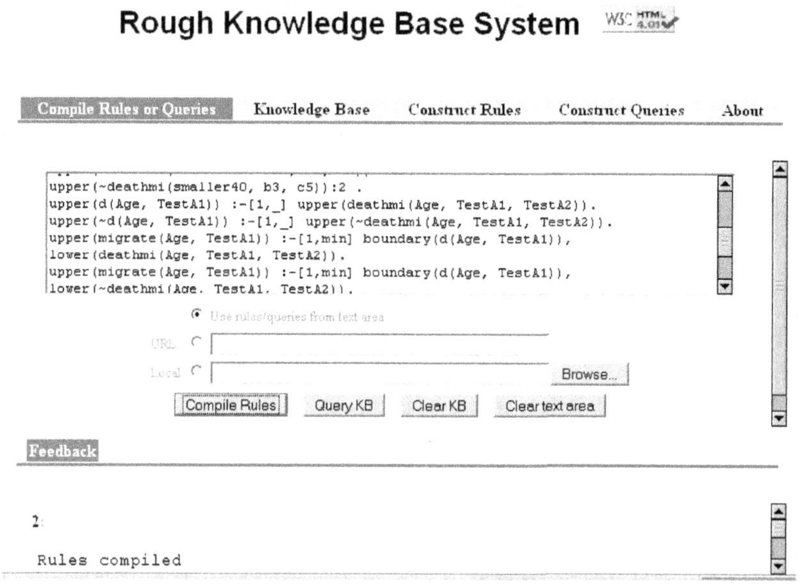

Rough Knowledge Base System

| Compile Rules or Queries | Knowledge Base | Construct Rules | Construct Queries | About |

```
upper(~deathmi(smaller40, b3, c5)):2 .
upper(d(Age, TestA1)) :-[1,_] upper(deathmi(Age, TestA1, TestA2)).
upper(~d(Age, TestA1)) :-[1,_] upper(~deathmi(Age, TestA1, TestA2)).
upper(migrate(Age, TestA1)) :-[1,min] boundary(d(Age, TestA1)),
lower(deathmi(Age, TestA1, TestA2)).
upper(migrate(Age, TestA1)) :-[1,min] boundary(d(Age, TestA1)),
lower(~deathmi(Age, TestA1, TestA2)).
```

    Use rules/queries from text area

    URL

    Local   Browse...

    Compile Rules   Query KB   Clear KB   Clear text area

Feedback

2:

Rules compiled

Fig. 3. The Rough Knowledge Base System

7.1   Concluding Remarks

Rough sets framework has two appealing aspects. First, it is a mathematical approach to deal with vague concepts. Second, rough set techniques can be used in data analysis to find patterns hidden in the data. The number of applications of rough sets to practical problems in different fields demonstrates the increasing interest in this framework and its applicability.

The first point above suggests that rough sets techniques can be generalized to knowledge bases. It is not uncommon that our knowledge about a concept bears contradictory information. For instance, one expert may state that a vehicle driven at medium speed in a wet road corresponds to a dangerous situation while another expert may consider that medium speed does not imply a dangerous traffic situation in general. Thus, the concept "dangerous traffic situation" cannot be defined precisely using the available knowledge. There will always be situations that are considered non-dangerous according the knowledge provided by one expert while the same situation is classified as dangerous when considering the criteria provided by the other expert. Rough set based techniques can be used to represent and deal with contradictory knowledge, as in the situation depicted previously. To this end we have proposed a language that caters for implicit definitions of rough sets.

Different regions (e.g. lower approximation, upper approximation or boundary region) of several rough sets can be combined to define a region of another rough set. In this way, existing rough sets can be used in the definition of other rough sets. This is achieved in the proposed language through the use of rough

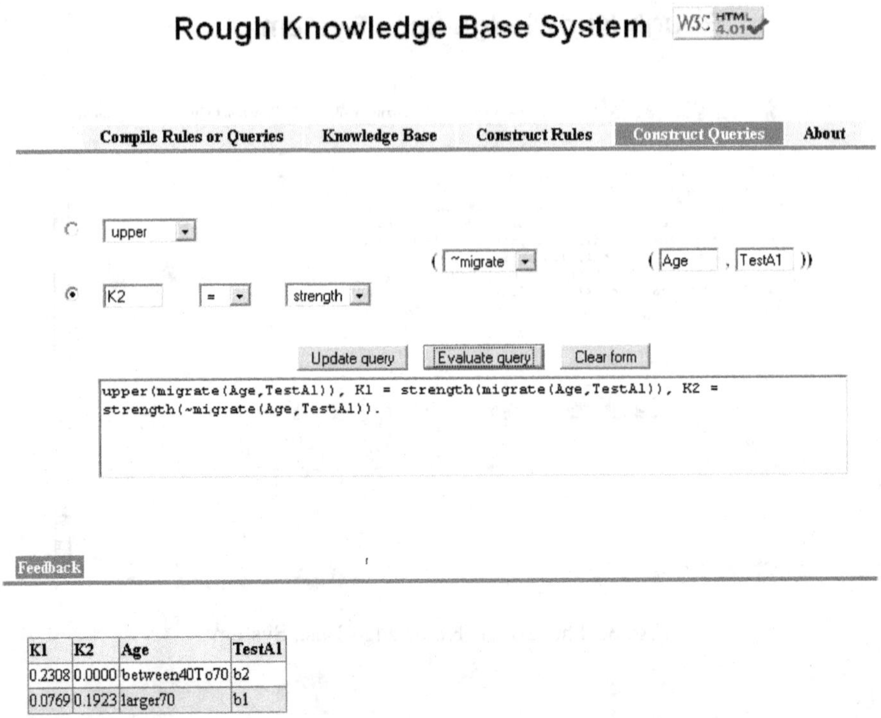

Rough Knowledge Base System   W3C HTML 4.01

| Compile Rules or Queries | Knowledge Base | Construct Rules | Construct Queries | About |

○ upper ▾

(~migrate ▾   (Age   , TestA1))

● K2   = ▾   strength ▾

Update query   Evaluate query   Clear form

```
upper(migrate(Age,TestA1)), K1 = strength(migrate(Age,TestA1)), K2 =
strength(~migrate(Age,TestA1)).
```

Feedback

K1	K2	Age	TestA1
0.2308	0.0000	between40To70	b2
0.0769	0.1923	larger70	b1

Fig. 4. The *RKBS* showing query Q_1 and its answer

clauses and rough facts, forming together a rough program. The main strengths of our language can be summarized as follows.

- The language captures and integrates in a uniform way vague knowledge with two possible sources: knowledge obtained directly from experimental data and encoded as rough facts; domain or expert knowledge expressed as rough rules. This contrasts with most of current rough set techniques that only allow definition of (vague) concepts to be obtained from experimental data.
- Several useful techniques and extensions to rough sets, reported in the literature [3,51], and implemented in an "ad hoc" way can be naturally expressed in our language.

Another important aspect of the work presented in this thesis is the definition of a query language to retrieve information about the defined rough sets and patterns implicit in the data.

The computational basis for reasoning with the rough sets defined in a rough program is a program transformation. Rough programs are compiled to extended logic programs whose semantics is captured by paraconsistent stable models. Systems like *Smodels* [43] or *dlv* [45] can then be readily used to run extended logic programs. The correctness of the proposed compilation technique has been proved.

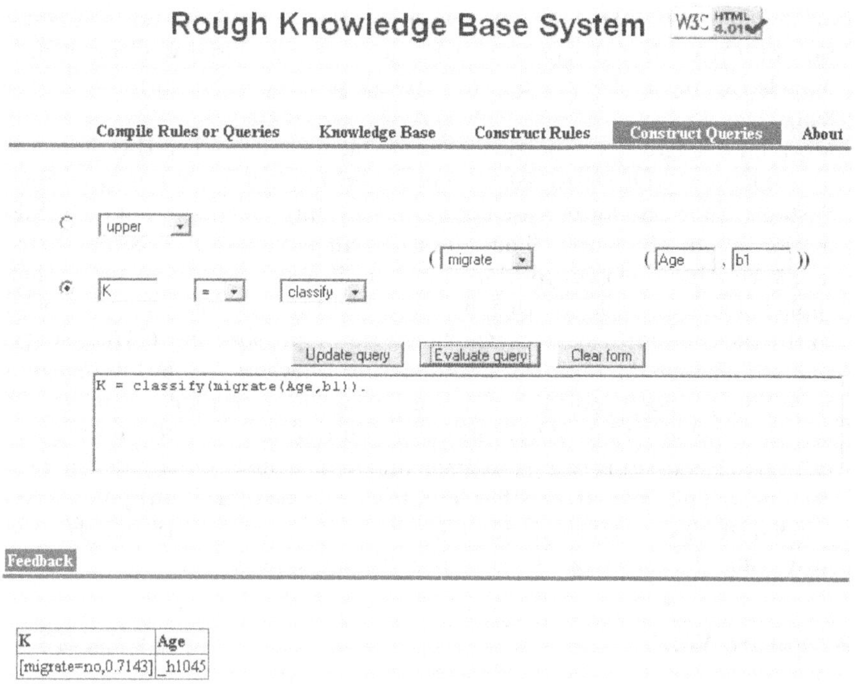

Rough Knowledge Base System   W3C HTML 4.01

Compile Rules or Queries   Knowledge Base   Construct Rules   **Construct Queries**   About

○ upper ▾

(migrate ▾)   (Age , b1))

● K   = ▾   classify ▾

Update query   Evaluate query   Clear form

K = classify(migrate(Age,b1)).

Feedback

K	Age
[migrate=no,0.7143]	h1045

Fig. 5. The *RKBS* showing query Q_2 and its answer

An extension of our language to quantitative measures has also been explored. Quantitative measures are particularly relevant in data mining applications. However, we have restricted this extension to non-recursive rough programs. We have implemented a system, called Rough Knowledge Base System, that can reason and answer queries about rough sets defined in this language. We also show that this extension allows to capture the variable precision rough set model [13] and to extend its application to implicitly defined rough sets.

To our knowledge, besides our work, only system *CAKE* [16] addresses the problem of defining implicitly rough sets. We present below a brief comparison of *CAKE* with the framework presented in this thesis.

– Our language distinguishes tuples for which there is no information available from tuples for which there is contradictory evidence. The latter case corresponds to tuples in the boundary region. System *CAKE* does not support this distinction: the boundary region includes tuples about which there is no information at all and tuples about which there is contradictory information. Hence, our language is based on a 4−valued logic while *CAKE* is based on a 3−valued logic.
– In our framework quantitative measures can be easily supported. This extension seems less obvious to achieve in *CAKE*.
– Another important difference is that *CAKE* only supports a restricted type of recursively defined rough sets, corresponding to stratified programs, while

our language supports any recursive (rough) program. However, this restriction of *CAKE* has the benefit of ensuring its tractability. Computing all models capturing the meaning of a rough program is an intractable problem.

- Most knowledge representation systems incorporate either the *open-world assumption* or the *closed-world assumption* in their reasoning procedures. Systems using the former assume that they may not have complete information about the world. Thus, information not known is assumed to be unknown. Under closed-world assumption, if a system cannot prove that a tuple belongs to a relation then it is assumed that the tuple does not belong to the relation. Both systems, the language we propose and *CAKE*, support reasoning under the *open-world assumption*. However, the latter also allows to apply the closed-world assumption locally in a particular context. This feature is achieved in *CAKE* through the use of *contextually closed queries* (CCQ) [62] consisting of a query itself and a particular context for evaluating the query. This context specifies minimization (maximization) policies to be applied to selected relations and a number of integrity constraints. However, answering queries using CCQs is co-NPtime complete [62].

7.2   Future Work

This section presents possible future directions of our research. We list below several aspects that can be improved and extensions of this work.

- We have discussed in section 6 an extension of our language supporting some basic quantitative measures. However, the declarative semantics of the language was not formally defined. Moreover, the compilation technique discussed only applies to non-recursive rough programs. Thus, we plan to formalize the declarative semantics of our extended language and investigate a computational technique that supports recursive rough programs with quantitative measures.
- We plan to investigate how the query language can be enriched. For instance, we would like to provide system support for formulating and testing hypothesis.
- More efficient implementation of our language is also one of our goals in the future. We plan to develop an implementation for recursive rough programs re-using the existing expertise in stable model systems, such as Smodels [43].
- We also plan to search for other concrete problems that can be formulated in our framework. To this end, we may take a closer look to rough mereology [63,64,65] applications. Mereology is a formal theory of parts and wholes. In rough mereology, "parthood" is a rough relation.
- A particular type of domain knowledge that often has to be considered when classifying objects of a universe is an ontology of decision classes. An interesting research direction is to investigate how ontologies can be represented in our framework. The problem of integrating learning algorithms based on rough set techniques with a gene ontology has been addressed in [66,6,67] to

predict gene function. Several operators to build approximations of decision classes are proposed. An open question is whether these operators could be easily encoded in our language or which extensions need to be considered to achieve that goal.

- This thesis presents foundation and implementation principles for a rule language able to support reasoning on incomplete and imprecise information. The necessity of rule languages for handling imprecise and incomplete Web data seems to be obvious. The research on this topic fits well with the objectives of the *EU FP6 Network of Excellence REWERSE* (`http://rewerse.net`) aiming at designing rule-based web reasoning languages. A topic of future research is deployment of the proposed language for web reasoning purposes.

Acknowledgements

I would like to express my sincere gratitude towards my supervisor, Jan Małuszynski, for his invaluable guidance and constant support.

At the department of Computer Science, at New University of Lisbon, my thanks goes to Carlos Viegas Damásio for his suggestions and the time taken to discuss many aspects of this work with me.

Notation Summary

Notation	Meaning
$|S|$	the number of elements in a set S
$A \setminus B$	set difference between A and B
$\mathcal{D} = (U, A, d)$	a decision table such that
	U is a set of objects, A is a set of condition attributes,
	and d is a decision attribute
R_A	the indiscernibility relation induced by
	a set of attributes A
R_A^*	the set of equivalence classes induced by R_A
$\overrightarrow{E^A}$	the tuple describing indiscernibility class E
(U, R_A)	an approximation space
$\overline{X}$	the upper approximation of rough set X
$\underline{X}$	the lower approximation of rough set X
$\underline{\overline{X}}$	the boundary region of rough set X
$cond(r) \rightarrow dec(r)$	a decision rule r
$cover(c)$	the set of objects satisfying condition c
$cover(r)$	$cover(cond(r))$
D_v	the set of objects having outcome v
	for the decision attribute
$red(r)$	a valued reduct for decision rule r

Notation	Meaning
$\kappa(A, B)$	the degree of functional dependency between the sets of attributes A and B
$red(A, B)$	a relative reduct of the set of attributes A w.r.t. $\kappa(A, B)$
Var	an alphabet of variable symbols
$Const$	an alphabet of constant symbols
$Pred$	an alphabet of predicate symbols
p/n	a predicate p with arity n
$\mathcal{I}$	(a rough) interpretation
$\mathcal{M}$	a (rough) model
X/c	a binding of a variable X to a constant c
θ	a substitution
$\neg A$	an explicit negated atom A
$not\ A$	a default negated atom A
$ground(\mathcal{P})$	a ground (rough) program
$H \colon\!\text{-}\ B.$	a (rough) clause
$\colon\!\text{-}\ B.$	an integrity constraint
$sem(\mathcal{P})$	the semantics of a (rough) program
$(\mathcal{Q}, \mathcal{P})$	a (rough) query to a (rough) program $\mathcal{P}$
$\psi_{\mathcal{I}}(\mathcal{P})$	the reduct of the ELP $\mathcal{P}$
$\Psi_{\mathcal{I}}(\mathcal{P})$	the ground rough program, obtained from $\mathcal{P}$, without lower approximations or testing literals in the body of its rough clauses
$\mathcal{I} \models l$	(rough) literal l is true in $\mathcal{I}$
$\mathcal{P} \models l$	(rough) program $\mathcal{P}$ implies (rough) literal l
$Q^{\mathcal{I}}$	the rough set denoted by predicate q in interpretation $\mathcal{I}$
$\langle t_1, \ldots, t_n \rangle$	a tuple whose attribute values are $t_1, \ldots, t_n$
t	a tuple
$[t]$	the indiscernibility class described by tuple t
$\langle t_1, \ldots, t_n \rangle : k$	k objects belonging to the indiscernibility class $[\langle t_1, \ldots, t_n \rangle]$ are positive examples
$\langle t_1, \ldots, t_n \rangle : k_1 : k_2$	k_1 (k_2) objects belonging to the indiscernibility class $[\langle t_1, \ldots, t_n \rangle]$ are positive (negative) examples
$q(\overline{X})$	$q(X_1, \ldots, X_1)$
$q(-, \overline{X})$	the first argument of predicate q is irrelevant
$u.1st$	if $u = \langle c_1, c_2 \rangle$ then $u.1st = c_1$
$u.2nd$	if $u = \langle c_1, c_2 \rangle$ then $u.2nd = c_2$
τ_1	the compilation function for rough clauses
τ_2	the compilation function for bodies of rough clauses belonging to rough programs without quantitative measures
τ_F	the compilation function for bodies of rough clauses in $RKBS$
τ_m	the compilation function for quantitative measures
τ_p	the compilation function for predicate symbols occurring in a rough program in $RKBS$
F	a commutative and associative binary function

References

1. Pawlak, Z.: Rough sets. International Journal of Information and Computer Science **11** (1982) 341–356
2. Nguyen, H.S., Nguyen, T.T., Skowron, A., Synak, P.: Knowledge discovery by rough set methods. In Callaos, N.C., ed.: Proc. of the International Conference on Information Systems Analysis and Synthesis (ISAS'96). (1996) 526–33
3. Komorowski, J., Øhrn, A.: Modelling prognostic power of cardiac tests using rough sets. Journal of Artificial Intelligence in Medicine **15** (1999) 167–191
4. Lazareck, L., Ramanna, S.: Classification of swallowing sound signals: A rough set approach. In Tsumoto, S., Słowiński, R., Grzymala-Busse, J.W., eds.: Proc. of the Fourth International Conference on Rough Sets and Current Trends in Computing (RSCTC'04). Number 3066 in LNAI, Springer-Verlag (2004) 679–684
5. Tay, F.E.H., Shen, L.: Economic and finantial prediction using rough sets model. European Journal of Operational Research **141** (2002) 641–659
6. Midelfart, H., Komorowski, J.: A rough set framework for learning in a directed acyclic graph. In Alpigini, J.J., Peters, J.F., Skowronek, J., Zhong, N., eds.: Proc. of the Third International Conference on Rough Sets and Current Trends in Computing. Volume 2475 of LNCS., Springer-Verlag (2002) 144–155
7. Pagliani, P.: Rough sets theory and logic-algebraic structures. In Orlowska, E., ed.: Incomplete Information: Rough Sets Analysis, Verlag-Physics (1997) 109–190
8. Yao, Y.Y., Lin, T.Y.: Generalizations of rough sets using modal logic. Journal of Intelligent Automation and Soft Computing **2** (1996) 103–120
9. Midelfart, H., Komorowski, J.: A rough set approach to inductive logic programming. In Ziarko, W., Yao, Y.Y., eds.: Proc. of the Second International Conference on Rough Sets and Current Trends in Computing (RSCTC'00). Volume 2005 of LNCS., Springer-Verlag (2000)
10. Wygralak, M.: Rough sets and fuzzy sets - some remarks on interrelations. Journal of Fuzzy Sets and Systems **29** (1989) 241–243
11. Dubois, D., Prade, H.: Putting rough sets and fuzzy sets together. In Slowinski, R., ed.: Handbook of Applications and Advances of the Rough Sets Theory, Kluwer Academic Publishers (1992) 204–232
12. Skowron, A., Stepaniuk, J.: Tolerance approximation spaces. Fundamenta Informaticae **27** (1996) 245–253
13. Ziarko, W.: Variable precision rough set model. Journal of Computer and Systems Science **46** (1993) 39–59
14. Øhrn, A., Komorowski, J.: ROSETTA: A rough set toolkit for analysis of data. In: Proc. of Third International Joint Conference on Information Sciences, Fifth International Workshop on Rough Sets and Soft Computing (RSSC'97). Volume 3., Durham, NC, USA (1997) 403–407
15. Ziarko, W., Fei, X.: VPRSM approach to WEB searching. In Alpigini, J., Peters, J., Skowron, A., Zhong, N., eds.: Proc. of the Third International Conference on Rough Sets and Current Trends in Computing, RSCTC'02. Number 2475 in LNAI, Springer-Verlag (2002) 514–521
16. Doherty, P., Łukaszewicz, W., Szałas, A.: CAKE: A Computer Aided Knowledge Engineering Technique. In van Harmelen, F., ed.: Proc. of the 15th European Conference on Artificial Intelligence, (ECAI'02), Amsterdam, IOS Press (2002) 220–224
17. Małuszyński, J., Vitória, A.: Defining rough sets by extended logic programs. In: On-Line Proc. of Paraconsistent Computational Logic Workshop (PCL'02). (2002) http://floc02.diku.dk/PCL/ and http://arxiv.org/list/cs.lo/0207#cs.lo/0207089.

18. Vitória, A., Małuszyński, J.: A logic programming framework for rough sets. In Alpigini, J., Peters, J., Skowron, A., Zhong, N., eds.: Proc. of the Third International Conference on Rough Sets and Current Trends in Computing (RSCTC'02). Number 2475 in LNAI, Springer-Verlag (2002) 205–212

19. Vitória, A., Damásio, C.V., Małuszyński, J.: Query answering for rough knowledge bases. In Wang, G., Liu, Q., Yao, Y., Skowron, A., eds.: Proc. of the Nineth International Conference on Rough Sets, Fuzzy Sets, Data Mining, and Granular Computing (RSFDGrC'03). Volume 2639 of LNAI., Springer-Verlag (2003) 197–204

20. Damásio, C.V., Pereira, L.M.: A survey of paraconsistent semantics for logic programs. In Gabbay, D.M., Smets, P., eds.: Handbook of Defeasible Reasoning and Uncertainty Management Systems. Volume 2., Kluwer Academic Publishers (1998) 241–320

21. Vitória, A., Damásio, C.V., Małuszyński, J.: From rough sets to rough knowledge bases. Fundamenta Informaticae **57** (2003) 215–246

22. Vitória, A., Damásio, C.V., Małuszyński, J.: Toward rough knowledge bases with quantitative measures. In Tsumoto, S., Słowiński, R., Komorowski, J., Grzymala-Busse, J.W., eds.: Proc. of the Fourth International Conference on Rough Sets and Current Trends in Computing (RSCTC'04). Volume 3066 of LNAI., Springer-Verlag (2004) 153–158

23. Andersson, R.:   Rough Knowledge Base System.   Available at http://www.ida.liu.se/rkbs (2004)

24. Pawlak, Z.: Rough sets. Theoretical Aspects of Reasoning about Data. Kluwer Academic Publishers, Dordrecht (1991)

25. Komorowski, J., Pawlak, Z., Polkowski, L., Skowron, A.: Rough sets: A tutorial. In Pal, S.K., Skowron, A., eds.: Rough Fuzzy Hybridization. A New Trend in Decision-Making, Springer-Verlag (1999) 3–98

26. Ziarko, W.: Rough set approaches for discovery of rules and attribute dependencies. In Kloesgen, W., Zytkow, J., eds.: Handbook of Data Mining and Knowledge Discovery, Oxford University Press (2002) 328–338

27. Slowinski, R., Vanderpooten, D.: Similarity relations as a basis for rough set approximations. In Wang, P.P., ed.: Advances in Machine Intelligence and Soft Computing. Volume 4. (1997) 17–33

28. Skowron, A., Polkowski, L.: Synthesis of decison systems from data tables. In Lin, T.Y., Cercone, N., eds.: Rough Sets and Data Mining Analysis of Imprecise Data, Kluwer Academic Publishers (1997) 259–300

29. Bazan, J.G.: Discovery of decision rules by matching new objects against data tables. In Polkowski, L., Skowron, A., eds.: Proc. of the First International Conference on Rough Sets and Current Trends in Computing (RSCTC'98). Volume 1424 of LNCS., Springer-Verlag (1998) 521–528

30. Stefanowski, J.: On rough set approaches to induction of decision rules. In Polkowski, L., Skowron, A., eds.: Rough Sets in Knowledge Discovery 1: Methodology and Applications, Springer-Verlag (1998) 501–529

31. Bazan, J.G.: Dynamic reducts and statistical inference. In: Proc. of the Sixth International Conference, Information Processing and Management of Uncertainty in Knowledge-Based Systems (IPMIU'96). Volume 3. (1996) 1147–1152

32. Lloyd, J.W.: Foundations of Logic Programming. Springer-Verlag (1987)

33. Nilsson, U., Małuszynski, J.: Logic, Programming and Prolog, 2nd edition. John Wiley & Sons, http://www.ida.liu.se/~ulfni/lpp/copyright.html (1995)

34. Baral, C.: Knowledge Representation, Reasoning and Declarative Problem Solving. Cambridge University Press (2003)

35. Brachman, R.J., Levesque, H.J.: Knowledge Representation and Reasoning. Elsevier (2004)
36. Pearce, D.: Answer sets and constructive logic, II: Extended logic programs and related non-monotonic formalisms. In Pereira, L.M., Nerode, A., eds.: Logic Programming and Nonmonotonic Reasoning - Proc. of the 2nd International Workshop, MIT Press (1993) 457–475
37. Sakama, C., Inoue, K.: Paraconsistent Stable Semantics for Extended Disjunctive Programs. Journal of Logic and Computation **5** (1995) 265–285
38. Gelfond, M., Lifschitz, V.: Logic programs with classical negation. In Warren, Szeredi, eds.: Proc. of the Seventh International Conference on Logic Programming, MIT Press (1990) 579–597
39. Gelfond, M., Lifschitz, V.: The stable model semantics for logic programming. In Kowalski, R.A., Bowen, K., eds.: Proc. of the Fifth International Logic Programming Conference and Symposium, Seattle, USA, MIT Press (1988) 1070–1080
40. Inoue, K., Sakama, C.: Negation as failure in the head. Journal of Logic Programming **35** (1998) 39–78
41. Alferes, J.J., Leite, J.A., Pereira, L.M., Przymusinska, H., Przymusiski, T.C.: Dynamic updates of non-monotonic knowledge bases. In: Journal of Logic Programming. Number 45, Elsevier (2000) 43–70
42. Niemelä, I., Simons, P.: Efficient implementation of the well-founded and stable model semantics. In Maher, M., ed.: Proc. of the Joint International Conference and Symposium on Logic Programming, Bonn, Germany, MIT Press (1996) 289–303
43. Simons, P.: Smodels system.
    (Available at http://www.tcs.hut.fi/Software/smodels/)
44. Eiter, T., Leone, N., Mateis, C., Pfeifer, G., Scarcello, F.: The KR system `dlv`: Progress report, comparisons and benchmarks. In Cohn, A.G., Schubert, L., Shapiro, S.C., eds.: KR'98: Principles of Knowledge Representation and Reasoning, San Francisco, California, Morgan Kaufmann (1998) 406–417
45. : The dlv project. (Available at http://www.dbai.tuwien.ac.at/proj/dlv/)
46. Deransart, P., Ed-Bali, A., Cervoni, L.: Prolog: The Standard Reference Manual. Springer-Verlag (1996)
47. Belnap, N.D.: A useful four-valued logic. In Epstein, G., Dunn, J.M., eds.: Modern Uses of Multiple-Valued Logic, Reidel Publishing Company (1977) 7–37
48. Belnap, N.D.: How computer should think. In Rydle, G., ed.: Contemporary Aspects of Philosophy, Oriel Press (1977) 30–56
49. Niemelä, I., Simons, P.: Smodels - an implementation of stable model and the well-founded semantics for normal logic programs. In Dix, J., Furbach, U., Nerode, A., eds.: Proc. of the Fourth International Conference on Logic Programming and Nonmonotonic Reasoning (LPNMR'97). Volume 1265 of LNAI., Springer-Verlag (1997) 420–429
50. Dantsin, E., Eiter, T., Gottlob, G., Voronkov, A.: Complexity and expressive power of logic programming. ACM Computing Surveys **33** (2001) 374–425
51. Ziarko, W.: Acquisition of Hierarchy-Structured Probabilistic Decision Tables and Rules from Data. In: Proc. of the World Congress on Computational Intelligence, Honolulu (2002)
52. Reiter, R.: A logic for deafult reasoning. Journal of Artificial Intelligence **13** (1980) 81–132
53. Andersson, R.: Implementation of a rough knowledge base system supporting quantitative measures. Master thesis, Linköping University, IDA (2004)

54. Zaniolo, C.: Key constraints and monotonic aggregates in deductive databases. In Kakas, A.C., Sadri, F., eds.: Computational Logic: Logic Programming and Beyond, Essays in Honour of Robert Kowalski. Volume 2408 of LNCS., Springer-Verlag (2002) 109–134

55. Faber, W., Leone, N., Pfeifer, G.: Recursive aggregates in disjunctive logic programs: Semantics and complexity. In Alferes, J.J., Leite, J.A., eds.: Proc. of the Nineth European Conference on Logics in Artificial Intelligence (JELIA'04). Volume 3229 of LNCS., Springer-Verlag (2004) 200–212

56. : Java. Sun Mycrosystems. (Available at http://java.sun.com/)

57. : XSB system. (Available at http://xsb.sourceforge.net/)

58. Chen, W., Swift, T., Warren, D.S.: Efficient top-down computation of queries under the well-founded semantics. Journal of Logic Programming **24** (1995) 161–199

59. Chen, W., Warren, D.S.: Tabled evaluation with delaying for general logic programs. Journal of the ACM **43** (1996) 20–74

60. Sagonas, K., Swift, T.: An abstract machine for tabled execution of fixed-order stratified logic programs. Journal of the ACM TOPLAS **20** (1998) 586–635

61. Swift, T.: Tabling for non-monotonic reasoning. Annals of Mathematics and Artifcial Intelligence **25** (1999) 201–240

62. Doherty, P., Kachniarz, J., Szałas, A.: Using contextually closed queries for local closed-world reasoning in rough knowledge databases. In Pal, S.K., Polkowski, L., Skowron, A., eds.: Rough-Neural Computing, Springer-Verlag (2004) 219–250

63. Polkowski, L., Skowron, A.: Rough mereology. In: Proc. of the Eighth International Symposium on Methodologies for Intelligent Systems, Springer-Verlag (1994) 85–94

64. Polkowski, L., Skowron, A.: Rough mereology: a new paradigm for approximate reasoning. International Journal of Approximate Reasoning **15** (1997) 333–4365

65. Polkowski, L., Skowron, A.: Rough mereological calculi of granules: A rough set approach to computation. Journal of Computational Intelligence **17** (2001) 472–492

66. Midelfart, H., Lægreid, A., Komorowski, J.: Classification of gene expression data in an ontology. In Crespo, J., Maojo, V., Martin, F., eds.: Proc. of the Second International Symposium on Medical Data Analysis (ISMDA'01). Volume 2199 of LNCS., Springer-Verlag (2001) 186–194

67. Midelfart, H.: Knowledge Discovery from cDNA Microarrays and a priori Knowledge. PhD thesis, Department of Computer and Information Science, Norwegian University of Science and Technology, NO-7491 Trondheim, Norway (2003)

Analogy-Based Reasoning in Classifier Construction

Arkadiusz Wojna

Institute of Informatics, Warsaw University,
Banacha 2, 02-097, Warsaw, Poland
wojna@mimuw.edu.pl

Abstract. Analogy-based reasoning methods in machine learning make it possible to reason about properties of objects on the basis of similarities between objects. A specific similarity based method is the k nearest neighbors (k-nn) classification algorithm. In the k-nn algorithm, a decision about a new object x is inferred on the basis of a fixed number k of the objects most similar to x in a given set of examples. The primary contribution of the dissertation is the introduction of two new classification models based on the k-nn algorithm.

The first model is a hybrid combination of the k-nn algorithm with rule induction. The proposed combination uses minimal consistent rules defined by local reducts of a set of examples. To make this combination possible the model of minimal consistent rules is generalized to a metric-dependent form. An effective polynomial algorithm implementing the classification model based on minimal consistent rules has been proposed by Bazan. We modify this algorithm in such a way that after addition of the modified algorithm to the k-nn algorithm the increase of the computation time is inconsiderable. For some tested classification problems the combined model was significantly more accurate than the classical k-nn classification algorithm.

For many real-life problems it is impossible to induce relevant global mathematical models from available sets of examples. The second model proposed in the dissertation is a method for dealing with such sets based on locally induced metrics. This method adapts the notion of similarity to the properties of a given test object. It makes it possible to select the correct decision in specific fragments of the space of objects. The method with local metrics improved significantly the classification accuracy of methods with global models in the hardest tested problems.

The important issues of quality and efficiency of the k-nn based methods are a similarity measure and the performance time in searching for the most similar objects in a given set of examples, respectively. In this dissertation both issues are studied in detail and some significant improvements are proposed for the similarity measures and for the search methods found in the literature.

Keywords: analogy-based reasoning, case-based reasoning, k nearest neighbors, similarity measure, distance based indexing, hybrid decision system, local metric induction.

J.F. Peters and A. Skowron (Eds.): Transactions on Rough Sets IV, LNCS 3700, pp. 277–374, 2005.
© Springer-Verlag Berlin Heidelberg 2005

1   Introduction

Decision-making as a human activity is often performed on different levels of abstraction. It includes both simple everyday decisions, such as selection of products while shopping, choice of itinerary to a workplace, and more compound decisions, e.g., in marking a student's work or in investments. Decisions are always made in the context of a current situation (i.e., the current state of the world) on the basis of the knowledge and experience acquired in the past. Computers support decision making. Several research directions have been developed to support computer-aided decision making. Among them are decision and game theory [57, 81], operational research [10], planning [28], control theory [67, 87], and machine learning [61]. The development of these directions has led to different methods of knowledge representation and reasoning about the real world for solving decision problems.

Decision-making is based on reasoning. There are different formal reasoning systems used by computers. Deductive reasoning [5] is based on the assumption that knowledge is represented and extended within a deductive system. This approach is very general and it encompasses a wide range of problems. However, real-life problems are usually very complex, and depend on many factors, some of them quite unpredictable. Deductive reasoning does not allow for such uncertainty. Therefore in machine learning another approach, called inductive reasoning [33, 50, 59], is used. Decision systems that implement inductive reasoning are based on the assumption that knowledge about a decision problem is given in the form of a set of examplary objects with known decisions. This set is called a training set. In the learning phase the system constructs a data model on the basis of the training set and then uses the constructed model to reason about the decisions for new objects called test objects. The most popular computational models used in inductive reasoning are: neural networks [15], decision trees [65], rule based systems [60], rough sets [63], bayesian networks [45], and analogy-based systems [68]. Inductive reasoning applied to large knowledge bases of objects made it possible to develop decision support systems for many areas of human activity, e.g., image, sound and handwriting recognition, medical and industrial diagnostics, credit decision making, fraud detection. Besides such general methods there are many specific methods dedicated to particular applications.

The goal of this dissertation is to present and analyze machine learning methods derived from the analogy-based reasoning paradigm [68], in particular, from case-based reasoning [3, 52]. Analogy-based reasoning reflects natural human reasoning that is based on the ability to associate concepts and facts by analogy. As in other inductive methods, we assume in case-based reasoning that a training set is given and reasoning about a new object is based on similar (analogous) objects from the training set.

Selection of a similarity measure among objects is an important component of this approach, which strongly affects the quality of reasoning. To construct a similarity measure and to compare objects we need to assume a certain fixed structure of objects. Most of data are collected in relational form: the objects are described by vectors of attribute values. Therefore, in the dissertation we assume this original structure of data. Numerous different metrics are used for such data [1, 14, 22, 51, 56, 77, 84, 88]. To construct such a metric one can use both general mathematical properties of the domains of attribute values and specific information encoded in the training data.

Case-based reasoning is more time-consuming than other inductive methods. However, the advance of hardware technology and the development of indexing methods for training examples [11, 29, 35, 43, 46, 62, 66, 78, 82] have made possible the application of case-based reasoning to real-life problems.

1.1   Results Presented in This Thesis

The research was conducted in two parallel directions. The first direction was based on the elaboration of reasoning methods and theoretical analysis of their quality and computational complexity. The second direction was focused on the implementation of the elaborated methods, and on experiments on real data followed by an analysis of experimental results. The quality of the methods developed was tested on data sets from the Internet data repository of the University of California at Irvine [16].

One of the widely used methods of case-based reasoning is the k nearest neighbors (k-nn) method [4, 23, 26, 31]. In the k-nn method the decision for a new object x is inferred from a fixed number k of the nearest neighbors of x in a training set. In the dissertation we present the following new methods and results related to the k-nn method:

1. A new metric for numerical attributes, called the Density Based Value Difference Metric (DBVDM) (Subsection 3.2),
2. An effective method for computing the distance between objects for the metrics WVDM [88] and DBVDM (Subsection 3.2),
3. Two attribute weighting algorithms (Subsections 3.4 and 3.5),
4. A new indexing structure and an effective searching method for the k nearest neighbors of a given test object (Section 4),
5. A classification model that combines the k-nn method with rule based filtering (Subsections 5.3 and 5.4),
6. The k-nn classification model based on locally induced metrics (Subsection 5.6).

Below we provide some detailed comments on the results of the dissertation.

Ad.(1). In case of the classical k-nn method is an important quality factor the selection of an appropriate similarity measure among objects[1, 2, 14, 22, 25, 51], [56, 77, 84, 85, 89, 88]. To define such a metric, in the first place, some general mathematical properties of the domains of attribute values can be used. The fundamental relation for comparing attribute values is the equality relation: for any pair of attribute values one can check if they are equal or not. Other relations on attribute values depend on the attribute type. In typical relational databases two types of attributes occur. Nominal attributes (e.g., color, shape, sex) are the most general. The values of such attributes can only be compared by the equality relation. The values of numerical attributes (e.g., size, age, temperature) are represented by real numbers. Numerical attributes provide more information about relations between values than nominal attributes, e.g., due to their linearly ordered structure and the existence of a measure of distance between values. The examples of metrics using only general relations on attribute values are the Hamming distance for nominal attributes and the l_p or the χ-square distance for numerical attributes.

However, in decision making such general relations on attribute values are not relevant, as they do not provide information about the relation between the values of attributes and the decision. Hence, an additional source of information, i.e., a training set, is used to construct a metric. By contrast to the properties of general relations on attribute values, this information depends on the problem to be solved and it helps to recognize which attributes and which of their properties are important in decision making for this particular problem. An example of such a data-dependent metric is provided by the Value Difference Metric (VDM) for nominal attributes. The VDM distance between two nominal values is defined on the basis of the distance between decision distributions for these two values in a given training set [77, 22]. Wilson and Martinez [88] proposed analogous metrics for numerical attributes: the Interpolated Value Difference Metric (IVDM) and the Windowed Value Difference Metric (WVDM). By analogy to VDM both metrics assign a decision distribution to each numerical value. To define such an assignment, for both metrics the objects whose values fall into a certain interval surrounding this value are used. The width of this interval is constant: it does not depend on the value. In many data sets the density of numerical values depends strongly on the values, and the constant width of the interval to be sampled can lead to the situation where for some values the sample obtained is not representative: it can contain either too few or too many objects.

In the dissertation we introduce the Density Based Value Difference Metric (DBVDM). In DBVDM the width of the interval to be sampled depends on the density of attribute values in a given training set. In this way we avoid the problem of having either too few or too many examples in the sample.

Ad.(2). The time required to compute the decision distribution for each numerical value by means of WVDM or DBVDM is linear with respect to the training set size. Hence, it is impractical to perform such a computation every time one needs the distance between two objects. In the dissertation we show that the decision distributions for all the values of a numerical attribute can be computed in total time $O(n \log n)$ (where n is the size of the given training set). This allows to compute the distance between two objects in logarithmic or even in constant time after preliminary conversion of the training set. This acceleration is indispensable if WVDM or DBVDM is to be applied to real-life data.

Ad.(3). Usually in real-life problems there are some factors that make attributes unequally important in decision making. The correlation of some attributes with the decision is stronger. Moreover, some attribute values are influenced by noise, which makes them less trustworthy than exact values of other attributes. Therefore, to ensure good similarity measure quality it is important to use attribute weights in its construction. Much research has been done on the development of algorithms for on-line optimization of attribute weights (i.e., training examples are processed sequentially and weights are modified after each example) [2, 48, 51, 69]. However, for real-life data sets the k-nn classification requires an advanced indexing method. We discuss this in more detail in the dissertation. In this case on-line algorithms are ineffective: indexing must be performed each time attribute weights are modified, i.e., after each example. Another disadvantage of on-line algorithms is that they are sensitive to the order of training examples.

Attribute weighting algorithms, used in practice, are batch algorithms with a small number of iterations, i.e., the algorithms that modify attribute weights only after having processed all the training examples. Lowe [56] and Wettschereck [84] have proposed such algorithms. Both algorithms use the conjugate gradient to optimize attribute weights, which means minimizing a certain error function based on the leave-one-out test on the training set. However, Lowe and Wettischereck's methods are applicable only to the specific weighted Euclidean metric.

In the dissertation we introduce two batch weighting algorithms assuming only that metrics are defined by a weighted linear combination of metrics for particular attributes. This assumption is less restrictive: attribute weighting can be thus applied to different metrics. The first algorithm proposed optimizes the distance to the objects classifying correctly in the leave-one-out test on the training set. The second algorithm optimizes classification accuracy in the leave-one-out test on the training set. We performed experiments consisting in the application of the proposed weighting methods to different types of metrics and in each case the weighting algorithms improved metric accuracy.

Ad.(4). Real-life data sets collected in electronic databases often consist of thousands or millions of records. To apply case-based queries to such large data tables some advanced metric-based indexing methods are required. These methods can be viewed as the extension of query methods expressed in the SQL language in case of relational databases where the role of similarity measure is taken over by indices and foreign keys, whereas similarity is measured by the distance between objects in an index and belonging the ones to the same set at grouping, respectively.

Metric-based indexing has attracted the interest of many researchers. Most of the methods developed minimize the number of I/O operations [9, 12, 13, 20, 43, 47], [55, 62, 66, 71, 83, 86]. However, the increase in RAM memory available in modern computers makes it possible to load and store quite large data sets in this fast-access memory and indexing methods based on this type of storage gain in importance. Efficiency of indexing methods of this type is determined by the average number of distance computations performed while searching a database for objects most similar to a query object. The first methods that reduces the number of distance computations have been proposed for the case of a vector space [11, 29]. They correspond to the specific Euclidean metric.

In the dissertation we consider different metrics defined both for numerical and nominal attributes and therefore we focus on more general indexing methods, such as BST [46], GHT [78], and GNAT [18]. GHT assumes that only a distance computing function is provided. BST and GNAT assume moreover that there is a procedure that computes center of an object set, which corresponds to computing the mean in a vector space. However, no assumptions about the properties of the center are used. In each of these two methods both the indexing and searching algorithms are correct for any definition of center. Such a definition affects only search efficiency.

In the most popular indexing scheme the indexing structure is constructed in the form of a tree. The construction uses the top-down strategy: in the beginning the whole training set is split into a number of smaller nodes and then each node obtained is recursively split into smaller nodes. Training objects are assigned to the leaves. All the three indexing methods (BST, GHT, and GNAT) follow this general scheme. One of the important components that affects the efficiency of such indexing trees is the node splitting procedure. BST, GHT, and GNAT use a single-step splitting procedure, i.e., the splitting algorithm selects criteria to distribute the objects from a parent node and then assigns the objects to child nodes according to these criteria. At each node this operation is performed once. In the dissertation we propose an indexing tree with an iterative k-means-like splitting procedure. Savaresi and Boley have shown that such a procedure has good theoretical splitting properties [70] and in the experiments we prove that the indexing tree with this iterative splitting procedure is more efficient than trees with a single-step procedure.

Searching in a tree-based indexing structure can be speeded up in the following way: the algorithm finds quickly the first candidates for the nearest neighbors and then it excludes branches that are recognized to contain no candidates closer than those previously found. Each of the three methods BST, GHT, and GNAT uses a different single mathematical criterion to exclude branches of the indexing tree. However, all the three criteria assume similar properties of the indexing tree. In the dissertation we propose a search algorithm that uses all the three criteria simultaneously. We show experimentally that for large data sets the combination of this new search algorithm with the iterative splitting based tree makes nearest neighbors searching up to several times more efficient than the methods BST, GHT, and GNAT. This new method allows us to apply the k-nn method to data with several hundred thousand training objects and for the largest tested data set it makes it possible to reduce the 1-nn search by 4000 times as compared with linear search.

Ad.(5). After defining a metric and choosing a method to speed up the search for similar objects, the last step is the selection of a classification model. The classical k-nn method finds a fixed number k of the nearest neighbors of a test object in the training set, assigns certain voting weights to these nearest neighbors and selects the decision with the greatest sum of voting weights.

The metric used to find the nearest neighbors is the same for each test object: it is induced globally from the training set. Real-life data are usually too complex to be accurately modeled by a global mathematical model. Therefore such a global metric can only be an approximation of similarity encoded in data and it can be inaccurate for specific objects. To ensure that the k nearest neighbors found for a test object are actually similar, a popular solution is to combine the k-nn method with another classification model.

A certain improvement in classification accuracy has been observed for models combining the k-nn approach with rule induction [25, 37, 54]. All these models use the approach typical for rule induction: they generate a certain set of rules a priori and then they apply these generated rules in the classification process. Computation of an appropriate set of rules is usually time-consuming: to select accurate rules algorithms need to evaluate certain qualitative measures for rules in relation to the training set.

In the dissertation we propose a classification model combining the k-nn with rule induction in such a way that after addition of the rule based component the increase of the performance time of the k-nn method is inconsiderable. The k-nn implements the lazy learning approach where computation is postponed till the moment of classification [6, 34]. We add rule induction to the k nearest neighbors in such a way that the combined model preserves lazy learning, i.e., rules are constructed in a lazy way at the moment of classification.

The combined model proposed in the dissertation is based on the set of all minimal consistent rules in the training set [74]. This set has good theoretical properties: it corresponds to the set of all the rules generated from all local reducts of the training set [94]. However, the number of all minimal consistent rules can be exponential with respect both to the number of attributes and to the training set size. Thus, it is practically impossible to generate them all. An effective lazy simulation of the classification based on the set of all minimal consistent rules for data with nominal attributes has been described by Bazan [6]. Instead of computing all minimal consistent rules a priori before classification the algorithm generates so called local rules at the moment of classification. Local rules have specific properties related to minimal consistent rules and, on the other hand, they can be effectively computed. This implies that classification based on the set of all minimal consistent rules can be simulated in polynomial time.

In the dissertation we introduce a metric-dependent generalization of the notions of minimal consistent rule and local rule. We prove that the model of rules assumed by Bazan [6] is a specific case of the proposed generalization where the metric is assumed to be the Hamming metric. We show that the generalized model has properties analogous to those of the original model: there is a relationship between generalized minimal consistent rules and generalized local rules that makes the application of Bazan's lazy algorithm to the generalized model possible.

The proposed metric-dependent generalization enables a combination of Bazan's lazy algorithm with the k-nn method. Using the properties of the generalized model we modify Bazan's algorithm in such a way that after addition of the modified algorithm to the k-nn the increase of the performance time is insignificant.

We show that the proposed rule-based extension of the k-nn is a sort of voting by the k nearest neighbors that can be naturally combined with any other voting system. It can be viewed as the rule based verification and selection of similar objects found by the k-nn classifier. The experiments performed show that the proposed rule-based voting gives the best classification accuracy when combined with a voting model where the nearest neighbors of a test object are assigned the inverse square distance weights. For some data sets the rule based extension added to the k-nn method decreases relatively the classification error by several tens of percent.

Ad.(6). The k-nn, other inductive learning methods and even hybrid combinations of these inductive methods are based on the induction of a mathematical model from training data and application of this model to reasoning about test objects. The induced data model remains invariant while reasoning about different test objects. For many real-life data it is impossible to induce relevant global models. This fact has been recently observed by researches in different areas, like data mining, statistics, multiagent systems [17, 75, 79]. The main reason is that phenomena described by real-life data are often too complex and we do not have sufficient knowledge in data to induce global models or a parameterized class of such models together with searching methods for the relevant global model in such a class.

In the dissertation we propose a method for dealing with such real-life data. The proposed method refers to another approach called transductive learning [79]. In this approach the classification algorithm uses the knowledge encoded in the training set, but it also uses knowledge about test objects in construction of classification models. This means that for different test objects different classification models are used. Application of transductive approach to problem solving is limited by longer performance time than in inductive learning, but the advance of hardware technology makes this approach applicable to real problems.

In the classical k-nn method the global, invariant model is the metric used to find the nearest neighbors of test objects. The metric definition is independent of the location of a test object, whereas the topology and the density of training objects in real data are usually not homogeneous. In the dissertation we propose a method for inducing a local metric for each test object and then this local metric is used to select the nearest neighbors. Local metric induction depends locally on the properties of the test object, therefore the notion of similarity can be adapted to these properties and the correct decision can be selected in specific distinctive fragments of the space of objects.

Such a local approach to the k-nn method has been already considered in literature [24, 32, 44]. However, all the methods described in literature are specific: they can be applied only to data from a vector space and they are based on local adaptation of a specific global metric in this space. In the dissertation we propose a method that requires a certain global metric but the global metric is used only for a preliminary selection of a set of training objects used to induce a local metric. This method is much more general: it makes the global metric and the local metric independent and it allows us to use any metric definition as a local metric.

In the experiments we show that the local metric induction method is helpful in the case of hard classification problems where the classification error of different global models remains high. For one of the tested data sets this method obtains the classification accuracy that has never been reported before in literature.

Partial results from the dissertation have been published and presented at the international conferences RSCTC, ECML and ICDM [8, 39, 38, 76, 91] and in the journal Fundamenta Informaticae [40, 92]. The methods described have been implemented and they are available in the form of a software library and in the system RSES [8, 73].

1.2   Organization of the Thesis

Section 2 introduces the reader to the problem of learning from data and to the evaluation method of learning accuracy (Subsections 2.1 and 2.2). It describes the basic model of analogy-based learning, the k-nn (Subsections 2.3–2.5), and it presents the experimental methodology used in the dissertation (Subsections 2.6 and 2.7).

Section 3 introduces different metrics induced from training data. It starts with the definition of VDM for nominal attributes (Subsection 3.1). Then, it describes three extensions of the VDM metric for numerical attributes: IVDM, WVDM and DBVDM, and an effective algorithm for computing the distance between objects for these metrics (Subsection 3.2). Next, two attribute weighting algorithms are presented: an algorithm that optimizes distance and an algorithm that optimizes classification accuracy (Subsections 3.3–3.5). Finally, experiments comparing accuracy of the described metrics and weighting methods are presented (Subsections 3.6–3.9).

Section 4 describes the indexing tree with the iterative splitting procedure (Subsections 4.2–4.4), and the nearest neighbors search method with three combined pruning criteria (Subsections 4.5 and 4.6). Moreover, It presents an experimental comparison of this search method with other methods known from the literature (Subsections 4.7 and 4.8).

In Section 5, first we describe the algorithm that estimates automatically the optimal number of neighbors k in the k-nn classifier (Subsection 5.1). The rest of the section is dedicated to two new classification models that use previously described components: the metrics, indexing and the estimation of the optimal k. First, the metric-based extension of rule induction and the combination of a rule based classification model with the k nearest neighbors method is described and compared experimentally with other known methods (Subsections 5.3–5.5). Next, the model with local metric induction is presented and evaluated experimentally (Subsections 5.6 and 5.7).

2   Basic Notions

In this section, we define formally the problem of concept learning from examples.

2.1   Learning a Concept from Examples

We assume that the target concept is defined over a universe of objects U^∞. The concept to be learnt is represented by a decision function $dec : U^\infty \to V_{dec}$. In the thesis we consider the situation, when the decision is discrete and finite $V_{dec} = \{d_1, \ldots d_m\}$. The value $dec(x) \in V_{dec}$ for an object $x \in U^\infty$ represents the category of the concept that the object x belongs to.

In the thesis we investigate the problem of decision learning from a set of examples. We assume that the target decision function $dec : U^\infty \to V_{dec}$ is unknown. Instead of this there is a finite set of training examples $U \subseteq U^\infty$ provided, and the decision values $dec(x)$ are available for the objects $x \in U$ only. The task is to provide an algorithmic method that learns a function (hypothesis) $h : U^\infty \to V_{dec}$ approximating the real decision function dec given only this set of training examples U.

The objects from the universe U^∞ are real objects. In the dissertation we assume that they are described by a set of n attributes $A = \{a_1, \ldots, a_n\}$. Each real object $x \in U^\infty$ is represented by the object that is a vector of values $(x_1, \ldots, x_n)$. Each value x_i is the value of the attribute a_i on this real object x. Each attribute $a_i \in A$ has its domain of values V_i and for each object representation $(x_1, \ldots, x_n)$ the values of the attributes belong to the corresponding domains: $x_i \in V_i$ for all $1 \leq i \leq n$. In other words, the space of object representations is defined as the product $\mathbb{X} = V_1 \times \ldots \times V_n$. The type of an attribute a_i is either numerical, if its values are comparable and can be represented by numbers $V_i \subseteq \mathbb{R}$ (e.g., age, temperature, height), or nominal, if its values are incomparable, i.e., if there is no linear order on V_i (e.g., color, sex, shape).

It is easy to learn a function that assigns the appropriate decision for each object in a training set $x \in U$. However, in most of decision learning problems a training set U is only a small sample of possible objects that can occur in real application and it is important to learn a hypothesis h that recognizes correctly as many objects as possible. The most desirable situation is to learn the hypothesis that is accurately the target function: $h(x) = dec(x)$ for all $x \in U^\infty$. Therefore the quality of a learning method depends on its ability to generalize information from examples rather than on its accuracy on the training set.

The problem is that the target function dec is usually unknown and the information about this function dec is restricted only to a set of examples. In such a situation a widely used method to compare different learning algorithms is to divide a given set of objects U into a training part U_{trn} and a test part U_{tst}, next, to apply learning algorithms to the training part U_{trn}, and finally, to measure accuracy of the induced hypothesis on the test set U_{tst} using the proportion of the correctly classified objects to all objects in the test set [61]:

$$accuracy(h) = \frac{|\{x \in U_{tst} : h(x) = dec(x)\}|}{|U_{tst}|}.$$

2.2   Learning as Concept Approximation in Rough Set Theory

The information available about each training object $x \in U_{trn}$ is restricted to the vector of attribute values $(x_1, \ldots, x_n)$ and the decision value $dec(x)$. This defines the indiscernibility relation $IND = \{(x, x') : \forall a_i \in A \; x_i = x'_i\}$. The indiscernibility relation IND is an equivalence relation and defines a partition in the set of the training objects U_{trn}. The equivalence class of an object $x \in U_{trn}$ is defined by $IND(x) = \{x' : xINDx'\}$. Each equivalence class contains the objects that are indiscernible by the values of the attributes from the set A. The pair (U_{trn}, IND) is called an approximation space over the set U_{trn} [63, 64].

Each decision category $d_j \in V_{dec}$ is associated with its decision class in U_{trn}: $Class(d_j) = \{x \in U_{trn} : dec(x) = d_j\}$. The approximation space $AS = (U_{trn}, IND)$ defines the lower and upper approximation for each decision class:

$$LOWER_{AS}(Class(d_j)) = \{x \in U_{trn} : IND(x) \subseteq Class(d_j)\}$$
$$UPPER_{AS}(Class(d_j)) = \{x \in U_{trn} : IND(x) \cap Class(d_j) \neq \emptyset\}$$

The problem of concept learning can be described as searching for an extension (U^∞, IND^∞) of the approximation space (U_{trn}, IND), relevant for approximation of the target concept dec. In such an extension each new object $x \in U^\infty$ provides an information $(x_1, \ldots, x_n) \in \mathbb{X}$ with semantics $IND^\infty(x) \subseteq U^\infty$. By $\|(x_1, \ldots, x_n)\|_{U_{trn}}$ and $\|(x_1, \ldots, x_n)\|_{U^\infty}$ we denote the semantics of the pattern $(x_1, \ldots, x_n)$ in U_{trn} and U^∞, respectively. Moreover, $\|(x_1, \ldots, x_n)\|_{U_{trn}} = IND(x)$ and $\|(x_1, \ldots, x_n)\|_{U^\infty} = IND^\infty(x)$.

In order to define the lower and upper approximation of $Class(d_j) \subseteq U^\infty$ using IND^∞ one should estimate the relationships between $IND^\infty(x)$ and $Class(d_l)$ for $l = 1, \ldots, m$.

In the dissertation two methods are used.

In the first method we estimate the relationships between $IND^\infty(x)$ and $Class(d_l)$ by:

1. selecting from U_{trn} the set $NN(x, k)$ of k nearest neighbors of x by using a distance function (metric) defined on patterns;
2. using the relationships between $\|(y_1, \ldots, y_n)\|_{U_{trn}}$ and $Class(d_l) \cap U_{trn}$ for $y \in NN(x, k)$ and $l = 1, \ldots, m$ to estimate the relationship between IND^∞ (x) and $Class(d_j)$.

One can also use another method for estimating the relationship between $IND^\infty(x)$ and $Class(d_j)$. Observe that the patterns from $\{(y_1, \ldots, y_n) : y \in U_{trn}\}$ are not enough general for matching arbitrary objects from U^∞. Hence, first, using a distance function we generalize the patterns $(y_1, \ldots, y_n)$ for $y \in U_{trn}$ to patterns $pattern(y)$ that are combinations of so called generalized descriptors $a_i \in W$, where $W \subseteq V_i$, with the semantics $\|a_i \in W\|_{U_{trn}} = \{y \in U_{trn} : y_i \in W\}$. The generalization preserves the following constraint: if $\|(y_1, \ldots, y_n)\|_{U_{trn}} \subseteq Class(d_l)$ then $\|pattern(y)\|_{U_{trn}} \subseteq Class(d_l)$. For a given $x \in U^\infty$ we select all *pattern(y)* that are matching x and we use the relationships between their semantics and $Class(d_l)$ for $l = 1, \ldots, m$ to estimate the relationship between $IND^\infty(x)$ and $Class(d_j)$.

Since in the considered problem of concept learning the only information about new objects to be classified is the vector of attribute values $(x_1, \ldots, x_n) \in \mathbb{X}$ the objects with the same value vector are indiscernible. Therefore searching for a hypothesis $h : U^\infty \to V_{dec}$ approximating the real function $dec : U^\infty \to V_{dec}$ is restricted to searching for a hypothesis of the form $h : \mathbb{X} \to V_{dec}$. To this end the space of object representations $\mathbb{X}$ is called for short the space of objects and we consider the problem of learning a hypothesis using this restricted form $h : \mathbb{X} \to V_{dec}$.

2.3 Metric in the Space of Objects

We assume that in the space of objects $\mathbb{X}$ a distance function $\rho : \mathbb{X}^2 \to \mathbb{R}$ is defined. The distance function ρ is assumed to satisfy the axioms of a pseudometric, i.e., for any objects $x, y, z \in \mathbb{X}$:

1. $\rho(x, y) \geq 0$ (positivity),
2. $\rho(x, x) = 0$ (reflexivity),
3. $\rho(x, y) = \rho(y, x)$ (symmetry),
4. $\rho(x, y) + \rho(y, z) \geq \rho(x, z)$ (triangular inequality).

The distance function ρ models the relation of similarity between objects. The properties of symmetry and triangular inequality are not necessary to model similarity but they are fundamental for the efficiency of the learning methods described in this thesis and for many other methods from the literature [9, 11, 12, 18, 19, 20, 29, 35, 36]. Sometimes the definition of a distance function satisfies the strict positivity: $x \neq y \Rightarrow \rho(x, y) > 0$. However, the strict positivity is not used by the distance based learning algorithms and a number of important distance measures like VDM [77] and the metrics proposed in this thesis do not satisfy this property.

In the l_p-norm based metric the distance between two objects $x=(x_1, \ldots, x_n)$, $y = (y_1, \ldots, y_n)$ is defined by

$$\rho(x, y) = \left(\sum_{i=1}^{n} \rho_i(x_i, y_i)^p \right)^{\frac{1}{p}}$$

where the metrics ρ_i are the distance functions defined for particular attributes $a_i \in A$.

Aggarwal et al. [1] have examined the meaningfulness of the concept of similarity in high-dimensional real value spaces investigating the effectiveness of the l_p-norm based metric in dependence on the value of the parameter p. They proved the following result:

Theorem 1. *For the uniform distribution of 2 points x, y in the cube $(0,1)^n$ with the norm l_p $(p \geq 1)$:*

$$\lim_{n \to \infty} E\left[\left(\frac{\max(\|x\|_p, \|y\|_p) - \min(\|x\|_p, \|y\|_p)}{\min(\|x\|_p, \|y\|_p)}\right)\sqrt{n}\right] = C\sqrt{\frac{1}{2p+1}}$$

where C is a positive constant and $\|\cdot\|_p$ denotes the standard norm in the space l_p.

It shows that the smaller p, the larger relative contrast is between the point closer to and the point farther from the beginning of the coordinate system. It indicates that the smaller p the more effective metric is induced from the l_p-norm. In the context of this result $p = 1$ is the optimal trade-off between the quality of the measure and its properties: $p = 1$ is the minimal index of the l_p-norm that preserves the triangular inequality. The fractional distance measures with $p < 1$ do not have this property.

On the basis of this result we assume the value $p = 1$ and in the thesis we explore the metrics that are defined as linear sum of metrics ρ_i for particular attributes $a_i \in A$:

$$\rho(x, y) = \sum_{i=1}^{n} \rho_i(x_i, y_i). \tag{1}$$

In the problem of learning from a set of examples U_{trn} the particular distance functions ρ_i are induced from a training set U_{trn}. It means that the metric definition depends on the provided examples and it is different for different data sets.

2.4   City-Block and Hamming Metric

In this subsection we introduce the definition of a basic metric that is widely used in the literature. This metric combines the city-block (Manhattan) distance for the values of numerical attributes and the Hamming distance for the values of nominal attributes.

The distance $\rho_i(x_i, y_i)$ between two values x_i, y_i of a numerical attribute a_i in the city-block distance is defined by

$$\rho_i(x_i, y_i) = |x_i - y_i|. \tag{2}$$

The scale of values for different domains of numerical attributes can be different. To make the distance measures for different numerical attributes equally significant it is better to use the normalized value difference. Two types of normalization are used. In the first one the difference is normalized with the range of the values of the attribute a_i

$$\rho_i(x_i, y_i) = \frac{|x_i - y_i|}{max_i - min_i}, \tag{3}$$

where $max_i = \max_{x \in U_{trn}} x_i$ and $min_i = \min_{x \in U_{trn}} x_i$ are the maximal and the minimal value of the attribute a_i in the training set U_{trn}. In the second type of normalization the value difference is normalized with the standard deviation of the values of the attribute a_i in the training set U_{trn}:

$$\rho_i(x_i, y_i) = \frac{|x_i - y_i|}{2\sigma_i}$$

where $\sigma_i = \sqrt{\frac{\sum_{x \in U_{trn}} (x_i - \mu_i)^2}{|U_{trn}|}}$ and $\mu_i = \frac{\sum_{x \in U_{trn}} x_i}{|U_{trn}|}$.

The distance $\rho_i(x_i, y_i)$ between two nominal values x_i, y_i in the Hamming distance is defined by the Kronecker delta:

$$\rho_i(x_i, y_i) = \begin{cases} 1 \text{ if } x_i \neq y_i \\ 0 \text{ if } x_i = y_i. \end{cases}$$

The combined city-block and Hamming metric sums the normalized value differences for numerical attributes and the values of Kronecker delta for nominal attributes. The normalization of numerical attributes with the range of values $max_i - min_i$ makes numerical and nominal attributes equally significant: the range of distances between values is $[0; 1]$ for each attribute. The only possible distance values for nominal attributes are the limiting values 0 and 1, whereas the normalized distance definition for numerical attributes can give any value between 0 and 1. It results from the type of an attribute: the domain of a nominal attribute is only a set of values and the only relation in this domain is the equality relation. The domain of a numerical attribute are the real numbers and this domain is much more informative: it has the structure of linear order and the natural metric, i.e., the absolute difference.

Below we define an important property of metrics related to numerical attributes:

Definition 2. *The metric ρ is consistent with the natural linear order of numerical values if and only if for each numerical attribute a_i and for each three real values $v_1 \leq v_2 \leq v_3$ the following conditions hold: $\rho_i(v_1, v_2) \leq \rho_i(v_1, v_3)$ and $\rho_i(v_2, v_3) \leq \rho_i(v_1, v_3)$.*

The values of a numerical attribute reflect usually a measure of a certain natural property of analyzed objects, e.g., size, age or measured quantities like temperature. Therefore, the natural linear order of numerical values helps often obtain useful information for measuring similarity between objects and the notion of metric consistency describes the metrics that preserve this linear order.

Fact 3. *The city-block metric is consistent with the natural linear order.*

Proof. The city-block metric depends linearly on the absolute difference as defined in Equation 2 or 3. Since the absolute difference is consistent with the natural linear order, the city-block metric is consistent too. □

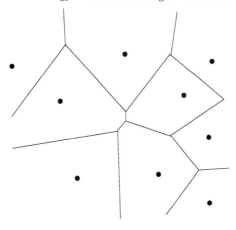

Fig. 1. The Voronoi diagram determined by examples on the Euclidean plane

2.5 K Nearest Neighbors as Analogy-Based Reasoning

One of the most popular algorithms in machine learning is the k nearest neighbors (k-nn). The predecessor of this method, the nearest neighbor algorithm (1-nn) [23], induces a metric ρ from the training set U_{trn}, e.g., the city-block and Hamming metric described in Subsection 2.4, and stores the whole training set U_{trn} in memory. Each test object x is classified by the 1-nn with the decision of the nearest training object $y_{nearest}$ from the training set U_{trn} according to the metric ρ:

$$y_{nearest} := \arg\min_{y \in U_{trn}} \rho(x, y),$$

$$dec_{1-nn}(x) := dec(y_{nearest}).$$

On the Euclidean plane (i.e., with the Euclidean metric) the regions of the points nearest to particular training examples constitute the Voronoi diagram (see Figure 1).

The k nearest neighbors is an extension of the nearest neighbor [26, 31]. Instead of the one nearest neighbor it uses the k nearest neighbors $NN(x, k)$ to select the decision for an object x to be classified. The object x is assigned with the most frequent decision among the k nearest neighbors:

$$dec_{k-nn}(x) := \arg\max_{d_j \in V_{dec}} |\{y \in NN(x, k) : dec(y) = d_j\}|. \tag{4}$$

Ties are broken arbitrary in favor of the decision d_j with the smallest index j or in favor of a randomly selected decision among the ties.

The k nearest neighbors method is a simple example of analogy-based reasoning. In this approach a reasoning system assumes that there is a database providing the complete information about examplary objects. When the system is asked about another object with an incomplete information it retrieves similar (analogous) objects from the database and the missing information is completed on the basis of the information about the retrieved objects.

In the k-nn the induced metric ρ plays the role of a similarity measure. The smaller the distance is between two objects, the more similar they are. It is important for the similarity measure to be defined in such a way that it uses only the information that is available both for the examplary objects in the database and for the object in the query. In the problem of decision learning it means that the metric uses only the values of the non-decision attributes.

2.6   Data Sets

The performance of the algorithms described in this dissertation is evaluated for a number of benchmark data sets. The data sets are obtained from the repository of University of California at Irvine [16]. This repository as the source of benchmark data sets is the most popular in the machine learning community and all the data sets selected to evaluate learning algorithms in this dissertation have been also used by other researchers. This ensures that the presented performance of algorithms can be compared to the performance of other methods from the literature.

To compare the accuracy of the learning models described in this dissertation (Section 3 and Section 5) 10 benchmark data sets were selected (see Table 1). All the selected sets are the data sets from UCI repository that have data objects represented as vectors of attributes values and have the size between a few thousand and several tens thousand of objects. This range of the data size was chosen because such data sets are small enough to perform multiple experiments for all the algorithms described in this dissertation and to measure their accuracy in a statistically significant way (see Subsection 2.7). The evaluation of these algorithms is based on the largest possible data sets since such data sets are usually provided in real-life problems.

To compare the efficiency of the indexing structures used to speedu up searching for the nearest neighbors (Section 4) all the 10 data sets from Table 1 were used again with 2 additional very large data sets (see Table 2). The size of the 2 additional data sets is several hundred thousand. The indexing and the searching

Table 1. The data sets used to evaluate accuracy of learning algorithms

Data set	Number of attributes	Types of attributes	Training set size	Test set size
segment	19	numeric	1 540	770
splice (DNA)	60	nominal	2 000	1 186
chess	36	nominal	2 131	1 065
satimage	36	numeric	4 435	2 000
mushroom	21	numeric	5 416	2 708
pendigits	16	numeric	7 494	3 498
nursery	8	nominal	8 640	4 320
letter	16	numeric	15 000	5 000
census94	13	numeric+nominal	30 160	15 062
shuttle	9	numeric	43 500	14 500

Table 2. The data sets used to evaluate efficiency of indexing structures

Data set	Number of attributes	Types of attributes	Training set size	Test set size
census94-95	40	numeric+nominal	199 523	99 762
covertype	12	numeric+nominal	387 308	193 704

process are less time consuming than some of the learning models. Therefore, larger data sets are possible to be tested. The 2 largest data sets illustrate the capabilities of the indexing methods described in the dissertation.

Each data set is split into a training and a test set. Some of the sets (*splice, satimage, pendigits, letter, census94, shuttle, census94-95*) are available in the repository with the original partition and this partition was used in the experiments. The remaining data sets (*segment, chess, mushroom, nursery, covertype*) was randomly split into a training and a test part with the split ratio 2 to 1. To make the results from different experiments comparable the random partition was done once for each data set and the same partition was used in all the performed experiments.

2.7   Experimental Evaluation of Learning Algorithms

Both in the learning models constructed from examples (Sections 3 and 5) and in the indexing structures (Section 4) described in the dissertation there are elements of non-determinism: some of the steps in these algorithms depend on selection of a random sample from a training set. Therefore the single test is not convincing about the superiority of one algorithm over another: difference between two results may be a randomness effect. Instead of the single test in each experiment a number of tests was performed for each data set and the average results are used to compare algorithms. Moreover, the Student's t-test [41, 30] is applied to measure statistical significance of difference between the average results of different algorithms.

The Student's t-test assumes that the goal is to compare two quantities being continuous random variables with normal distribution. A group of sample values is provided for each quantity to be compared. In the dissertation these quantities are either the accuracy of learning algorithms measured on the test set (see Subsection 2.1) or the efficiency of the indexing and searching algorithm measured by the number of basic operations performed.

There are the paired and the unpaired Student's t-test. The paired t-test is used where there is a meaningful one-to-one correspondence between the values in the first and in the second group of sample values to be compared. In our experiments the results obtained in particular tests are independent. In such a situation the unpaired version of the Student's t-test is appropriate.

Another type of distinction between different tests depends on the information one needs to obtain from a test. The one-tailed t-test is used if one needs to know whether one quantity is greater or less than another one. The two-tailed t-test is used if the direction of the difference is not important, i.e., the infor-

Table 3. The Student's t-test probabilities

$df \setminus \alpha$	90%	95%	97.5%	99%	99.5%
1	3.078	6.314	12.706	31.821	63.657
2	1.886	2.920	4.303	6.965	9.925
3	1.638	2.353	3.182	4.541	5.841
4	1.533	2.132	2.776	3.747	4.604
5	1.476	2.015	2.571	3.365	4.032
6	1.440	1.943	2.447	3.143	3.707
7	1.415	1.895	2.365	2.998	3.499
8	1.397	1.860	2.306	2.896	3.355
9	1.383	1.833	2.262	2.821	3.250
10	1.372	1.812	2.228	2.764	3.169

mation whether two quantities differ or not is required only. In our experiments the information about the direction of difference (i.e., whether one algorithm is better or worse than another one) is crucial so we use the one-tailed unpaired Student's t-test.

Let X_1 and X_2 be continuous random variables and let p be a number of values sampled for each variable X_i. In the Student's t-test only the means $E(X_1), E(X_2)$ and the standard deviations $\sigma(X_1), \sigma(X_2)$ are used to measure statistical significance of difference between the variables. First, the value of t is to be calculated:

$$t = \frac{E(X_1) - E(X_2)}{\sqrt{\frac{\sigma(X_1)^2 + \sigma(X_2)^2}{p}}}.$$

Next, the degree of freedom df is to be calculated:

$$df = 2(p - 1).$$

Now the level of statistical significance can be checked in the table of the t-test probabilities (see Table 3). The row with the calculated degree of freedom df is to be used. If the calculated value of t is greater than the critical value of t given in the table then X_1 is greater than X_2 with the level of significance α given in the header of the column. The level of significance α means that X_1 is greater than X_2 with the probability α.

3   Metrics Induced from Examples

This section explores metrics induced from examples.

3.1   Joint City-Block and Value Difference Metric

Subsection 2.4 provides the metric definition that combines the city-block metric for numerical attributes and the Hamming metric for nominal attributes. In this subsection we focus on nominal attributes.

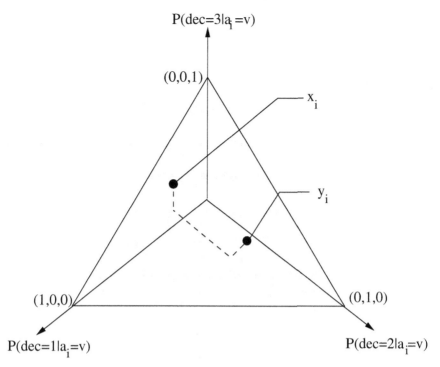

Fig. 2. An example: the Value Difference Metric for the three decision values $V_{dec} = \{1, 2, 3\}$. The distance between two nominal values x_i, y_i corresponds to the length of the dashed line.

The definition of the Hamming metric uses only the relation of equality in the domain of values of a nominal attribute. This is the only relation that can be assumed in general about nominal attributes. This relation carries often insufficient information, in particular it is much less informative than the structure of the domains for numerical attributes where the values have the structure of linear order with a distance measure between the values.

Although in general one can assume nothing more than equality relation on nominal values, in the problem of learning from examples the goal is to induce a classification model from examples assuming that a problem and data are fixed. It means that in the process of classification model induction the information encoded in the database of examples should be used. In the k nearest neighbors method this database can be used to extract meaningful information about relation between values of each nominal attribute and to construct a metric.

This fact has been used first by Stanfill and Waltz who defined a measure to compare the values of a nominal attribute [77]. The definition of this measure, called the Value Difference Metric (VDM), is valid only for the problem of learning from examples. It defines how much the values of a nominal attribute $a_i \in A$ differ in relation to the decision dec. More precisely, the VDM metric estimates the conditional decision probability $P(dec = d_j | a_i = v)$ given a nominal value v and uses the estimated decision probabilities to compare nominal values. The VDM distance between two nominal values x_i, y_i is defined by the difference between the estimated decision probabilities $P(dec = d_j | a_i = x_i)$, $P(dec = d_j | a_i = x_i)$ corresponding to the values x_i, y_i (see Figure 2):

$$\rho_i(x_i, y_i) = \sum_{d_j \in V_{dec}} |P(dec = d_j | a_i = x_i) - P(dec = d_j | a_i = y_i)|. \qquad (5)$$

The estimation of the decision probability $P(dec = d_j | a_i = v)$ is done from the training set U_{trn}. For each value v, it is defined by the decision distribution in the set of all the training objects that have the value of the nominal attribute a_i equal to v:

$$P_{VDM}(dec = d_j | a_i = v) = \frac{|\{x \in U_{trn} : dec(x) = d_j \wedge x_i = v\}|}{|\{x \in U_{trn} : x_i = v\}|}.$$

From Equation 5 and the definition of $P_{VDM}(dec = d_j | a_i = v)$ one can see that the more similar the correlations between each of two nominal values $x_i, y_i \in V_i$ and the decisions $d_1, \ldots, d_m \in V_{dec}$ in the training set of examples U_{trn} are the smaller the distance in Equation 5 is between x_i and y_i. Different variants of this metric were used in many applications [14, 22, 77].

To define a complete metric the Value Difference Metric needs to be combined with another distance function for numerical attributes. For each pair of possible data objects $x, y \in \mathbb{X}$ the following condition $\rho_i(x_i, y_i) \leq 2$ is satisfied for any nominal attribute $a_i \in A$. It means that the range of possible distances for the values of nominal attributes in the Value Difference Metric is $[0; 2]$. It corresponds well to the city-block distance for a numerical attribute a_i normalized by the range of the values of this attribute in the training set U_{trn} (see Subsection 2.4):

$$\rho_i(x_i, y_i) = \frac{|x_i - y_i|}{max_i - min_i}.$$

The range of this normalized city-block metric is $[0; 1]$ for the objects in the training set U_{trn}. In the test set U_{tst} this range can be exceeded but it happens very rarely in practice. The most important property is that the ranges of such a normalized numerical metric and the VDM metric are of the same order.

The above described combination of the distance functions for nominal and numerical attributes was proposed by Domingos [25]. The experimental results described in Subsection 3.7 and 3.9 prove that this combination is more effective than the same normalized city-block metric combined with the Hamming metric.

3.2   Extensions of Value Difference Metric for Numerical Attributes

The normalized city-block metric used in the previous subsection to define the joint metric uses information from the training set: it normalizes the difference between two numerical values v_1, v_2 by the range of the values of a numerical attribute $max_i - min_i$ in the training set. However, it defines the distance between values of the numerical attribute on the basis of the information about this attribute only, whereas the distance definition for nominal attributes makes use of the correlation between the nominal values of an attribute and the decision values. Since this approach improves the effectiveness of metrics for nominal attributes (see Subsection 3.7) analogous solutions has been investigated for numerical attributes.

Wilson and Martinez proposed two analogous distance definitions. In the Interpolated Value Difference Metric (IVDM) [88, 89] it is assumed that the range of values $[min_i; max_i]$ of a numerical attribute a_i in a training set is discretized into s equal-width intervals. To determine the value of s they use the heuristic value

$$s = \max\left(|V_{dec}|, 5\right).$$

The width of such a discretized interval is:

$$w_i = \frac{max_i - min_i}{s}.$$

In each interval $I_p = [min_i + (p-1) \cdot w_i; min_i + p \cdot w_i]$, where $0 \le p \le s+1$, the midpoint mid_p and the decision distribution $P(dec = d_j | a_i \in I_p)$ are defined by

$$mid_p = min_i + (p - \frac{1}{2}) \cdot w_i,$$

$$P(dec = d_j | a_i \in I_p) = \begin{cases} 0 & \text{if } p = 0 \text{ or } p = s+1 \\ \frac{|\{x \in U_{trn} : dec(x) = d_j \wedge x_i \in I_p\}|}{|\{x \in U_{trn} : x_i \in I_p\}|} & \text{if } 1 \le p \le s. \end{cases}$$

To determine the decision distribution in the IVDM metric for a given numerical value v the two neighboring intervals are defined by

$$I(v) = \max\{p \le s+1 : mid_p \le v \vee p = 0\},$$

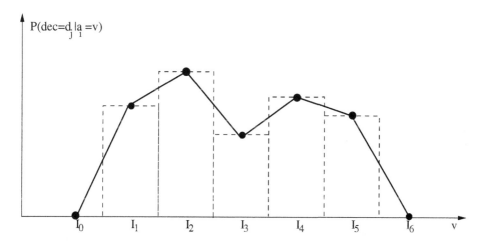

Fig. 3. An example of the interpolated decision distribution for a single decision d_j with the number of intervals $s = 5$

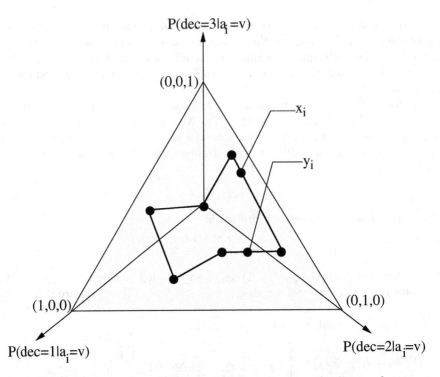

Fig. 4. The Interpolated Value Difference Metric: to measure the distance between two numerical values x_i, y_i the decision distribution for each value is interpolated between the decision distributions in the midpoints of the two neighboring intervals.

$$\overline{I(v)} = \min\{p \geq 0 : mid_p \geq v \lor p = s + 1\}.$$

If v is out of the range $[mid_0; mid_{s+1}]$ the interval indices are set either to zero: $\underline{I(v)} = \overline{I(v)} = 0$ or to $s + 1$: $\underline{I(v)} = \overline{I(v)} = s + 1$, and the null distribution is assigned to v. If v lies in the range $[mid_0; mid_{s+1}]$ there are two cases. If $\underline{I(v)}$ and $\overline{I(v)}$ are equal the value v is exactly the midpoint of the interval $I_{\underline{I(v)}} = I_{\overline{I(v)}}$ and the decision distribution from this interval $P(dec = d_j | a_i \in I_{\underline{I(v)}})$ is used to compare v with other numerical values. Otherwise, the decision distribution for the value v is interpolated between the two neighboring intervals $I_{\underline{I(v)}}$ and $I_{\overline{I(v)}}$. The weights of the interpolation are proportional to the distances to the midpoints of the neighboring intervals (see Figure 3):

$$P_{IVDM}(dec = d_j | a_i = v) =$$
$$P(dec = d_j | a_i \in I_{\underline{I(v)}}) \cdot \frac{mid_{\overline{I(v)}} - v}{w_i} + P(dec = d_j | a_i \in I_{\overline{I(v)}}) \cdot \frac{v - mid_{\underline{I(v)}}}{w_i}.$$

The decision distributions for the values of a numerical attribute correspond to the broken line in the space of decision distributions in Figure 4. The dimension of this space is equal to the number of decisions $m = |V_{dec}|$. To define the IVDM metric these decision distributions for numerical values are used by analogy to the decision distributions for nominal values of nominal attributes the VDM metric. The IVDM distance between two numerical values is defined by Equation 5 as equal to the city-block distance between the two corresponding distributions in the space of decision distributions.

The IVDM metric can be explained by means of sampling the value of $P(dec = d_j | a_i \in I_p)$ at the midpoint mid_p of each discretized interval $[mid_p - \frac{w_i}{2}; mid_p + \frac{w_i}{2}]$. Then the IVDM metric interpolates between these sampled points to provide a continuous approximation of the decision probability $P(dec = d_j | a_i = v)$ for the whole range of values of the attribute a_i.

The IVDM metric is computationally effective. The limits of the range of values min_i, max_i, the interval width w_i and the decision distributions in the discretized intervals $I_0, \ldots, I_{s+1}$ for all attributes can be computed in linear time $O(|U_{trn}| |A|)$. The cost of the single distance computation is also linear $O(|A| |V_{dec}|)$: the two neighboring intervals of a value v can be determined in a constant time by the evaluation of the expressions:

$$
\underline{I(v)} = \begin{cases} 0 & \text{if } v < min_i - \frac{w_i}{2} \\ s+1 & \text{if } v > max_i + \frac{w_i}{2} \\ \left\lfloor \frac{v - min_i + \frac{w_i}{2}}{w_i} \right\rfloor & \text{if } v \in [min_i - \frac{w_i}{2}; max_i + \frac{w_i}{2}], \end{cases}
$$

$$
\overline{I(v)} = \begin{cases} 0 & \text{if } v < min_i - \frac{w_i}{2} \\ s+1 & \text{if } v > max_i + \frac{w_i}{2} \\ \left\lceil \frac{v - min_i + \frac{w_i}{2}}{w_i} \right\rceil & \text{if } v \in [min_i - \frac{w_i}{2}; max_i + \frac{w_i}{2}]. \end{cases}
$$

and the interpolation of two decision distributions can be computed in $O(|V_{dec}|)$.

Another extension of the VDM metric proposed by Wilson and Martinez is the Windowed Value Difference Metric (WVDM) [88]. It replaces the linear interpolation from the IVDM metric by sampling for each numerical value. The interval width w_i is used only to define the size of the window around the value to be sampled. For a given value v the conditional decision probability $P(dec = d_j | a_i = v)$ is estimated by sampling in the interval $[v - \frac{w_i}{2}; v + \frac{w_i}{2}]$ that v is the midpoint in:

$$
P_{WVDM}(dec = d_j | a_i = v) =
$$
$$
\begin{cases} 0 & \text{if } v \leq min_i - \frac{w_i}{2} \text{ or } v \geq max_i + \frac{w_i}{2} \\ \frac{|\{x \in U_{trn} : dec(x)=d_j \wedge |x_i - v| \leq \frac{w_i}{2}\}|}{|\{x \in U_{trn} : |x_i - v| \leq \frac{w_i}{2}\}|} & \text{if } v \in [min_i - \frac{w_i}{2}; max_i + \frac{w_i}{2}]. \end{cases}
$$

The WVDM metric locates each value v to be estimated in the midpoint of the interval to be sampled and in this way it provides a closer approximation of the conditional decision probability $P(dec = d_j | a_i = v)$ than the IVDM metric. However, the size of the window is constant. In many problems the density of numerical values is not constant and the relation of being similar between two numerical values depends on the range where these two numerical values occur. It means that the same difference between two numerical values has different meaning in different ranges of the attribute values. For example, the meaning of the temperature difference of the one Celsius degree for the concept of water freezing is different for the temperatures over 20 degrees and for the temperatures close to zero.

Moreover, in some ranges of the values of a numerical attribute the sample from the training set can be sparse and the set of the training objects falling into a window of the width w_i may be insufficiently representative to estimate correctly the decision probability. In the extreme case the sample window can even contain no training objects.

To avoid this problem we propose the Density Based Value Difference Metric (DBVDM) that is a modification of the WVDM metric. In the DBVDM metric the size of the window to be sampled depends on the density of the attribute values in the training set. The constant parameter of the window is the number of the values from the training set falling into the window rather than its width. To estimate the conditional decision probability $P(dec = d_j | a_i = v)$ for a given value v of a numerical attribute a_i the DBVDM metric uses the vicinity set of the value v that contains a fixed number n of objects with the nearest values of the attribute a_i. Let $w_i(v)$ be such a value that

$$\left| \left\{ x \in U_{trn} : |v - x_i| < \frac{w_i(v)}{2} \right\} \right| \leq n \text{ and } \left| \left\{ x \in U_{trn} : |v - x_i| \leq \frac{w_i(v)}{2} \right\} \right| \geq n.$$

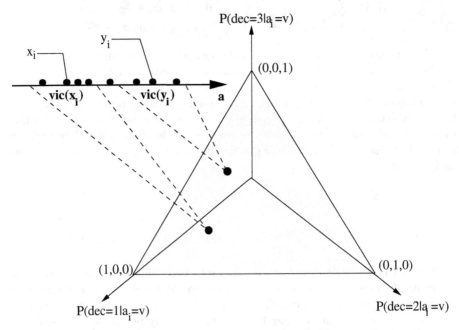

Fig. 5. The Density Based Value Difference Metric: The decision distributions for x_i, y_i are sampled from the windows $vic(x_i), vic(y_i)$ around x_i and y_i, repsectively, with a constant number of values in a training set

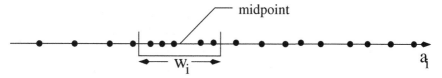

Fig. 6. A window with the midpoint ascending in the domain of values of a numerical attribute a_i

The value $w_i(v)$ is equal to the size of the window around v dependent on the value v. The decision probability in the DBVDM metric for the value v is defined as in the WVDM metric. However, it uses this flexible window size $w_i(v)$ (see Figure 5):

$$P_{DBVDM}(dec = d_j | a_i = v) = \frac{\left| \left\{ x \in U_{trn} : dec(x) = d_j \wedge |x_i - v| \le \frac{w_i(v)}{2} \right\} \right|}{\left| \left\{ x \in U_{trn} : |x_i - v| \le \frac{w_i(v)}{2} \right\} \right|}.$$

The DBVDM metric uses the sample size n as the invariable parameter of the procedure estimating the decision probability at each point v. If the value n is selected reasonably the estimation of the decision probability avoids the problem of having either too few or too many examples in the sample. We performed a number of preliminary experiments and we observed that the value $n = 200$ was large enough to provide representative samples for all data sets and increasing the parameter n above 200 did not improve the classification accuracy.

The WVDM and the DBVDM metric are much more computationally complex than the IVDM metric. The basic approach where the estimation of the decision probability for two numerical values v_1, v_2 to be compared is performed during distance computation is expensive: it requires to scan the whole windows around v_1 and v_2 at each distance computation. We propose another solution where the decision probabilities for all values of a numerical attribute are estimated from a training set a priori before any distance is computed.

Theorem 4. *For both metrics WVDM and DBVDM the range of values of a numerical attribute can be effectively divided into $2 \cdot |U_{trn}| + 1$ or less intervals in such a way that the estimated decision probability in each interval is constant.*

Proof. Consider a window in the domain of real values moving in such a way that the midpoint of this window is ascending (see Figure 6). In case of the WVDM metric the window has the fixed size w_i. All the windows with the midpoint $v \in \left(-\infty; min_i - \frac{w_i}{2}\right)$ contain no training objects. While the midpoint of the window is ascending in the interval $\left[min_i - \frac{w_i}{2}; max_i + \frac{w_i}{2}\right]$ the contents of the window changes every time when the lower or the upper limit of the window meets a value from the training set U_{trn}. The number of different values in U_{trn} is at most $|U_{trn}|$. Hence, each of the two window limits can meet a new value at most $|U_{trn}|$ times. Hence, the contents of the window can change at most $2 \cdot |U_{trn}|$ times. Since the decision probability estimation is constant if the contents of the window does not change there are at most $2 \cdot |U_{trn}| + 1$ intervals each with constant decision probability.

In the DBVDM metric at the beginning the window contains a fixed number of training objects with the lowest values of the numerical attribute to be considered. Formally, while the midpoint of the window is ascending in the range $\left(-\infty; min_i + \frac{w_i(min_i)}{2}\right)$ the upper limit of the window is constantly equal to $min_i + w_i(min)$ and the lower limit is ascending. Consider the midpoint ascending in the interval

$\left[min_i + \frac{w_i(min_i)}{2}; max_i - \frac{w_i(max_i)}{2}\right]$. In DBVDM the size of the window is changing but one of the two limits of the window is constant. If the lower limit has recently met an object from the training set then it is constant and the upper limit is ascending. If the upper limit meets an object it becomes constant and the lower limit starts to ascend. This repeats until the upper limit of the window crosses the maximum value max_i. Hence, as in WVDM, the contents of the window can change at most $2 \cdot |U_{trn}|$ times and the domain of numerical values can be divided into $2 \cdot |U_{trn}| + 1$ intervals each with constant decision probability.   □

Given the list of the objects from the training set sorted in the ascending order of the values of a numerical attribute a_i the proof provides a linear procedure for finding the intervals with constant decision probability. The sorting cost dominates therefore the decision probabilities for all the values of all the attributes can be estimated in $O(|A| |U_{trn}| \log |U_{trn}|)$ time. To compute the distance between two objects one needs to find the appropriate interval for each numerical value in these objects. A single interval can be found with the binary search in $O(\log |U_{trn}|)$ time. Hence, the cost of a single distance computation is $O(|A| |V_{dec}| \log |U_{trn}|)$. If the same objects are used to compute many distances the intervals corresponding to the attribute values can be found once and the pointers to these intervals can be saved.

All the metrics presented in this subsection: IVDM, WVDM and DBVDM use the information about the correlation between the numerical values and the decision from the training set. However, contrary to the city-block metric none of those three metrics is consistent with the natural linear order of numerical values (see Definition 2). Summing up, the metrics IVDM, WVDM and DBVDM are based more than the city-block metric on the information included in training data and less on the general properties of numerical attributes.

3.3   Weighting Attributes in Metrics

In the previous subsections we used the distance defined by Equation 1 without attribute weighting. This definition treats all attributes as equally important. However, there are numerous factors that make attributes unequally significant for classification in most real-life data sets. For example:

- some attributes can be strongly correlated with the decision while other attributes can be independent of the decision,
- more than one attribute can correspond to the same information, hence, taking one attribute into consideration can make other attributes redundant,

– some attributes can contain noise in values, which makes them less trust-worthy than attributes with the exact information.

Therefore, in many applications attribute weighting has a significant impact on the classification accuracy of the k-nn method [2, 51, 56, 85]. To improve the quality of the metrics described in Subsection 3.2 we also use attribute weighting and we replace the non-weighted distance definition from Equation 1 with the weighted version:

$$\rho(x, y) = \sum_{i=1}^{n} w_i \cdot \rho_i(x_i, y_i). \tag{6}$$

In the dissertation we combine attribute weighting with linear metrics. As we substantiated in Subsection 2.3 the linear metric is the optimal trade-off between the quality of the measure and its properties.

Weighting methods can be categorized along several dimensions [85]. The main criterion for distinction depends on whether a weighting algorithm com-putes the weights once following a pre-existing model or uses feedback from performance of a metric to improve weights iteratively. The latter approach has an advantage over the former one: the search for weight settings is guided by estimation how well those settings perform. Thus, attribute weights are adjusted to data more than in case of a fixed, pre-existing model. In this dissertation we propose the weighting methods that incorporate performance feedback.

The next distinction among algorithms searching in a weight space depends on the form of a single step in an algorithm. The algorithms fall into two categories:

– on-line algorithms: training examples are processed sequentially and the weights are modified after each example; usually the weights are modified in such a way that the distance to nearby examples from the same class is de-creased and the distance to nearby examples from other classes is increased,
– batch algorithms: the weights are modified after processing either the whole training set or a selected sample from the training set.

Online algorithms change weights much more often than batch algorithms so they require much less examples to process. However, for large data sets both online and batch algorithms are too expensive and an advanced indexing method must be applied (see Section 4). In such a case online algorithms are impractical: indexing must be performed every time when weights are modified, in online algorithms it is after each example. Therefore we focus our research on batch methods. Batch algorithms have the additional advantage: online algorithms are sensitive to an order of training examples, whereas batch algorithms are not.

Lowe [56] and Wettschereck [84] have proposed such batch algorithms using performance feedback. Both algorithms use the conjugate gradient to optimize attribute weights in order to minimize a certain error function based on the leave-one-out test on a training set. However, Lowe and Wettischereck's methods are applicable only to the specific weighted Euclidean metric. To make it possible to apply attribute weighting to different metrics we propose and test two batch

---

Algorithm 1. Attribute weighting algorithm optimizing distance

---

```
nearest(x) - the nearest neighbor of x in the sample S_trn
```

for each attribute $w_i := 1.0$
$modifier := 0.9$
$convergence := 0.9$
repeat l times
    $S_{trn} :=$ `a random training sample from` U_{trn}
    $S_{tst} :=$ `a random test sample from` U_{trn}
    $MR := \dfrac{\sum_{x \in S_{tst} : dec(x) \neq dec(nearest(x))} \rho(x, nearest(x))}{\sum_{x \in S_{tst}} \rho(x, nearest(x))}$
    for each attribute a_i
        $MR(a_i) := \dfrac{\sum_{x \in S_{tst} : dec(x) \neq dec(nearest(x))} \rho_i(x_i, nearest(x)_i)}{\sum_{x \in S_{tst}} \rho_i(x_i, nearest(x)_i)}$
    for each attribute a_i
        if $MR(a_i) > MR$ then $w_i := w_i + modifier$
    $modifier := modifier \cdot convergence$

---

methods based on less restrictive assumptions. They assume only that metrics are defined by the linear combination of metrics for particular attributes as in Equation 6. The first proposed method optimizes distance to the objects classifying correctly in a training set and the second one optimizes classification accuracy in a training set.

A general scheme of those algorithms is the following: they start with the initial weights $w_i := 1$, and iteratively improve the weights. At each iteration the algorithms use the distance definition from Equation 6 with the weights w_i from the previous iteration.

3.4   Attribute Weighting Method Optimizing Distance

Algorithm 1 presents the weighting method optimizing distance. At each iteration the algorithm selects a random training and a random test samples S_{trn} and S_{tst}, classifies each test object x from S_{tst} with its nearest neighbor in S_{trn} and computes the global misclassification ratio MR and the misclassification ratio $MR(a_i)$ for each attribute a_i. The misclassification ratio is the ratio between the sums of the distances to the nearest neighbors $\rho(x, nearest(x))$ for the incorrectly classified objects and for all training objects, respectively. Attributes with greater misclassification ratio $MR(a_i)$ than others have a larger share in the distance between incorrectly classified objects and their nearest neighbors. All attributes a_i that have the misclassification ratio $MR(a_i)$ higher than the global misclassification ratio MR have the weights w_i increased.

If the misclassification ratio $MR(a_i)$ of an attribute a_i is large then the distance between incorrectly classified objects and their nearest neighbors is influenced by the attribute a_i more than the distance between correctly classified objects and their nearest neighbors. The goal of weight modification is

Algorithm 2. Attribute weighting algorithm optimizing classification accuracy

$nearest(x)$ - the nearest neighbor of x with the same decision
in the sample S_{trn}
$\overline{nearest}(x)$ - the nearest neighbor of x with a different decision
in the sample S_{trn}

for each attribute $w_i := 1.0$
$modifier := 0.9$
$convergence := 0.9$
repeat l **times**
    $S_{trn} :=$ a random training sample from U_{trn}
    $S_{tst} :=$ a random test sample from U_{trn}
    $correct := \left|\{x : \rho(x, nearest(x)) \le \rho(x, \overline{nearest}(x))\}\right|$
    for each attribute a_i
        $correct(a_i) := \left|\{x : \rho_i(x_i, nearest(x)_i) \le \rho_i(x_i, \overline{nearest}(x)_i)\}\right|$
    for each attribute a_i
        if $correct(a_i) > correct$ **then** $w_i := w_i + modifier$
    $modifier := modifier \cdot convergence$

to replace incorrectly classifying nearest neighbors without affecting correctly classifying nearest neighbors. Increasing the weights of attributes with the large misclassification ratio gives a greater chance to reach this goal than increasing the weights of attributes with the small misclassification ratio.

In order to make the procedure convergable the coefficient $modifier$ used to modify the weights is decreased at each iteration of the algorithm. We performed a number of preliminary experiments to determine the appropriate number of iterations l. It is important to balance between the optimality of the final weights and the time of computation. For all tested data sets we observed that increasing the number of iterations l above 20 did not improve the results significantly and on the other hand the time of computations with $l = 20$ is still acceptable for all sets. Therefore in all further experiments we set the number of iterations to $l = 20$.

3.5   Attribute Weighting Method Optimizing Classification Accuracy

Algorithm 2 presents the weighting method optimizing classification accuracy. At each iteration the algorithm selects a random training and a random test samples S_{trn} and S_{tst} and for each test object x from S_{tst} it finds the nearest neighbor $nearest(x)$ with the same decision and the nearest neighbor $\overline{nearest}(x)$ with a different decision in S_{trn}. Then for each attribute a_i it counts two numbers. The first number $correct$ is the number of objects that are correctly classified with their nearest neighbors according to the total distance ρ, i.e., the objects for which the nearest object with the correct decision $nearest(x)$ is closer than the nearest object with a wrong decision $\overline{nearest}(x)$. The second number $correct(a_i)$ is the number of objects for which the component $\rho_i(x_i, nearest(x)_i)$ related to the attribute a_i in the distance to the correct nearest neighbor $\rho(x, nearest(x))$ is less than the corresponding component $\rho_i(x_i, \overline{nearest}(x)_i)$ in the distance to the wrong nearest neighbor $\rho(x, \overline{nearest}(x))$. If the number of objects correctly classified by a particular attribute a_i ($correct(a_i)$) is greater than the number of objects correctly classified by the total distance ($correct$), the weight for this attribute w_i is increased. Like in the previous weighting algorithm to make the procedure convergable the coefficient $modifier$ used to modify the weights is decreased at each iteration and the number of iterations is set to $l = 20$ in all experiments.

3.6   Experiments

In the next subsections we compare the performance of the k nearest neighbors method for the metrics and the weighting methods described in the previous subsections. We compare the Hamming metric (Subsection 2.4) and the Value Difference Metric (Subsection 3.1) for nominal attributes and the city-block metric (Subsection 2.4), the Interpolated Value Difference Metric and the Density Based Value Difference Metric (Subsection 3.2) for numerical attributes. Comparison between the Interpolated and the Windowed Value Difference Metric (Subsection 3.2) was presented in [88] and the authors reported that there was no significant difference between both metrics. Since the interpolated version is more efficient it was chosen to be compared in this dissertation. As the attribute weighting models we compare the algorithm optimizing distance, the algorithm optimizing classification accuracy and the model without weighting, i.e., all weights are equal $w_i := 1$.

To compare the metrics and the weighting methods we performed a number of experiments for the 10 benchmark data sets presented in Table 1. Each data set was partitioned into a training and a test set as described in Subsection 2.7 and the test set was classified by the training set with the k nearest neighbors method. Each data set was tested 5 times with the same partition and the average classification error is used for comparison. To compare accuracy we present the classification error for $k = 1$ and for k with the best accuracy for each data set. The results for k with the best accuracy are computed in the following way. In each test the classification error was computed for each value of k in the range $1 \leq k \leq 200$ and the smallest error among all k was chosen to compute the average error from 5 tests. It means that for the same data set the results of particular tests can correspond to different values of k.

3.7   Results for Data with Nominal Attributes Only

First, we compare the metrics and the weighting methods for data only with nominal attributes. Among the 10 described data sets there are 3 sets that contain only nominal attributes: *chess, nursery* and *splice*.

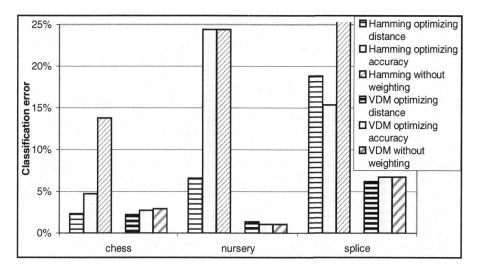

Fig. 7. The average classification error of the 1-nn for the two metrics: Hamming metric and VDM and for the three weighting models: Optimizing distance, optimizing classification accuracy and without weighting

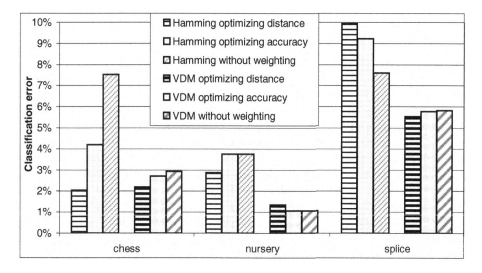

Fig. 8. The average classification error of the k-nn with the best k for the two metrics: Hamming metric and VDM and for the three weighting models: Optimizing distance, optimizing classification accuracy and without weighting

Figure 7 presents the average classification error of the nearest neighbor method for those 3 data sets. The graph presents the results for the two metrics and for the three weighting methods.

The results for the VDM metric are distinctly better than for the Hamming metric therefore we focus our attention on comparison of the weighting methods for the VDM metric.

The differences between the weighting methods are much smaller but the Student's t-test (see Subsection 2.7) indicates that they are still significant. In case of the data set *chess* the method optimizing distance outperforms the two others the maximum confidence level 99.5%. In case of the data set *nursery* the weighting does not help: the algorithm optimizing classification accuracy gives exactly the same result as without weighting and the algorithm optimizing distance gives a worse result with the confidence level 99.5%. In case of the data set *splice* the algorithm optimizing distance has again the lowest classification error but the statistical significance of the difference is only 97.5%.

Figure 8 presents the average classification error for the best value of k. As in case of the 1-nn, the results for the VDM metric are much better than for the Hamming metric so we compare the weighting methods for the VDM metric.

The table below presents the average value of k with the smallest classification error for particular data sets.

Metric	Hamming			VDM		
Weighting	optimizing distance	optimizing accuracy	none	optimizing distance	optimizing accuracy	none
chess	2.6	1.8	3	1	1	1
nursery	11.6	13	13	1	1	1
splice	152.4	78.4	158	8.2	7	7

In case of the data sets *chess* and *nursery* the average value of the best k for all weighting models for the metric VDM is 1 what means that in all tests the smallest error was obtained for $k = 1$. Hence, for those two data sets and for all the weighting models the average classification error for the best k is equal to the average classification error for $k = 1$ presented before. For *chess* the weighting method optimizing distance outperformed the others with the maximum confidence level 99.5% and for *nursery* the model without weighting provided exactly the same results like the weighting optimizing classification accuracy and both models outperformed the weighting optimizing distance also with the maximum confidence level 99.5%. In case of the data set *splice* the average values of the best k for all the weighting models are greater than 1 so the results are slightly different. In case of 1-nn the weighting optimizing distance is the best but only with the confidence level 97.5% whereas in case of the best k the weighting optimizing distance is the best with the maximum confidence level 99.5%.

The results for data sets with nominal attributes show clearly that the VDM metric is more accurate than the Hamming metric. In case of the Hamming metric the properties of the domain of values of a nominal attribute are only used (the equality relation), whereas in the case of the VDM metric the information contained in a training set is also used. In the latter case a structure of a metric is learnt from a set of values of an attribute in a training set. In comparison to the equality relation such a structure is much richer and it allows to adapt the VDM metric more accurately to data than in the case of the Hamming metric.

The comparison between the weighting methods is not unilateral. However, in most cases the method optimizing distance works best and in case when it loses (for the data set *nursery*) the difference is not so large: the error 1.38% of the weighting optimizing distance in comparison to the error 1.07% of the remaining methods.

3.8   Results for Data with Numerical Attributes Only

In this subsection we present the performance analysis of the metric and weighting models for data only with numerical attributes. There are 6 data sets that contain only numerical attributes: *letter, mushroom, pendigits, satimage, segment* and *shuttle*. All the tests for the data set *mushroom* gave the error 0% and all the tests for the data set *shuttle* gave an error not greater than 0.1%. These two data sets are very easy and the classification results for them can not be a reliable basis for comparison of different metrics and weighting methods. Therefore we exclude those two sets from analysis and we focus on the 4 remaining data sets: *letter, pendigits, satimage* and *segment*.

Figure 9 presents the average classification error of the nearest neighbor method for those 4 data sets. The graph presents the results for the three metrics and for the three weighting methods. First we compare again the metrics. The results are not so unilateral as in case of data with nominal attributes. The table below presents statistical significance of the differences in accuracy between the tested metrics.

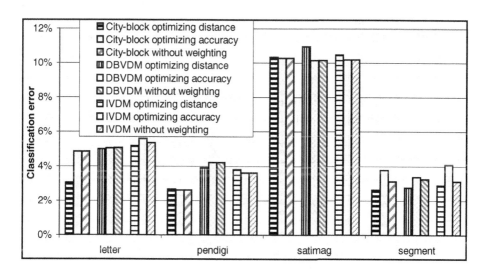

Fig. 9. The average classification error of the 1-nn for the three metrics: The city-block metric, DBVDM and IVDM and for the three weighting models: Optimizing distance, optimizing classification accuracy and without weighting

Weighting	optimizing distance	optimizing accuracy	none
letter	City-block 99.5%	City-block 99.5%	City-block 99.5%
pendigits	City-block 99.5%	City-block 99.5%	City-block 99.5%
satimage	City-block 90%	DBVDM 99,5%	DBVDM 99,5%
segment	City-block ¡90% (from DBVDM) 90% (from IVDM)	DBVDM 99.5%	City-block & IVDM 99.5%

Each cell in the table presents the metric (or metrics) that the best classification accuracy was obtained for, and explains the confidence level of the difference between this best metric and the other tested metrics for the data set given in the row header and with the weighting method given in the column header. For example, the cell on the crossing of the first row and the first column states that for the data set *letter* with the weighting method optimizing distance the best accuracy was obtained by the city-block metric and the probability that the city-block metric outperforms the others is 99.5%.

The results from the table indicate that the city-block metric wins in most cases, especially when combined with the weighting method optimizing distance. In this case the city-block metric is never worse: for *letter* and *pendigits* it wins with the maximum confidence level 99.5% and for *satimage* and *segment* the classification accuracy for all metrics is similar. In combination with the two other weighting methods the results are not unilateral but still the city-block metric dominates.

If we consider the value of k with the smallest classification error, in each test for the tree data sets: *letter*, *pendigits* and *satimage* it is usually greater than 1. The table below presents the average value of the best k for particular data sets:

Metric	City-block			DBVDM			IVDM		
Weighting	opt. dist.	opt. acc.	none	opt. dist.	opt. acc.	none	opt. dist.	opt. acc.	none
letter	1	5	5	3	2.6	3	1.4	1	1
pendigits	4.6	4	4	3.2	4	4	3.8	4	4
satimage	4.2	3	3	4.6	3	3	3.8	3	3
segment	1.6	1	1	1	1	1	1.4	1	1

Since in tests the best value of k was often greater than 1 the results are different from the case of $k = 1$. Figure 10 presents the average classification error for the best value of k and in the table below we present the winning metric (or metrics) and the confidence level of the difference between the winning metric and the others for the results at Figure 10:

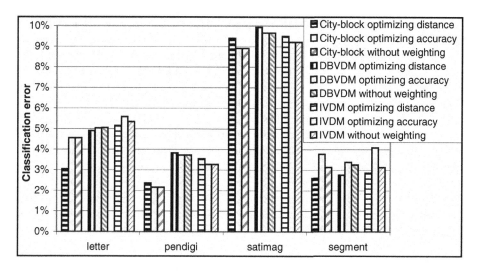

Fig. 10. The average classification error of the k-nn with the best k for the three metrics: The city-block metric, DBVDM and IVDM metric and for the three weighting models: Optimizing distance, optimizing classification accuracy and without weighting

Weighting	optimizing distance	optimizing accuracy	none
letter	City-block 99.5%	City-block 99.5%	City-block 99.5%
pendigits	City-block 99.5%	City-block 99.5%	City-block 99.5%
satimage	City-block ¡90% (from IVDM) 99.5% (from DBVDM)	City-block 99.5%	City-block 99.5%
segment	City-block ¡90% (from DBVDM) 90% (from IVDM)	DBVDM 99.5%	City-block & IVDM 99.5%

The results are even more unilateral than for the case of $k = 1$. The city-block metric loses only in one case: for the data set *segment* when combined with the weighting method optimizing classification accuracy.

The general conclusion is that the city-block metric is the best for data with numerical attributes and up to now different attempts to replace it with metrics induced from data like the VDM metric for nominal attributes are unsuccessful. This conclusion for numerical data is opposite to the analogous conclusion for nominal data. Like the Hamming metric the city-block metric uses mainly the properties of the domain of values of a numerical attribute. The probable reason for the opposite observation is that the properties of a numerical attribute are much more informative than the equality relation in case of a nominal attribute. In many cases the natural linear order in the domains of numerical attributes corresponds well with the properties of objects important for a decision attribute and the information provided in this linear order is rich enough to work well in the city-block metric. Therefore, it is difficult to construct a better metric from training data. The proposed metrics: DBVDM and IVDM are not consistent with the natural linear order of numerical values. The presented results show that this order is important for reasoning from numerical attributes.

Now, we compare accuracy of the weighting models. The results are presented in Figures 9 and 10 by means of the graphs used for comparison of metrics. The results are different for the different data sets and metrics. However, the city-block metric appeared to be generally the best so we focus on this metric. The table below presents the winning weighting method (or methods) and the confidence level of the difference between this winning method and the others in case of the 1-nn classification and in case of the classification with the best k (using the city-block metric):

k	$k = 1$	the best k
letter	optimizing distance 99.5%	optimizing distance 99.5%
pendigits	optimizing acc. & none ¡90%	optimizing acc. & none 99.5%
satimage	optimizing acc. & none ¡90%	optimizing acc. & none 99.5%
segment	optimizing distance 99% (from none) 99.5% (from optimizing acc.)	optimizing distance 99.5%

The results are not unilateral but we show that the method optimizing distance dominates for the city-block metric. For this metric the results of the method optimizing accuracy differs from the results without weighting only in case of the data set *segment*: the model without weighting provides a better classification. Then it is enough to compare the method optimizing distance to the model without weighting. For $k = 1$, in the cases where the method optimizing distance wins, the statistical significance of the difference is quite large: at least 99%, and the error reduction is also large: from 4.85% to 3.05% for *letter* (37% of the relative difference) and from 3.13% to 2.63% for *segment* (16% of the relative difference) whereas in cases when the method optimizing distance loses the difference is statistically insignificant and relatively very small: 2% of the relative difference for *pendigits* and 0.5% for *satimage*. For the best k all the differences are statistically significant with the maximum confidence level 99.5% but the relative differences are still in favour of the method optimizing distance: for *letter* and *segment* the reduction in error is similar to the case of $k = 1$ (33% and 17% respectively) and for *pendigits* and *satimage* the opposite relative differences in error are only 8% and 5% respectively.

The conclusion is that for the city-block metric it pays to apply the method optimizing distance because a gain in case of improvement can be much larger than a loss in case of worsening. For the two other metrics the results of the weighting methods are more similar and the method optimizing distance does not have the same advantage as in case of the city-block metric.

3.9   Results for Data with Numerical and Nominal Attributes

In this subsection we present analysis of the performance of the metrics and the weighting models for data with both nominal and numerical attributes. There is only one such a data set: *census94*. It is the most difficult data set among all the tested sets: the classification accuracy obtained for *census94* by different classification algorithms from the literature is the lowest [40, 53].

Figure 11 presents the average classification error of the 1-nn for the data set *census94* for all the combinations of the four joint metrics: the Hamming with the city-block, the VDM with the city-block, the VDM with the DBVDM and the VDM with the IVDM metric and the three weighting models: optimizing distance, optimizing accuracy and without weighting. The results are surprising: the best combination is the VDM metric for nominal attributes with the DB-VDM metric for numerical attributes. The same is in the analogous classification results of the *k*-nn with the best *k* presented at Figure 12.

Generally, the combinations of the VDM metric with its extensions for numerical attributes: DBVDM and IVDM work better than with the city-block metric. In a sense it is contradictory to the results for data only with numerical attributes. The possible explanation is that the data *census94* are more difficult and the information contained in the properties of the domain of numerical attributes does not correspond directly to the decision. The metrics DBVDM and IVDM are more flexible, they can learn from a training set more than the city-block metric and in case of such difficult data they can adapt more accurately to data.

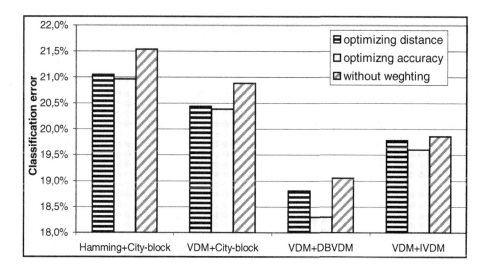

Fig. 11. The average classification error of the 1-nn for the four joint metrics: Hamming with the city-block metric, VDM with the city-block metric, VDM with DBVDM and VDM with IVDM and for the three weighting models: Optimizing distance, optimizing classification accuracy and without weighting, obtained for the data set *census94*

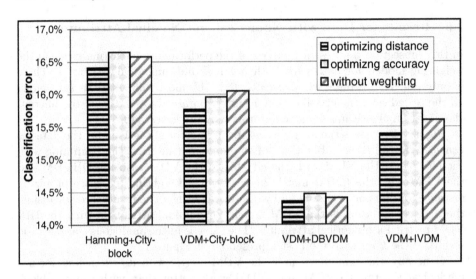

Fig. 12. The average classification error of the k-nn with the best k for the four joint metrics: Hamming with the city-block metric, VDM with the city-block metric, VDM with DBVDM and VDM with IVDM and for the three weighting models: Optimizing distance, optimizing classification accuracy and without weighting, obtained for the data set *census94*

In case of the simpler data sets from Subsection 3.8 the experimental results do not indicate clearly that one of the two metrics DBVDM or IVDM dominates. In case of the data set *census94* the difference between the DBVDM and the IVDM metric is more visible in favour of the DBVDM metric: the classification accuracy of VDM joint with DBVDM is always at least 1% better than the accuracy of VDM joint with IVDM.

Now, we compare the weighting models. The results for $k = 1$ and for the best k are quite different. For $k = 1$ the method optimizing classification accuracy gives the best classification in combination with all the metrics, whereas for the best k the method optimizing distance gives the best accuracy also for all metrics. It is related to the fact that the value of k with the best accuracy is always large for the data set *census94*. The table below presents the average of the best k for all the combinations of the metrics and the weighting methods:

	optimizing distance	optimizing accuracy	none
Hamming+City-block	43.4	23	45
VDM+City-block	34.2	23	27
VDM+DBVDM	84.2	61	83
VDM+IVDM	41a	31	41

The difference in classification between the method optimizing distance and the method optimizing accuracy is not large: in all cases it is below 3% of the relative classification error. The advantage of the method optimizing distance is that in all cases it gives a better result than the model without weighting whereas the method optimizing accuracy is sometimes worse.

3.10 Summary

In this section we have presented the following new methods:

- a Density Based Value Difference Metric for numerical attributes (Subsection 3.2): as distinguished from other metrics of this type from the literature [89, 88] the estimation of the decision probability in DBVDM depends on the density of values,
- an effective method for computing the distance between objects for the metrics WVDM [88] and DBVDM (Subsection 3.2),
- the two attribute weighting batch algorithms using performance feedback, applicable to the whole class of linear metrics: the first one optimizes distance in a training set (Subsection 3.4) and the second one optimizes classification accuracy in a training set (Subsection 3.5).

The experimental results presented in Subsections 3.7, 3.8 and 3.9 lead to the following final conclusions. For nominal attributes the general properties of the domain of values are poor and the information contained in training data is much richer and, therefore, it is important for classification accuracy to incorporate the information from training data into a metric. Hence, a good solution for nominal attributes is the Value Difference Metric described in Subsection 3.1. For numerical attributes the situation is different. The natural linear order provided in the properties of numerical attributes is an important, powerful source of information about objects and in most cases the city-block metric consistent with this natural linear order outperforms the metrics that do not regard this order so strictly. However, the results in Subsection 3.9 show that in cases where data are difficult and the relation between numerical attributes and a decision is not immediate the information contained in data can be still important for the accuracy of a metric. In this case the best classification accuracy has been obtained with use of the DBVDM metric.

In summary, the combination of the VDM metric for nominal attributes and the city-block metric for numerical attributes gives generally the best accuracy and we choose this metric to use in further research: on methods accelerating k nearest neighbors search described in Section 4 and on more advanced metric-based classification models described in Section 5. Since the DBVDM metric was the best for the most difficult data set and in a few other cases, in some experiments in Section 5 we use also the combination of the VDM and the DBVDM metric for comparison.

Comparison of the weighting models does not indicate a particular method to be generally better than others but weighting attributes in a metric usually improves classification accuracy so we decided to choose one for further experiments. Since both for nominal and numerical data the weighting algorithm optimizing distance seems to dominate this one is chosen to be always applied in all further experiments.

Some of the presented metrics and weighting algorithms have been included in the system RSES [8, 73]. The system provides different tools to analyze data, in particular the k nearest neighbors classifier. Two of the presented metrics: the joint VDM and city-block metric and the joint VDM and DBVDM metric and all the three presented weighting models are available in this classifier. They are implemented exactly as described in this section and some of the described parameters of the metrics and the weighting algorithms are available to be set by a user.

4   Distance-Based Indexing and Searching for k Nearest Neighbors

Distance-based indexing and the problem of searching for k nearest neighbors is investigated in this section.

4.1   Problem of Searching for k Nearest Neighbors

In this section we consider the efficiency problem of the k-nn classifier described in Subsection 2.5. For a long time k-nn was not used in real-life applications due to its large computational complexity. However, the development of methods accelerating searching and the technology advance in recent decade made it possible to apply the method to numerous domains like spatial databases, text information retrieval, image, audio and video recognition, DNA and protein sequence matching, planning, and time series matching (e.g., in stock market prognosis and weather forecasting) [3, 80].

The main time-consuming operation in the k-nn classifier is the distance-based searching for k nearest neighbors of a given query object. Distance-based searching is an extension of the exact searching widely used in text and database applications. It is assumed that a distance measure ρ is defined in a space of objects $\mathbb{X}$ and the problem is to find the set $NN(x, k)$ of k objects from a given training set $U_{trn} \subseteq \mathbb{X}$ that are nearest to a given query object x.

We restrict our consideration to application of k-nn for object classification. It requires fast access to data therefore we concentrate on the case when data are kept in the main memory. With growing size of the main memory in data servers this case attracts more and more attention of people working in different application areas.

The basic approach to searching for k nearest neighbors in a training set is to compute the distance from a query object to each data object in the training set and to select the objects with the smallest distances. The computational cost of finding the nearest neighbors from U_{trn} to all queries in a test set U_{tst} is $O(|U_{tst}| |U_{trn}|)$. In many applications the size of a database is large (e.g., several hundred thousand objects) and the cost $O(|U_{tst}| |U_{trn}|)$ is not acceptable. This problem is an important issue in many applications therefore it is the subject of the great interest among researchers and practitioners and a considerable effort has been made to accelerate searching techniques and a number of indexing

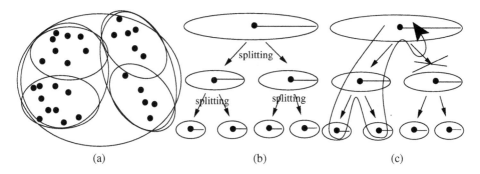

Fig. 13. Indexing and searching in a data set: (a) the hierarchical structure of data clusters (b) the indexing tree with the nodes corresponding to data clusters (c) search pruning in the indexing tree

methods both general and for particular applications have been developed. The most popular idea of indexing is a top-down scheme introduced by Fukunaga and Narendra [35]. It splits the whole training set into clusters in such a way that each cluster contains objects from the same region of a data space (Figure 13a). Each cluster has a compact representation that allows to check quickly whether the cluster can contain the nearest neighbors of a query object (Figure 13b). Instead of comparing a query object directly with each data object first it is compared against the whole regions. If a region is recognized not to contain the nearest neighbors it is discarded from searching. In this way the number of distance computations, and in consequence the performance time, are considerably reduced (Figure 13c).

In the literature one can find indexing methods based on the bottom-up scheme like Ward's clustering [82]. It was recognized that bottom-up constructions lead to a very good performance but instead of reducing the computational cost those bottom-up methods transfer it only from searching to indexing, i.e., searching is much faster but indexing has the $O(|U_{trn}|^2)$ complexity. Hence, this approach is too expensive for most of applications and the top-down scheme has remained the most popular in practice. In the dissertation we focus on the top-down scheme.

An important issue for indexing method construction are the initial assumptions made about a data space. Different models are considered in the literature. The first one assumes that data objects are represented by vectors from a vector space. This model is applicable to databases with numerical attributes or with complex multimedia objects transformable to vectors. It makes it possible to use the algebraic operations on objects in an indexing algorithm: summation and scaling, and construct new objects, e.g., the mean of a set of objects. However, not all databases fit to the model of a vector space. In the dissertation we consider data with both numerical and nominal attributes. The domains of nominal attributes do not have the structural properties used in the model with a vector space. The indexing methods for such data use only a small subset of the properties available in a vector space. Moreover, there are data not based on feature vectors, e.g., texts with the editing distance, DNA or time dependent sequences or plans. The structure of such objects is very specific and for such data the model is limited only by the distance axioms defined in Subsection 2.3. They are sufficient for the k nearest neighbors classification (see Subsection 2.5) therefore many indexing methods in the literature assume the distance model based only on these axioms.

Since a great part of applications is associated with structural databases and multimedia objects transformed to feature vectors a number of indexing techniques have been developed for vector spaces (e.g.,, quad-trees [29] and k-d trees [11]). The cost of a distance computing operation between two vectors is usually low so the methods such as grid-files [62], k-d-b tree [66], R-tree [43] and its variants R^+-tree [71] and R^*-tree [9] were focused on optimizing the number of I/O operations. The above techniques work well for low dimensional problems, but the performance degrades rapidly with increasing dimensionality. This phenomenon called the dimensional curse have been theoretically substantiated by Beyer et al. [13]. They proved that under certain reasonable assumptions the ratio of the distances to the nearest and the farthest neighbor converges to 1 while increasing the dimension of the vector space. To avoid the problem some specialized methods for high-dimensional spaces have been proposed: X-trees [12], SR-trees [47], TV-trees [55] and VA-files [83].

All the above tree based methods are based on regions in the shape of hypercubes so application of these methods is strictly limited to vector spaces. However, a large number of databases with other kinds of distance measures have raised an increase of interest in general distance-based indexing methods. An exhaustive overview of indexing methods for metric spaces is contained in [19]. SS-tree [86] uses a more general clustering scheme with spheres instead of rectangles as bounding regions but it is still limited to vector spaces because it uses the mean as the center of a cluster. A general distance-based indexing scheme is used in BST [46] and GHT [78]. Both trees have the same construction but different search pruning criteria are used. GNAT [18], SS-tree [86] and M-tree [20] are specialized versions of the BST/GHT tree. To balance the tree GNAT determines separately the number of child nodes for each node. As the splitting procedure GNAT uses the algorithm that selects the previously computed number of centers from a sample and assigns the objects from the parent node to the nearest centers. SS-tree and M-tree are focused on optimizing the number of I/O operations. They maintain a structure of nodes similar to B-trees and assume the dynamic growth of the database. Clustering in M-tree is similar to the clustering algorithm in SS-tree but M-tree uses either a random or a sampled set of the centers instead of the means. Thus, it uses only the distance function and is applicable to any metric space.

All the above mentioned indexing structures from the literature use a one-step clustering procedure to split a node in the tree. Such a procedure selects a number of cluster centers among objects in the given node and assigns each data object from the node to the nearest center. Moreover, the described searching methods use always a single search pruning criterion to accelerate searching in

Algorithm 3. The indexing schema

k - the splitting degree of the tree nodes
$root$ - the top node with all the training data objects from U_{trn}
$priorityQueue$ - the priority queue of leaf nodes used
                 for the selection of the next node to be split

$priorityQueue := \{root\}$
repeat
    $parent :=$ the next node from $priorityQueue$ to be split
    $splitCluster(parent, k)$
    add k child nodes of $parent$ to $priorityQueue$
until the number of nodes in $priorityQueue \geq \frac{1}{5}|U_{trn}|$

---

an indexing structure. In the next subsections we propose a new method that uses an iterative clustering procedure to split nodes while indexing instead of the one-step procedure and combines three search pruning criteria from BST, GHT and GNAT into one.

We present three versions of this method, depending on the model of data. The first version is appropriate for the model of a vector space, i.e., for data only with numerical attributes. The second variant is appropriate for the model of data considered in the dissertation, i.e., for data with both numerical and nominal attributes. It depends on the metric used to measure distance between data object too. Since the joint city-block and the Value Difference Metric provides the best classification accuracy in the experiments from Section 3 we present the version of indexing that assumes this metric to be used. As the third solution we propose the algorithm based on the most general assumption that only a distance measure is available for indexing.

4.2 Indexing Tree with Center Based Partition of Nodes

Most of the distance based indexing methods reported in the literature [11, 29, 43], [66, 71] and all the methods presented in the paper are based on a tree-like data structure. Algorithm 3 presents the general indexing scheme introduced by Fukunaga and Narendra [35]. All indexing algorithms presented in the paper fit to this scheme. It starts with the whole training data set U_{trn} and splits recursively the data objects into a fixed number k of smaller clusters. The main features that distinguish different indexing trees are the splitting degree of tree nodes k, the splitting procedure $splitCluster$ and the pruning criteria used in the search process.

Algorithm 3 assumes that the splitting degree k is the same for all nodes in the tree. An exception to this assumption is Brin's method GNAT [18] that balances the tree by selecting the degree for a node proportional to the number of data objects contained in the node. However, on the ground of experiments Brin concluded that a good balance was not crucial for the performance of the tree. In Subsection 4.4 we present the results that confirm this observation.

Algorithm 4. The iterative k-centers splitting procedure $splitCluster(objects, k)$

```
objects - a collection of data objects to be split
            into k clusters
Cl₁,...,Clₖ - partition of data objects from objects
            into a set of clusters
centers - the centers of the clusters Cl₁,...,Clₖ
prevCenters - the centers of clusters
              from the last but one iteration
getCenter(Clⱼ) - the procedure computing
              the center of the cluster Clⱼ


repeat
    centers := select k initial seeds c₁,...,cₖ from objects
    for each x ∈ objects
        assign x to the cluster Clⱼ
        with the nearest center cⱼ ∈ centers
    prevCenters := centers
    centers := ∅
    for each cluster Clⱼ
        cⱼ := getCenter(Clⱼ)
        add cⱼ to centers
until prevCenters = centers
```

We have assumed that the algorithm stops when the number of leaf nodes exceeds $\frac{1}{5}$ of the size of the training set $|U_{trn}|$, in other words when the average size of the leaf nodes is 5. It reflects the trade-off between the optimality of a search process and the memory requirements. To make the search process effective the splitting procedure $splitCluster$ has the natural property that data objects that are close each to other are assigned to the same child node. Thus, small nodes at the bottom layer of the tree have usually very close objects, and splitting such nodes until singletons are obtained and applying search pruning criteria to such small nodes do not save many distance comparisons. On the other hand, in our implementation the memory usage for the node representation is 2-3 times larger than for the data object representation so the model with the number of leaf nodes equal to $\frac{1}{5}$ of the number of data objects does not increase memory requirements as significantly as the model where nodes are split until the leafs are singletons and the number of all tree nodes is almost twice as the size of the training data set U_{trn}.

Algorithm 4 presents the iterative splitting procedure $splitCluster(objects, k)$ that generalizes the k-means algorithm. Initially, it selects k objects as the centers $c_1, \ldots, c_k$ of clusters. Then it assigns each object x to the cluster with the nearest center and computes the new centers $c_1, \ldots, c_k$. This assignment procedure is iterated until the same set of centers is obtained in two subsequent iterations.

The procedure $getCenter(\cdot)$ computes the center of a cluster of objects. Except for this procedure the presented indexing structure preserves the generality: it uses only the notion of distance. The indexing structure is correct for any definition of the procedure $getCenter(\cdot)$. However, the efficiency of searching in this structure depends strongly on how the centers of clusters are defined. Therefore we propose different definitions of the centers, depending on the information about the type of a space of objects.

In case of a vector space we propose the means as the centers of clusters:

$$getCenter(Cl) := \frac{\sum_{x \in Cl} x}{|Cl|}$$

In this case Algorithm 4 becomes the well known k-means procedure. Boley and Savaresi have proved the following property of the 2-means algorithm:

Theorem 5. *[70] If a data set is an infinite set of data points uniformly distributed in a 2-dimensional ellipsoid with the semi-axes of the length 1 and a $(0 < a < 1)$ the 2-means iterative procedure with random selection of initial centers has 2 convergence points: one is locally stable and one is locally unstable. The splitting hyperplanes corresponding to the convergence points pass through the center of the ellipsoid and are orthogonal to the main axes of the ellipsoid. In the stable convergence point the splitting hyperplane is orthogonal to the largest axis (see Figure 14).*

This theorem shows that in an infinite theoretical model the 2-means procedure with random selection of initial centers converges in a sense to the optimal

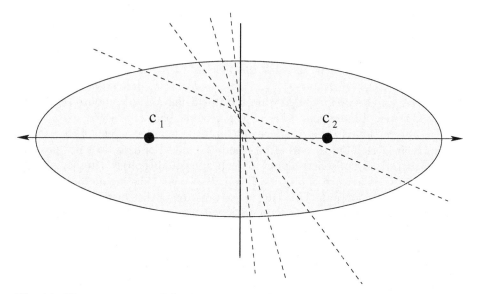

Fig. 14. The convergence of the 2-means procedure to the locally stable partition for data distributed uniformly in an ellipse; the splitting line is orthogonal to the largest axis of the ellipse

partition of data, which may substantiate good splitting properties of this procedure in practice and explain the good experimental performance of the tree based on the 2-means splitting procedure presented in Subsection 4.7.

In the dissertation, we consider data with both numerical and nominal attributes. In Section 3 the joint city-block and VDM metric was proved to provide the best classification accuracy. Therefore, for data with both types of attributes we present the definition of the center of a cluster $getCenter(Cl)$ that assumes this metric to be used for measuring the distance between objects.

The numerical attributes constitute a vector space. Therefore, as in the first version, we propose the mean to be the center value for each numerical attribute a_i:

$$getCenter(Cl)_i := \frac{\sum_{x \in Cl} x_i}{|Cl|}.$$

In case of a nominal attribute the domain of values does not provide the operations of summation and division and the only general property of nominal values is the equality relation. Therefore, as in the problem of metric definition, to define the center of a set of nominal values we use the information encoded in data. Since the centers of clusters are used only by the operation of the distance computation, it is enough to define how the center of nominal values is represented in Equation 5 defining the Value Difference Metric. This equation does not use values directly but it uses the decision probability estimation $P_{VDM}(dec = d_j | a_i = v)$ for each nominal value v. Therefore, for each nominal attribute a_i it is enough to define the analogous decision probability estimation for the center of a set of nominal values:

$$P_{VDM}(dec = d_j | a_i = getCenter(Cl)_i) := \frac{\sum_{x \in Cl} P_{VDM}(dec = d_j | a_i = x_i)}{|Cl|}.$$

This definition of the decision probability estimation for the centers of clusters is correct, because it satisfies the axioms of probability: $P_{VDM}(dec = d_j | getCenter(Cl)_i) \geq 0$ and $\sum_{d_j \in V_{dec}} P_{VDM}(dec = d_j | getCenter(Cl)_i) = 1$. The indexing and searching algorithm use this definition to compute the VDM distance between centers and other objects from a space of objects.

The last version of the procedure $getCenter(\cdot)$ is independent of the metric definition. It is useful in the situation where the model of data does not provide the information how to construct new objects and training objects in U_{trn} are the only objects from a space of objects $\mathbb{X}$ available for an indexing method. In this general case we propose the following approximation of the cluster center. When a cluster Cl contains one or two data objects it selects any of them as the center of Cl. Otherwise the algorithm constructs a sample S that contains the center used to assign objects in the previous iteration of the procedure $splitCluster$ and randomly selected $\max(3, \lfloor \sqrt{|Cl|} \rfloor)$ other objects from Cl. Then it computes the distances among all pairs of objects from S, and selects the object in S that minimizes the second moment of the distance ρ in S, as the new center of Cl:

$$getCenter(Cl) := \arg \min_{x \in S} E\left(\rho(x, y)^2\right).$$

In this way it selects the center from S that minimizes the variance of S. The assumption that the center from the previous iteration is included into the sample S in the next iteration makes it possible to use the previous center in the next center selection. It provides a chance for the stopping condition to be satisfied at each iteration and saves a significant number of unnecessary iterations.

The choice of the value $\max(3, \left\lfloor \sqrt{|Cl|} \right\rfloor) + 1$ as the size of the sample S in this center selection algorithm is strictly related to its complexity. A single iteration of the algorithm requires $|S|^2$ distance computations: it computes the distance among all pairs of objects in S. Since the size of the sample S is $O(|Cl|^{\frac{1}{2}})$ the computational cost of a single iteration remains linear with respect to the cluster size $|Cl|$, and thus, it is comparable to the case of the k-means procedure used for vector spaces.

The last algorithm selects the approximate centers for clusters among objects belonging to these clusters. Therefore, in the next subsections we call it the k-approximate-centers algorithm.

The discussion and experimental analysis related to selection of initial centers in Algorithm 4 and the degree of nodes of the presented indexing structure are presented in the next two subsections.

4.3   Selection of Initial Centers

One can consider three general approaches for selection of the initial centers for clusters in the procedure *splitCluster* (see Algorithm 4) known from the literature: random, sampled [18] and exhaustive. The description of BST [46] and GHT [78] is quite general and either it does not specify any particular selection of initial centers or it assumes a simple random model. M- [20] and SS-trees [86] are the dynamic structures and the splitting procedures assume that they operate on an existing inner node of a tree and they have access only to the information contained in a node to be split. While splitting a non-leaf node the algorithm does not have access to all data objects from the subtree of the node so the splitting procedures from M- and SS-trees are incomparable to the presented iterative procedures.

To select the initial centers in GNAT [18] a random sample of the size $3k$ is drown from a set of data objects to be clustered and the initial k centers are picked from this sample. First, the algorithm picks one of the sample data objects at random. Then it picks the sample point that is the farthest from this one. Next, it picks the sample point that is the farthest from these two, i.e., the minimum distance from the two previously picked seeds is the greatest one among all unpicked sample objects. Finally, it picks the point that is the farthest from these three and so on until there are k data points picked.

In the dissertation, we propose yet another method for selecting initial k centers presented in Algorithm 5. It is similar to GNAT's method but it selects the first center more carefully and for selection of the others it uses the whole set to be clustered instead of a sample. Therefore we call this algorithm the

Algorithm 5. The global algorithm for selection of initial centers in the procedure *splitCluster*

$c := getCenter(Cl)$
$c_1 := \arg\max_{x \in Cl} \rho(c, x)$
for $j := 2$ to k
    $c_j := \arg\max_{x \in Cl} \min_{1 \leq l \leq j-1} \rho(c_l, x)$

---

global selection of the farthest objects. First, the algorithm computes the center c of the whole set to be clustered Cl. As the first seed c_1 it picks the object that is the farthest from the center c of the whole data set. Then it repeats selection of the farthest objects as in GNAT, but from the whole set Cl instead of from a sample. The algorithm can be performed in $O(|Cl| k)$ time: it requires to store the minimal distance to selected centers $\min_{1 \leq l \leq j-1} \rho(c_l, x)$ for each object $x \in Cl$ and to update these minimal distances after selection of each next center. For small values of k this cost is acceptable.

One can consider the exhaustive procedure that checks all k-sets among objects to be clustered as the sets of k centers and selects the best one according to a predefined quality measure. However, the computational cost of this method $O(|Cl|^k)$ does not allow us to use it in practice.

Figure 15 presents the performance of the search algorithm for three different seeding procedures used in the k-means based indexing trees with $k = 2$, $k = 3$ and $k = 5$: a simple random procedure, GNAT's sampled selection of the farthest objects and the global selection of the farthest objects described above. The experiments have been performed for the joint city-block and VDM metric with the representation of the center of a cluster extended to nominal attributes as described in the previous subsection, and for the searching algorithm described in Subsections 4.5 and 4.6. All 12 benchmark data sets presented in Tables 1 and 2 have been tested in the following way: the training part of a data set have been indexed with the k-means based indexing tree (once for each combination of $k \in \{2, 3, 5\}$ and the three seeding procedures), and for each object in a test set the two searches have been performed in each indexing tree: for the 1 nearest neighbor and for the 100 nearest neighbors of the test object. At each search the number of distance computations has been counted. The graphs present the average number of distance computations for the whole test set in the 1-nn and the 100-nn search.

The results indicate that the indexing trees with all three methods have comparable performance what may be explained with the good theoretical convergence property of the k-means algorithm formulated in Theorem 5. However, for a few larger data sets: *census94*, *census94-95*, *covertype* and *letter* the difference between the global and the two other selection methods is noticeable. In particular, the largest difference is for the data set *census94-95*, e.g., in case of the 2-means based indexing tree the global method takes only 65% of the time of the sampled method and 60% of the time of the random method (as presented at the two upper graphs at Figure 15).

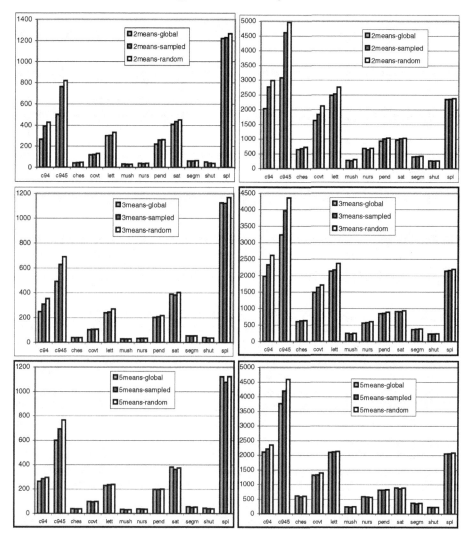

Fig. 15. The average number of distance computations per single object in 1-nn search (the left graphs) and 100-nn search (the right graphs) with the use of 2-means based, 3-means based and 5-means based indexing trees, and with the three different methods of initial center selection: Globally farthest, sampled farthest and random

Summing up, the global method seems to have a little advantage over the others and we decide to use this one in further experiments described in the next subsections.

4.4   Degree of the Indexing Tree

In order to analyze the performance of the k-means based indexing trees, in dependence on the degree of nodes k, we have performed experiments for 8

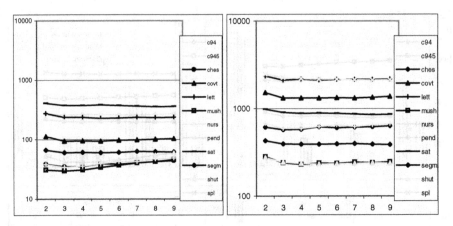

Fig. 16. The average number of distance computations per single object in 1-nn search (the left graph) and 100-nn (the right graph) with the use of the k-means based indexing trees with k in the range $2 \le k \le 9$

successive values of k ranging from 2 to 9. Figure 16 presents the performance graphs for particular data sets. As it is shown they are quite stable in the range of tested values except for the value 2 and different values of k have the best performance for particular data sets. For the 1-nn search 7 data sets have the best performance at $k = 3$, 1 at $k = 4$ and 2 at $k = 5$ and $k = 8$. For 100-nn search 4 data sets have the best performance at $k = 3$, 2 at $k = 4, 5$ and 8 and 1 at $k = 7$ and 9. These statistics indicate that the best performance is for small values of k (but greater than 2). Assignment of k to 3, 4 or 5 ensures almost optimal performance.

In the literature the splitting degree of tree nodes is usually assumed to be constant over all nodes in a tree. The exception to this rule is the GNAT structure [18] that attempts to balance the size of branches by choosing different splitting degrees for nodes. It assumes a fixed k to be the average splitting degree of nodes and applies the following procedure to construct a tree. The top node is assigned the degree k. Then each of its child nodes is assigned the degree proportional to the number of data points contained in this child node (with a certain minimum and maximum) so that the average degree of all the child nodes is equal to the global degree k. This process works recursively so that the child nodes of each node have the average degree equal to k. In his experiments Brin set the minimum of the degree to 2 and the maximum to $\min(5k, 200)$. On the basis of experiments he reported that good balance was not crucial for the performance of the tree.

We have implemented this balancing procedure too. In case of $k = 2$ the value 2 is both the average and the minimal possible value of the splitting degree in the k-means balanced indexing tree so the balancing procedure assigns the degree 2 to all nodes and it behaves identically as in case of the constant degree 2. Hence, the comparison of the balanced and the constant degree selections makes sense for the value of k greater than 2. Figure 17 presents the comparison between the

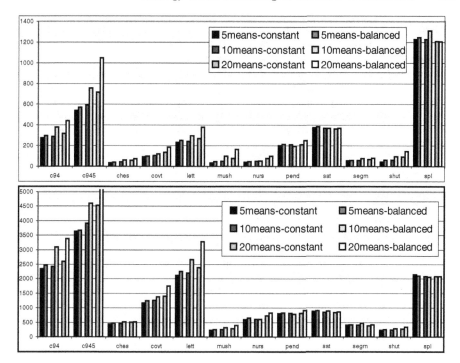

Fig. 17. The average number of distance computations per single object in 1-nn search (the upper graph) and in 100-nn search (the lower graph) with the use of the k-means based indexing trees with constant and with balanced degrees of tree nodes; for each data set the first pair of columns represents the performance for the constant and for the balanced degree $k = 5$, the second pair represents the performance for $k = 10$ and the third one for $k = 20$

k-means based balanced trees where k is the average degree of child nodes and the corresponding k-means trees with the constant degree k. The results show that the balancing procedure does not improve performance of the tree with a constant degree and in many experiments searching in the tree with a constant degree is even faster. It indicates that in order to make profit from balancing more sophisticated procedures are required. Up to now it is not known whether there is a balancing policy with acceptable computational complexity having a significant advantage over the non-balanced structures.

4.5 Searching in the Indexing Tree

In this subsection we present Algorithm 6 that is a general searching schema finding a fixed number k of data objects nearest to the query q [35]. The algorithm traverses the indexing tree rooted at *root* in the depth-first order. In *nearestQueue* it stores the nearest data objects, maximally k, from already visited nodes. At each tree node n the algorithm checks with pruning criteria

---

Algorithm 6. Searching schema

$root$ - the root node of the indexing tree to be searched
$nodeStack$ - the stack of nodes to be searched
$nearestQueue$ - the queue of the data objects nearest to q
    sorted according to the distance ρ
$discard(n :node, q :query, r_q :range)$ - the procedure checking whether
    pruning criteria apply to a node n while searching
    the neighbors of q in the distance less or equal r_q

$nodeStack := \{root\}$
repeat
    $n :=$ pull the top node from $nodeStack$
    $r_q := \max_{x \in nearestQueue} \rho(q, x)$
    if $|nearestQueue| < k$ or not $discard(n, q, r_q)$
        if n is a leaf
            for each data object $x \in n$
                if $|nearestQueue| < k$ then add x to $nearestQueue$
                else
                    check x against the farthest
                    object $y \in nearestQueue$
                    and replace y with x if $\rho(q, x) < \rho(q, y)$
        else
            push the child nodes of n to $nodeStack$ in the decreasing
            order of the distance of the the child node centers
            to the query q (the nearest on the top)
until $nodeStack$ is empty
return $nearestQueue$

---

whether n should be visited, i.e., whether n can contain an object that is closer to the query q than any previously found nearest neighbor from $nearestQueue$. If so and the node n is a leaf, it compares each data object $x \in n$ against data objects in $nearestQueue$ and replaces the farthest object y from $nearestQueue$ by x, if x is closer to the query q than y. In case where the node n is an inner node it adds the child nodes of n to $nodeStack$ to be visited in the future.

The important issue for efficiency of the algorithm is the selection of a heuristic procedure determining the order of visiting child nodes. The child nodes of the same parent node are visited always in the increasing order of the distance between the center of a child node and the query q, i.e., the child node with the nearest center is visited first and the child node with the farthest center is visited last. The closer center of a child node is to the query q the closer objects to the query are contained in this node. If the nodes with the nearest centers are visited first, it is more probable to find near neighbors quickly and to limit the range of search r_q to a small radius. Thus, more nodes are discarded during further search.

In Subsection 4.6 different node pruning criteria for the function *discard* are described and compared.

4.6   Optimization of Searching in the Indexing Tree

Algorithm 6 presents the searching procedure that uses search pruning criteria to discard nodes while traversing the indexing tree. If the algorithm finds the first k objects and inserts them to *nearestQueue* it starts to check with the procedure $discard(n, q, r_q)$ whether subsequent visited nodes can contain objects closer to the query q than any previously found neighbor from *nearestQueue*. The algorithm does it in the following way: it stores the current search radius r_q defined as the distance $\rho(q, y)$ between the query q and the farthest from q object $y \in nearestQueue$ and for each visited node it checks whether the node can contain an object x such that $\rho(q, x) < r_q$.

The definition of the k nearest neighbor classification model from Subsection 2.5 does not use the axioms of metric from Subsection 2.3. Those axioms are not required for the model but they serve two other purposes. In the first place, they represent mathematically the natural properties of the notion of analogy. Second, all the metric axioms are necessary for correctness of search pruning rules described in the literature [18, 20, 46, 78, 86]. In this subsection we describe all these pruning rules and we propose a combination of the presented rules into one rule.

The most common search pruning criterion applied in BST [46], SS-tree [86] and M-tree [20] uses the covering radius (Figure 18a). Each node n of the indexing tree keeps the center c_n computed with the function $getCenter(n)$ and the covering radius r_n:

$$r_n := \max_{x \in n} \rho(c_n, x).$$

A node n is discarded from searching if the intersection between the ball around q containing all nearest neighbors from *nearestQueue* and the ball containing all members of the node n is empty:

$$\rho(c_n, q) > r_q + r_n.$$

Uhlmann has proposed another criterion for his Generalized-Hyperplane Tree (GHT) [78]. The important assumption for correctness of this criterion is that at the end of the splitting procedure (see Algorithm 4) each object from a parent node is assigned to the child node with the nearest center. It is ensured with the stopping condition: the splitting procedure stops if the centers from the last and the last but one iteration are the same. The procedure returns the object assignment to the centers from the last but one iteration and this stopping condition makes this assignment appropriate for the final centers too.

Uhlmann's criterion uses the hyperplanes separating the child nodes of the same parent (Figure 18b). A node n_i is discarded if there is a brother node n_j of n_i (another child node of the same parent node as n_i) such that the whole query ball is placed beyond the hyperplane separating n_i and n_j (midperpendicular to the segment connecting the centers c_{n_i} and c_{n_j}) on the side of the brother node n_j:

$$\rho(c_{n_i}, q) - r_q > \rho(c_{n_j}, q) + r_q.$$

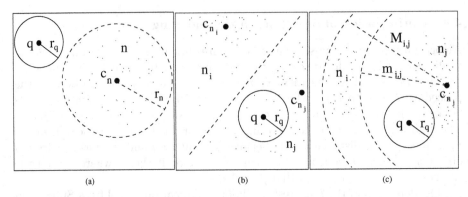

Fig. 18. The three search pruning criteria: (a) the covering radius from BST (b) the hyperplane cut from GHT (c) the rings-based from GNAT

The third pruning criterion used in Brin's GNAT tree [18] is also based on a mutual relation among brother nodes but it is more complex (Figure 18c). If the degree of a tree node is k then each child node n_i keeps the minimum $m_{i,1}, \ldots, m_{i,k}$ and the maximum $M_{i,1}, \ldots, M_{i,k}$ distances between its elements and the centers $c_{n_1}, \ldots, c_{n_k}$ of the remaining brother nodes:

$$m_{i,j} = \min_{x \in n_i} \rho(c_{n_j}, x),$$

$$M_{i,j} = \max_{x \in n_i} \rho(c_{n_j}, x).$$

The node n_i is discarded if there is a brother node n_j such that the query ball is entirely placed outside the ring around the center of n_j containing all members of n_i:

$$\text{either } \rho(c_{n_j}, q) + r_q < m_{i,j} \text{ or } \rho(c_{n_j}, q) - r_q > M_{i,j}.$$

The covering radius and the hyperplane criterion require only to store the center c_n and the covering radius r_n in each node n. The criterion based on the rings requires more memory: each node stores the $2(k-1)$ distances to the centers of the brother nodes.

All the three described criteria are based on the notion of the center of a node. The hyperplane based criterion requires moreover the object assignment condition to be satisfied but it is ensured with the stopping condition of the splitting procedure. Hence, all the three criteria can be applied simultaneously to the indexing structure described in Subsection 4.2, and in this dissertation we propose their combination as the complex criterion for acceleration of the nearest neighbors search.

Figure 19 presents the experimental comparison of the performance for all possible combinations of the three criteria. In a single form the most effective criterion is the covering radius, the least effective is the hyperplane criterion and the differences in performance among all three criteria are significant. In case of the 100-nn search the covering radius alone is almost as powerful as all the three criteria. Addition of the two remaining criteria does not increase

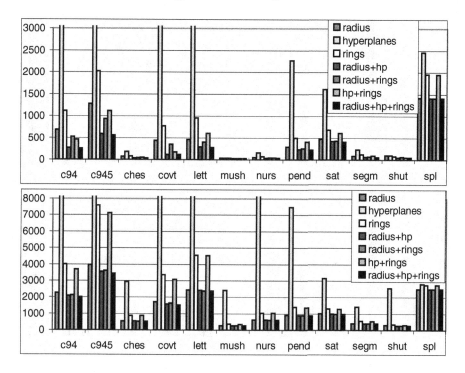

Fig. 19. The average number of distance computations per single object in 1-nn search (the upper graph) and 100-nn search (the lower graph) with the use of the 2-means based indexing tree, and with all the possible combinations of the three search pruning criteria: The covering radius, the hyperplanes and the rings

the performance. The different behavior is observed in case of the 1-nn search: none of them is comparable to the case where all the three criteria are applied. Both the covering radius and the hyperplane cut are crucial for the performance and only the rings based criterion can be removed with no significance loss in the performance.

The presented results indicate that the combination of the different criteria improves the performance of the k-nn search with a single criterion at least for small values of k. On the other hand, in both cases of the 1-nn and the 100-nn search addition of the memory consuming criterion based on rings does not improve the combination of the two remaining criteria. This result may suggest that the covering radius and the hyperplanes provide the optimal pruning combination and there is no need to search for a more sophisticated pruning mechanism.

4.7   Analysis of Searching Cost in the Indexing Tree

The most interesting question is how much the search process profits from the additional cost due to the iterative splitting procedure presented in Algorithm

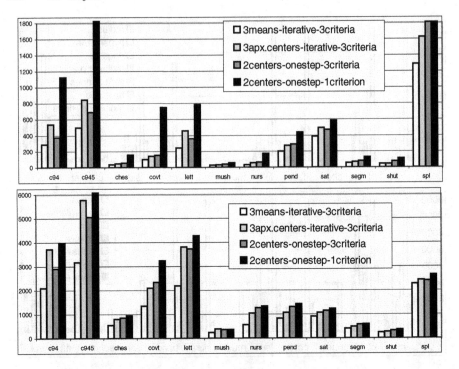

Fig. 20. The average number of distance computations per single object in 1-nn search (the upper graph) and 100-nn search (the lower graph) with the use of the indexing trees with the iterative 3-means, with the iterative 3-approximate-centers, and with the one-step 2-centers splitting procedure, this last tree in two search variants: With the combination of the 3 search pruning criteria and with the single covering radius criterion

4 and the combined search pruning criterion from the previous subsection in comparison to the case with the one-step procedure and a single pruning criterion. The iterative procedure selects initial centers, assigns the data objects to be split to the nearest centers and computes new centers of clusters. Then, the assignment of the data objects to the centers and computation of the new cluster centers is iterated as long as the same set of the cluster centers is generated in two subsequent iterations. The one-step procedure works as in the other indexing trees BST, GHT, GNAT, SS-tree and M-tree. It stops after the first iteration and uses the initial centers as the final ones. The globally farthest data objects are used as the set of the initial centers both in the iterative and in the non-iterative splitting procedure.

Figure 20 presents the cost of searching in the trees with the iterative k-means, with the iterative k-approximate-centers and with the one-step k-centers splitting procedure. The results both for the trees with the iterative procedures and for the first tree with the one-step procedure are obtained with the use of the combination of all the three search pruning criteria. The fourth column at each data set presents the performance of the one-step based tree with the single covering radius criterion. We chose this criterion for comparison since it had the best performance among all the three tested criteria (see Subsection 4.6). Except for a single case we have observed that the performance of the one-step based trees deteriorates while increasing k (it has been checked for $k = 2, 3, 5$ and 7). Then for comparison we have selected the most competitive value $k = 2$ (the exception was the 100-nn search in the data set *splice*, the case $k = 5$ has provided the best performance, and hence, this case has been presented at the graph instead of $k = 2$). In case of both iterative procedures the value $k = 3$ was used since it is one of the most optimal values (see Subsection 4.4).

While comparing the performance of the iterative 3-means (the first column) and the one-step 2-centers (the third column) procedures the profit from applying the iterative procedure is noticeable. In case of the 1-nn search the savings range from 20% (*satimage*) to 50% (*nursery*), in case of the 100-nn search the savings are similar to the 1-nn case, except for a single data set *splice* where the saving is 5%. These results indicate that replacing the one-step procedure with the iterative 3-means procedure can improve the performance even twice.

The comparison between the third and the fourth column presents the profit for the tree with the one-step procedure only from the application of the combined search pruning criterion instead of the single one. For the 1-nn search the combined criterion outperforms significantly the single one, in particular for the largest data sets (*census94, census94-95, covertype*) the acceleration reaches up to several times. For the 100-nn search the difference is not so large but it is still noticeable. These results show that for the tree with the one-step splitting procedure the complex criterion is crucial for the performance of the tree. In case of the tree with the k-means splitting procedure the results are different, i.e., the difference in performance between the single covering radius and the combined criteria is much smaller (see Subsection 4.6). It indicates that the iterative k-means procedure has very good splitting properties and the choice of the search pruning criterion for this case is not so crucial as for the non-iterative case.

The comparison between the first and the fourth columns shows that the tree with the 3-means splitting procedure and the complex search pruning criterion is always at least several tens percent more effective than the tree with the one-step procedure and a single criterion. In case of the 1-nn search the former tree is usually even several times more effective than the latter one.

We obtain different conclusions while comparing the iterative k-approximate-centers (the second column) and the one-step (the third column) procedures. Although for most of data sets the iterative procedure outperforms the non-iterative one, the differences in the performance are usually insignificant and for the three large data sets (*census94, census94-95, letter*) the performance of the iterative procedure is even worse than the performance of the non-iterative one. These results indicate that in case of the tree with the k-approximate-centers the profit in the performance is mainly due to the complex search criterion. Since the only feature that differentiates the k-means and the k-approximate-centers procedures is how the algorithm selects and represents the center of a cluster of data objects this feature seems to be an important issue for the performance of indexing trees.

Uhlmann has introduced another type of an indexing structure: the vantage point tree [78]. It is the binary tree constructed in such a way that at each node the data objects are split with the use of the spherical cut. Given a node n the splitting algorithm picks an object $p \in n$, called the vantage point, and computes the median radius M, i.e., half of the data objects from n fall inside the ball centered at the vantage point p with the radius M and half of them fall outside this ball. The objects inside the ball $\{x \in n : \rho(p, x) \leq M\}$ are inserted into the left branch of the node n and the objects outside the ball $\{x \in n : \rho(p, x) > M\}$ are inserted into the right branch. The vantage point tree is balanced and the construction takes $O(n \log n)$ time in the worst case. While searching for the nearest neighbors of a query q the branch with objects inside the ball is pruned if $M + \rho(q, x_{nearest}) < \rho(p, q)$ and the branch with the objects outside the ball is pruned if $M - \rho(q, x_{nearest}) > \rho(p, q)$. Yianilos has described an implementation of the vantage point tree with sampled selection of the vantage point [95]. For the experimental comparison we have implemented this structure as described by Yianilos.

Figure 21 presents the comparison of the performance of the trees with the 2-means, with the 2-approximate-centers and with the vantage point splitting procedure (since the vantage point tree is a binary tree we use $k = 2$ in all the tested trees to make them comparable). The result are presented only

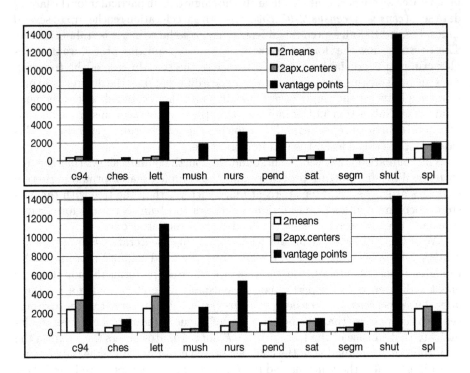

Fig. 21. The average number of distance computations in 1-nn search (the upper graph) and 100-nn search (the lower graph) with the use of the indexing trees with the k-means, with the k-approximate-centers and with the vantage point splitting procedure

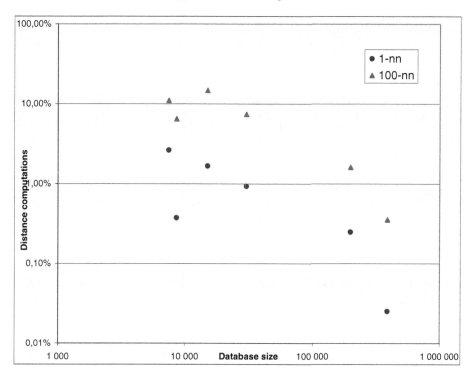

Fig. 22. The average reduction of the number of distance computations in relation to linear search obtained by the 3-means based indexing tree with the 3 search pruning criteria presented in percentage terms in dependence on the training set size

for 10 data sets because for the 2 largest data sets: *census94-95* and *cover-type* the experiments with the vantage point tree takes too much time. The results show a large advantage of the trees based on the centers over the tree based on the vantage points. It indicates that the center based representation of tree nodes provides better search pruning properties than the vantage point based representation.

An interesting question is how much of the searching cost the indexing tree with the iterative splitting procedure and the three pruning criteria reduces in relation to the linear scan of a training set. Figure 22 presents the average reduction of the linear search cost for training sets of the different sizes (for the six largest data sets from all 12 sets in Tables 1 and 2). A particularly advanced acceleration level has been reached for the two largest data sets. The size of the data set *covertype* is almost 400 thousand, whereas the average number of distance comparisons per single object (the fourth column set from the left at Figure 20) is less than 100 for the 1-nn search and close to 1300 for the 100-nn search. It means that the 3-means based tree reduces the cost of searching 4000 times in case of the 1-nn search and 300 times in case of the 100-nn search. For the second largest data set *census94-95* (the second column set from the left at Figure 20, the size almost 200 thousand) the reductions in cost are 400

times and 60 times, respectively. This good performance has been reached both
due to the improved splitting procedure and due to the use of the complex
search criterion.

4.8   Comparison of Searching and Indexing Cost

The results from the previous subsection have proved that the k-means based
indexing tree is a good accelerator of searching for the nearest neighbors. The
question arises whether the cost of constructing a tree is not too large in comparison to the cost of searching.

Figure 23 presents the comparison between the number of computed distances
per single object in the indexing process (in other words the average cost of
indexing a single object) and the average number of the distances computed in
the 100-nn search.

The results for the k-means and for the k-approximate centers procedure are
similar. For $k = 2$ they are quite optimistic, for all the tested data sets except
shuttle the average cost of indexing a single object is several times lower than
the average cost of searching for the 100 nearest neighbors of a single object.

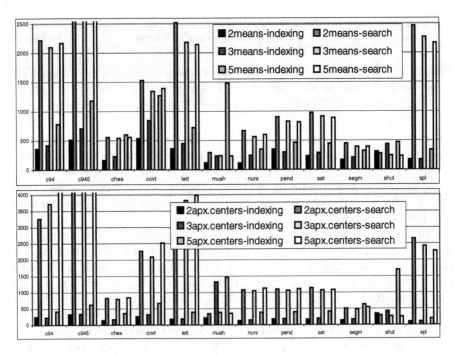

Fig. 23. The average number of distance computations per single object in the indexing
algorithm and in 100-nn search, with the use of the indexing trees with the k-means (the
upper graph) and with the k-approximate-centers (the lower graph) splitting procedure.
For each data set the first pair of columns represents the costs of indexing and searching
for $k = 2$, the second pair represents these costs for $k = 3$ and the third one for $k = 5$.

It means that if the size of the training and the test set are of the same order the main workload remains on the side of the search process. For the data sets *shuttle* and *mushroom* the differences in the cost are smaller but it results from the fact that the search process is more effective for these two data sets than for the others.

The situation changes to worse while increasing the degree k. In case of $k = 5$ the cost of indexing for the five data sets: *chess, covertype, mushroom, segment* and *shuttle* is at least comparable and is sometimes higher than the cost of searching. It has been mentioned in Subsection 4.4 that the computational cost of searching is stable for $k \geq 3$. On the other hand, the cost of indexing increases significantly while increasing the degree k. It means that the larger degree k the lower profit from applying the advanced indexing structure is. The results from this subsection and Subsection 4.4 indicate that the best trade-off between the indexing cost and the search performance is obtained for $k = 3$. Increasing the value of k more increases the cost of indexing with no profit from searching.

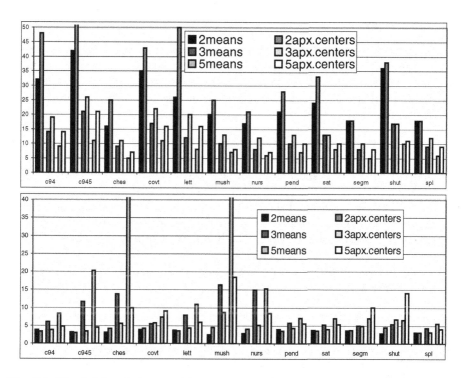

Fig. 24. The height of the tree (the upper graph) and the average number of iterations in the splitting procedure (the lower graph) in the k-means based and in the k-approximate-centers based indexing tree. For each data set the first pair of columns represents the height (at the upper graph) and the iterations (at the lower graph) for the 2-means based and the 2-approximate-centers based tree, the second pair represents these quantities for $k = 3$ and the third one for $k = 5$.

We have analyzed the case of the 100-nn search. In many application, e.g., while searching for the optimal value of k or for some topological properties in a data set, there is the need to search for a large number of the nearest neighbors. In this case the presented trees keep the appropriate balance between the costs of construction and searching. The results for the 1-nn case do not lead to such an unequivocal conclusion. The usefulness of the presented structures for queries with a small k depends more on the specific properties of a data set and on the number of queries to be performed.

The upper graph at Figure 24 provides some information about the shape of the indexing trees. The fact that the height of the trees, i.e., the distance between the root and the deepest leaf, exceeds rarely 25, indicates that the shape of the trees is quite balanced: they do not contain very long thin branches. The lower graph presents the average number of iterations in the splitting procedures. In many experiments, especially for $k = 2$, this number does not exceed 5 what indicates that the construction cost in case of the tree with the iterative splitting procedure is only a few times larger than in case of the tree with the non-iterative procedure.

4.9   Summary

In this section we analyze the different properties of distance based indexing algorithms with center based splitting procedures and search for the optimal parameters of the indexing algorithm. As the result we introduce the following new methods to be used in construction of an indexing structure and in the search process:

- the iterative procedure for splitting nodes in an indexing tree (Subsection 4.2) generalizing the k-means algorithm; the procedure has been presented in two versions: the specific case where the weighted joint city-block and VDM metric is used and the general case of any metric,
- the method for selection of the initial centers in the node splitting procedure (Subsection 4.3); in this method, as distinguished from Brin's method [18], the initial centers are searched globally among all the objects from a given tree node instead of in a sample,
- a complex search pruning criterion combining three different single criteria (Subsection 4.6).

We have compared the three methods for selection of the initial centers in the iterative splitting procedure: random, sampled and global methods. Savaresi and Boley have reported that the 2-means algorithm has good convergence properties [70] so the selection of the initial centers is not very important. This result has been confirmed by the experimental results for most of the tested data sets. However, it was obtained for an infinite theoretical model and there are real-life data sets where the global selection of the initial centers gives a little better performance than the other two methods. We have also observed that the performance of the indexing trees of different splitting degrees is comparable except for the tree of the degree $k = 2$ that has a little worse performance than the trees of the degrees $k \geq 3$. On the other hand, the cost of indexing increases significantly while increasing the degree k. These observations lead to the conclusion that the degree $k = 3$ is the optimal trade-off between the performance of the search process and the cost of indexing.

We have compared the significance of the three different search pruning criteria using the center based representation of tree nodes. Two of criteria are based on the covering radius and on the separating hyperplanes, and the third criterion is based on rings that require more information to be stored at the tree nodes. The experimental results indicate that the most effective criterion is the covering radius. In searching for the 100 nearest neighbors this single criterion is as efficient as all the three criteria combined together. In the case of the 1-nn search none of the tree criteria alone is comparable to all the three criteria used simultaneously, but the combination of the two: the covering radius and the hyperplane criterion is. These results indicate that the two simple criteria define the optimal combination and there is no need to search for a more sophisticated mechanism like the rings based criterion.

The center based indexing trees outperform the vantage point trees. However, the performance of the center based tree still depends much on how the center of a set of data objects is constructed or selected. While comparing the iterative k-means algorithm to the non-iterative one the advantage of the former one is noticeable but the k-means algorithm is applicable only to vector spaces. As a general solution we have proposed the approximate centers that replace the means by centers selected from a sample of objects. Although there is some evidence that the approximate centers perform a little better than the non-iterative centers the difference does not seem to be significant. The gap between the performance of the means and the approximate centers is much larger. These observations indicate that the representation of the center of a set of data objects is crucial for the effectiveness of the center based search pruning and we find the problem of the center selection an important issue for future research.

The experimental results show that the tree with the iterative 3-means splitting procedure and the combined search pruning criteria is up to several times more effective than the one-step based tree with a single criterion. A particularly advanced acceleration level in comparison to the linear search has been reached in case of the largest data sets. The presented structure has reduced the 1-nn search cost 4000 times in case of the data set *covertype* and 400 times in case of the data set *census94-95*. During the 100-nn search the reductions of the performance cost are 300 and 60 times, respectively. These results show the great capability of k-nn based methods in applications to large databases.

It is known that bottom-up constructions give a very good performance but such an immediate construction requires $O(n^2)$ time. Brin, in conclusions of [18], has considered the iterative transformation of the tree from a top-down construction to a bottom-up construction in such a way that at each iteration the tree is constructed with the use of the structure from the previous iteration. Such an approach can result in an indexing structure that reflects more topological properties of a data set than a tree constructed by the top-down method. We find it interesting to instantiate this idea.

The presented indexing and searching method is also described in [91, 92]. It was implemented with the programming language Java and it is used to accelerate the k nearest neighbors classifier in the system RSES [8, 73].

5   Neighborhood-Based Classification Methods

Neighborhood-based classification methods are investigated in this section.

5.1   Estimating the Optimal Neighborhood Size

In the experiments we noticed that the accuracy of the k-nn classifier depends significantly on the number k and different k are appropriate for different data sets. Therefore it is important to estimate the optimal value of k before classification and in this subsection we consider this problem.

Since the optimal value k depends on the data set, we present an algorithm that estimates this optimal value from a training set. The idea is that the leave-one-out classification is applied to the training set in the range of values $1 \leq k \leq k_{max}$ and k with the best leave-one-out accuracy is chosen to be used for a test set. Applying it directly requires repeating the leave-one-out estimation k_{max} times. However, we emulate this process in time comparable to the single leave-one-out test for k equal to the maximum possible value $k = k_{max}$. Algorithm 7 implements this idea.

The function $getClassificationVector(x, k_{max})$ returns the decision of the k-nn classifier for a given object x for the subsequent values of k in the range $1 \leq k \leq k_{max}$. After calling this function for all training objects $x \in U_{trn}$ the algorithm compares the total accuracy of different values k for the whole set U_{trn} and it selects the value k with the maximum accuracy.

At the beginning the function $getClassificationVector(x, k_{max})$ finds the k_{max} training objects from $U_{trn} \setminus \{x\}$ that are nearest to the object x. This is the most time-consuming operation in this function and performing it once instead of for each value $1 \leq k \leq k_{max}$ saves a significant amount of performance time. Then the function counts the votes for particular decisions for successive values of k and for each k it stores the most frequent decision as the result of the classification of the object x. In this way it implements the majority voting model but by analogy one can implement other voting models of the k-nn classifier.

To find the k_{max} nearest training objects from $U_{trn} \setminus \{x\}$ one can use the indexing and searching method described in Section 4 with the small modification: the searching algorithm ignores the object x during search and it does not add it to the set of the nearest neighbors.

The setting $k_{max} = 100$ makes the algorithm efficient enough to apply it to large data sets and we use this setting in further experiments. The maximum

Algorithm 7. The function $findOptimalK$ estimating the optimal value k from a training set U_{trn} in the range $1 \leq k \leq k_{max}$

```
function  findOptimalK(k_max)
    for each  x ∈ U_trn
        A_x := getClassificationVector(x, k_max)
    return  arg max_{1≤k≤k_max} |{x ∈ U_trn : A_x[k] = dec(x)}|

function  getClassificationVector(x, k_max)
    n_1, …, n_{k_max} := the sequence of the k_max nearest neighbors of x
                sorted in the increasing order of the distance to x
    for each  d_j ∈ V_dec  votes[d_j] := 0
    mostFrequentDec :=arg max_{d_j∈V_dec} |{x ∈ U_trn : dec(x) = d_j}|
    for  k := 1 to  k_max
        votes[dec(n_k)] := votes[dec(n_k)] + 1
        if  votes[dec(n_k)] > votes[mostFrequentDec]
            then  mostFrequentDec := dec(n_k)
        A_x[k] := mostFrequentDec
    return  A_x
```

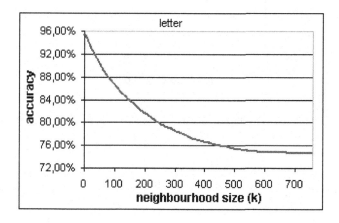

Fig. 25. The classification accuracy for the data set *letter* in dependence on the parameter k

possible value of k_{max} is the size of the training set $|U_{trn}|$. The interesting question is how much the setting $k_{max} = 100$ affects the classification results. To answer this question the following experiment was performed for the data sets from Table 1: for the smallest two data sets: *chess* and *splice* the k-nn accuracy was computed for all possible values of k and for the 8 remaining data sets accuracy was computed for all values k with the maximum value $k_{max} = 500$. For each data set the classification accuracy was measured with the leave-one-out method applied to the training set.

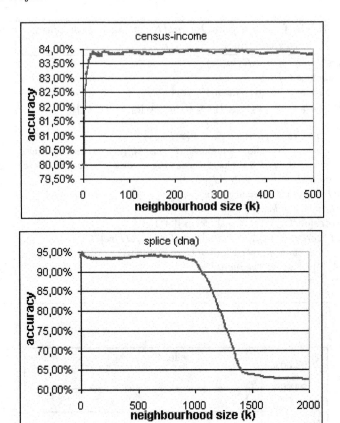

Fig. 26. The classification accuracy for the data sets *census94* and *splice* in dependence on the parameter k

For 8 of the tested data sets (all the sets except *census94* and *splice*) the maximum accuracy was obtained for small values of k (always ≤ 5) and while increasing k the accuracy was significantly falling down (see, e.g., Figure 25).

The dependence between the accuracy and the value k for the two remaining data sets is presented at Figure 26. For the data set *splice* the accuracy remains stable in a wide range of k, at least for the whole range $1 \leq k \leq 1000$, and it starts to fall down for $k > 1000$. However, the maximum accuracy was obtained at the beginning of this wide range: for $k = 15$. In case of the data set *census94* accuracy becomes stable for $k \geq 20$ and it remains stable to the maximum tested value $k = 500$. We observed that the maximum accuracy was obtained for $k = 256$ but the difference to the accuracy for the best k in the range $1 \leq k \leq 100$ was insignificant: accuracy for $k = 24$ was only 0.04% lower than for $k = 256$.

The conclusion is that for all the tested data sets the accuracy reaches a value close to the maximum for a certain small value k. Then either it starts quickly to fall or it remains stable for a wide range of k, but then the fluctuations in the accuracy are very small. The conclusion is that accuracy close to the maximum can be always found in the range $1 \leq k \leq 100$. Therefore, $k_{max} = 100$ provides a good balance between the optimality of the results and the searching time and we assume this setting in further experiments.

5.2 Voting by k Nearest Neighbors

During the classification of any test object x in the k-nn classifier the k nearest neighbors of x vote for different decisions and the classifier chooses the best decision according to a certain voting model.

The most common majority voting model [31] is given in Equation 4. This model assigns the same weight to each object in the set of the k nearest neighbors $NN(x, k)$.

In the literature there are a number of other voting models that take into consideration the distances from the neighbors to the test object x [27, 72]. It has been argued that for a finite training set U_{trn} the distance weighted voting models can outperform the majority voting model [4, 58, 84, 93, 96].

Dudani [27] proposed the inverse distance weight where the weight of a neighbor vote is inversely proportional to the distance from this neighbor to the test object x. In this way closer neighbors are more important for classification than farther neighbors. In the dissertation we consider the modified version of Dudani's model, the inverse square distance weights:

$$dec_{weighted-knn}(x) := \arg \max_{d_j \in V_{dec}} \sum_{y \in NN(x,k):dec(y)=d_j} \frac{1}{\rho(x,y)^2}. \qquad (7)$$

In the above model the weight of any neighbor vote is inversely proportional to the square of the distance from this neighbor to the test object x. It makes the weights more diversified than in Dudani's model.

Empirical comparison of the two voting models: with the equal weights and with the inverse square distance weights is discussed in Subsection 5.5.

5.3 Metric Based Generalization of Lazy Rule Induction

In this subsection we consider another approach to learning from examples based on rule induction. Rule induction is one of the most popular approaches in machine learning [7, 21, 42, 60, 74]. Our goal is to combine the k nearest neighbors method with rule induction.

The k-nn model implements the lazy learning approach [6, 34]. In this approach the model is assumed to be induced at the moment of classification. For the k-nn it is implemented in a natural way because the k nearest neighbors are searched in relation to a test object. In this subsection we consider a lazy approach to rule based classification. Bazan [6] has proposed an effective lazy rule induction algorithm for data with nominal attributes. In this subsection, we make the first step towards combining Bazan's algorithm with the k nearest neighbors method: we extend this algorithm to the case of data with both numerical and nominal attributes. The extension uses the assumption that a linear weighted metric from Equation 6 is provided for data. We show that the proposed extension generalizes Bazan's method.

The main feature of rule based classifiers is the set of rules used for classifying objects.

Definition 6. *A rule consists of a premise and a consequent:*

$$a_{i_1} = v_1 \wedge \ldots \wedge a_{i_p} = v_p \Rightarrow dec = d_j.$$

The premise is conjunction of attribute conditions and the consequent indicates a decision value. A rule is said to cover an example $x = (x_1, \ldots, x_n)$, and vice versa, the example x is said to match the rule, if all the attribute conditions in the premise of the rule are satisfied by the object values: $x_{i_1} = v_1, \ldots, x_{i_p} = v_p$. The consequent $dec = d_j$ denotes the decision value that is assigned to an object if it matches the rule.

In rule based classifiers a set of rules is induced from a training set. The important properties of rules are consistency and minimality [74].

Definition 7. *A rule $a_{i_1} = v_1 \wedge \ldots \wedge a_{i_p} = v_p \Rightarrow dec = d_j$ is consistent with a training set U_{trn} if for each object $x \in U_{trn}$ matching the rule the decision of the rule is correct, i.e., $dec(x) = d_j$.*

The notion of consistency describes the rules that classify correctly all the covered objects in a given training set.

Definition 8. *A consistent rule $a_{i_1} = v_1 \wedge \ldots \wedge a_{i_p} = v_p \Rightarrow dec = d_j$ is minimal in a training set U_{trn} if for each proper subset of conditions occurring in the premise of this rule $C \subset \{a_{i_1} = v_1, \ldots, a_{i_p} = v_p\}$ the rule built from these conditions, i.e., $\bigwedge C \Rightarrow dec = d_j$ is inconsistent with the training set U_{trn}.*

The notion of minimality selects the consistent rules of the minimum length in terms of the number of conditions in the premise of a rule. These rules maximize also the set of covered objects in a training set.

The complete set of all minimal consistent rules has good theoretical properties: it corresponds to the set of all rules generated from all local reducts of a given training set [94]. However, the number of all minimal consistent rules can be exponential in relation both to the number of attributes $|A|$ and to the training set size $|U_{trn}|$ and computing all minimal consistent rules is often infeasible [90]. Therefore many rule induction algorithms are based on a smaller set of rules [7, 42].

However, in the dissertation we consider a rule based classification model that allows us to classify objects on the basis of the set of all minimal consistent rules without computing them explicitly. The decision for each object to be classified is computed using the rules covering the object. Usually in a given set of rules they are not mutually exclusive and more than one rule can cover a test object. Therefore a certain model of voting by rules is applied to resolve conflicts between the covering rules with different decisions.

Algorithm 8. Algorithm $decision_{local-rules}(x)$ classifying a given test object x based on lazy induction of local rules

---

```
for each d_j ∈ V_dec  support[d_j] := ∅
for each  y ∈ U_trn
    if r_local(x, y) is consistent with U_trn then
        support[dec(y)] := support[dec(y)] ∪ {y}
return arg max_{d_j ∈ V_dec} |support[d_j]|
```

Definition 9. *The* support *of a rule* $a_{i_1} = v_1 \wedge \ldots \wedge a_{i_p} = v_p \Rightarrow dec = d_j$ *in a training set* U_{trn} *is the set of all the objects from* U_{trn} *matching the rule and with the same decision* d_j:

$$support(a_{i_1} = v_1 \wedge \ldots \wedge a_{i_p} = v_p \Rightarrow dec = d_j) =$$
$$\{x = (x_1, \ldots, x_n) \in U_{trn} : x_{i_1} = v_1 \wedge \ldots \wedge x_{i_p} = v_p \wedge dec(x) = d_j\}.$$

In the dissertation, we focus on the commonly used rule based classification model using the notion of the rule support:

$$dec_{rules}(x, R) := \arg \max_{d_j \in V_{dec}} \left| \bigcup_{\alpha \Rightarrow dec = d_j \in R: \, x \text{ satisfies } \alpha} support(\alpha \Rightarrow dec = d_j) \right|.$$
(8)

where R is a given set of rules used by the classifier. This model computes the support set for each rule $r \in R$ covering a test object x and then it select the decision with the greatest total number of the supporting objects.

Algorithm 8 presents Bazan's lazy rule induction algorithm $decision_{local-rules}$ [6] that simulates this rule support based classifier dec_{rules} where R is the complete set of all minimal consistent rules. The algorithm was designed originally only for data with nominal attributes and it is based on the following notion of a local rule:

Definition 10. *The* local rule *for a given pair of a test object* x *and a training object* $y \in U_{trn}$ *is the rule* $r_{local}(x, y)$ *defined by*

$$\bigwedge_{a_i: y_i = x_i} a_i = y_i \Rightarrow dec = dec(y).$$

The conditions in the premise of the local rule $r_{local}(x, y)$ are chosen in such a way that both the test object x and the training object y match the rule and the rule is maximally specific relative to the matching condition. This is opposite to the definition of a minimal consistent rule where the premise of a rule is minimally specific. However, there is the following relation between minimal consistent rules and local rules:

Fact 11. *[6] The premise of a local rule* $r_{local}(x, y)$ *for a test object* x *and a training object* $y \in U_{trn}$ *implies the premise of a certain minimal consistent rule if and only if the local rule* $r_{local}(x, y)$ *is consistent with the training set* U_{trn}.

It means that if a local rule is consistent with a training set then it can be generalized to a certain minimal consistent rule covering both the test and the training object and this property is used to compute the support set of minimal consistent rules matched by a test object in Algorithm 8. Instead of computing all minimal consistent rules covering a given test object x to be classified the algorithm generates the local rules spanned by the object x and each training object $y \in U_{trn}$, and next, it checks the consistency of each local rule against the training set U_{trn}. If the local rule $r_{local}(x, y)$ is consistent with the training set, then the object y supports a certain minimal consistent rule and the algorithm uses y to vote. Hence, the following conclusion can be drawn:

Corollary 12. *[6] The classification result of the rule support based classifier from Equation 8 with the set R of all minimal consistent rules and the lazy local rule induction classifier (Algorithm 8) is the same for each test object x:*

$$dec_{rules}(x, R) = decision_{local-rules}(x).$$

To check the consistency of a local rule $r_{local}(x, y)$ with the training set U_{trn} the algorithm checks for each object $z \in U_{trn}$ with the decision different from y: $dec(z) \neq dec(y)$ whether z matches the local rule. Hence, the time complexity of the lazy rule induction algorithm for a single test object is $O(|U_{trn}|^2 |A|)$ and the classification of the whole test set U_{tst} has the time complexity $O(|U_{trn}|^2 |U_{tst}| |A|)$. It means that lazy induction of rules reduces the exponential time complexity of the rule based classifier to the polynomial time. This makes it possible to apply this algorithm in practice.

The original version of the algorithm was proposed for data only with nominal attributes and it uses equality as the only form of conditions on attributes in the premise of a rule (see Definition 6). We generalize this approach to data with both nominal and numerical attributes and with a metric ρ defined by linear combination of metrics for particular attributes (see Equation 6). Equality as the condition in the premise of the rule from Definition 6 represents selection of attribute values, in this case always a single value. We replace equality conditions with a more general metric based form of conditions. This form allows us to select more than one attribute value in a single attribute condition, and thus, to obtain more general rules.

First, we define the generalized versions of the notions of rule and consistency.

Definition 13. *A generalized rule consists of a premise and a consequent:*

$$\rho_{i_1}(a_{i_1}, v_1) \leq r_1 \wedge \ldots \wedge \rho_{i_p}(a_{i_p}, v_p) < r_p \Rightarrow dec = d_j.$$

Each condition $\rho_{i_j}(a_{i_j}, v_j) \leq r_j$ or $\rho_{i_j}(a_{i_j}, v_j) < r_j$ in the premise of the generalized rule is described as the range of acceptable values of a given attribute a_{i_q} around a given value v_q. The range is specified by the distance function ρ_{i_q} that is the component of the total distance ρ and by the threshold r_q.

The definition of rule consistency with a training set for the generalized rules is analogous to Definition 7.

Definition 14. *A consistent generalized rule* $\rho_{i_1}(a_{i_1}, v_1) < r_1 \wedge \ldots \wedge \rho_{i_p}(a_{i_p}, v_p)$
$< r_p \Rightarrow dec = d_j$ *is* minimal *in a training set* U_{trn} *if for each attribute*
$a_{i_q} \in \{a_{i_1}, \ldots, a_{i_p}\}$ *occurring in the premise of the generalized rule the rule*
$\rho_{i_1}(a_{i_1}, v_1) < r_1 \wedge \ldots \wedge \rho_{i_q}(a_{i_q}, v_q) \leq r_q \wedge \ldots \wedge \rho_{i_p}(a_{i_p}, v_p) < r_p \Rightarrow dec = d_j$ *with*
the enlarged range of acceptable values on this attribute (obtained by replacing
$<$ *by* $\leq$ *in the condition of the original rule) is inconsistent with the training*
set U_{trn}.

Observe, that each condition in the premise of a minimal consistent generalized rule is always a strict inequality. It results from the assumption that a
training set U_{trn} is finite.

For the generalized version of the classifier based on the set of all minimal consistent rules we use the notion of a generalized rule center.

Definition 15. *An object* $(x_1, \ldots, x_n)$ *is* center *of the rule from Definition 13*
if for each attribute a_{i_j} *from its premise we have* $x_{i_j} = v_j$.

Observe, that a rule can have many centers if there are attributes that do not
occur in the premise of the rule.

In the generalized rule support based classification model the support is
counted using the set R equal to the set of all generalized minimal consistent
rules centered at a test object x:

$$decision_{gen-rules}(x, R) :=$$

$$\arg\max_{d_j \in V_{dec}} \left| \bigcup_{\alpha \Rightarrow dec = d_j \in R:\, x \text{ is a center of } \alpha \Rightarrow dec = d_j} support(\alpha \Rightarrow dec = d_j) \right| .(9)$$

Although in the generalized version we consider only minimal consistent rules
centered at a test object the number of these rules can be exponential as in the
non-generalized version:

Fact 16. *For arbitrary large set of attributes* A *there is a training set* U_{trn} *and*
a test object x *such that the number of minimal consistent rules centered at* x *is*
exponential with respect both to the number of attributes $|A|$ *and to the size of*
the training set $|U_{trn}|$.

Proof. We assume that the number of attributes $n = |A|$ is even and the decision
is binary: $V_{dec} = \{0, 1\}$. In the proof we use any linear metric from Equation
6 to define distance between attribute values and we assume only that each
attribute has at least two different values, let us assume that $\{0, 1\} \subseteq V_i$. We
define the training set U_{trn} consisting of $\frac{n}{2} + 1$ objects. The first object x^0 has
all the attribute values and the decision value equal to 0: $x_i^0 = 0$, $dec(x^0) = 0$.
Any object x^j from the remaining $\frac{n}{2}$ objects $x^1, \ldots, x^{\frac{n}{2}}$ has the two values of
neighboring attributes and the decision value equal to 1: $x_{2j-1}^j = x_{2j}^j = 1$,
$dec(x^j) = 1$ and the remaining attributes have the value equal to 0: $x_i^j = 0$
$(i \neq 2j-1, 2j)$. Consider the object $x = x^0$ and minimal consistent rules centered

at x. To exclude each of the training object x^j a minimal consistent rule contains the condition with exactly one of the two attributes that have the value 1 in the object x^j. On the other hand the rule can contain the condition with any attribute from each pair a_{2j-1}, a_{2j}. It means the for each selection function $sel : \{1, \ldots, \frac{n}{2}\} \to \{0, 1\}$ there is the corresponding minimal consistent rule:

$$\bigwedge_{1 \leq j \leq \frac{n}{2}} p_{2j-sel(j)} \left(a_{2j-sel(j)}, x^0_{2j-sel(j)} \right) < p_{2j-sel(j)}(0, 1) \Rightarrow dec = 0.$$

Each of the above rules is unique. Hence, the number of minimal consistent rules centered at x is $2^{\frac{n}{2}}$ and we obtain the following exponential relation between this number of minimal consistent rules denoted by R_x and the number of attributes and the training set size:

$$R_x = 2^{|U_{trn}|-1} = (\sqrt{2})^{|A|}. \qquad \square$$

Since it is impossible to enumerate all generalized minimal consistent rules in practice we propose to simulate the generalized rule support based classification model from Equation 9 by analogy to Algorithm 8. First, we introduce the definition of a generalized local rule analogous to Definition 10. The conditions in a generalized local rule are chosen in such a way that both the test and the training object match the rule and the conditions are maximally specific.

Definition 17. *The* generalized local rule *for a given pair of a test object x and a training object $y \in U_{trn}$ is the rule $r_{gen-local}(x, y)$:*

$$\bigwedge_{a_i \in A} \rho_i(a_i, x_i) \leq \rho_i(y_i, x_i) \Rightarrow dec = dec(y).$$

For each attribute a_i the range of acceptable values in the corresponding condition of the generalized local rule is defined as the set of values whose distance to the attribute value x_i in the test object is less or equal to the distance from the attribute value y_i in the training object to x_i.

First, we identify the relation between the original and the generalized notion of a local rule. Let us consider the case where to define the generalized rules the Hamming metric described in Subsection 2.4 is used for all the attributes, both the nominal and the numerical ones.

Fact 18. *For the Hamming metric the notion of the generalized local rule $r_{gen-local}(x, y)$ in Definition 17 is equivalent to the notion of the local rule $r_{local}(x, y)$ in Definition 10.*

Proof. Consider a single attribute a_i. If the values of x and y on this attribute are equal $x_i = y_i$ the corresponding condition in the local rule $r_{local}(x, y)$ has the form of equality $a_i = y_i$. The attribute distance in the Hamming metric between two equal values is 0 so the corresponding condition in the generalized local rule has the form $\rho_i(a_i, x_i) \leq 0$. The distance between two attribute values in the

Hamming metric is 0 if and only if these two value are equal. Hence, in case of $x_i = y_i$ the corresponding conditions $a_i = y_i$ and $\rho_i(a_i, x_i) \leq 0$ in the local and in the generalized local rule, respectively, are equivalent.

If the values of x and y on the attribute a_i are different $x_i \neq y_i$ the condition corresponding to the attribute a_i does not occur in the local rule $r_{local}(x, y)$. This means that the local rule accepts all values on this attribute. In the generalized local rule $r_{gen-local}(x, y)$ the corresponding condition has the form $\rho_i(a_i, x_i) \leq 1$. But in the Hamming metric the attribute distance between two values is always either 0 or 1 so the condition $\rho_i(a_i, x_i) \leq 1$ is satisfied for all the values of the attribute a_i too.

Hence, for each attribute the corresponding conditions in the local rule $r_{local}(x, y)$ and in the generalized rule $r_{gen-local}(x, y)$ with the Hamming metric are equivalent so the whole premises of these two rules are equivalent too.   □

Now we present an example how the presented generalization works for the case of a non-trivial metric. We consider the joint city-block and VDM metric defined in Subsection 3.1. Let us assume that the following training set is provided:

Object	Age (A)	Weight (W)	Sex (S)	BloodGroup (BG)	Diagnosis
y_1	35	90	M	A	Sick
y_2	40	65	F	AB	Sick
y_3	45	68	F	AB	Healthy
y_4	40	70	M	AB	Healthy
y_5	45	75	M	B	Sick
y_6	35	70	F	B	Healthy
y_7	45	70	M	0	Healthy

Age and *Weight* are the numerical attributes and *Sex* and *BloodGroup* are the nominal attributes. We consider the following test object:

Object	Age (A)	Weight (W)	Sex (S)	BloodGroup (BG)	Diagnosis
x_1	50	72	F	A	?

For the attribute *BloodGroup* there are 4 possible values: *A*, *AB*, *B* and *0*. To construct the generalized local rules for x_1 we need to compute the attribute distance from A to each other value:

$$\rho_{BG}(A, A) = 0,$$
$$\rho_{BG}(A, AB) = |P(Diagn = Healthy|A) - P(Diagn = Healthy|AB)| -$$
$$- |P(Diagn = Sick|A) - P(Diagn = Sick|AB)| =$$
$$= \left|0 - \frac{2}{3}\right| - \left|1 - \frac{1}{3}\right| = \frac{4}{3},$$
$$\rho_{BG}(A, B) = |P(Diagn = Healthy|A) - P(Diagn = Healthy|B)| -$$
$$- |P(Diagn = Sick|A) - P(Diagn = Sick|B)| =$$
$$= \left|0 - \frac{1}{2}\right| - \left|1 - \frac{1}{2}\right| = 1,$$

$$\rho_{BG}(A,0) = |P(Diagn = Healthy|A) - P(Diagn = Healthy|0)| -$$
$$- |P(Diagn = Sick|A) - P(Diagn = Sick|0)| =$$
$$= |0 - 1| - |1 - 0| = 2.$$

Consider the generalized local rule $r_{gen-local}(x_1, y_1)$. Since the objects x_1 and y_1 have the same value A on the attribute $BloodGroup$, the local rule accepts only this value on the attribute $BloodGroup$:

$$A \in [35; 65] \wedge W \in [54; 90] \wedge BG = A \Rightarrow Diagn = Sick.$$

No other training object except for y_1 satisfies the premise of this rule so it is consistent and it can be extended to a minimal consistent rule, e.g.,

$$BG = A \Rightarrow Diagn = Sick.$$

If we consider the generalized local rule $r_{gen-local}(x_1, y_2)$ for the objects x_1 and y_2, the distance between the values of x_1 and y_2 on the attribute $BloodGroup$ is $\rho_{BG}(A, AB) = \frac{4}{3}$. It makes the three values A, AB and B be accepted in the rule $r_{gen-local}(x_1, y_2)$ on the attribute $BloodGroup$:

$$A \in [40; 60] \wedge W \in [65; 79] \wedge S = F \wedge BG \in \{A, AB, B\} \Rightarrow Diagn = Sick.$$

Now we obtain the inconsistent rule because, e.g., the object y_3 satisfies the premise of this rule and it has the inconsistent decision $Diagn = Healthy$.

The most important property of the presented generalization is the relation between generalized minimal consistent rules and generalized local rules analogous to Fact 11.

Theorem 19. *The premise of the generalized local rule $r_{gen-local}(x, y)$ for a test object x and a training object $y \in U_{trn}$ implies the premise of a certain generalized minimal consistent rule centered at x if and only if the generalized local rule $r_{local}(x, y)$ is consistent with the training set U_{trn}.*

Proof. First, we show that if the generalized local rule $r_{gen-local}(x, y)$ is consistent with the training set U_{trn} it can be extended to the generalized minimal rule centered at x. We define the sequence of rules $r^0, \ldots, r^n$ in the following way. The first rule in th sequence is the local rule $r^0 = r_{gen-local(x,y)}$. To define each next rule r_i we assume that the previous rule r_{i-1}:

$$\bigwedge_{1 \le j < i} \rho_j(a_j, x_j) < M_j \bigwedge_{i \le j \le n} \rho_j(a_j, x_j) \le \rho_j(y_j, x_j) \Rightarrow dec = dec(y).$$

is consistent with th training set U_{trn} and the first $i-1$ conditions of the rule r_{i-1} are maximally general, i.e., replacing any strong inequality $\rho_j(a_j, x_j) < M_j$ for $j < i$ by the weak makes this rule inconsistent. Let S_i be the set of all the object that satisfy the premise of the rule r_{i-1} with the condition on the attribute a_i removed:

$$S_i = \{z \in U_{trn} : z \text{ satisfies } \bigwedge_{1 \le j < i} \rho_j(a_j, x_j) < M_j \bigwedge_{i < j \le n} \rho_j(a_j, x_j) \le \rho_j(y_j, x_j)\}.$$

In the rule r_i the i-th condition is maximally extended in such way that the rule remains consistent. It means that the range of acceptable values for the attribute a_i in the rule r_i has to be not larger than the attribute distance from x to any object in S_i with a decision different from $dec(y)$. If S_i does not contain an object with a decision different from $dec(y)$ the range remains unlimited:

$$M_i = \begin{cases} \infty & \text{if } \forall z \in S_i \, dec(z) = dec(y) \\ \min\{\rho_i(z_i, x_i) : z \in S_i \wedge dec(z) \neq dec(y)\} & \text{otherwise.} \end{cases}$$

(10)

If we limit the range of values on the attribute a_i in the rule r_i by the M_i with the strong inequality in the condition:

$$\bigwedge_{1 \leq j < i} \rho_j(a_j, x_j) < M_j \wedge \rho_i(a_i, x_i) < M_i \bigwedge_{i < j \leq n} \rho_j(a_j, x_j) \leq \rho_j(y_j, x_j) \Rightarrow dec = dec(y)$$

then it ensures that the rule r_i remains consistent. On the other hand, the value of M_i in Equation 10 has been chosen in such a way that replacing the strong inequality by the weak inequality or replacing the range by a value larger than M_i causes the situation where a certain object with a decision different from $dec(y)$ satisfies the condition on the attribute a_i and the whole premise of the rule r_i, i.e., the rule r_i becomes inconsistent.

Since r_{i-1} was consistent the range M_i is greater than the range for the attribute a_i in the rule r_{i-1}: $M_i > \rho(y_i, x_i)$. Hence, the ranges for the previous attributes $M_1, \ldots, M_{i-1}$ remain maximal in the rule r_i: widening of one of these ranges in the rule r_{i-1} makes an inconsistent object match r_{i-1} and the same happens for the rule r_i.

By induction the last rule $r_n : \bigwedge_{1 \leq j \leq n} \rho_j(a_j, x_j) < M_j \Rightarrow dec = dec(y)$ in the defined sequence is consistent too and all the conditions are maximally general. Then r_n is consistent and minimal. Since the premise of each rule r_{i-1} implies the premise of the next rule r_i in the sequence and the relation of implication is transitive the first rule r_0 that is the generalized local rule $r_{gen-local}(x, y)$ of the objects x, y implies the last rule r_n that is a minimal consistent rule. Thus we have proved the theorem for the case when the generalized local rule is consistent.

In case where the generalized local rule $r_{gen-local}(x, y)$ is inconsistent each rule centered at x implied by $r_{gen-local}(x, y)$ covers all objects covered by $r_{gen-local}(x, y)$, in particular it covers an object causing inconsistency. Hence, each rule implied by $r_{gen-local}(x, y)$ is inconsistent too. $\square$

The above theorem allows to define an effective generalized version of the local rule based algorithm simulating the rule support based classifier (see Algorithm 8). Algorithm 9 works in the same way as the non-generalized version. Instead of computing all the generalized minimal consistent rules centered at a given test object x to be classified the algorithm generates the generalized local rules spanned by the object x and each training object $y \in U_{trn}$ and then checks consistency of each local rule against the training set U_{trn}. The time complexity of the generalized lazy rule induction algorithm is the same as the complexity of

Algorithm 9. Algorithm $decision_{gen-local-rules}(x)$ classifying a given test object x based on lazy induction of the generalized local rules

---

```
for each dⱼ ∈ V_dec  support[dⱼ] := ∅
for each y ∈ U_trn
    if r_gen-local(x, y) is consistent with U_trn then
        support[dec(y)] := support[dec(y)] ∪ {y}
return arg max_{dⱼ∈V_dec} |support[dⱼ]|
```

---

the non-generalized version: $O(|U_{trn}|^2 |U_{tst}| |A|)$. Theorem 19 ensures that the algorithm counts only those objects that are covered by a certain generalized minimal consistent rule centered at x. Hence, we obtain the final conclusion.

Corollary 20. *The classification result of the generalized rule support based classifier from Equation 9 with the set R of all the generalized minimal consistent rules centered at x and the generalized lazy local rule induction classifier (Algorithm 9) is the same for each each test object x:*

$$decision_{gen-rules}(x, R) = decision_{gen-local-rules}(x).$$

In this way we extended the effective lazy rule induction algorithm for data with nominal attributes to the case of data with both nominal and numerical attributes and with linear weighted distance provided.

5.4   Combination of k Nearest Neighbors with Generalized Lazy Rule Induction

In this subsection we consider an approach from multistrategy learning, i.e., a method combining more than one different approaches. We examine the combination of the k nearest neighbors method with rule induction. In the literature there is a number of different solutions combining these two methods [25, 37, 54]. Contrary to the other solutions combining k-nn with rule induction we propose the algorithm that preserves lazy learning, i.e., rules are constructed in lazy way at the moment of classification like the nearest neighbors. The proposed combination uses the metric based generalization of rules described in the previous subsection.

For each test object x Algorithm 9 looks over all the training examples $y \in U_{trn}$ during construction of the support sets $support[d_j]$. Instead of that we can limit the set of the considered examples to the set of the k nearest neighbors of x. The intuition is that training examples far from a test object x are less relevant for classification than closer objects. Therefore in the algorithm combining the two approaches we use the modified definition of the rule support, depending on a test object x:

Definition 21. *The k-support of the generalized rule $\alpha \Rightarrow dec = d_j$ for a test object x is the set:*

$$k - support(x, \alpha \Rightarrow dec = d_j) = \{y \in NN(x, k) : y \text{ matches } \alpha \wedge dec(x) = d_j\}.$$

The k-support of the rule contains only those objects from the original support set that belong to the set of the k nearest neighbors.

Now, we define the classification model that combines the k-nn method with rule induction by using the k-supports of the rules:

$$decision_{knn-rules}(x, R) := \arg\max_{d_j \in V_{dec}} \left| \bigcup_{r \in R: \, r \text{ centered in } x} k-support(x, r) \right|. \quad (11)$$

In the above model R is the set of all generalized minimal consistent rules. The difference between the generalized rule support based classifier $decision_{gen-rules}$ from Equation 9 and the combined classifier $decision_{knn-rules}$ is that the combined classifier counts only those objects supporting minimal consistent rules that belong to the set of the k nearest neighbors.

The form of the definition of the combined classifier $decision_{knn-rules}$ in Equation 11 presents the difference between the combined classifier and the pure rule based classifiers described in the previous subsection. Now, we consider the combined classifier from the point of view of the k nearest neighbors method.

Fact 22. *The combined classifier $decision_{knn-rules}$ can be defined by the equivalent formula:*

$$decision_{knn-rules}(x, R) := \arg\max_{d_i \in V_{dec}} \sum_{y \in NN(x,k): dec(y)=d_i} \delta(y, R) \quad (12)$$

where the value of $\delta(y)$ is defined by

$$\delta(y) := \begin{cases} 1 \text{ if } \exists r \in R \text{ centered in } x \text{ supported by } y \\ 0 \text{ otherwise} \end{cases}$$

and R is the set of all generalized minimal consistent rules for the training set U_{trn}.

The above fact shows that the combined classifier presented in this subsection can be considered as a special sort of the k nearest neighbors method: it can be viewed as the k-nn classifier with the specific rule based zero-one voting model. Such a zero-one voting model is a sort of filtering: it excludes some of the k nearest neighbors from voting. Such a voting model can be easily combined with other voting models, e.g., with the inverse square distance weights defined in Equation 7:

$$dec_{weighted-knn-rules}(x, R) := \arg\max_{d_i \in V_{dec}} \sum_{y \in NN(x,k): dec(y)=d_i} \frac{\delta(y, R)}{\rho(x, y)^2}. \quad (13)$$

As for the generalized rule support classifier we propose an effective algorithm simulating the combined classifier $decision_{knn-rules}$ based on the generalized local rules. The operation of consistency checking for a single local rule

Algorithm 10. Algorithm $decision_{gen-local-knn-rules}(x)$ simulating the classifier $decision_{knn-rules}(x)$ with lazy induction of the generalized local rules

```
for each dⱼ ∈ V_dec  support[dⱼ] := ∅
neighbor₁,...,neighbor_k := the k nearest neighbors of x
                sorted from the nearest to the farthest object
for each i := 1 to k
    if r_gen-local(x, neighborᵢ) is consistent
    with neighbor₁,...,neighborᵢ₋₁ then
        support[dec(neighborᵢ)] :=support[dec(neighborᵢ)]∪{neighborᵢ}
return arg max_{dⱼ∈V_dec} |support[dⱼ]|
```

in Algorithm 9 takes $O(|U_{trn}||A|)$ time. If the linear weighted distance from Equation 6 is used we can use the following fact to accelerate the consistency checking operation in the local rule based algorithm for the combined classifier $decision_{knn-rules}$:

Fact 23. *For each training object $z \in U_{trn}$ matching a generalized local rule $r_{gen-local}(x,y)$ based on a linear weighted distance ρ the distance between the objects x and z is not greater than the distance between the objects x and y:*

$$\rho(x,z) \leq \rho(x,y).$$

Proof. The generalized local rule $r_{gen-local}(x,y)$ for a test object $x = (x_1,\ldots, x_n)$ and a training object $y = (y_1,\ldots,y_n)$ has the form

$$\bigwedge_{a_i \in A} \rho_i(a_i,x_i) \leq \rho_i(y_i,x_i) \Rightarrow dec = dec(y).$$

If $z = (z_1,\ldots,z_n)$ matches the rule then it satisfies the premise of this rule. It means that for each attribute $a_i \in A$ the attribute value z_i satisfies the following condition: $\rho_i(z_i,x_i) \leq \rho_i(y_i,x_i)$. Hence, we obtain that the distance between the objects x and z is not greater than the distance between the objects x and y:

$$\rho(x,z) = \sum_{a_i \in A} w_i\rho_i(z_i,x_i) \leq \sum_{a_i \in A} w_i\rho_i(y_i,x_i) = \rho(x,y). \qquad \square$$

The above fact proves that to check consistency of a local rule $r_{gen-local}(x,y)$ with a training set U_{trn} it is enough to check only those objects from the training set U_{trn} that are closer to x than the object y.

Algorithm 10 presents the lazy algorithm simulating the classifier $decision_{knn-rules}(x,R)$ combining the k nearest neighbors method with rule induction. The algorithm follows the scheme of the generalized local rule based algorithm described in the previous subsection (see Algorithm 9). There are two differences. First, only the k nearest neighbors of a test object x are allowed to vote for decisions. Second, the consistency checking operation for each local rule $r_{gen-local}(x,y)$ checks only those objects from the training set U_{trn} that are closer to x than the object y. Thus the time complexity of the consistency checking operation for a single neighbor is $O(k\,|A|)$. For a single test object the consistency checking operation is performed once for each of the k nearest neighbors. Hence, the cost of consistency checking in the whole procedure testing a single object is $O(k^2\,|A|)$. In practice, it takes less time than searching for the k nearest neighbors. In this way we have obtained an important property of the proposed combination: addition of the rule induction to the k nearest neighbors algorithm does not lengthen significantly the performance time of the k-nn method.

Algorithm 10 simulates the classifier $decision_{knn-rules}(x, R)$ correctly only if the distances from a test object x to training objects are different. To omit this assumption the algorithm requires two small changes. First, the procedure searching for the k nearest neighbors of x returns all objects that are equally distant from x as the k-th nearest neighbor of x. It means that sometimes the algorithm considers more than k nearest neighbors of x. Second, in the procedure checking consistency of a rule $r_{gen-local}(x, neighbor_i)$ the algorithm checks also all the neighbors $neighbor_{i+1}, \ldots, neighbor_{i+l}$ that are equally distant from x as the neighbor $neighbor_i$.

5.5   Experimental Results for Different Voting Models

In this subsection we compare the performance of the k-nn method with different voting models described in the previous subsections. Four voting models are compared: the majority voting model with equal weights defined in Equation 4, the inverse square distance weights (see Equation 7), the zero-one voting model using generalized minimal consistent rules to filter objects (see Equation 12) and the combination of the last two methods, i.e., the inverse square distance weights with rule based filtering (see Equation 13).

On the basis of the results from Section 3 we choose two most effective metrics to be used in the experiments. The first tested metric is the joint city-block and VDM metric (see Subsection 3.1) with the attribute weighting method optimizing distance (see Subsection 3.4) and the second tested metric is the joint DBVDM and VDM metric (see Subsection 3.2) with the same attribute weighting method.

To compare the voting models we performed a number of experiments for the 10 benchmark data sets presented in Table 1. Each data set was partitioned into a training and a test set as described in Subsection 2.7. For each data set and for each voting model the k nearest neighbors method was trained and tested 5 times for the same partition of the data set and the average classification error was calculated for comparison. In each test first the metric was induced from the training set, then the optimal value of k was estimated from the training set in the range $1 \leq k \leq 200$ with Algorithm 7 and finally, the test part of a data set was tested with the k nearest neighbor method for the previously estimated value of k.

Both in case of the joint city-block and VDM metric and in case of the joint DBVDM and VDM metric all the tests for the data set *mushroom* gave the error 0% and all the tests for the data set *shuttle* gave an error not greater than 0.1%.

Since the classification error for these two data sets is always very small it does not provide reliable results to compare different voting models and we focus on the 8 remaining data sets: *census94, chess, letter, nursery, pendigits, satimage, segment* and *splice*.

First, we consider the weighted joint city-block and VDM metric. The table below presents the average value of the estimation of the optimal k for particular voting models with this metric:

Data set	equal weights	sqr. inv. dist. weights	equal weights & rule based filter.	sqr. inv. dist. weights & rule based filter.
census94	30.6	119.4	88.8	181
chess	1	1	1.4	49.2
letter	1	4	1	5.6
nursery	5.4	15.4	19.4	16.8
pendigits	2.2	3.6	2.2	4.2
satimage	2	5.2	2	5.6
segment	1	2.4	1	5.6
splice	1.8	3	2.6	3.6

The estimation of the optimal k for the models with the inverse square distance weights is usually larger than for the models with equal weights. It indicates that the most important objects for classification of a test object are the nearest neighbors but a number of farther objects can provide useful information too. The information from farther objects should only be considered less important than information from the nearest objects.

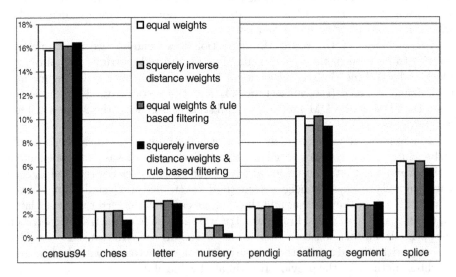

Fig. 27. The average classification error of the k-nn with the optimal k estimated from the training set for the joint city-block and VDM metric with the attribute weighting method optimizing distance

Figure 27 presents the average classification error of the k nearest neighbor method for the weighted joint city-block and VDM metric. The table below presents the best voting model for this metric and the confidence level of the differences between the best voting model and the others (see Subsection 2.7) for particular data sets:

The data set	The winning voting model	The confidence level
census94	equal weights	99.5%
chess	sqr. inv. distance weights & rule based filtering	99.5%
letter	sqr. inv. distance weights & rule based filtering	90% (from sqr. inv. dist.) 99.5% (from the two remaining)
nursery	sqr. inv. distance weights & rule based filtering	99.5%
pendigits	sqr. inv. distance weights & rule based filtering	¡90% (from sqr. inv. dist.) 90% (from the two remaining)
satimage	sqr. inv. distance weights & rule based filtering	90% (from sqr. inv. dist.) 99.5% (from the two remaining)
segment	equal weights both with and without rule based filtering	¡90% (from sqr. inv. dist.) 95% (from sqr. inv. dist. & rule based filter.)
splice	sqr. inv. distance weights & rule based filtering	90% (from sqr. inv. dist.) 99.5% (from the two remaining)

Figure 27 and the above table indicate that the voting model combining the inverse square distance weights with rule based filtering is the best: it gives the smallest error for six of the eight data sets: *chess, letter, nursery, pendigits, satimage* and *splice*. The most noticeable reduction in error is obtained for the two data sets: in case of the data set *nursery* the combined voting model gives the 0.3% error in comparison to the 1.57% error of the pure majority voting model with equal weights and in case of the data set *chess* the combined model gives the 1.46% error in comparison to the 2.24% error of the majority model.

While comparing the second and the third column in Figure 27 for each of the six data sets where the combined model is the best the model with the inverse square distance weights alone provides always a smaller error than the model with the equal weights and rule based filtering. This observation indicates that the inverse square distance weights are a more important component for the accuracy of the combined voting model than the rule based filtering. It is also confirmed by the fact that the difference between the combined voting model and the two models with equal weights (with and without rule based filtering) has almost always the maximum confidence level whereas the difference between the inverse square distance weights with and without rule based filtering has often a low confidence level. However, in case of the data sets *nursery* and *chess*, where the reductions in error by the combined voting model are largest, both components of this model contribute significantly to such good results.

Now, we consider the second metric: the weighted joint DBVDM and VDM metric. Since for data with nominal attributes this metric is equivalent to the weighted joint city-block and VDM metric we present the result only for data that contain numerical attributes. The table below presents the average value of the estimation of the optimal k for particular voting models with the weighted joint DBVDM and VDM metric:

Data set	equal weights	sqr. inv. dist. weights	equal weights & rule based filter.	sqr. inv. dist. weights & rule based filter.
census94	40.2	168.4	128.2	183
letter	1	4.4	1	6
pendigits	1	3.8	1	5.2
satimage	3.6	4	3.6	3.8
segment	1	1	1	4.8

We can make the same observation as in case of the weighted joint city-block and VDM metric: the voting models with the inverse square distance weights make use of the distance-dependent weights and they use more objects to vote than the models with equal weights.

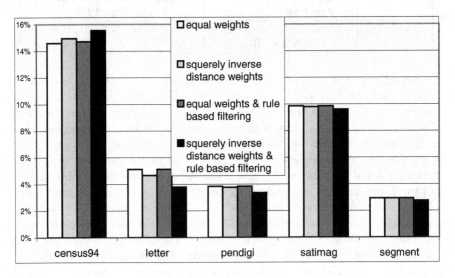

Fig. 28. The average classification error of the k-nn with the optimal k estimated from the training set for the joint DBVDM and VDM metric with the attribute weighting method optimizing distance

Figure 28 presents the average classification error of the k nearest neighbor method for the weighted joint DBVDM and VDM metric. The table below presents the best voting model for this metric and the confidence level of the difference between the best voting model and the others, for particular data sets:

The data set	The winning voting model	The confidence level
census94	equal weights	97.5% (from eq. weights (with rule based filter.) 99.5% (from the two remaining)
letter	inv. sqr. distance weights & rule based filtering	99.5%
pendigits	sqr. inv. distance weights & rule based filtering	99.5%
satimage	sqr. inv. distance weights & rule based filtering	95% (from sqr. inv. dist.) 99.5% (from the two remaining)
segment	sqr. inv. distance weights & rule based filtering	90%

As in case of the weighted joint city-block and VDM metric the results indicate that the voting model combining the inverse square distance weights with rule based filtering is the best: it gives the smallest error for all the data sets except for *census94*. The results are usually worse than for the weighted joint city-block and VDM metric. An interesting observation is that each of the two components of the combined metric: the inverse square distance weights and rule based filtering alone gives very small improvement (compare the second and the third column for each data set in Figure 28) and only the combination of these two components gives more noticeable reduction of the classification error.

The final conclusion from the presented results is that the voting model combining the inverse square distance weights with rule based filtering gives generally the best classification accuracy. It indicates that the significance of the information for a test object provided by the nearest neighbors correlates with the distance of the nearest neighbors to the test object and it is helpful to use this correlation. Distance measure and grouping of objects by rules are the two different sorts of similarity models and the application of rule based filtering to the nearest neighbors is a sort of combination of these two models. The neighbors selected for voting in such a combined method are similar to a test object according to both models, which gives more certainty that these neighbors are appropriate for decision making.

The above general conclusion does not fit to the results for the data set *census94*. This is related with the specificity of this data set. The estimated value of the optimal k for the data set *census94* is always much larger than for the other data sets. In case of such a large neighborhood the models with the inverse square distance weights are not enough accurate to improve the classification results and more accurate voting model should be searched.

5.6 K Nearest Neighbors with Local Metric Induction

All the variants of the k nearest neighbors method presented in the dissertation up to now and all other machine learning methods based on inductive concept learning: rule based systems, decision trees, neural networks, bayesian networks

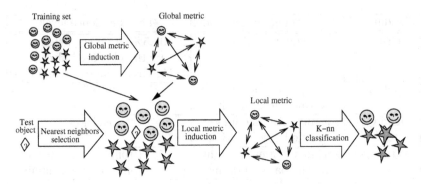

Fig. 29. K-nn classification with local metrics

and rough sets [59, 61, 63] induce a mathematical model from training data and apply this model to reasoning about test objects. The induced model of data remains invariant for different test objects. For many real-life data it is not possible to induce relevant global models. This fact has been recently observed by researches from different areas like data mining, statistics, multiagent systems [17, 75, 79]. The main reason is that phenomena described by real-life data are often too complex and we do not have sufficient knowledge to induce global models or a parameterized class of such models together with feasible searching methods for the relevant global model in such a class. Developing methods for dealing with such real-life data is a challenge.

In this subsection we propose a step toward developing of such methods. We propose a classification model that is composed of two steps. For a given test object x, first, a local model dependent on x is induced, and next, this model is used to classify x.

To apply this idea we extend the classical k-nn classification model described in Subsection 2.5. The classical k-nn induces a global metric ρ from the training set U_{trn}, and next, for each test object x it uses this induced metric ρ to find the k nearest neighbors of x and it computes a decision from the decisions of these k neighbors. We propose a new algorithm extending the classical k-nn with one additional intermediate step (see Figure 29). First, it induces a global metric ρ like in the classical k-nn method but this global metric ρ is used only in preliminary elimination of objects not relevant for classifying x. For each test object x the extended algorithm selects a neighborhood of x according to the global metric ρ and it induces a local metric ρ^x based only on the selected neighborhood. Local metric induction is a step to build a model that depends locally on the properties of the test object x. The final k nearest neighbors that are used to make a decision for the test object x are selected according to the locally induced metric.

A local approach to the k-nn method has been already considered in the literature. However, all the methods described in the literature apply only to data with numerical attributes and they assume always a specific metric to be defined.

Friedman proposed a method that combines the k-nn method with recursive partitioning used in decision trees [32]. For each test object the method starts with the whole training set and it constructs a sequence of partitions. Each partition eliminates a number of training objects. In this way after the last partition a small set of k objects remains to be used for classification. To make a single partition the algorithm selects the partition with the greatest decision discernibility.

The algorithm proposed by Hastie and Tibshirani [44] starts with the Euclidean metric and for each test object it iteratively changes the weights of attributes. At each iteration it selects a neighborhood of a test object and it applies local discriminant analysis to shrink the distance in the direction parallel to the boundary between decision classes. Finally, it selects the k nearest neighbors according to the locally transformed metric.

Domeniconi and Gunopulos use a similar idea but they use support vector machines instead of local discriminant analysis to determine class boundaries and to shrink the distance [24]. Support vectors can be computed during the learning phase what makes this approach much more efficient in comparison to local discriminant analysis.

As opposed to the above three methods our method proposed in this subsection is general: it assumes only that a procedure for metric induction from a set of objects is provided.

Algorithm 11. The k nearest neighbors algorithm $decision_{local-knn}(x)$ with local metric induction

---

```
ρ - the global metric induced once
      from the whole training set U_trn
l - the size of the neighborhood
      used for local metric induction
k_opt - the optimal value of k estimated
      from the training set U_trn (k_opt ≤ l)

NN(x,l) := the set of l nearest neighbors of x from U_trn
            according to the global metric ρ
ρ^x := the local metric induced from the neighborhood NN(x,l)
NN_local(x,k_opt) := the set of k_opt nearest neighbors of x
            from NN(x,l) according to the local metric ρ^x
return arg max_{d_j ∈ V_dec} |{y ∈ NN_local(x,k_opt) : dec(y) = d_j}|
```

In the learning phase our extended method induces a global metric ρ and estimates the optimal value k_{opt} of nearest neighbors to be used for classification. This phase is analogous to the classical k-nn.

Algorithm 11 presents the classification of a single query object x by the method extended with local metric induction. First, the algorithm selects the l nearest neighbors $NN(x,l)$ of x from the training set U_{trn} according to the global metric ρ. Next, it induces a local metric ρ^x using only the selected neighborhood $NN(x,l)$. After that the algorithm selects the set $NN_{local}(x,k_{opt})$ of the nearest neighbors of x from the previously selected neighborhood $NN(x,l)$ according to this local metric ρ^x . Then, the selected set $NN_{local}(x, k_{opt})$ is used to compute the decision $decision_{local-knn}(x)$ that is returned as the final result for the query object x.

Both for the global and for the local metric definition the algorithm can use any metric induction procedure. Moreover, different metrics can be used in the global and in the local step.

The neighborhood size l is the parameter of the extended method. To improve classification accuracy this value should be large, usually at least of an order of several hundred objects. To accelerate the selection of a large number of nearest neighbors from a training set we use the indexing tree with the iterative 3-means splitting procedure and the combined search pruning criteria described in Section 4.

The optimal value k_{opt} is estimated from a training set within the range $1 \leq k \leq l$ with the use of the same efficient procedure as in case of the classical k-nn presented in Algorithm 7 in Subsection 5.1. However, the estimation process uses Algorithm 11 as the classification procedure instead of the classical k-nn classification procedure from Equation 4. This is the only difference between the learning phases of the classical and the extended method.

In Algorithm 11 we use the most popular majority voting model with equal weights. However, as in the classical k-nn method any voting model can be used in the method with local metric induction.

The classical k-nn is a lazy method: it induces a global metric and it performs the rest of computation at the moment of classification. The algorithm proposed in this subsection extends this idea: it repeats the metric induction at the moment of classification. The proposed extension allows us to use the local properties of data topology in the neighborhood of a test object and to adjust the metric definition to these local properties.

5.7   Comparison of k-nn with Global and with Local Metric

In this subsection we compare the performance of the k nearest neighbors method with the local metric induction described in the previous subsection and the performance of the classical k-nn method. We compare the classical k-nn and the extended k-nn with the three different values of the neighborhood size l: 100, 200 and 500.

To compare the methods we tested the 10 benchmark data sets presented in Table 1. As in all the previous experiments described in the dissertation each data set was partitioned into a training and a test set as described in Subsection 2.7. Then training and testing for each data set and for each classification method was performed 5 times for the same partition of the data set and the average classification error was calculated for comparison. For the classical k-nn method the optimal value k_{opt} was estimated in the range $1 \leq k_{opt} \leq 200$ and for the extended method for each of the values of l: 100, 200 and 500 the optimal value k_{opt} was estimated in the range $1 \leq k_{opt} \leq l$.

The two most effective global metrics were tested: the joint city-block and VDM metric (see Subsection 3.1) with the attribute weighting method optimiz-

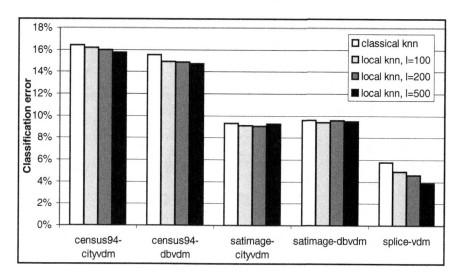

Fig. 30. The average classification error of the classical and the extended k-nn with the two metrics: The weighted joint city-block and VDM metric and the weighted joint DBVDM and VDM metric as the global metric, and with the weighted joint city-block and VDM metric as the local metric, obtained for the three different neighborhood sizes: 100, 200 and 500

ing distance (see Subsection 3.4) and the joint DBVDM and VDM metric (see Subsection 3.2) with the same attribute weighting method. Using the weighted joint DBVDM and VDM metric as the local metric makes the k nearest neighbors impractical: the performance time of k-nn with this metric becomes too long. Therefore the weighted joint city-block and VDM metric was always used as the local metric. Since the voting model combining the inverse square distance weights with rule based filtering provides generally the best classification accuracy (see Subsection 5.5) we apply this voting model in the experiment.

For the seven of the 10 tested data sets: *chess, letter, mushroom, nursery, pendigits, segment* and *shuttle* the classical k-nn method with the combined voting model obtained the classification accuracy over 97% (see Subsection 5.5). Such a good accuracy is hard to improve and for these seven data sets the k nearest neighbors with local metric induction does not provide better results or the improvement in accuracy is insignificant. Therefore we focus on the three most difficult data sets: *census94* (16.44% error by the weighted joint city-block and VDM metric and 15.54% error by the weighted joint DBVDM and VDM metric), *satimage* (9.33% error by the weighted city-block metric and 9.6% error by the weighted DBVDM metric) and *splice* (5.77% error by the weighted VDM metric).

Figure 30 presents the average classification errors of the classical and the extended k-nn method with the two types of the global metric. For the three presented data sets the extended method provides always better classification accuracy than the classical method. The table below presents the confidence

level of the difference between each of the extended method and the classical
k-nn (see Subsection 2.7) for each data set:

Data set Global metric	local k-nn (l=100) vs. classical k-nn	local k-nn (l=200) vs. classical k-nn	local k-nn (l=500) vs. classical k-nn
census94 city-block & VDM	99.5%	99.5%	99.5%
census94 DBVDM & VDM	99.5%	99.5%	99.5%
satimage city-block	95%	99%	¡90%
satimage DBVDM	¡90%	¡90%	¡90%
splice VDM	99.5%	99.5%	99.5%

The best improvement was obtained for the data set *splice*. The difference
between the extended and the classical k-nn has always the maximum confidence
level and in the best case of $l = 500$ the extended method reduced the classification error from 5.77% to 3.86%. This result is particularly noteworthy: the
author has never met such a good result in the literature for this data set. For
the data set *census94* the difference between the extended and the classical k-nn
has always the maximum confidence level too. For this data set the improvement
by the extended method is not so large but it is still noticeable: in the best case
of $l = 500$ for the weighted joint DBVDM and VDM metric the classification
error was reduced from 15.54% to 14.74%. The least effect of applying local metric induction one can observed for the data set *satimage*: only in one case the
statistical significance of the difference between the extended and the classical
method is trustworthy: in case of $l = 200$ for the weighted city-block metric the
classification error was reduced from 9.33% to 9.07%.

An interesting observation is that the largest improvement was obtained for
data only with nominal attributes and the smallest improvement was obtained for
data only with numerical attributes. It correlates with the fact that the general
properties of the domain of values of nominal attributes are poor and the methods
for data with nominal attributes learn mainly from the information encoded in
data. Hence, a metric induced globally from the whole training set and a metric
induced locally from a neighborhood of a test object can differ significantly, the
local metric can adapt strongly to local properties of the neighborhood and thus the
possibility of improving accuracy by the local metric is large. In case of numerical
attributes there are a structure of linear order and a distance measure defined in
the set of values. In many cases this structure corresponds well with the properties
of objects important for the decision attribute. The weighted city-block metric is
consistent with this structure and it is often enough to apply this metric in order to
obtain almost optimal classification accuracy. Therefore improving a global metric
by local metric induction in case of data with numerical attributes is much more
difficult than in case of data with nominal attributes.

An interesting issue is the dependence between the classification accuracy and the neighborhood size l used to induce a local metric. For the two data sets: *census94* and *splice* where the improvement by local metric induction is significant the best results was obtained for the maximum tested neighborhood size $l = 500$. It indicates that an important factor for the quality of a local metric is the representativeness of the sample used for metric induction and it is important to balance between the locality and the representativeness of the neighborhood used to induce a local metric.

5.8 Summary

In this section we have introduced two new classification models based on the k nearest neighbors:

- k nearest neighbors method combined with rule based filtering of the nearest neighbors,
- k nearest neighbors method based on a locally induced metric.

In the beginning, we have presented the algorithm estimating the optimal value of k from training data. The algorithm allows us to set automatically an appropriate value of k.

Then we have considered different voting models known from the literature. The most popular is the majority voting model where all the k nearest neighbors are weighted with equal weights. The distance based voting model replaces equal weights by the inverse square distance weights. The classification accuracy of the distance based voting model is better than the majority voting model what reflects the fact that the significance of the information for a test object provided by the nearest neighbors correlates with the distance of the nearest neighbors to the test object and it is helpful to use this correlation.

The first new model introduced in this section adds rule based filtering of the k nearest neighbors to the classical k-nn method. As the origin we took Bazan's lazy algorithm simulating effectively the classification model based on all minimal consistent rules for data with nominal attributes and we generalized the equality based model of minimal consistent rules to the metric based model. Next, we adapted Bazan's algorithm to the metric based model of minimal consistent rules, and finally, we attached this generalized rule based classification model to the k nearest neighbors in the form of rule based filtering of the k nearest neighbors. An important property of the proposed combination is that the addition of rule based filtering does not change essentially the performance time of the k nearest neighbors method. The experimental results show that the application of rule based filtering improves the classification accuracy especially when combined with the voting model with the inverse square distance weights. It indicates that rule based filtering improves selection of objects for reasoning about a test object.

The estimation of the optimal value of k and all the described voting models are available in the k nearest neighbors classifier in the system RSES [8, 73].

The classifier makes it possible to choose between the model with equal weights and the model with the inverse square distance weights, and optionally, it allows us to apply rule based filtering to the k nearest neighbors for each test object.

As the second method we proposed a new classification model that is an extension of the classical k-nn classification algorithm. The extended method induces a different metric for each test object using local information in the neighborhood of an object. The k-nn model with a local metric corresponds to the idea of transductive reasoning [79]. The transductive approach assumes that a classification model should depend on the objects to be classified and it should be adapted according to the properties of these objects. The presented extension of the k-nn algorithm implements transduction: the local metric induction adapts the metric definition to the local topology of data in the neighborhood of an object to be classified.

The experimental results show that the local approach is particularly useful in the case of hard problems. If the classification error of the methods based on global models remains large a significant improvement can be obtained with the local approach.

An important problem related to the k nearest neighbors method with local metric induction is that the local metric induction for each test object is a time-consuming step. As a future work we consider the extension of data representation in such a way that the algorithm can use the same local metric for similar test objects.

6   Conclusions

In the dissertation we have developed different classification models based on the k-nn method and we have evaluated them against real-life data sets.

Among the k-nn voting models based on the global metric the most accurate model can be obtained by the method combining the inverse square distance weights with the nearest neighbors filtering performed by means of the set of minimal consistent rules. The assignment of the inverse square distance weights to the nearest neighbors votes reflects the fact that the more similar a test object is to a training object, the more significant is for the test object the information provided by the training object. The rule-based filtering method introduced in the dissertation makes it possible to construct an alternative model that combined with the k-nn method enables verification of objects recognized by the k-nn as similar and the rejection of the objects that are not confirmed to be similar by the rule based model. The proposed rule-based extension is independent of the metric and does not increase the performance time of the classical k-nn method. Therefore it can be applied whenever the classical k-nn is applicable.

We compared different metrics in the k-nn classification model. In general, the best metrics are the normalized city-block metric for numerical attributes and the Value Difference Metric for nominal attributes, both combined with attribute weighting. For nominal attributes there is no mathematical structure in the domain of values, therefore the Value Difference Metric uses the information encoded in the training data to measure similarity between nominal values. Domains with numerical values have the structure of linear order and a distance measure consistent with this linear order. These properties reflect usually the natural relationship among the values of the numerical attribute and this information is often sufficient to define an accurate similarity measure for the values of a numerical attribute. However, there are decision problems where the natural metric on numerical values does not reflect directly the differences between decision values. For such data the correlation between the attribute values and the decision encoded in training objects is the information more useful than the general structure of numerical values. Hence, solutions analogous to the Value Difference Metric for nominal attributes are more accurate. We have analyzed three numerical metrics of this type: IVDM, WVDM, and DBVDM. They estimate decision probabilities for particular numerical values in the training set and use these estimations to define distance between values. In DBVDM the sample for decision probability estimation is chosen more carefully than in IVDM and WVDM and it gives the most accurate classification among these three metrics.

There are hard classification problems where the relationship between attributes and the decision is complex and it is impossible to induce a sufficiently accurate global model from data. For such data the method with local model induction is a better solution. The algorithm yields a separate local decision model for each test object. This approach allows us to adapt each local model to the properties of a test object and thus to obtain a more accurate classification than in the case of the global approach.

To apply metric-based classification models to large databases a method that would speed up the nearest neighbors search is indispensable. The extension of the indexing and searching methods described in literature, i.e., the iterative splitting based tree with three search pruning criteria, as proposed in the dissertation, allows us to use the k-nn method to data sets with several hundred thousand objects.

The results presented in the dissertation do not exhaust all the aspects of case-based reasoning from data. The following extensions can be considered.

Experiments with different metrics for numerical data have proved that the city-block metric is more accurate than metrics that do not preserve consistency with the natural linear order of the real numbers. However, such a metric uses the natural metric of real numbers and the training set is used marginally to modify this natural metric. An interesting issue is to construct and investigate metrics preserving the natural order of numerical values and to use training data to differentiate the value of the distance in dependence on the range of values to be compared.

A more general problem related to metric construction is that the induction of an accurate metric only from data without additional knowledge is impossible for more complex decision problems. Therefore the development of methods for inducing similarity models based on interaction with a human expert acquires particular importance.

Another possible continuation is related to the induction of local classification models. The method presented in the dissertation induces a separate model for each test object. Such a solution is computationally much more expensive than methods based on the global model. Another possible solution is to use the common sense assumption that a local model can be relevant for a fragment of a space of objects. Such an approach has been already used in data mining [49] where transactions are partitioned into groups and a specific optimization function is defined for each group. By analogy, one can partition training objects into groups of similar objects and construct one local classification model for each group. This approach integrated with an indexing structure could be comparable to the global k-nn method in terms of efficiency.

Acknowledgments

The paper is the full version of my phd dissertation supervised by Andrzej Skowron, approved in May 2005 by Warsaw University, Faculty of Mathematics, Informatics and Mechanics.

I wish to express my gratitude to my supervisor Professor Andrzej Skowron for leading me towards important research issues and for his guidance in my research work. His attitude to me has been invariably kind and gentle, and Professor has always enabled me to pursue my own research endeavors. I am also indebted to him for making such great efforts to thoroughly revise this dissertation. I am happy that I have been given the opportunity to work with Professor Andrzej Skowron.

I thank my wife, Ewa. Her sincere love, support and forbearance helped me very much in writing this dissertation.

I am also grateful to many other people who contributed to my PhD dissertation. Dr Marcin Szczuka has supported me with hardware and software necessary for performing experiments. The pleasant and fruitful cooperation with Grzegorz Góra helped me determine my research objectives. Long discussions with dr Jan Bazan, Rafal Latkowski, and Michal Mikolajczyk, on the design and implementation of the programming library *rseslib* allowed me to make it useful for other researchers. Professor James Peters, dr Wiktor Bartol and Joanna Kuźnicka revised thoroughly the language of the dissertation.

Financial support for this research has been provided by the grants 8 T11C 009 19, 8 T11C 025 19, 4 T11C 040 24 and 3 T11C 002 26 from Ministry of Scientific Research and Information Technology of the Republic of Poland.

References

1. Ch. C. Aggarwal, A. Hinneburg, and D. A. Keim. On the surprising behaviour of distance metrics in high dimensional space. In *Proceedings of the Eighth Internatinal Conference on Database Theory*, pages 420–434, London, UK, 2001.
2. D. W. Aha. Tolerating noisy, irrelevant and novel attributes in instance-based learning algorithms. *International Journal of Man-Machine Studies*, 36:267–287, 1992.

3. D. W. Aha. The omnipresence of case-based reasoning in science and applications. *Knowledge-Based Systems*, 11(5-6):261–273, 1998.

4. D. W. Aha, D. Kibler, and M. K. Albert. Instance-based learning algorithms. *Machine Learning*, 6:37–66, 1991.

5. K. Ajdukiewicz. *Logika Pragmatyczna*. PWN, Warszawa, 1974.

6. J. G. Bazan. Discovery of decision rules by matching new objects against data tables. In *Proceedings of the First International Conference on Rough Sets and Current Trends in Computing*, volume 1424 of *Lectures Notes in Artificial Intelligence*, pages 521–528, Warsaw, Poland, 1998. Springer-Verlag.

7. J. G. Bazan and M. Szczuka. RSES and RSESlib - a collection of tools for rough set computations. In *Proceedings of the Second International Conference on Rough Sets and Current Trends in Computing*, volume 2005 of *Lectures Notes in Artificial Intelligence*, pages 106–113, Banff, Canada, 2000. Springer-Verlag.

8. J. G. Bazan, M. Szczuka, A. G. Wojna, and M. Wojnarski. On the evolution of Rough Set Exploration System. In *Proceedings of the Fourth International Conference on Rough Sets and Current Trends in Computing*, volume 3066 of *Lectures Notes in Artificial Intelligence*, pages 592–601, Uppsala, Sweden, 2004. Springer-Verlag.

9. N. Beckmann, H. P. Kriegel, R. Schneider, and B. Seeger. The $R^\star$-tree: an efficient and robust access method for points and rectangles. In *Proceedings of the 1990 ACM SIGMOD International Conference on Management of Data*, pages 322–331, Atlantic City, NJ, 1990.

10. R. E. Bellman. *Dynamic Programming*. Princeton University Press, Princeton, New Jersey, 1957.

11. J. L. Bentley. Multidimensional binary search trees used for associative searching. *Communications of the ACM*, 18(9):509–517, 1975.

12. S. Berchtold, D. Keim, and H. P. Kriegel. The X-tree: an index structure for high dimensional data. In *Proceedings of the Twenty Second International Conference on Very Large Databases*, pages 28–39, 1996.

13. K. S. Beyer, J. Goldstein, R. Ramakrishnan, and U. Shaft. When is "nearest neighbor" meaningful? In *Proceedings of the Seventh International Conference on Database Theory*, pages 217–235, Jerusalem, Israel, 1999.

14. Y. Biberman. A context similarity measure. In *Proceedings of the Ninth European Conference on Machine Learning*, pages 49–63, Catania, Italy, 1994.

15. C. M. Bishop. *Neural Networks for Pattern Recognition*. Oxford University Press, Oxford, England, 1996.

16. C. L. Blake and C. J. Merz. UCI repository of machine learning databases. http://www.ics.uci.edu/~mlearn/MLRepository.html, Department of Information and Computer Science, University of California, Irvine, CA, 1998.

17. L. Breiman. Statistical modeling - the two cultures. *Statistical Science*, 16(3):199–231, 2001.

18. S. Brin. Near neighbor search in large metric spaces. In *Proceedings of the Twenty First International Conference on Very Large Databases*, pages 574–584, 1995.

19. E. Chavez, G. Navarro, R. Baeza-Yates, and J. L. Marroquin. Searching in metric spaces. Technical Report TR/DCC-99-3, Department of Computer Science, University of Chile, 1999.

20. P. Ciaccia, M. Patella, and P. Zezula. M-tree: an efficient access method for similarity search in metric spaces. In *Proceedings of the Twenty Third International Conference on Very Large Databases*, pages 426–435, 1997.

21. P. Clark and T. Niblett. The CN2 induction algorithm. *Machine Learning*, 3:261–284, 1989.

22. S. Cost and S. Salzberg. A weighted nearest neighbor algorithm for learning with symbolic features. *Machine Learning*, 10:57–78, 1993.
23. T. M. Cover and P. E. Hart. Nearest neighbor pattern classification. *IEEE Transactions on Information Theory*, 13:21–27, 1967.
24. C. Domeniconi and D. Gunopulos. Efficient local flexible nearest neighbor classification. In *Proceedings of the Second SIAM International Conference on Data Mining*, 2002.
25. P. Domingos. Unifying instance-based and rule-based induction. *Machine Learning*, 24(2):141–168, 1996.
26. R. O. Duda and P. E. Hart. *Pattern Classification and Scene Analysis*. Wiley, New York, NY, 1973.
27. S. Dudani. The distance-weighted k-nearest-neighbor rule. *IEEE Transactions on Systems, Man and Cybernetics*, 6:325–327, 1976.
28. R. E. Fikes and N. J. Nilsson. STRIPS: A new approach to the application of theorem proving to problem solving. *Artificial Intelligence*, 2(3-4):189–208, 1971.
29. R. Finkel and J. Bentley. Quad-trees: a data structure for retrieval and composite keys. *ACTA Informatica*, 4(1):1–9, 1974.
30. R. A. Fisher. Applications of "student"s' distribution. *Metron*, 5:3–17, 1925.
31. E. Fix and J. L. Hodges. Discriminary analysis, non-parametric discrimination: Consistency properties. Technical Report 4, USAF School of Aviation and Medicine, Randolph Air Field, 1951.
32. J. Friedman. Flexible metric nearest neighbor classification. Technical Report 113, Department of Statistics, Stanford University, CA, 1994.
33. J. Friedman, T. Hastie, and R. Tibshirani. *The Elements of Statistical Learning*. Springer, New York, NY, 2001.
34. J. H. Friedman, R. Kohavi, and Y. Yun. Lazy decision trees. In *Proceedings of the Thirteenth National Conference on Artificial Intelligence*, pages 717–724, Cambridge, 1996.
35. K. Fukunaga and P. M. Narendra. A branch and bound algorithm for computing k-nearest neighbors. *IEEE Transactions on Computers*, 24(7):750–753, 1975.
36. V. Gaede and O. Gunther. Multidimensional access methods. *ACM Computing Surveys*, 30(2):170–231, 1998.
37. A. R. Golding and P. S. Rosenbloom. Improving accuracy by combining rule-based and case-based reasoning. *Artificial Intelligence*, 87(1-2):215–254, 1996.
38. G. Góra and A. G. Wojna. Local attribute value grouping for lazy rule induction. In *Proceedings of the Third International Conference on Rough Sets and Current Trends in Computing*, volume 2475 of *Lectures Notes in Artificial Intelligence*, pages 405–412, Penn State Great Valley, PA, 2002. Springer-Verlag.
39. G. Góra and A. G. Wojna. RIONA: a classifier combining rule induction and k-nn method with automated selection of optimal neighbourhood. In *Proceedings of the Thirteenth European Conference on Machine Learning*, volume 2430 of *Lectures Notes in Artificial Intelligence*, pages 111–123, Helsinki, Finland, 2002. Springer-Verlag.
40. G. Góra and A. G. Wojna. RIONA: a new classification system combining rule induction and instance-based learning. *Fundamenta Informaticae*, 51(4):369–390, 2002.
41. "Student" (W. S. Gosset). The probable error of a mean. *Biometrika*, 6:1–25, 1908.
42. J. W. Grzymala-Busse. LERS - a system for learning from examples based on rough sets. In R. Slowinski, editor, *Intelligent Decision Support, Handbook of Applications and Advances of the Rough Sets Theory*, pages 3–18. Kluwer Academic Publishers, Dordrecht, Boston, London, 1992.

43. A. Guttman. R-trees: a dynamic index structure for spatial searching. In *Proceedings of the 1984 ACM SIGMOD International Conference on Management of Data*, pages 47–57, Boston, MA, 1984.

44. T. Hastie and R. Tibshirani. Discriminant adaptive nearest neighbor classification. *IEEE Transactions on Pattern Analysis and Machine Intelligence*, 18(6):607–616, 1996.

45. F. V. Jensen. *An Introduction to Bayesian Networks*. Springer Verlag, New York, 1996.

46. I. Kalantari and G. McDonald. A data structure and an algorithm for the nearest point problem. *IEEE Transactions on Software Engineering*, 9(5):631–634, 1983.

47. N. Katayama and S. Satoh. The SR-tree: an index structure for high dimensional nearest neighbor queries. In *Proceedings of the 1997 ACM SIGMOD International Conference on Management of Data*, pages 369–380, Tucson, Arizona, 1997.

48. K. Kira and L. A. Rendell. A practical approach to feature selection. In *Proceedings of the Ninth International Conference on Machine Learning*, pages 249–256, Aberdeen, Scotland, 1992. Morgan Kaufmann.

49. J. Kleinberg, Ch. Papadimitriou, and P. Raghavan. Segmentation problems. *Journal of the ACM*, 51(2):263–280, 2004.

50. W. Klösgen and J. M. Żytkow, editors. *Handbook of Data Mining and Knowledge Discovery*. Oxford University Press, Inc., New York, NY, USA, 2002.

51. I. Kononenko. Estimating attributes: Analysis and extensions of RELIEF. In *Proceedings of the Seventh European Conference on Machine Learning*, volume 784 of *Lectures Notes in Artificial Intelligence*, pages 171–182, Catania, Italy, 1994. Springer-Verlag.

52. D. B. Leake, editor. *Case-Based Reasoning: Experiences, Lessons and Future Directions*. AAAI Press/MIT Press, 1996.

53. J. Li, G. Dong, K. Ramamohanarao, and L. Wong. DeEPs: a new instance-based discovery and classification system. *Machine Learning*, 2003. to appear.

54. J. Li, K. Ramamohanarao, and G. Dong. Combining the strength of pattern frequency and distance for classification. In *Proceedings of the Fifth Pacific-Asia Conference on Knowledge Discovery and Data Mining*, pages 455–466, Hong Kong, 2001.

55. K. I. Lin, H. V. Jagadish, and C. Faloustos. The TV-tree: an index structure for high dimensional data. *VLDB Journal*, 3(4):517–542, 1994.

56. D. Lowe. Similarity metric learning for a variable kernel classifier. *Neural Computation*, 7:72–85, 1995.

57. D. R. Luce and H. Raiffa. *Games and Decisions*. Wiley, New York, 1957.

58. J. E. S. Macleod, A. Luk, and D. M. Titterington. A re-examination of the distance-weighted k-nearest-neighbor classification rule. *IEEE Transactions on Systems, Man and Cybernetics*, 17(4):689–696, 1987.

59. R. S. Michalski. A theory and methodology of inductive learning. *Artificial Intelligence*, 20:111–161, 1983.

60. R. S. Michalski, I. Mozetic, J. Hong, and H. Lavrac. The multi-purpose incremental learning system AQ15 and its testing application to three medical domains. In *Proceedings of the Fifth National Conference on Artificial Intelligence*, pages 1041–1045, 1986.

61. T. M. Mitchell. *Machine Learning*. McGraw-Hill, Portland, 1997.

62. J. Nievergelt, H. Hinterberger, and K. Sevcik. The grid file: an adaptable symmetric multikey file structure. *ACM Transactions on Database Systems*, 9(1):38–71, 1984.

63. Z. Pawlak. *Rough Sets - Theoretical Aspects of Reasoning about Data*. Kluwer Academic Publishers, Dordrecht, 1991.

64. L. Polkowski and A. Skowron. Synthesis of decision systems from data tables. In T. Y. Lin and N. Cercone, editors, *Rough Sets and Data Mining: Analysis of Imprecise Data*, pages 259–299. Kluwer Academic Publishers, Dordrecht, 1997.

65. J. R. Quinlan. *C4.5: Programs for Machine Learning*. Morgan Kaufmann, San Mateo, CA, 1993.

66. J. Robinson. The K-D-B-tree: a search structure for large multi-dimensional dynamic indexes. In *Proceedings of the 1981 ACM SIGMOD International Conference on Management of Data*, pages 10–18, New York, 1981.

67. A. Rosenblueth, N. Wiener, and J. Bigelow. Behavior, purpose, and teleology. *Philosophy of Science*, 10:18–24, 1943.

68. S. J. Russell. *Use of Knowledge in Analogy and Induction*. Morgan Kaufmann, 1989.

69. S. Salzberg. A nearest hyperrectangle learning method. *Machine Learning*, 2:229–246, 1991.

70. S. M. Savaresi and D. L. Boley. On the performance of bisecting K-means and PDDP. In *Proceedings of the First SIAM International Conference on Data Mining*, pages 1–14, Chicago, USA, 2001.

71. T. Sellis, N. Roussopoulos, and C. Faloustos. The R+-tree: a dynamic index for multi-dimensional objects. In *Proceedings of the Thirteenth International Conference on Very Large Databases*, pages 574–584, 1987.

72. R. N. Shepard. Toward a universal law of generalization for psychological science. *Science*, 237:1317–1323, 1987.

73. A. Skowron et al. Rough set exploration system. http://logic.mimuw.edu.pl/~rses, Institute of Mathematics, Warsaw University, Poland.

74. A. Skowron and C. Rauszer. The discernibility matrices and functions in information systems. In R. Slowinski, editor, *Intelligent Decision Support, Handbook of Applications and Advances of the Rough Sets Theory*, pages 331–362. Kluwer Academic Publishers, Dordrecht, 1992.

75. A. Skowron and J. Stepaniuk. Information granules and rough-neural computing. In *Rough-Neural Computing: Techniques for Computing with Words*, Cognitive Technologies, pages 43–84. Springer-Verlag, Heidelberg, Germany, 2003.

76. A. Skowron and A. G. Wojna. K nearest neighbors classification with local induction of the simple value difference metric. In *Proceedings of the Fourth International Conference on Rough Sets and Current Trends in Computing*, volume 3066 of *Lectures Notes in Artificial Intelligence*, pages 229–234, Uppsala, Sweden, 2004. Springer-Verlag.

77. C. Stanfill and D. Waltz. Toward memory-based reasoning. *Communications of the ACM*, 29(12):1213–1228, 1986.

78. J. Uhlmann. Satisfying general proximity/similarity queries with metric trees. *Information Processing Letters*, 40(4):175–179, 1991.

79. V. Vapnik. *Statistical Learning Theory*. Wiley, Chichester, GB, 1998.

80. M. Veloso. *Planning and Learning by Analogical Reasoning*. Springer, 1994.

81. J. von Neumann and O. Morgenstern. *Theory of Games and Economic Behavior*. Princeton University Press, Princeton, New Jersey, 1944.

82. J. Ward, Jr. Hierarchical grouping to optimize an objective function. *Journal of the American Statistical Association*, 58:236–244, 1963.

83. R. Weber, H. J. Schek, and S. Blott. A quantitative analysis and performance study for similarity-search methods in high-dimensional spaces. In *Proceedings of the Twenty Fourth International Conference on Very Large Databases*, pages 194–205, 1998.

84. D. Wettschereck. *A Study of Distance-Based Machine Learning Algorithms.* PhD thesis, Oregon State University, 1994.

85. D. Wettschereck, D. W. Aha, and T. Mohri. A review and empirical evaluation of feature weighting methods for a class of lazy learning algorithms. *Artificial Intelligence Review*, 11:273–314, 1997.

86. D. A. White and R. Jain. Similarity indexing with the SS-tree. In *Proceedings of the Twelve International Conference on Data Engineering*, pages 516–523, New Orleans, USA, 1996.

87. N. Wiener. *Cybernetics.* Wiley, New York, 1948.

88. D. R. Wilson and T. R. Martinez. Improved heterogeneous distance functions. *Journal of Artificial Intelligence Research*, 6:1–34, 1997.

89. D. R. Wilson and T. R. Martinez. An integrated instance-based learning algorithm. *Computational Intelligence*, 16(1):1–28, 2000.

90. A. G. Wojna. Adaptacyjne definiowanie funkcji boolowskich z przykladow. Master's thesis, Warsaw University, 2000.

91. A. G. Wojna. Center-based indexing for nearest neighbors search. In *Proceedings of the Third IEEE International Conference on Data Mining*, pages 681–684, Melbourne, Florida, USA, 2003. IEEE Computer Society Press.

92. A. G. Wojna. Center-based indexing in vector and metric spaces. *Fundamenta Informaticae*, 56(3):285–310, 2003.

93. D. Wolpert. Constructing a generalizer superior to NETtalk via meithematical theory of generalization. *Neural Networks*, 3:445–452, 1989.

94. J. Wróblewski. Covering with reducts - a fast algorithm for rule generation. In *Proceedings of the First International Conference on Rough Sets and Current Trends in Computing*, volume 1424 of *Lectures Notes in Artificial Intelligence*, pages 402–407, Warsaw, Poland, 1998. Springer-Verlag.

95. P. N. Yianilos. Data structures and algorithms for nearest neighbor search in general metric spaces. In *Proceedings of the Fourth Annual ACM/SIGACT-SIAM Symposium on Discrete Algorithms*, pages 311–321, Austin, Texas, 1993.

96. J. Zavrel. An empirical re-examination of weighted voting for k-nn. In *Proceedings of the Seventh Belgian-Dutch Conference on Machine Learning*, pages 139–148, Tilburg, The Netherlands, 1997.

Appendix. List of Symbols Used in the Dissertation

$|\ldots|$ — size of a set

$\|\ldots\|_p$ — norm of a vector in the space l_p

μ_i — mean of a numerical attribute a_i in a training set U_{trn}

ρ — distance function $\mathbb{X}^2 \to \mathbb{R}$

ρ_i — distance function $\mathbb{R}^2 \to \mathbb{R}$ defined for the values of an attribute a_i

σ_i — standard deviation of a numerical attribute a_i in a training set U_{trn}

$\sigma(X)$ — standard deviation of a continuous variable X

a_i — single attribute

A — set of attributes

c_j — center of the j-th cluster Cl_j in a node splitting procedure

Cl_j — j-th cluster of data objects in a node splitting procedure

dec — decision function $\mathbb{X} \to V_{dec}$ to be learnt by classifiers

df — degree of freedom in the Student's t-test

d_j — single decision value

$E(X)$ — expected value of a continuous variable X

I_p — p-th interval at discretization of a numerical attribute for the IVDM metric

$\overline{I}(v)$ — index of the upper neighboring interval of a value v for the IVDM metric

$\underline{I}(v)$ — index of the lower neighboring interval of a value v for the IVDM metric

k — number of nearest neighbors in the k-nn classifier

k_{max} — upper limit of the range of values examined by the procedure estimating the optimal k

l — number of iterations in attribute weighting algorithms

max_i — maximum value of an attribute a_i in a training set U_{trn}

mid_p — midpoint of the interval I_p

min_i — minimum value of an attribute a_i in a training set U_{trn}

MR — global misclassification ratio

$MR(a_i)$ — misclassification ratio for an attribute a_i

$NN(x, k)$ — set of the k nearest neighbors of a data object x in a training set U_{trn}

$P(dec = d_j | a_i = v)$ — conditional decision probability given a value v of an attribute a_i

$P(dec = d_j | a_i \in I)$ — conditional decision probability given an interval of values I of an attribute a_i

$P_{DBVDM}(dec = d_j | a_i = v)$ — estimated conditional decision probability in the DBVDM metric

$P_{IVDM}(dec = d_j | a_i = v)$ — estimated conditional decision probability in the IVDM metric

$P_{VDM}(dec = d_j | a_i = v)$ — estimated conditional decision probability in the VDM metric

$P_{WVDM}(dec = d_j | a_i = v)$ — estimated conditional decision probability in the WVDM metric

$r_{local(x,y)}$ — local rule for a pair of a test object x and a training object $y \in U_{trn}$

$r_{gen-local(x,y)}$ — generalized local rule for a pair of a test object x and a training object $y \in U_{trn}$

R_x — number of generalized minimal consistent rules centered at x

$\mathbb{R}$ — set of real numbers

s — number of intervals at discretization of a numerical attribute for the IVDM metric

$support(r)$ — set of all objects in U_{trn} matching the rule r

t — value t in the Student's t-test

U_{trn} — training set

U_{tst} — test set

V_{dec} — set of decision values

V_i — domain of values of the attribute a_i

x_i — value of an attribute a_i in a data object $x \in \mathbb{X}$

$\mathbb{X}$ — space of data objects, domain of learning

Author Index

Lecture Notes in Computer Science

For information about Vols. 1–3680

please contact your bookseller or Springer